Advances in Soil Science

SOIL MANAGEMENT AND GREENHOUSE EFFECT

Edited by

R. Lal
John Kimble
Elissa Levine
B. A. Stewart

CRC LEWIS PUBLISHERS

Boca Raton London Tokyo

Library of Congress Cataloging-in-Publication Data

Catalog record available from the Library of Congress.

Lewis Publishers is an imprint of CRC Press

International Standard Book Number 1-56670-117-1
Printed in the United States of America 1 2 3 4 5 6 7 8 9 0
Printed on acid-free paper

Preface

Global change and its causes and effects in relation to natural and anthropogenic activity have been the recent focus of concern within the scientific community. An important component of this issue is the role management of soils plays in contributing as a source or sink of carbon in the environment. Soil management extends across agricultural, urban, and natural environments and is a critical player in controlling carbon dynamics. Management also must be considered in the context of policy and economics in order to ensure the well-being of society over the long term.

The chapters in this book are presented to emphasize the importance of managing soils properly with an awareness of their effect on global change and specifically, the greenhouse effect. We have chosen to publish these manuscripts in the series *Advances in Soil Science* because of its proven record of disseminating knowledge about soils to a large international community. This volume will provide the scientific community with valuable information about how soil management will affect carbon dynamics, and how soils will behave within different ecosystems. Issues which deal with policy options and their affects on soil management and decisions which need to be made with regards to the best utilization of the pedosphere (soil resources) are also addressed.

This volume is an integrated effort funded by the United States Department of Agriculture's Global Change Office, Soil Conservation Service (SCS), and Forest Service (FS), the Environmental Protection Agency (EPA), the National Aeronautics and Space Administration (NASA), and The Ohio State University to bring together the ideas of many different scientists on the topic of *Soil Management and Greenhouse Effect*. The information here is an attempt to address the gaps in our knowledge of the role of soil management and policy options in global change and, at the same time, to present a *state of the art* compendium of our present knowledge on these issues. There are still questions regarding the role that soils play as a sink or source of carbon in the global carbon cycle. This book addresses our current knowledge of this role and identifies gaps in our understanding.

The editors would like to thank the authors for their efforts in documenting what they know about *Soil Management and Greenhouse Effect* which will greatly assist us in moving forward in soil science and environmental research. We have enjoyed working with the authors and have appreciated their stimulating contributions. The authors have described past changes and made predictions of what can be expected in the future. Their efforts will greatly help others appreciate the important role that soils play in potential global change. We would also like to thank the staff of Lewis Publishers and Kïrsten Stuart of the SCS in Lincoln, Nebraska for all their efforts in completing this volume so that the scientific community would have timely information on this important topic.

The Editors

About the Editors:

Dr. R. Lal is a Professor of Soil Science in the School of Natural Resources at The Ohio State University, Columbus, Ohio. Prior to joining Ohio State in 1987, he served as a soil scientist for 18 years at The International Institute of Tropical Agriculture, Ibadan, Nigeria. Prof. Lal is a fellow of the Soil Science Society of America, American Society of Agronomy, and The Third World Academy of Sciences. He is recipient of both the International Soil Science Award, and the Soil Science Applied Research Award of the Soil Science Society of America.

Dr. John Kimble is a Research Soil Scientist at the USDA Soil Conservation Service National Soil Survey Laboratory in Lincoln, Nebraska. For the past 4 years, Dr. Kimble has managed the Global Change project of the Soil Conservation Service, and has worked for the last 13 years with US Agency for International Development projects dealing with soil related problems in more than 40 developing countries. He is a member of the American Society of Agronomy, the Soil Science Society of America, the International Soil Science Society, and the International Humic Substances Society.

Dr. Elissa Levine is a Soil Scientist in the Biospheric Sciences Branch at the NASA Goddard Space Flight Center since September, 1986. Prior to her employment at NASA, Dr. Levine served as a Resident Research Associate sponsored by the National Academy of Sciences, National Research Council. She was also a Post-Doctoral Fellow in the Plant and Soils Department at the University of Connecticut. Dr. Levine is a member of the American Society of Agronomy, the Soil Science Society of America, Society of Soil Scientists of Southern New England, Gamma Sigma Delta (Agricultural Honor Society), and Sigma Delta Epsilon (Graduate Women in Science).

Dr. B.A. Stewart is a Distinguished Professor of Soil Science, and Director of the Dryland Agriculture Institute at West Texas A&M University, Canyon, Texas. Prior to joining West Texas A&M University in 1993, he was Director of the USDA Conservation and Production Research Laboratory, Bushland, Texas. Dr. Stewart is past president of the Soil Science Society of America, and was a member of the 1990-93 Committee of Long Range Soil and Water Policy, National Research Council, National Academy of Sciences. He is a Fellow of the Soil Science Society of America, American Society of Agronomy, Soil and Water Conservation Society, a recipient of the USDA Superior Service Award, and a recipient of the Hugh Hammond Bennett Award by the Soil and Water Conservation Society.

Contributors

R.R. Allmaras, USDA Agricultural Research Service, University of Minnesota, 439 Borlaug Hall, 1991 Upper Buford Circle, St. Paul, Minnesota 55108.

Denis A. Angers, Agriculture Canada, Station de Recherches, 2560 Boulevard Hochelaga, Sainte-Foy, Quebec G1V 2J3, Canada.

T.O. Barnwell, Jr., Environmental Research Laboratory, U.S. Environmental Protection Agency, Athens, Georgia 30613.

Sandra Brown, Department of Forestry, University of Illinois, Urbana, Illinois 61801.

Aziz Bouzaher, Environmental Division, The World Bank, 1818 H Street, N.W., Washington, D.C. 20433.

J. Mac Callaway, Environmental Services Group, RCG/Hagler Bailly, Inc., Boulder, Colorado 80301.

C.A. Cambardella, National Soil Tilth Laboratory, USDA Agricultural Research Service, Ames, Iowa 50011.

L.J. Cihacek, Department of Soil Science, North Dakota State University, Fargo, North Dakota 58105-5638.

C.E. Clapp, USDA Agricultural Research Service, University of Minnesota, 439 Borlaug Hall, 1991 Upper Buford Circle, St. Paul, Minnesota 55108.

Richard G. Cline, Forest Environment Research Staff, USDA Forest Service, Washington, D.C., 20013.

V.L. Cochran, USDA Agricultural Research Service, 309 O'Neill Building, Fairbanks, Alaska 99775.

C.V. Cole, USDA Agricultural Research Service, Natural Resource Ecology Laboratory, Colorado State University, Fort Collins, Colorado 80523.

C.R. Crozier, Wetland Biogeochemistry Institute, Louisiana State University, Baton Rouge, Louisiana 70803-7511.

A.S. Donigian, Jr., AQUA TERRA Consultants, Mountain View, California 94043.

John M. Duxbury, Department of Soil, Crop, and Atmospheric Sciences, Cornell University, Bradfield and Emerson Halls, Ithaca, New York 14853.

D.J. Eckert, Department of Agronomy, Ohio State University, 2021 Coffey Road, Columbus, Ohio 43210-1085.

Robert L. Edmonds, College of Forest Resources, University of Washington, Seattle, Washington 98195.

Edward T. Elliott, Natural Resource Ecology Laboratory, Colorado State University, Fort Collins, Colorado 80523.

N.R. Fausey, USDA Agricultural Research Service, Soil Drainage Research Unit, Ohio State University, Columbus, Ohio 43210.

G.G. Gaston, Department of Civil Engineering, Oregon State University, Corvallis, Oregon 97331-2302.

D.J. Greenland, Department of Soil Science, University of Reading, United Kingdom.

Peter M. Groffman, Institute of Ecosystem Studies, Box AB, Millbrook, New York 12545.

L.A. Harding, Natural Resource Ecology Laboratory, Colorado State University, Fort Collins, Colorado 80523.

Phyllis Henderson, Biological Sciences Center, Desert Research Institute, University of Nevada, Reno, Nevada 89501.

Derald J. Holtkamp, Resource and Environmental Policy Division, Center for Agricultural and Rural Development, Iowa State University, Ames, Iowa 50011.

D.R. Huggins, University of Minnesota, Southwest Minnesota Experiment Station, Lamberton, Minnesota 56152.

R.C. Izaurralde, Department of Soil Science, University of Alberta, Edmonton, Alberta T6G 2E3, Canada.

R.B. Jackson, IV, Environmental Research Laboratory, U.S. Environmental Protection Agency, Athens, Georgia 30613.

Dale W. Johnson, Biological Sciences Center, Desert Research Institute, University of Nevada, Reno, Nevada 89501.

Mark G. Johnson ManTech Environmental Technology Inc., U.S. EPA Environmental Research Laboratory, 200 S.W. 35th Street, Corvallis, Oregon 97333.

N.G. Juma, Department of Soil Science, University of Alberta, 4-42 Earth Science Building, Edmonton, T6G 2E3, Canada.

Katsumi Kumagai, Yamagata Prefecture Agricultural Experiment Station, Yamagata 990-02, Japan.

J. Kimble, National Soil Survey Center, Soil Conservation Service, Federal Building, Room 152, 100 Centennial Mall North, Lincoln, Nebraska 68508-3866.

T.P. Kolchugina, Department of Civil Engineering, Oregon State University, Corvallis, Oregon 97331-2302.

R. Lal, Department of Agronomy, Ohio State University, 2021 Coffey Road, Columbus, Ohio 43210-1085.

J.A. Lamb, University of Minnesota, Department of Soil Science, 439 Borlaug Hall, 1991 Upper Buford Circle, St. Paul, Minnesota 55108.

Changsheng Li, Institute for the Study of Earth, Oceans, and Space, University of New Hampshire, Durham, New Hampshire 03824.

T.J. Logan, Department of Agronomy, Ohio State University, 2021 Coffey Road, Columbus, Ohio 43210-1085.

S.S. Malhi, Agriculture Canada Research Station, Lacombe, Alberta T0C 1S0, Canada.

Bruce A. McCarl, Department of Agricultural Economics, Texas A&M University, College Station, Texas 77840.

Mark J. McDonnell, University of Connecticut, Bartlett Arboretum, 151 Brookdale Road, Stamford, Connecticut 06903.

A.K. Metherell, AgResearch, c/o Soil Science Department, P.O. Box 84, Lincoln University, New Zealand.

Katsuyuki Minami, National Institute of Agro-Environmental Sciences, Tsukuba 305, Japan.

William J. Mitsch, School of Natural Resources, The Ohio State University, 2021 Coffey Road, Columbus, Ohio 43210.

A.R. Mosier, USDA Agricultural Research Service, 301 S. Howes Street, Fort Collins, Colorado 80522.

M. Nyborg, Department of Soil Science, University of Alberta, Edmonton, Alberta T6G 2E3, Canada.

W.H. Patrick, Jr. Wetland Biogeochemistry Institute, Louisiana State University, Baton Rouge, Louisiana 70803-7511.

W.J. Partan, Natural Resource Ecology Laboratory, Colorado State University, Fort Collins, Colorado 80523.

A.S. Patwardhan, AQUA TERRA Consultants, Mountain View, California 94043.

Keigh Paustian, Natural Resource Ecology Laboratory, Colorado State University, Fort Collins, Colorado 80523.

G.A. Peterson, Department of Agronomy, Colorado State University, Fort Collins, Colorado 80523.

Steward T.A. Pickett, Institute of Ecosystem Studies, Box AB, Millbrook, New York 12545.

Richard V. Pouyat, U.S. Forest Service, Northeastern Forest Experiment Station, c/o SUNY College of Environmental Science and Forestry, 5 Moon Library, Syracuse, New York 13210.

Randall Reese, Resource and Environmental Policy Division, Center for Agricultural and Rural Development, Iowa State University, Ames, Iowa 50011.

G. Philip Robertson, Department of Crop and Soil Sciences, Michigan State University, East Lansing, Michigan 48824.

A.L. Rowell, Computer Sciences Corporation, Athens, Georgia 30613.

V.A. Rozhkov, Dokuchaev Soil Institute, Moscow, Pyzhevskyi per., 7, Russia 109017.

Gregory A. Ruark, Forest Environment Research Staff, USDA Forest Service, Washington, D.C., 20013.

R. Neil Sampson, American Forests, 1516 P Street NW, Washington, D.C. 20005.

S.F. Schlentner, USDA Agricultural Research Service, 309 O'Neill Building, Fairbanks, Alaska 99775.

Jason Shogren, Department of Economics, Iowa State University, Ames, Iowa 50011.

A.Z. Shwidenko, International Institute for Applied Systems Analysis, Laxenburg, Austria A-2361.

Thomas G. Siccama, School of Forestry and Environmental Studies, 370 Prospect Street, Yale University, New Haven, Connecticut 06511.

Whendee L. Silver, School of Forestry and Environmental Studies, 370 Prospect Street, Yale University, New Haven, Connecticut 06511.

E.D. Solberg, Soil and Crop Management Branch, Alberta Agriculture, Food and Rural Development, Edmonton, Alberta T6H 5T6, Canada.

E.B. Sparrow, USDA Agricultural Research Service, 309 O'Neill Building, Fairbanks, Alaska 99775.

B.A. Stewart, Dryland Agriculture Institute, West Texas A&M University, Canyon, Texas 79016.

Joel P. Tilley, School of Forestry and Environmental Studies, 370 Prospect Street, Yale University, New Haven, Connecticut 06511.

Haruo Tsuruta, National Institute of Agro-Environmental Sciences, Tsukuba 305, Japan.

G. Uehara, Department of Agronomy and Soil Science, University of Hawaii at Manoa, Honolulu, Hawaii 96822.

M.G. Ulmer, Department of Soil Science, Norgh Dakota State University, Fargo, North Dakota 58105-5638.

T.S. Vinson, Department of Civil Engineering, Oregon State University, Corvallis, Oregon 97331-2302.

Daniel J. Vogt, School of Forestry and Environmental Studies, 370 Prospect Street, Yale University, New Haven, Connecticut 06511.

Kristiina A. Vogt, School of Forestry and Environmental Studies, 370 Prospect Street, Yale University, New Haven, Connecticut 06511.

R. Paul Voroney, Department of Land Resource Science, University of Guelph, Guelph, Ontario N1G 2W1, Canada.

Z.P. Wang, Wetland Biogeochemistry Institute, Louisiana State University, Baton Rouge, Louisiana 70803-7511.

K.B. Weinrich, Computer Sciences Corporation, Athens, Georgia 30613.

Xinyuan Wu, School of Natural Resources, The Ohio State University, 2021 Coffey Road, Columbus, Ohio 43210.

Kazuyuki Yagi, National Institute of Agro-Environmental Sciences, Tsukuba 305, Japan.

Wayne C. Zipperer, U.S. Forest Service, Northeastern Forest Experiment Station, c/o SUNY College of Environmental Science and Forestry, 5 Moon Library, Syracuse, New York 13210.

Contents

C. Forest Ecosystems

D. Cold Ecosystems

E. Wetlands and Rice Paddies

F. Arid Lands

G. Tropical Ecosystems

H. Policy Issues

I. Managing Soils for Mitigating Greenhouse Effect

J. Research and Development Priorities

CHAPTER 1

World Soils as a Source or Sink for Radiatively-Active Gases

R. Lal, J. Kimble, and B.A. Stewart

I. Introduction

Terrestrial ecosystems, comprising world biota and soils, play a major role in regulating atmospheric concentrations of radiatively-active gases. Oceans comprise the largest reservoir for carbon and may play a major role in buffering the atmospheric concentration of CO_2. The geological reservoir has become active only within the last two centuries since humans began mining fossil fuel.

Distribution of carbon pool among various reservoirs, based on estimates by Post et al. (1990) and Sundquist (1993), are shown in Figure 1. In addition to being a major carbon pool, world soils influence atmospheric concentration of CO_2 and other radiatively-active gases through their effects on biota. Estimates of annual fluxes of carbon among various pools shown in Table 1 indicate a net imbalance of about 1.8 $\pm$ 1.4 gigaton C yr^{-1}. This imbalance is attributed to the terrestrial ecosystems among which world soils are the major component. Land use and soil management can influence both eflux and influx of C between soil and the atmosphere, and of several other radiatively active gases e.g., N_2O, NO_x, CH_4, and CFCs.

I. Soil Degradation and Greenhouse Gas Emissions

Soil is a principal medium for plant growth, and about 99% of the human food supply comes from soil and related ecosystems. Soil degradation, reduction in productive capacity due to misuse and mismanagement aggravated by harsh environmental conditions is rapidly increasing because of increasing demand on the finite extent, and fragile nature of soil resources. Worldwide, about 2 billion ha of soil have been degraded to some extent (ISRIC, 1991-92; WRI, 1992-93), and there is a widespread land hunger especially in densely-populated Asia (Pimentel et al., 1994). Land hunger leads to cultivation of unsuitable and marginal lands for food crop production or grazing that further exacerbates soil degradation and feeds the vicious cycle. In addition to decreasing food production, soil degradation accentuates environmental problems. In developed economies, soil degradation leads to increased reliance on agricultural chemicals (e.g., fertilizers, pesticides) and irrigation to off-set adverse effects on productivity. These chemicals, especially fertilizers, may accentuate greenhouse gas emissions from soil-related processes. In developing economies, soil degradation leads to deforestation and expansion of agricultural activities to marginal lands in ecologically-sensitive ecoregions (Myers, 1989). Deforestation and cultivation of grasslands are among the major causes of greenhouse gas emissions (Houghton, 1994).

Depending on the interactive effects among biophysical and socioeconomic factors, there are three principal types of soil degradation i.e., physical, chemical, and biological. Soil physical degradation is set-in-motion by decline in structural attributes that lead to crusting, compaction, low infiltration rate, high runoff rate, and accelerated soil erosion (Figure 2). Worldwide 83 x 10^6 ha or 4% of degraded soils have been subjected to some degree of physical degradation (Oldeman et al., 1990). Accelerated soil erosion is

ISBN 1-56670-117-1/95/$0.00+.50

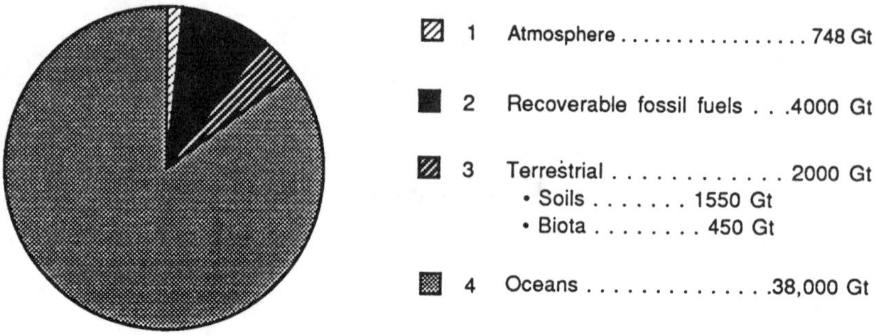

Figure 1. 1988-Estimates of carbon pool among various reservoirs.

Table 1. Carbon exchange among various reservoirs

Flux		Reservoir	Rate (gigaton C yr^{-1})
Efflux to the atmosphere	(i)	Fossil fuel burning	5.3
	(ii)	Land use	0.6-2.6
	(iii)	Plant respiration	40-60
	(iv)	Residue decay	50-60
		Sub-total	95.9-127.9
Influx from the atmosphere	(i)	Photosynthesis	100-120
	(ii)	Ocean uptake	1.6-2.4
		Sub-total	101.6-122.4
Imbalance (eflux-influx)			1.8 ± 1.4

(Calculated from Post et al., 1990; Sundquist, 1993.)

the most severe form of physical degradation, and about 80% of the world's agricultural land presumably suffers moderate to severe erosion and 10% slight to moderate erosion (Speth, 1994). Worldwide 1094 x 10^6 ha or 56% of degraded soils are prone to erosion by water and 548 x 10^6 ha or 28% of degraded soils are prone to wind erosion (Oldeman et al., 1990). Chemical degradation leads to loss of bases from the soil solum and to acidification. Nutrient imbalance, deficiency of some nutrients and toxicity of others, is a major problem of low productivity in vast areas of the tropics and sub-tropics. Worldwide 240 x 10^6 ha or 12% of degraded soils have been subjected to some degree of chemical degradation (Oldeman et al., 1990). Soil biological degradation, reduction in soil organic carbon content and in biomass carbon with concomitant effects on activity and species diversity of soil fauna and flora, has a major effect on greenhouse gas emissions from soil. These soil degradative processes are driven by socio-economic and political factors and in turn affect flux of radiatively-active gases from soils (Figure 2). Soil degradative processes, and especially the biological degradation may drastically influence eflux of CO_2, N_2O, NO_x and CFCs and decrease influx of CH_4 (Figure 2).

III. Soil Management and Carbon Dynamics

The magnitude and type of greenhouse gas emissions from soil degradative processes depend on land use, cropping systems, and soil management. Soil management affects carbon dynamics and gaseous emissions through its influence on soil properties and processes (Figure 3). Principal soil properties affected by soil management are soil moisture and temperature regimes, aeration, and aggregation. Micro-aggregation can

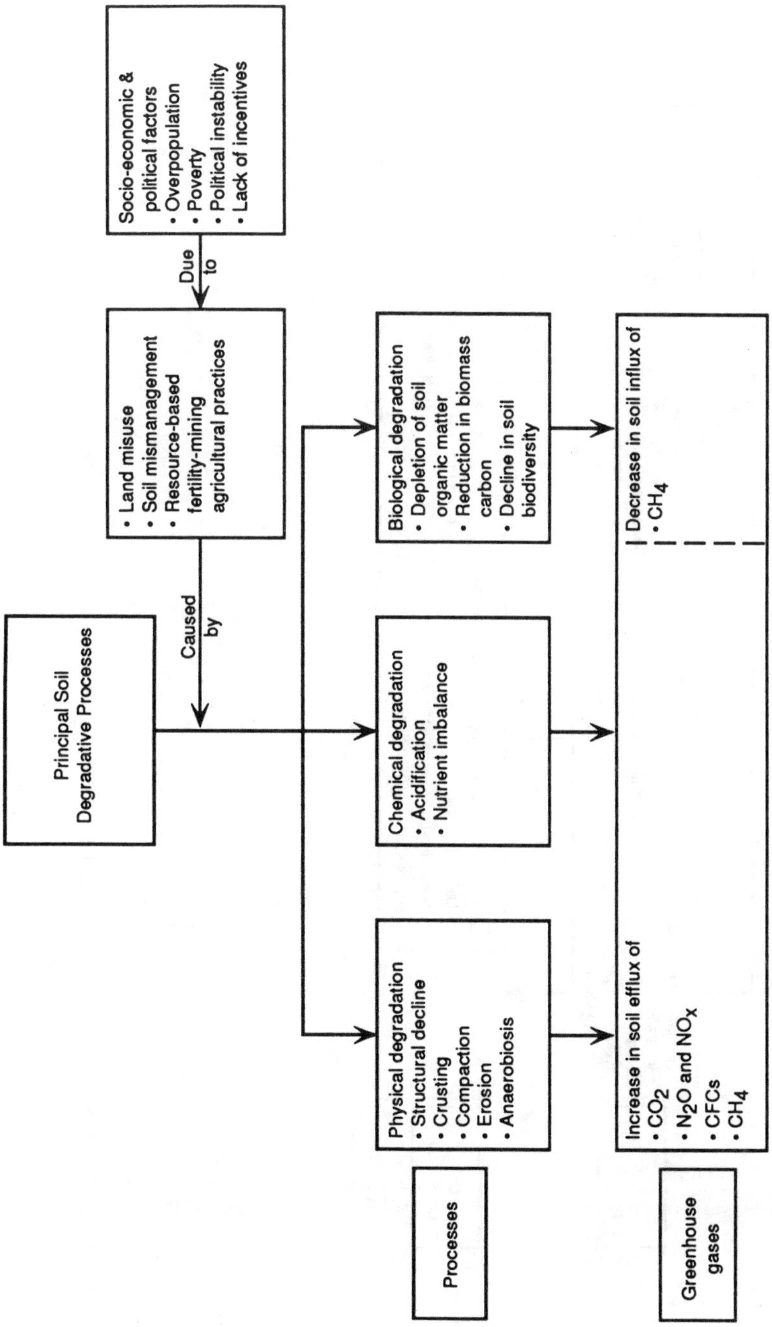

Figure 2. Interactive effects of socio-economic and biophysical factors on soil degradation and greenhouse gas emissions.

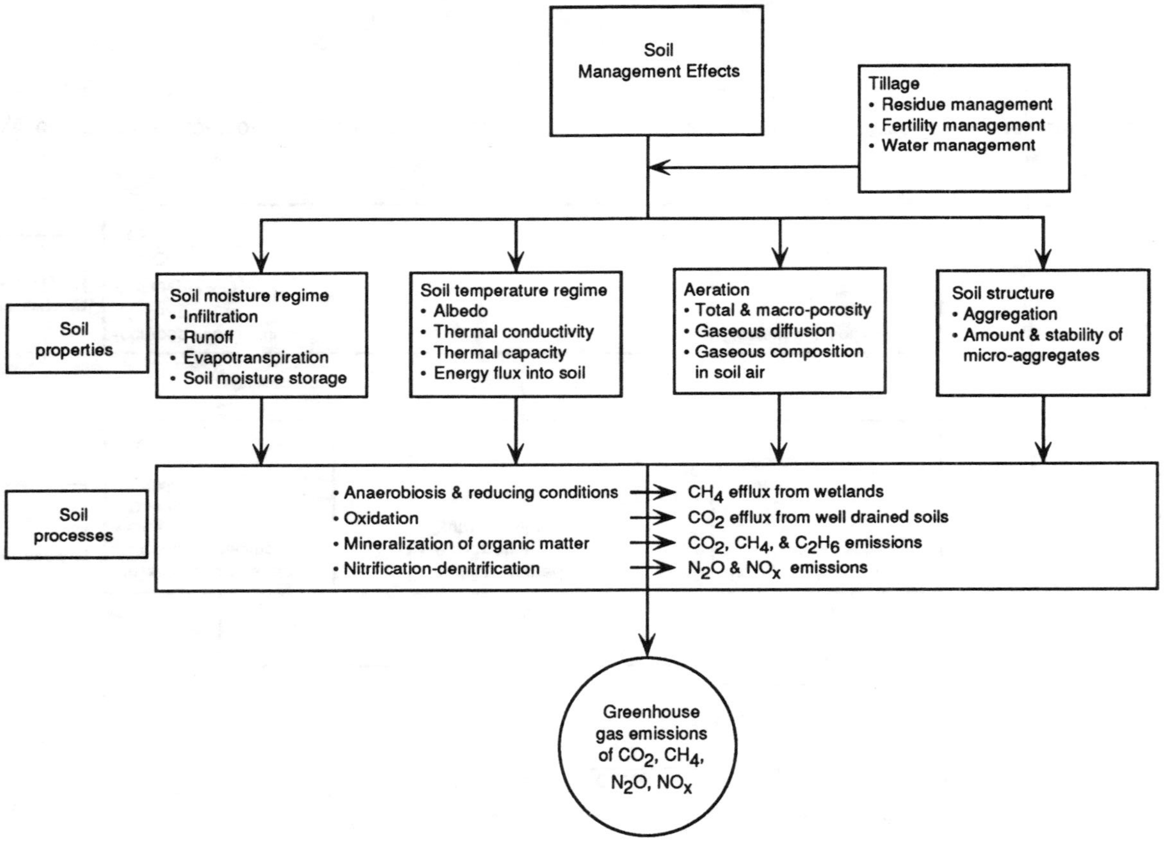

Figure 3. Soil management effects on greenhouse gas emissions from soil-related processes.

be grossly affected by soil management with strong effects on release of carbon hitherto inaccessible even to microbes.

Mineralization of carbon and nitrogen in soil follow a similar pattern. A simple model to predict the rate of change of C (and N) in soil is shown in equation 1 (Stevenson, 1982):

$$\frac{dC}{dt} = a - kC \qquad \textbf{(1)}$$

where k is the decomposition constant, C (or N) is the carbon content of a given mass of soil at time t, and a is the accretion constant reflecting the amount of C added to the soil through agricultural operations e.g., crop residue return, application of compost and farmyard manure, dung from grazing cattle, etc. It is the magnitude of the term (a - kC) that determines whether soil is a sink (sequesteration) or source (efflux) of carbon. Magnitude of the decomposition constant k depends on land use, cropping systems, soil and crop management (Lal, 1989; Figure 4). All other factors remaining the same, the magnitude of decomposition constant k is generally more for tropical than temperate environment. Furthermore, the magnitude is low for natural ecosystems, and for agricultural practices based on science-based systems of soil and crop management involving judicious off-farm inputs and conservation tillage involving crop residue return. The magnitude of k is also low for soil restorative measures that enhance soil fertility. In contrast, the magnitude of k is high for practices that involve deforestation, expansion of agricultural activities to ecologically-sensitive ecoregions, biomass burning, cultivation of marginal lands, resource-based and low-input agriculture, and plow-based methods of seedbed preparation that accentuate extremes of soil moisture and temperature regimes, and systems that exacerbate soil degradative processes (Figure 4).

Soil tillage is an important tool to regulate decomposition constant k. While plow-based tillage may increase k, conservation tillage systems decrease k. Kern and Johnson (1991) estimated that the organic carbon content of soils in the contiguous United States to 30 cm depth is 5,304 to 8,654 Tg C with 1,710 to 2,831 Tg C at 0 to 8 cm depth, and 1,383 to 2,240 Tg C at 8 to 15 cm depth. Maintaining the 1990 levels of conventional tillage (73% of arable land) until 2020 would result in loss of 46 to 78 Tg C. Increase in conservation tillage area from 27% in 1990 to 57% by 2020 would result in loss of 27 to 45 Tg C. However, increase in conservation tillage to 76% of the land by 2020 would result in loss of 14 to 24 Tg C. Widespread adoption of soil conserving systems, especially in the tropics and sub-tropics, can lead to substantial reduction in loss of soil organic carbon and emission of greenhouse gases.

IV. Restoration of Degraded Lands

Reliable estimates of the extent of global soil degradation do not exist. The UNEP (1986) estimated that about 2 billion ha of once biologically productive land has been rendered unproductive through irreversible degradation. A report by ISRIC (1991-92) estimates that worldwide about 2×10^9 ha of land has been degraded to some extent. From a global perspective, restoration of degraded lands offers a tremendous potential to sequester atmospheric carbon. Restoration of biological productivity of these lands would enhance their soil organic carbon content and render them as an effective terrestrial carbon sink. If soil organic carbon content of 2×10^9 ha of land for the top 0-10 cm layer could be increased by 0.01% yr^{-1}, it would amount to C influx of 250 Tg yr^{-1} assuming soil bulk density of 1.25 Mg m^{-3}. This influx is 2 to 2.5 times the annual C fixation of 100 to 120 Tg by world biota through photosynthesis (Post et al., 1990)

Restoration of degraded lands requires an adoption of world soil policy at the global level. To be effective, it may be implemented through the United Nations as a World Soil Charter on Restoration of Degraded Lands for Carbon Sequesteration. Restorative practices would involve large-scale afforestation, application of fertilizers and soil amendments, construction of runoff management and erosion control devices, ban against deforestation and biomass burning, and incentives for adoption of conservation tillage based on return of crop residues to the soil and frequent use of cover crops and planted fallows in food crop rotations. Improvements of grasslands and pastures is equally crucial to carbon sequesteration. Controlled grazing at low stocking rate, seeding improved pasture species, and using chemicals and soil amendments are important considerations in restoration of degraded pastures.

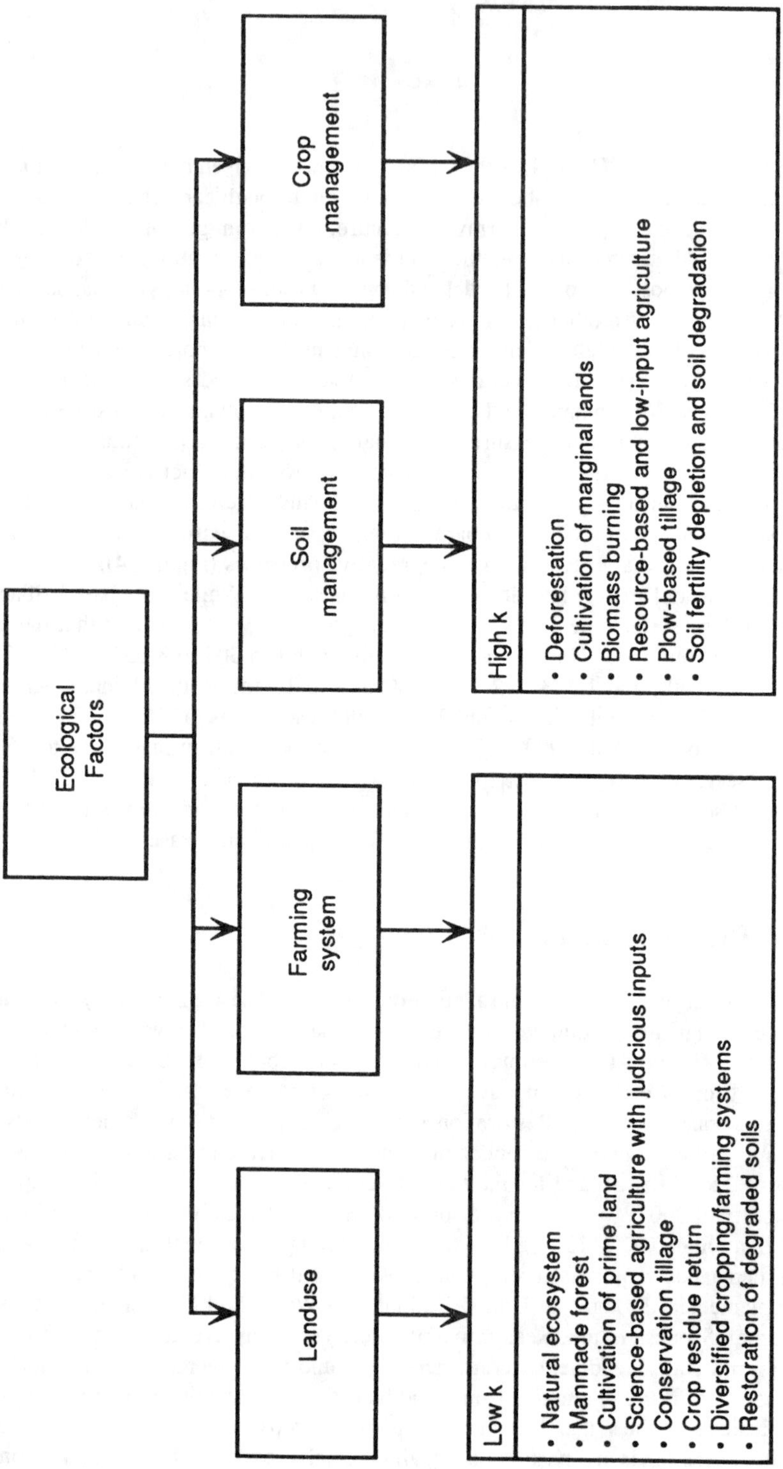

Figure 4. Factors affecting decomposition constant k of soil organic carbon.

V. Conclusions

World soil misuse and soil mismanagement set in motion a wide range of soil degradative processes. Soil degradation by physical, chemical, and biological processes render soil as a major source of CO_2, CH_4, N_2O, NO_x and other radiatively-active gases. Carbon emission from soil-related processes is enhanced by mineralization of soil organic matter leading to crusting, compaction, anaerobiosis, erosion, leaching, acidification, and decline in soil biodiversity.

Agricultural practices that enhance gaseous emissions from soil include deforestation, biomass burning, plow-based soil tillage, cultivation of marginal lands, and resource-based or low-input agriculture. In contrast, world soils can be used as a major sink for atmospheric carbon by enhancing soil organic carbon content or decreasing the rate of its mineralization. Adoption of soil-enhancing agricultural practices would increase soil organic carbon content or retard the rate of its depletion. These practices include conservation tillage, runoff management and erosion control, use of prime agricultural land on science-based inputs, and use of improved crops and cultivars. There is a need to develop institutional support and identify policy incentives for farmers in developing economies to adopt these agricultural practices. Furthermore, carbon sequesteration in world soils on a global scale can be accelerated by restoration of degraded lands. It is important that world soil policy be developed and implemented through the United Nations for using world's degraded soils as a sink for atmospheric carbon.

References

Houghton, R.A., 1994. Changes in the storage of terrestrial carbon since 1850. In: Lal et al. (eds.) *Soils and Global Change.* Lewis Publishers, Chelsea, MI.

ISRIC, 1991-1992. Bi-annual report. ISRIC, Wageningen, The Netherlands.

Kern, J.S. and M.G. Johnson, 1991. The impact of conservation tillage use on soil and atmospheric carbon in the contiguous United States. Man Tech Environmental Technology, Inc., USEPA-Environmental Research Laboratory, Corvallis, Oregon.

Lal, R., 1989. Soil as a potential source or sink of carbon in relation to the greenhouse effect. In: Greenhouse Gas Emissions from Agricultural Systems. Vol. 2-Appendix, USEPA, Washington, D.C.

Myers, N., 1989. Deforestation rates in tropical forests and their climatic implications. Friends of the Earth report, London.

Oldeman, L.R., R.T.A. Hakkeling, and W.G. Sombrock, 1990. World map of the status of human-induced soil degradation. An explanatory note. ISRIC, Wageningen, The Netherlands.

Pimentel, D., R. Harman, M. Pacenza, J. Pecarsky, and M. Pimentel, 1994. Natural resources and an optimum human population. *Population and Environment* 15 (5): 347-369.

Post, W.M., T-H. Peng, W.R. Emanuel, A.W. King, V.H. Dale, and D.L. DeAngelis, 1990. The global carbon cycle. *American Scientist* 78: 312-326.

Speth, J.G., 1994. Towards an effective and operational international convention on desertification. International Convention on Desertification, United Nations' International Negotiating Committee, New York.

Stevenson, F.J., 1982. *Humus chemistry: genesis, composition, reactions.* J. Wiley & Sons, New York, 443 pp.

Sundquist, E.T., 1993. The global carbon dioxide budget. *Science* 259: 934-941.

UNEP, 1986. Farming systems principles for improved food production and the control of soil degradation in arid and semi-arid tropics. ICRISAT, Hyderabad, India.

World Resources Institute, 1992-93. Towards sustainable development. WRI, Washington, D.C., 367 pp.

CHAPTER 2

Land Use and Soil Carbon in Different Agroecological Zones

D.J. Greenland

I. Introduction

The quantity of carbon in soils is substantial. It is larger than the total atmosphere and biosphere store, and may be equal to as much as a quarter of the total of fossil fuel reserves. It is also subject to considerable change as land use changes. The importance of these changes is not only to the global carbon balance, but also to the present and future potential of the soil to produce sufficient food and fiber to feed and clothe the world and to meet the demand for wood for fuel, building, and other domestic purposes. There can be no doubts about the importance of an accurate estimate of the store of carbon in the soil,and the rate at which it is changing, and may change in the future as land use changes. The significance is well illustrated by the suggestion of Wallace et al. (1990) that soil organic matter could be increased at a rate equal to the present rate of carbon accumulation in the atmosphere, and so largely neutralise the greenhouse effect. This paper examines this suggestion in terms of current knowledge of soil carbon levels, and the dynamics of soil organic matter changes.

II. The Quantity of Soil Carbon - Problems of Determination and Current Estimates

As Franzmeier et al. (1985) noted some years back, "Of the various C reservoirs interacting with the atmosphere.....soil C is the largest, but its mass is the least certain." The statement remains true and before considering the dynamics of soil carbon it is necessary to put the available evidence within the context of uncertainties in the measurement of soil carbon.

The quantity of carbon in the soil has to be determined by analysis of samples of different soils for the amount of carbon contained per unit weight of soil. This proportion by weight has then to be converted to a volume basis by determination of the bulk density of the soil, and then the area and depth of soil to which those measurements refer estimated. The measurements for individual soils have then to be integrated on a global basis. At each stage there is a considerable margin of error.

A. Analytical Errors.

The most commonly used method for determination of soil carbon is by wet dichromate oxidation. From the time of its introduction this method has been known to underestimate the carbon content of the soil when compared to values obtained by dry combustion (Walkley, 1947.) If it is used without refluxing the discrepancy may be anything between 10 and 50 per cent; with refluxing the error should be less than 10%

ISBN 1-56670-117-1/95/$0.00+.50

(Kalembasa and Jenkinson, 1973; Metson et al., 1979.) Empirical correction factors of 1.1 to 1.3 are commonly employed with the method. Nevertheless its simplicity, and the fact that it determines the readily oxidised material, which is thought to be that which is most biologically accessible,has enabled it to maintain its position as the most widely used method.

Dry combustion analysis may also underestimate soil carbon. Kalembasa and Jenkinson (1973) found a small increase in carbon recovery on increasing the standard combustion temperature from 700 to 1000 degrees Celsius, and Tabatabai and Bremner (1970) and Carr (1973) found higher recoveries when the combustion was done at the temperatures around 1600 degrees attained in an induction furnace. While the carbon not oxidized by dichromate may mostly be carbon not involved in biological reactions, the carbon not converted to carbon dioxidein normal combustion analysis may originally have been in any organic or inorganic form. Incomplete combustion at 800 degrees Celsius is believed to arise from the formation of silicon carbides and other compounds which are only oxidized at the temperatures achieved in the induction furnace. The discrepancies are usually much larger for subsoil carbon than for carbon in surface soils.

These analytical errors mean that there may be a systematic underestimate of total soil carbon,of the order of 10 to 30 per cent, in many of the analyses reported for surface soils, and rather greater discrepancies for subsoils.

B. Problems in Bulk Density Determination

Determinations of the bulk density of the relevent soil depth can also involve a systematic error. This mostly arises when laboratory rather than field determinations are made. Laboratory samples are usually relatively small clods or aggregates. While their bulk density can be measured accurately, the figure obtained will usually be too high as the inter-clod or inter-aggregate spaces are ignored. The effect is a systematic overestimate of the bulk density of the whole soil horizon being studied. In this case the error is probably below 10%, and in the opposite direction as far as total carbon in the soil is concerned to the systematic error in the chemical determination.

C. Depth and Time Scale Problems

A further systematic error may arise due to omission of all or part of the subsoil carbon. Many soils contain significant amounts of carbon in the subsoil. Although carbon dating has shown that subsoil carbon in temperate zone soils is usually of greater age than that in surface soils (Table 1a) the limited data from tropical soils suggests that there is a considerably faster turnover of organic matter in the subsoil as well as in the surface soil (Table 1b,c.) Detwiler (1985) has suggested that on clearing tropical forests little change is produced in the soil below about 60 cms. This conclusion requires further verification. There is known to be substantial root growth and other biological activity well below such depth, and in many tropical soils well below 1 meter,the depth often used to assess total global soil carbon.

Changes in soil carbon at depths below 1m are undoubtedly slower than in surface soils. However in the context of global change the time scale must be reckoned in much larger units than are used in agricultural studies. Lugo and Brown (1986) believe that the normal time periods used to estimate achievement of a steady state in forest soils (around 100 years) are generally too short.

It is very desirable that a better basis for determining to what depth soil measurements for carbon balance studies should be made, and the appropriate time intervals to be considered, is established.

D. Sampling and Extrapolation Problems

Finally and most importantly there are the sampling and extrapolation errors involved in relating the carbon in the horizon of the particular soil sampled to the area of similar soils from which it was taken, and integrating the information on a global basis.

Soil samples are usually collected for survey purposes from "typical pedons" and no systematic attempt is made to determine the variability of soil properties on a spatial basis (Arnold and Wilding, 1991). Random sampling errors for composite soil cores taken from plots used in most agronomic studies are known to be of the order of 10 to 15% for surface soils on plots established on a relatively uniform area

Table 1. Radiocarbon age of organic matter in soil (a) collected in the last century from the unmanured plot of Broadbalk field, Rothamsted, UK, and collected recently from rice soils in (b) Thailand and (c) the Philippines

Sampling depth	Organic carbon, %	Equivalent age, years
(a)		
0 - 23	0.94	1,400
23 - 46	0.61	2,000
46 - 69	0.47	3,700
114 - 137	0.24	9,900
206 - 229	0.20	12,100
(b)		
14 - 18	-	700
26 - 44	-	800
54 - 66	-	1,100
78 - 90	-	1,100
(c)		
22 - 28	-	500
54 - 66	-	800
78 - 88	-	1,000

(Data for (a) from Jenkinson, 1975; data for (b) and (c) approximated from the report of Dr. Becker-Heidmann, quoted by Scharpenseel, 1993.)

of land where all the soil belongs to one series. The error in using a profile sample will be very much larger. This is well illustrated by the data of Kimble et al. (1990) who compared the organic carbon contents of 2,715 soil profiles according to soil order, and showed that for five orders the coefficients of variation ranged from 42.4 to 70.09 per cent.

Differences between soils of different soil orders, or of different groups within an order, may be as great as or greater than the differences due to different land use. Land use will mostly dominate the more variable content of carbon in the surface soil, while the amount of carbon in the whole soil profile will be determined by the soil characteristics reflected in the soil classification.

E. Estimates of Total Global Soil Carbon

Clearly any estimate of global soil carbon must therefore be accepted with considerable reservations. Nevertheless there have been several estimates published, and these are rather consistent (Table 2) in arriving at a global figure between 1 and 3Pg. More details about the derivation of these estimates are given by Esser (1989) and Bouwman (1990), and in the papers cited in those publications.

III. Dynamics of Soil Organic Carbon.

There have been several attempts at estimating the extent to which anthropogenic factors can influence the quantity of global organic carbon, including the effects on phytomass as well as soil carbon. However before the value of these attempts can be assessed it is desirable to consider the basic factors controlling the dynamics of organic carbon in the soil.

A. Additions of Organic Material.

Of dominant importance is the quantity of organic material returned to the soil. Most is of course from vegetation growing on the soil. However it may also be deliberately added, and a very small quantity will usually be contributed by the action of autotrophic organisms.

Table 2. Estimates of global soil carbon (in thousands of millions of tonnes)

Reference	Amount
Ajtay et al., 1979	1,800
Bohn, 1976, 1982	2,200
Post et al., 1982	700 to 2,900
Brown and Lugo, 1982	1,500
Bouwman, 1990	1,700
Esser, 1990	1,500
Kimble, Eswaran, and Cook, 1990	1,000
Wallace, Wallace, and Cha, 1990	3,000
Scharpenseel, 1993	1,800 to 2,000

As well as the amount of organic carbon added the kind of material can be important. Leaf, root and woody material are well known to decompose at different rates,the lignified material decomposing more slowly than other organic matter. Other inherent characteristics of the organic materials added to the soil,and of the compounds formed in their decomposition, will effect the rate and extent of the decomposition process.

Jenkinson and Rayner (1977), Emanuel et al. (1984) and Parton et al. (1988) have all found it necessary to postulate several organic matter "compartments" where decomposition occurs at different rates to enable their models of organic matter change to fit experimental data and the data on organic matter age as determined by carbon dating.

Jenkinson, Adams, and Wild (1991) have recently calculated the annual input of carbon to each of 18 "life zones" (Table 3). They used a model of soil carbon dynamics in which added plant material was divided into "readily decomposable plant material" (DPM) and "resistant plant material" (RPM). They took the DPM/RPM ratio to be 0.25 for forests, 0.43 for tropical woodland and savanna, 0.5 for deserts, tundra and steppe, and 0.67 for cultivated land.

B. Decomposition Processes

The material which reaches the soil will be immediately subject to decomposition processes associated with the activities of the soil fauna and flora. The rate at which the added material decomposes is determined not only by its composition but also by intrinsic soil properties, and temperature and moisture conditions. Probably of most general importance is aeration, which is largely dependent on drainage.

In temperate regions in welldrained and aerated soils the oxidative decomposers are active and break down added organic materials rather easily, while in poorly aerated or anaerobic soils they act more slowly, and anaerobic decomposition is a much slower process. Thus poorly drained soils will tend to have greater amounts of organic carbon than well drained soils. In extreme cases of anaerobism muck and peat soils develop.

In the tropics anaerobic decomposers can operate at rates more comparable to the aerobic decomposers and differences are smaller. This has particular relevence for the soils in which rice is produced.

Soil acidity slows down the rate of decomposition (Ayanaba and Jenkinson, 1990; Greenland et al., 1992) by its effects on soil organisms. Soil texture influences soil aeration, and hence the rate of organic matter decomposition, through its influence on soil structure and air and water movement. Clay content, and clay type, may also influence decomposition rates through adsorption and occlusion of carbon in soil pores not accessible to the main decomposing organisms, and the formation of stable complexes between aluminum sites and the organic compounds (Wada, 1980; Virakornphanich et al., 1988). This type of complex formation between active hydrous oxides and humic materials is believed to account for the exceptionally high organic matter levels in many Andisols (Parfitt and Clayden, 1991) and the higher than expected levels in many Oxisols and Rhodustalfs (Greenland et al., 1992).

"Parent material" is sometimes stated to be an inherent factor determining the rate of organic matter decomposition in the soil (see e.g. Franzmeier et al., 1985). It does so partly by determining soil texture and mineralogy, and also by determining soil characteristics related to productivity and biological activity such as acidity and nutrient availability. Temperature also has a critically important influence on biological activity in the soil, and hence on the rate at which organic matter is decomposed.

Table 3. Stock of soil carbon and calculated annual input of carbon for soils in different "life zones," assuming all soils have 20% clay and carbon content of each life zone represents the equilibrium level for the existing climate and vegetation type

Life zone	Carbon stock C, Gt	Carbon input A, GT/an	A/C %
Tundra	191	0.9	0.47
Boreal desert	20	0.1	0.5
Cool desert	43	0.9	2.1
Warm desert	20	0.6	3.0
Tropical desert bush	2	0.1	5.0
Cool temperate steppe	120	2.7	2.3
Temperate thorn steppe	30	1.8	6.0
Tropical woodland and savanna	129	11.5	8.9
Boreal moist forest	49	0.8	1.6
Boreal wet forest	133	4.7	3.5
Temperate forest cool	43	3.1	7.2
Temperate forest warm	61	7.1	11.6
Tropical forest very dry	22	1.7	7.7
Tropical forest dry	24	1.1	4.6
Tropical forest moist	60	13.2	22.0
Tropical forest wet	78	15.3	19.6
Cultivated land	167	10.2	6.1

(From Jenkinson et al., 1991.)

In most instances agricultural practices will lead to lower rates of addition of organic matter to the soil, as well as higher rates of decomposition, through cultivation, drainage, and crop removal. On the other hand addition of fertilisers and manures which lead to greater crop yields and larger root systems and more crop residues will lead to greater levels of soil organic matter. While drainage causes more rapid decomposition of organic materials, irrigation will increase the amount of organic matter added to the soil primarily by stimulating plant growth.

C. The Equilibrium Concept

For any given soil and land management practice an equilibrium level of organic carbon will be attained if the practice is continued for a sufficiently long period. At equilibrium inputs are equal to outputs. For this to be true it is necessary that the intrinsic factors effecting the organic matter content remain constant. This is in fact seldom exactly correct, although a valuable approximation. Soil physical conditions for instance change with organic matter content, and as they improve productivity increases and so does the rate of addition of organic matter, further increasing the build-up of soil carbon.

Many experimental studies have shown that for a given land use the amount of organic matter in the soil tends to an equilibrium value. These studies have also shown that the simple relationship

$$dC/dt = A - kC$$

(where C is the soil carbon content, t is time, A is the addition of carbon to the system in unit time, and k a constant which measures the amount of carbon lost from the system in unit time) gives a reasonable approximation to the measured values. The agreement is better if organic nitrogen rather than carbon is used as the measure of soil organic matter, as there is almost always a more rapid initial loss of carbon as the C/N ratio narrows from the values typical of plant material to those of the soil (Jenkinson, 1990).

When additions equal losses

$$dC/dt = 0, \text{ and } C = A/k.$$

Thus if the annual addition of carbon to the soil is known, and the rate of decomposition of carbon in the soil, the equilibrium carbon content can be calculated. Alternatively, if a system has attained equilibrium, and the decomposition constant can be calculated, for instance using 14C data, the carbon addition can be calculated (Jenkinson et al., 1992.) Current models are considerably more sophisticated than those based on simple first order reaction equations, although the basic principles are still the same (see e.g. Jenkinson and Rayner, 1977; Van Veen and Paul, 1981; Parton et al., 1983; Emanuel et al., 1984; Jenkinson et al., 1987; Parton et al., 1988, and Verberne et al., 1990). In these models carbon is compartmentalised into fractions decomposing at different rates, but still according to first order reaction kinetics, and factors such as temperature, moisture content and clay content taken into account.

D. Other Factors Affecting Soil Carbon

Parton et al. (1988) recognise the importance of simultaneous additions of nitrogen, phosphorus, and sulfur to the soil in controlling the level of soil organic matter. It is well established by observation that the ratio of the quantity of carbon in a soil to that of nitrogen, phosphorus and sulfur remains approximately constant, whatever the level of organic matter in the soil. Given that P and S contents are intrinsic soil properties,they may set a limit to the level of soil organic matter which can be attained, if the levels of these elements are particularly low (Tate and Salcedo, 1988). More commonly, nitrogen is the element most likely to be limiting, so that the equilibrium level is set by the amount of nitrogen fixed by or added to the system.

In addition to oxidation of organic matter losses of carbon from the soil can also occur through erosion and from pedogenic carbonate due to soil acidification. Erosion can have major effects on plant growth and thereby reduce the return of organic matter to the soil. At the present time there is inadequate information available to estimate the significance of the effect, although attempts are currently being made to model the process (Bouwman, 1989; Biot, 1991). Loss of soil by erosion also means that carbon is removed from the point of loss. The eroded soil is however deposited somewhere so that there may be no escape of carbon to the atmosphere.

Carbonates are widely distributed in soils. They may be a source of carbon dioxide loss to the atmosphere as soil acidification occurs. Carbonate can also be precipitated in soils under appropriate pedogenic conditions. Again there is inadequate information to quantify the net effect on a global basis.

Neither of these latter mechanisms of carbon loss will be considered further in this paper, although their importance in some situations must be recognised.

IV. Changing Levels of Soil Carbon.

A. Land Use and Soil Carbon

Given the above background to the levels and amounts of carbon and the factors which control the levels in different soils, and recognizing the considerable limitations on the data, is it possible to reach any conclusions about how changes in land use are likely to affect the global soil carbon balance?

As noted above, the amount of carbon in the soil is determined in part by intrinsic soil properties and in part by land use and the inputs of organic matter to the soil. Climate also has a major effect through its control of soil moisture and temperature, and hence plant productivity and organic matter decomposition rates. Thus to arrive at any estimate of how soil carbon is likely to change requires that assessments are based on soil, land use, and climate characteristics of any particular region. In addition it is necessary that an assessment is made of the relation of existing soil carbon levels to equilibrium levels for the current and any proposed future land use.

Esser (1990) has made what is probably the most comprehensive attempt to relate past global soil carbon changes to land use. His data are derived from the FAO Soil Map of the World, the vegetation atlas of Schmithusen (1976), and the land use atlas of the Instituto Geographico de Agostini (1969, 1971, 1973.)

Based on his data Scharpenseel (1993) believes that 30 to 60 Pg of carbon have been lost from cultivated soils in the past 100 years. However Bertram (1986) estimated on the basis of 13C analyses that there had been a net soil gain of 60 Pg of carbon in the past 50 years. The discrepancy could be due to the movement of carbon from surface soils to deeper horizons by eluviation with clay particles and by podsolisation.

B. Models and the Prediction of Soil Carbon Changes

To predict future changes will require the use of models, related to different forms of land use. As noted above, several quite sophisticated models are now available to predict future levels. The models have been developed where a good database exists - the Jenkinson and Rayner model used the data provided by the field trials at Rothamsted which had been continued for more than 100 years and soil carbon determined at frequent intervals. The "Century" model of Parton et al. (1988) has been verified using the large volume of information available in the central grasslands region of the United States (Parton et al., 1989; Cole et al., 1989; Burke et al., 1989).

The challenge now is to analyse existing data and to collect further data relevant to a much wider range of soils, climates, and land use.

The relative areas of the major land use zones as used for the Global Assessment of Soil Degradation (GLASOD) are given in Table 4. An indication of how global soil and biomass carbon relate to land use is shown diagrammatically in Figure 1 (Goudriaan, 1993). Arable land is a little more than one tenth of the vegetated area,and changes in soil carbon due to changes in agricultural practices will only affect a corresponding proportion of the total store of soil carbon, or about 200Pg. Thus an annual increase of about 1.5% in the amount of carbon in agricultural soils would be needed to offset the current annual rate of increase of 3Pg of carbon as carbon dioxide in the atmosphere (Wallace et al., 1990).

C. Scope for Carbon Increase, Temperate Zone, Arable Soils.

Average yields of wheat in Europe and maize in North America in recent years have exceeded 5Mg/ha. The quantity of crop residues will be similar or rather larger and contain about 40% of carbon. Root residues and exudates will add a further amount of carbon to the soil. Thus if the residues are incorporated and most retained in the soil it is possible for those parts of the temperate areas producing high crop yields to make a sufficient contribution to the carbon sequestration requirement. However it is likely that most arable soils in the temperate zone are currently cultivated at or near the equilibrium level for the agricultural system practiced. The Rothamsted data (Jenkinson, 1991; Jenkinson et al., 1992) show that in spite of crop yields of the order of 6 to 7 Mg/ha soil carbon has reached a steady state only a little above the equilibrium level established for the unmanured plot (Figure 2).

In the Rothamsted experiments the crop residues other than roots and stubble are not returned to the soil. The observed increase must therefore arise from the roots and root slough and exudate and any litter fall during the life of the crop. Jenkinson et al. (1992) have recently shown that carbon added in this way may be calculated from the apparent steady state conditions established after many years of consistent land use, together with data for decomposition rates assessed from 14C turnover and measurement of biomass carbon.

The results (Table 5) show that the annual addition on the unmanured continuous wheat plot was 1.3 Mg/ha, and on the NPK plot was 1.7 Mg/ha. The removals of carbon in grain and straw were 0.93 and 3.51 Mg/ha bringing the Net Primary Production (NPP) to 2.2 and 5.2 Mg/ha respectively. However because the soil has attained the equilibrium state for the system of management currently practiced, sequestration of carbon is zero. Addition of straw to the soil would of course increase the soil carbon level. On the NPK plot this would almost exactly double the carbon input, and in time lead to a level of soil carbon almost twice the existing equilibrium level, assuming that the rate of decomposition was not significantly affected by the straw addition.

Until the steady state is again approached this could represent an annual carbon gain of the order of 1%. The areas in temperate regions where increasing crop yields and straw return are likely to lead to significantly higher equilibrium levels may be as much as a quarter of the agricultural land of the world, and hence such practices merit strong encouragement. Increases in soil carbon will also have occurred in recent years through changes from conventional plow and harrow cultivation to minimum tillage practices. The "set aside" policy now pursued in Europe and the United States may also lead to increasing organic

Table 4. Areas in millions of ha of agricultural land, permanent pasture, and forest and woodland, and the portions of these areas affected by human-induced soil degradation

	Agricultural land			Permanent pasture			Forest and woodland		
	Total	Degraded	%	Total	Degraded	%	Total	Degraded	%
Africa	187	121	65	793	243	31	683	130	19
Asia	536	206	38	978	197	20	1,273	344	27
S. America	142	64	45	478	68	14	896	112	13
C. America	38	28	74	94	10	11	66	25	38
N. America	236	63	26	274	29	11	621	4	1
Europe	287	72	25	156	54	35	353	92	26
Oceania	49	8	16	439	84	19	156	12	8
World	1,475	562	38	3,212	685	21	4,048	719	18

(From Oldeman et al., 1990.)

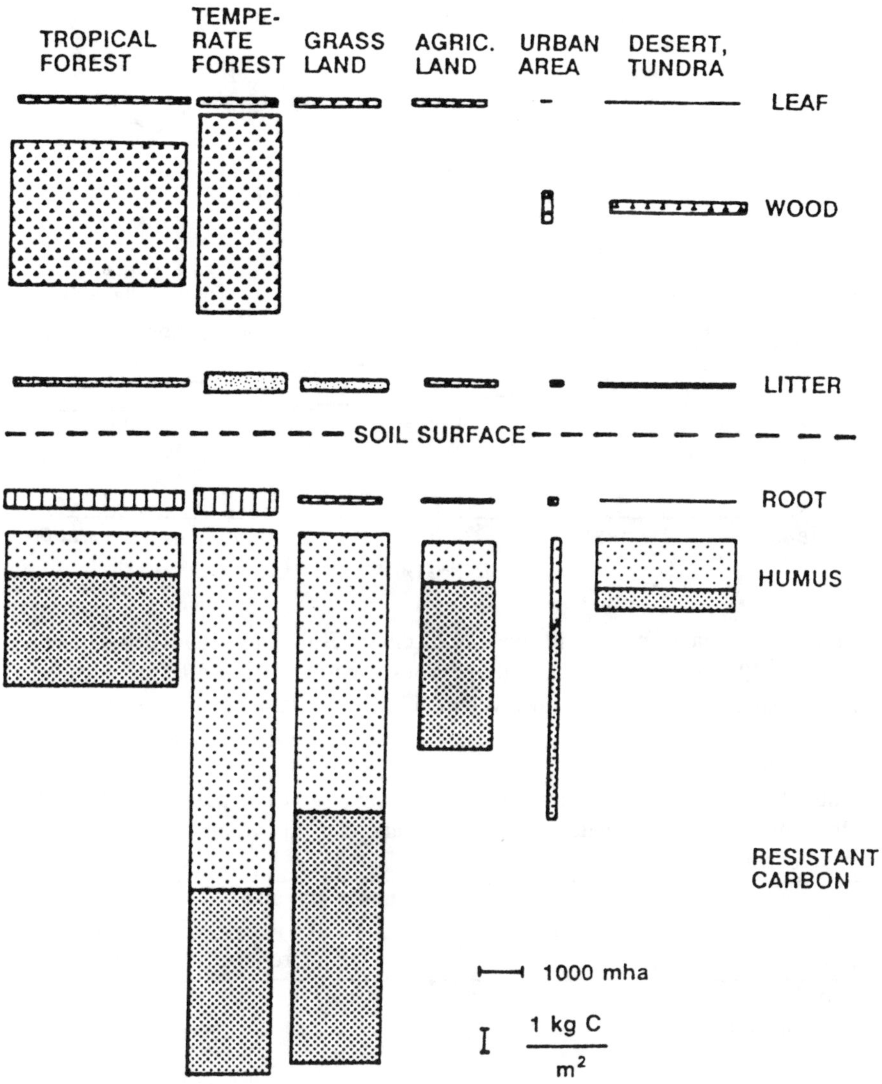

Figure 1. Carbon in the soil and land cover in different ecological zones. (From Goudriaan, 1993.)

matter levels in those soils no longer cultivated, provided that they are allowed to develop a vegetative cover.

Organic farming methods are unlikely to be economic for the majority of farmers. Thus it is to be expected that in temperate areas the increases in carbon sequestration are likely to arise from straw retention, minimum tillage practices, and yield optimization. Results from the Sanborn plots in Missouri (Wagner, 1989) support these conclusions, although they, like pasture-wheat rotation trials in Australia (Greenland, 1971; Russell, 1986), confirm the value of a legume in the management system.

D. Scope for Carbon Increase, Tropical Zone, Arable Soils

The dynamics of organic matter in tropical soils has attracted much interest and created much argument for many years (Greenland et al., 1992.) The topic has also been the subject of a recent major conference (Coleman et al., 1989.) The interest has arisen because of the importance of organic matter to the productivity of many soils in the tropics, and the rapidity with which readily observable changes occur.

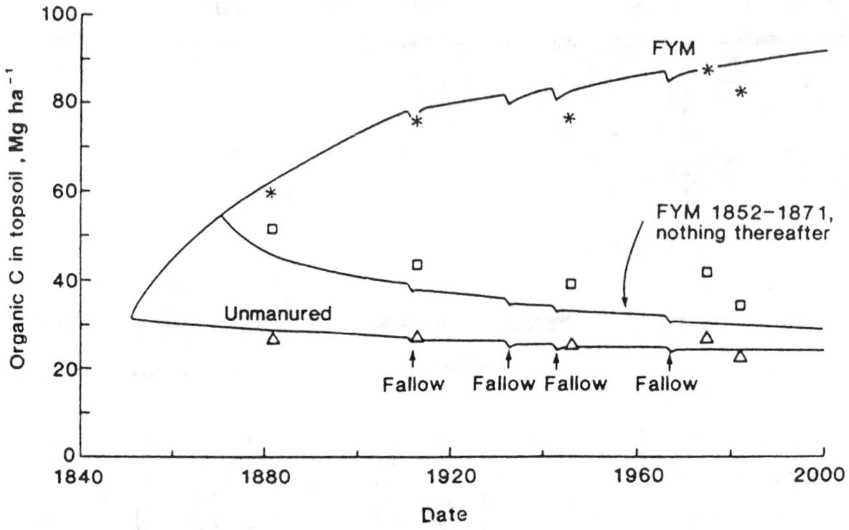

Figure 2. Organic carbon in the top 23 cm of Broadbalk field, Rothamsted, U.K., 1840-1990. Symbols are observations, lines are derived from the current Rothamsted organic matter turnover model. (From Jenkinson, 1991.)

Table 5. Net primary production (NPP) as tonnes of carbon per ha per year at Rothamsted, U.K., estimated from ^{14}C data and the soil carbon content at equilibrium

Site	Carbon returned to soil	Carbon accumulated in vegetation/crop	Net primary production
Broadbalk Wilderness	3.5	1.27	4.8
Geescroft Wilderness	2.5	0.84	3.3
Park grass	3.0	0.97	4.0
Broadbalk wheat, unmanured	1.3	0.93	2.2
Broadbalk wheat, N,P,K	1.7	3.51	5.2

(From Jenkinson et al., 1992.)

1. Moist and Wet Forest Regions.

Soil organic matter levels under tropical rain forest are maintained at relatively high levels by the large amounts of carbon in litter and other returns (UNESCO, 1978). Breaking the cycle by clearing the forest for agricultural or other uses leads to a rapid loss of soil carbon as the new levels of carbon addition are usually much lower than those from the forest. In these areas crop yields average less than 2 Mg/ha and in spite of intercropping carbon returns to the soil are likely to be much less than under forest canopy. It is also true that some form of shifting cultivation has been practiced in most forested areas of the tropics, ınd "fallow" periods are much less than the 40 years or more needed to restore a level of organic matter close to equilibrium.

Many of the soils in these areas have also been significantly degraded (Oldeman et al., 1990). Thus if the return of organic matter to the soil can be increased by increasing crop yields, and if minimum tillage, intercropping, and agroforestry methods are used to add organic matter and minimize soil degradation, it is likely that substantially greater amounts of carbon could be sequestered. Just how much is a question that is difficult to answer at present for lack of appropriate long term trials.

2. Savanna and Drier Forest Regions

The majority of soils in the tropical savanna regions are of low organic matter content, partly due to the general practice of burning during the dry season, partly due to the low inherent fertility of many of the soils (Jones and Wild, 1975; Pieri, 1992) and partly due to soil erosion (Oldeman et al., 1990). The carbon levels are probably below the equilibrium levels to be expected under more productive management of the soils, whether for arable crops or improved pastures. Thus the introduction of improved production systems in these areas should lead to increases in soil carbon levels.

3. Tropical Wetlands Used for Rice

Of the 800 mha. of land cultivated annually in the tropics about 100 mha. are used to produce rice, and of this 80 mha. are used for wetland rice (Figure 3). Although they only represent 6% of the total agricultural land in the world, they are a rather special case. This is because the decomposition of much of the added organic material occurs under anaerobic conditions. This not only tends to slow down decomposition but also leads to the formation of hydrocarbons rather than only carbon dioxide. They may be a potential sink for some carbon, although their limited extent means that they cannot assume major importance. The 40 mha in the temperate and subtropical region are more likely to be a significant carbon sink than the tropical region ricelands because of their lower rates of organic matter decomposition.

For the rice production in the tropical wetlands it is probably much more important that organic additions be minimized to reduce methane production than inputs be increased to enable more carbon to be stored in the soil. Addition of organic matter to wetland rice soils not only provides more carbon as a potential source of methane and other hydrocarbons, but is also instrumental in lowering the redox potential of the soil to the level at which hydrocarbons can be formed.

E. Scope for Carbon Increase in Soils under Tropical Forest and Grassland

In most of the remaining 2 billion hectares of forest vegetation in the tropics the standing vegetation has developed and accumulated sufficient nutrients at least to maintain itself within a closed nutrient cycle. Nutrients are largely cycled through litter back to the soil. Until the forest is cut down for agricultural or other purposes there is a sizeable return of organic matter, and a satisfactory equilibrium level of soil carbon will be established. This process has been thoroughly studied, although most data refer to specific sites and it is not easy to know how well the sites represent wider areas. In most instances it is also uncertain whether the soil carbon levels represent equilibrium (Lugo and Brown, 1986). However unless there has been serious forest interference in the past so that the level of organic matter is below the equilibrium level typical of forested conditions it is unlikely that much can be done to increase carbon sequestration.

As noted above the soils of many tropical grassland areas are degraded (or desertified) so that if a more productive vegetative cover could be established and maintained a considerable increase in soil carbon would be possible. Australian experience has shown that use of a legume species with fertilizers carrying P and S will increase soil carbon levels (Russell, 1986). The same methods, plus control of burning, have been shown to be applicable to tropical grasslands. There is thus real potential in these areas to sequester much carbon in the soil. The wider use of the methods is likely to be restricted for economic reasons, unless a market exists for the animal products which are raised on the pastures.

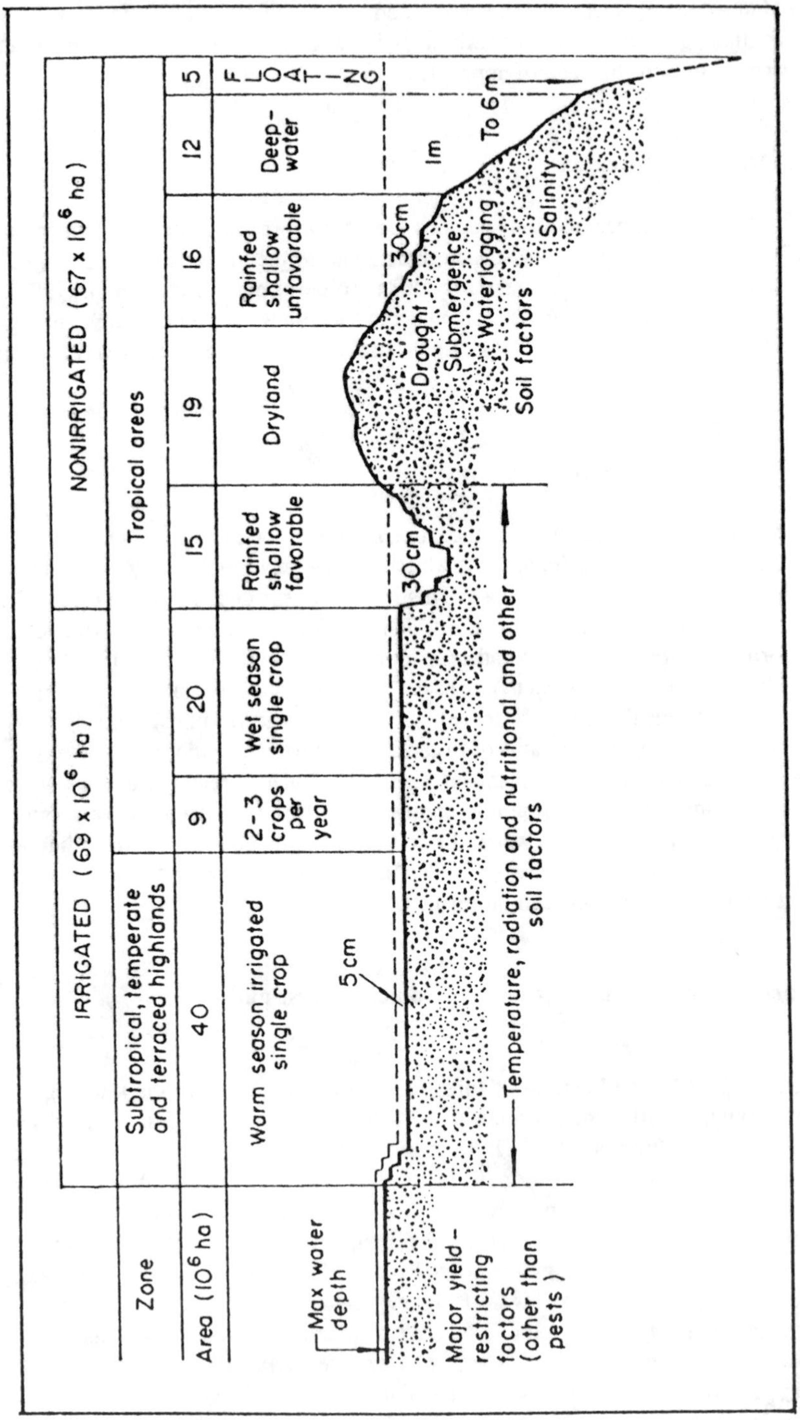

Figure 3. Areas of land used for different systems of rice cultivation. (From IRRI, 1984.)

V. Summary and Conclusions

Several estimates indicate that total global soil carbon is of the order of 2,000 Pg. The current annual rate of accumulation of carbon as carbon dioxide in the atmosphere is of the order of 3Pg. Thus if the soil content of carbon increases by an average of 0.1-0.2% per annum additions from fossil fuel burning can be largely offset. Is it feasible to increase soil carbon to this extent?

The carbon content of the great majority of soils falls between 2 and 40 kg/m^2, so that 0.2% represents an addition of 0.04 to 0.40 Mg/ha. Given that a cereal crop usually produces residues containing 0.5 to 5.0 Mg/ha, it is at least reasonable to consider the possibility, particularly for the soils of lower initial carbon content.

Equilibrium levels of soil carbon are determined primarily by the ratio of the rate of addition of organic material to the soil to the rate at which it decomposes. The rates of carbon addition and loss depend on the type of land use to which the soil is subjected. The rate of sequestration (retention in the soil) depends upon how far the soil is below its equilibrium level for the type of land management being practised.

If a figure is to be estimated for the potential amount of carbon which can be sequestered in the soil it is necessary that an approximation be made as to how existing levels of soil carbon relate to equilibrium levels for different systems of soil management.

Much is now known about carbon levels in different soil types, and of the processes which affect the carbon level. There are also several models available which enable the net effects of the operation of the different processes to be integrated over time. Long term experiments on well defined soils are necessary to enable the validity of the models to be assessed for a range of different soils in different agroecological zones.

Rice soils are a special case. When organic matter is incorporated much is evolved as methane. Thus rice soils are not appropriate to use as sites to sequester carbon.

Sufficient is now known of the processes of carbon gains and losses from soils for sophisticated models to be developed which can predict changes for the environments and types of land use in the areas for which they were developed. It is also likely that they can be used more widely for predicting soil carbon levels under different land use systems. Before any decision can be reached about their wider use they need to be tested and refined using data from carefully conducted studies on land use in different ecological zones. The data to be used in the models and the ways in which the information obtained must be extrapolated to refer to wider areas needs to be critically assessed in relation to the errors involved at each step. Inadequate attention has usually been given to the need to assess errors when levels of soil carbon and future projections are reported. Further research on the sources of error in soil carbon contents is needed, as well as more data. Data is particularly needed in the tropics where it appears probable that greater carbon sequestration is possible.

Detwiler (1986) estimates that at present these areas are responsible for the release of about 0.1 Pg of carbon annually. If the present level of carbon return from crops and organic additions to the soil were to be increased from its present level of around 0.5 Mg/ha to 3 or 4 Mg/ha, which is quite feasible, at least when using agroforestry methods, more than 2Pg will enter the soil, of which a significant proportion is likely to be retained, the amount depending on how the existing soil carbon level relates to the equilibrium level for the more productive farming system.

Most opportunity to increase carbon sequestration on arable land appears to exist in the tropics,where many soils are somewhat degraded, and substantial increases in the carbon inputs to the soil might be achieved by improved farming systems (Goreau, 1990).

Unfortunately only limited information of the type required to substantiate and quantify these conclusions, collected over an adequate time period, exists at present. It is important that a number of sites be established and funding secured on a long term basis to ensure that the necessary information is obtained. Many grasslands may be well below their equilibrium levels, particularly in the tropics, and so offer an opportunity for greater carbon sequestration. This will usually require restrictions on burning and the development of the grassland for animal production by introducing improved grasses and legumes, and raising the nutrient status of the soils.

In the temperate zone improvements in crop production practices on arable land could lead to increased carbon sequestration, but as most soils are farmed at or near the equilibrium level representative of high input farming systems the increases are unlikely to be large.

There is an urgent research need to establish for a range of well defined soils what are the potential equilibrium levels of soil organic matter under well defined and economically productive management systems. To do so will require several well monitored long term experiments. It will also be necessary to establish on a properly quantified basis how present soil carbon levels relate to the equilibrium level under the farming system currently used, as well as under improved management systems.

Acknowledgement

The author is grateful to Dr. David Jenkinson for many helpful comments and suggestions, and for his review of the manuscript.

References

Ajtay, G.L., P. Ketner, and P. Duvigneaud. 1979. Terrestrial primary production and phytomass. p.129-181. In: B.Bolin, E.T. Degens, S. Kempe, and P. Ketner (eds.), *The global carbon cycle.* (SCOPE Report no. 13.) J. Wiley & Sons, Chichester, U.K.

Arnold, R.W. and L.P. Wilding. 1991. The need to quantify spatial variability. p. 1-8. In: M.J. Mausbach and L.P. Wilding (eds.), *Spatial variability of soils and landforms.* Soil Sci. Soc. Amer. Inc., Madison, Wis.

Ayanaba, A. and D.S. Jenkinson. 1990. Decomposition of carbon-14 labelled ryegrass and maize under tropical conditions. *Soil Sci. Soc. Am. J.* 54:112-115.

Bertram, H.G. 1986. Zur Rolle des Bodens im globalen Kohlenstoffzyklus. Veroffentlichungen der Naturforschenden Gesellschaft zu Emden v. 1814. 8:3-93.

Biot, Y. 1991. Forecasting future production. p 593-611. In: J. Dumanski, E. Pushparajah, M. Latham, and R. Myers (eds.) Evaluation for sustainable land management in the developing world, vol. 2. IBSRAM, Bangkok, Thailand.

Bohn, H.L. 1976. Estimates of organic carbon in world soils. *Soil Sci. Soc. Am. J.* 40:468-470.

Bohn, H.L. 1982. Estimates of organic carbon in world soils: II. *Soil Sci. Soc. Am .J.* 46:1118-1119.

Bouwman, A.F. 1989. Modelling soil organic matter decomposition and rainfall erosion in two tropical soils after forest clearing for permanent agriculture. *Land Degradation and Rehabilitation* 1:125-140.

Bouwman, A.F. (ed.) 1990. *Soils and the Greenhouse Effect.* J. Wiley & Sons, Chichester, U.K. 575 pp.

Brown, S.and A.E. Lugo. 1982. The storage and production of organic matter in tropical forests and their role in the global carbon cycle. *Biotropica.* 14:161-187.

Burke, I.C., C.M. Yonker, W.J. Parton, C.V. Cole, K. Flach, and D.S. Schimel. 1989. Texture, climate and cultivation effects on soil organic matter content in U.S. grassland soils. *Soil Sci. Soc. Am. J.* 53:800-805.

Carr, C.E. 1973. Gravimetric determination of soil carbon using the Leco induction furnace. *J. Sci. Food Agric.* 24:1091-1095.

Cole, C.V., J.W.B. Stewart, D.S. Ojima, W.J. Parton, and D.S. Schimel. 1989. Modelling land use effects of soil organic matter dynamics in the North American Great Plains. *Developments in Plant and Soil Sciences.* 39:89-98.

Coleman, D.C., J.M. Oades, and G. Uehara (eds.), 1989. *Dynamics of soil organic matter in tropical ecosystems.* University of Hawaii Press, Honolulu, HI, USA.

Detwiler, R.P. 1985. Land use change and the global carbon cycle: the role of tropical soils. *Biogeochemistry* 2:67-93.

Emanuel, W.R., G.G. Killough, W.M. Post, and H.H. Shugart. 1984. Modelling terrestrial ecosystems in the global carbon cycle with shifts in carbon storage capacity by land use change. *Ecology* 65:970- 983.

Esser, G. 1989. Global land use changes from 1860 to 1980 and future projections to 2500. *Ecological Modelling* 44:307-316.

Esser, G. 1990. Modelling global terrestrial sources and sinks of carbon dioxide with special reference to soil organic matter. p. 247-261. In: A.F.Bouwman (ed.), *Soils and the Greenhouse Effect.* J. Wiley & Sons. Chichester, U.K.

Franzmeier, D.P., G.D. Lemme, and R.J. Miles. 1985. Organic carbon in soils of the north-central United States. *Soil Sci. Soc. Am. J.* 49:702-708.

Goreau, T.J. 1990. Balancing atmospheric carbon dioxide. *Ambio* 19:230-236.

Goudriaan, J. 1990. In "At Global Change - Do Soils Matter?" International Society of Soil Science and ISRIC, Wageningen, Netherlands.

Greenland, D.J. 1971. Changes in the nitrogen status and physical condition of soils under pasture. *Soils and Fertilisers* 34:237-251.

Greenland, D.J., A. Wild, and D. Phillips. 1992. Organic matter in soils of the tropics: from myths to complex reality. p. 17-34. In: R. Lal and P.A. Sanchez (eds.), *Myths and science of soils in the tropics.* Am. Soc. Agron., Madison, Wisconsin.

IRRI. 1984. Terminology for Rice growing Environments. International Rice Research Institute, Los Banos, Philippines.

Istituto Geographico de Agostini. 1969, 1971, 1973. World Atlas of Agriculture. Novara, Italy.

Jenkinson, D.S. 1975. The turnover of organic matter in agricultural soils. *Reports Welsh Soils Discussion Group.* 16:91-105.

Jenkinson, D.S. 1990. The turnover of organic carbon and nitrogen in soil. *Phil. Trans. Roy. Soc. London, B.* 329:361-368.

Jenkinson, D.S. 1991. The Rothamsted long-term experiments: are they still of use? *Agron. J.* 83:2-10.

Jenkinson D.S. and J. Rayner. 1977. The turnover of soil organic matter in some of the Rothamsted Classical Experiments. *Soil Sci.* 123:298-305.

Jenkinson, D.S., P.B.S. Hart, J.H. Rayner, and C.L. Parry. 1987. Modelling the turnover of organic matter in long-term experiments at Rothamsted. *INTECOL Bull.* 15:1-8.

Jenkinson, D.S., D.E. Adams, and A. Wild. 1991. Model estimates of carbon dioxide emissions from soil in response to global warming. *Nature* 351:304-306.

Jenkinson, D.S., D.D. Harkness, E.D. Vance, D.E. Adams, and A.F. Harrison. 1992. Calculating net primary production and annual input of organic matter to soil from the amount and radiocarbon content of soil organic matter. *Soil Biol. Biochem.* 24:295-308.

Jones, M.J. and A. Wild. 1975. Soils of the West African savanna. p. 246. Tech. Comm. no. 55, CAB international, Wallingford, U.K.

Kalembasa, S.J. and D.S. Jenkinson. 1973. A comparative study of titrimetric and gravimetric methods for the determination of organic carbon in soil. *J. Sci. Food Agric.* 24:1085-1090.

Kimble, J.M., H. Eswaran, and T. Cook. 1990. Organic carbon on a volume basis in tropical and temperate soils. *Trans. Int. Congr. Soil. Sci. Kyoto* 5:248-253.

Lugo, A.E. and S. Brown. 1986. Steady state terrestrial ecosystems and the global carbon cycle. *Vegetation* 68:83-90.

Metson, A.J., L.C. Blakemore, and D.A. Rhoades. 1979. Methods for the determination of soil organic carbon: a review, and application to New Zealand soils. *New Zealand J. Sci.* 22:205-228.

Oldeman, L.R., R.T.A. Hakkeling, and W.G. Sombroek. 1990. World map of the status of human-induced soil degradation:an explanatory note. ISRIC, Wageningen, Netherlands and UNEP. Kenya. 27 pp and 3 maps.

Parfitt, R.L. and B. Clayden. 1991. Andisols - the development of a new order in soil taxonomy. *Geoderma* 49:181-198.

Parton, W.J., D.W.Anderson, C.V.Cole, and J.W.B. Stewart. 1983. Simulation of soil organic matter formation and mineralisation in semi-arid ecosystems. p. 533-550. In: R.R. Lowrance (ed.), Nutrient cycling in agricultural ecosystems. Univ. Georgia, Spec. Pub. Vol.23.

Parton, W.J., J.W.B. Stewart, and C.V. Cole. 1988. Dynamics of C, N, P and S in grassland soils: a model. *Biogeochemistry* 5:109-131.

Parton, W.J., C.V. Cole, J.W.B. Stewart, D.S.Ojima, and D.S. Schimel. 1989. Simulating regional patterns soil C, N, and P dynamics in the U.S. central grasslands region. *Developments in Plant and Soil Sciences* 39:99-108.

Pieri, C.J.M.G. 1992. *Fertility of soils: a future for farming in West Africa.* Spriger Verlag, Berlin. 348 pp.

Post, W.M., W.R. Emanuel, P.J. Zinke, and A.G. Stangenberger. 1982. Soil carbon pools and wild life zones. *Nature* 298:156-159.

Post, W.M., T.H. Peng, W.R.Emanuel, A.W. King, V.H. Dale, and D.L. DeAngelis. 1990. The global carbon cycle. *American Scientist* 78:310-326.

Russell, J.S. 1986. Improved pastures. p. 374-396. In: J.S. Russell and R.F. Isbell (eds.), *Australian Soils: the Human Impact.* University of Queensland Press, St.Lucia, Queensland, Australia.

Scharpenseel, H.W. 1993. Sustainable land use in the light of resilience to soil organic matter fluctuations. In: D.J. Greenland and I. Szabolcs (eds.), *Soil Resilience and Sustainable Land Use*. CAB-International, Wallingford, U.K.

Schmithusen, J. 1976. Atlas zur Biogeographie. Meyers Grosser Physischer Weltatlas. Mannheim, Germany.

Tabatabai, M.A. and J.M. Bremner. 1970. Use of the LECO 70-second carbon analyser for total carbon analysis of soils. *Soil Sci. Soc. Amer. Proc.* 34:608-610.

Tate, K.R. and I. Salcedo. 1988. Phosphorus control of soil organic matter accumulation and cycling. *Biogeochemistry* 5:99-107.

UNESCO. 1978. Tropical forest ecosystems: a state of knowledge report prepared by UNESCO/-UNEP/FAO. Natural Resources Research XIV. UNESCO, Paris.

Veen, J.A. van and E.A. Paul. 1981. Organic carbon dynamics in grassland soils. 1. Background information and carbon simulation. *Can. J. Soil Sci.* 61:185-201.

Verberne, E.L.J., J. Hassink, P. de Willigen, J.J.R. Groot, J.A. van Veen, P. de Willigen, and J.A. van Veen. 1990. Modelling organic matter dynamics in different soils. *Neth. J. Agric. Sci.* 38:221-238.

Virakornphanich, P.S., I. Wada, and K. Wada. 1988. Metal-humus complexes in A horizons of Thai and Korean red and yellow soils. *J. Soil Sci.* 39:529-538.

Wada, K. 1980. Mineralogical characteristics of Andisols. p. 87-108. In: B.K.G. Theng (ed.), *Soils with Variable Charge*. New Zealand Society of Soil Science, Lower Hutt, New Zealand.

Wagner, G.H. 1989. Lessons in soil organic matter from Sanborn Field. Special Report, College of Agriculture. University of Missouri. Special Report 415:64-70.

Walkley, A. 1947. A critical examination of a rapid method for determining organic carbon in soils - effect of variations in digestion conditions and of inorganic soil constituents. *Soil Science* 63:251-264.

Wallace, A.G., A. Wallace, and J.W. Cha. 1990. Soil organic matter and the global carbon cycle. *J. Plant Nut.* 13:459-466.

CHAPTER 3

Carbon Pools, Fluxes, and Sequestration Potential in Soils of the Former Soviet Union

T.P. Kolchugina, T.S. Vinson, G.G. Gaston, V.A. Rozhkov, and A.Z. Shwidenko

I. Introduction

General circulation models (GCMs) (Schneider, 1990; Washington and Mitchel, 1984; Manabe and Wetherald, 1987; Hansen et al., 1984; Schlesinger and Zhao, 1989) predict that a continuous increase in the atmospheric concentration of carbon dioxide (CO_2) and other greenhouse gases (e.g. methane (CH_4)) will cause climate warming (IPCC, 1990). World temperatures may be 5°C warmer by the end of the twenty-first century. Climate warming may be more pronounced in the Northern Hemisphere (Etkin, 1990) with "... as much as 10° to 20°C warming locally near the ice sheets ..." (Schneider, 1990).

The long-term ecological consequences of the change in the chemical composition of the atmosphere are not fully understood. An increase in plant productivity with warming may occur due to a longer growing season, enhanced soil aeration and nutrient availability (Chapin and Shaver, 1981; Billings et al., 1982; Chapin, 1984; Rastetter et al., 1991; Shaver et al., 1992). Plant productivity may increase by 55% in response to a 10% increase in the growing season (Miller et al., 1976). Global warming is also likely to increase rates of soil respiration (Dixon and Turner, 1991; Raich and Schlesinger, 1992). The effect of climate warming on the net carbon flux between terrestrial ecosystems and the atmosphere depends upon the balance between accumulation and decomposition processes.

The global reservoir of carbon stored in terrestrial soils was estimated at 2,020 Pg C (Kobak, 1988). Soil respiration of terrestrial ecosystems is estimated at 50-75 Pg C yr^{-1} on the global scale (Houghton and Woodwell, 1989; Schlesinger, 1977). Soil respiration results from the autotrophic respiration of plants (root respiration), and respiration of heterotrophic organisms (decomposition of plant detritus, e.g., litter, dead roots, stems, and soil organic matter).

Conversion of lands to agricultural production causes a sharp decrease in carbon stored in soil (Haas et al., 1957; Hobbs and Brown, 1965; Houghton et al., 1983; Mann, 1986; Kobak et al., 1987). The release of carbon from soil to the atmosphere during the last two centuries was estimated at 10 to 40 Pg C as a result of conversion of forest land to agriculture (Bolin, 1977). When soils are cultivated, the surface layers (0-30 cm) often lose 20-50% of their carbon (Schlesinger, 1984). Soils of the arable lands in the former Soviet Union (FSU) lost 24 to 50% since the onset of agriculture (Kononova, 1984; Rozanov et al., 1989; Gaston et al., 1993). The number of microorganisms which participate in mineralization of soil organic matter increases dramatically in arable land compared to undisturbed land soils (Kobak, 1988). Most of the carbon lost from disturbed soils is lost through increased oxidation rather than erosion (Schlesinger, 1977). It has been estimated that approximately half of the carbon transported from agricultural soils by erosion is eventually transferred to the atmosphere (Kern and Johnson, 1991).

It is difficult to evaluate the exact amount of organic matter removed from agricultural lands as a result of water and wind erosion. It was estimated that over the past century world soils lost 384 Pg of humus, which is equivalent to 230 Pg C (Gorshkov et al., 1980). Assuming 2,020 Pg C in the global carbon pool

ISBN 1-56670-117-1/95/$0.00+.50

(Kobak, 1988), the average annual loss from water and wind erosion is approximately 0.11% of the initial soil organic carbon content.

A number of management strategies have great potential to conserve carbon in agricultural soils. Ridge tillage, contour cropping, stubble mulching, and shelter belts all serve to reduce the loss of carbon from agricultural areas from soil erosion (Barnwell et al., 1992; IPCC, 1992). Shelter belts increase biodiversity and improve microclimate. Shelter belts also increase available water content and decrease rates of wind and water erosion in soils within protected areas (Vorobyov, 1985). When considering options for additional storage of atmospheric carbon in agricultural soils, fewer management strategies are available. Growing cover crops for green manure, irrigation of dry lands, addition of carbon from outside sources, and conservation tillage have all been suggested as options to sequester additional atmospheric carbon in agricultural soils. Kern and Johnson (1991) indicate that of the possible conservation tillage options "... only no-till significantly results in increased [carbon] sequestration."

The arable land in the FSU is approximately 210 Mha (10^6 ha) (US Department of Agriculture, 1990; Gaston et al., 1993; Vinson and Kolchugina, 1993). The FSU sources report the arable land at 227.7 Mha (USSR State Committee for the Protection of Nature, 1989). The arable land is almost twice the cultivated area of the United States (120 Mha), and over four times the area of Canadian arable lands (46 Mha).

At present 5 Mha of shelter belts protect 40 Mha of arable lands in the FSU. The rate of establishment of forest shelter belts in the FSU has decreased over the past few decades (Anonymous, 1991). Collective farms would not set aside lands for forest plantations. It is estimated that an additional 18 Mha of shelter belts should be established in the FSU to protect 200 Mha of arable lands and pastures (Rozhkov and Shwidenko, 1992, unpublished data). The benefit of shelter belt protection is especially feasible in dry climatic regions such as Middle Asia, the Northern Caspian Region of Russia, and Kazhakhstan (Anonymous, 1991). Seventy-one million hectares of land may be afforested including 18 Mha of shelter belts, 35 Ma of lands severely disturbed by water and wind erosion, 15 Mha of gullies, and 3 Mha of heavily polluted lands.

Projections of the future terrestrial carbon balance in soils of the FSU under climate warming can be made after the present carbon pools and fluxes are estimated. Analyses of ecosystems feedback to warming may be based on assumptions and may involve the creation of scenarios under which changes in carbon cycling are considered (Schneider 1989).

II. Purpose

The purpose of the study reported herein was to assess (1) soil carbon pools in the FSU, (2) current and post climate warming biogenic carbon emissions in the FSU, and (3) management options to conserve and sequester carbon.

III. Methodology

A. Carbon Pools and Fluxes

The soil carbon pools and rates of carbon accumulation/emissions were assessed as components of the biogenic carbon budget of the FSU (Vinson and Kolchugina, 1993). The biogenic carbon cycle consists of a combination of pools and fluxes. The pools are carbon stores in soil and vegetation, including live vegetation (i.e., *phytomass*) and plant detritus (i.e., *mortmass* and *litter*). In the present study, the term mortmass was used to describe coarse above-ground and below-ground woody debris. The term litter was used to define the upper soil layer comprised of fine woody debris and leaves that are not completely decomposed.

The effluxes are carbon emissions resulting from plant respiration and decomposition of organic matter. The processes of formation of new organic matter in soil and vegetation (i.e., *humus*, *foliage formation* and *NPP*) represent carbon influxes. The NPP equals the difference between *gross photosynthesis* (GPP) and *respiration of autotrophic organisms* (R_A). The R_A amounts to 44 to 52% (48% on average) of GPP. Root respiration (R_{Ar}) comprises one-third of the R_A (Kobak, 1988).

Carbon fluxes can be measured or calculated. The rates of soil respiration were measured in many of the world's ecosystems (Kobak, 1988; Makarov, 1988; Raich and Schlesinger, 1992). When carbon effluxes are measured, the contribution from different processes cannot be distinguished. For example, when the soil carbon efflux is measured, it is difficult to distinguish between effluxes resulting from R_{Ar} and R_H (decomposition of litter, below-ground mortmass and soil organic matter). The quantitative method allows one to separate fluxes (Kolchugina and Vinson, 1993a,b; Vinson and Kolchugina, 1993).

To calculate pools and fluxes of biogenic carbon the geographic area within which carbon may be quantified must be isolated. The term *ecoregion* was applied to the boundaries and areal extent of the geographic area. The term *ecosystem* was applied to the combination of certain soil-vegetation formations within an ecoregion. The term *biome* was applied to the complex of ecosystems within a climatic belt or subbelt.

The carbon cycle parameters may be expressed in terms of carbon content (for pools) or rate (for influxes or effluxes) ha^{-1} for a variety of soil-vegetation complexes. If the soil-vegetation complexes which comprise an ecosystem are related to the natural attributes identified on maps which are used to isolate ecoregions, then the carbon budget for an ecoregion can be established simply by multiplying the area of the ecoregion (in ha) by the carbon content(s) and flux(es). The carbon contents and fluxes for all the ecoregions may be summed to arrive at the carbon budget for a larger region, biome, or nation (Vinson and Kolchugina, 1993).

B. Isolation of Ecoregions

About 95% of the territory of the FSU, including Russia, Ukraine, Belorussia, Moldavia, Kazakhstan, and the Baltic states, was categorized by the soil-vegetation type of the ecosystem, the presence of peatlands and cultivation intensity. Maps containing information on the distribution of zonal soil-vegetation associations within the FSU (Ryabchikov, 1988), distribution of peatlands (Isachenko, 1988) and cultivation intensity of arable lands (Cherdantsev, 1961) were digitized and computer-superimposed with a geographical information system (GIS) (Burrough, 1986). The map with the distribution of soil-vegetation associations (Ryabchikov, 1988) provided the basis for ecoregion isolation.

The ecosystems presented by Ryabchikov (1988) were aggregated into nine biomes. The *polar desert* biome included areas covered by ice and stony barrens. The *tundra* biome included herbaceous and shrub tundra formations of polar and subpolar belts. The *forest-tundra/sparse taiga* biome included forest ecosystems within the subpolar climatic belt and northern areas of the boreal climatic belts. This biome unified ecosystems with sparse forest cover. The *taiga* biome included ecosystems within the boreal climatic belt with light- or dark-crown coniferous forest vegetation of northern, middle, and southern subzones. The *mixed-deciduous forest* biome unified mixed (coniferous-broadleaf or small-leaf) or broadleaf forests within the temperate climatic belt. The *forest-steppe* biome included mixed coniferous-deciduous forests and grasslands of the temperate climatic belt. The *steppe* biome included grasslands. The *desert-semidesert* biome included shrub-grass and shrub-tree desert formations of temperate and subtropical belts. The *subtropical woodland* biome included mountainous formations of the subtropical belt.

C. Data Bases for Natural Carbon Cycle Parameters

Bazilevich's (1986) data base on carbon accumulation in vegetation formations of the FSU (total and portions, e.g., woody parts, above- and below-ground parts, green parts, of phytomass, mortmass, and NPP) was associated with the ecosystems presented by Ryabchikov (1988). Kobak's (1988) soil data base (total and stable soil organic matter content and rate of accumulation) for more than 40 soil types of the polar, boreal, temperate, and tropical belts was used to characterize the soil component of the carbon cycle.

D. Integration of Ecoregion Areas and Carbon Data Bases

Carbon pools and fluxes for natural ecosystems in the FSU were estimated by integrating the carbon data bases and the GIS analysis results (hectare data), using commercially available spreadsheet software (Microsoft Corporation, 1991). Data on carbon content of peatlands were integrated with hectare data of peatlands within biomes. The carbon pools and influxes for the biomes in the FSU were obtained by summing ecosystem contributions assuming that (1) forests totally cover the area of ecosystems (excluding arable land) within forest-tundra/sparse taiga, taiga and mixed-deciduous forest biomes; and (2) forests occupy one-half of the area (excluding arable land) of the forest-steppe and subtropical woodland biomes.

E. Estimation of Carbon Effluxes under Present Climate

Carbon effluxes under the present climate were calculated from carbon influxes (i.e., NPP, and rates of foliage and soil organic matter formation) using the methodology presented by Kolchugina and Vinson (1993a,b) and Vinson and Kolchugina (1993). It was assumed that (1) all ecosystems were initially in an equilibrium state (NPP equals R_H), (2) the carbon pools (i.e., phytomass, mortmass, litter, and soil) were constant, (3) the litter pool was mainly formed from foliage, and (4) most of the soil organic matter was formed from litter. Following these assumptions, the carbon efflux from mortmass decomposition was assumed to be equal to mortmass production. In turn, mortmass production was assumed to be equal to phytomass production (NPP and production of different parts of plants). The carbon efflux from litter decomposition was calculated as the difference between foliage formation (green-assimilating parts production) and total humus formation. The carbon efflux from soil organic matter decomposition was calculated as the difference between total and stable humus formation. The carbon efflux from R_{Ar} was calculated from NPP, assuming that R_{Ar} comprises one-third of the total R_A, and R_A comprises 48% (on average) of the GPP; NPP equals the difference between GPP and R_A. The sum of R_{Ar} and R_H (below-ground mortmass, litter, and soil organic matter decomposition) was compared with field measurements of the surface soil carbon efflux (Kobak, 1988).

F. Estimation of Changes in Carbon Fluxes under Climate Warming

The increase in decomposition rates under climate warming was estimated based on the $Q_{10} = 2.4$ value reported by Raich and Schlesinger (1992). It was assumed that the air temperature would increase by 4°, 3°, and 2°C in the polar desert to forest-tundra/sparse taiga zone, the taiga to forest-steppe zone, and the steppe to desert-semidesert zone, respectively. Further, it was assumed that the average soil temperature increase would be half of the air temperature increase. An increase in above-ground mortmass decomposition was estimated under the air temperature increase scenario. An increase in litter, below-ground mortmass, and soil organic matter decomposition rates was estimated under the soil temperature increase scenario.

With climate warming plant productivity may increase by 55% under favorable environmental conditions (Shaver et al. 1992; Kolchugina and Vinson, 1993b). Increased decomposition may be accompanied by loss of nutrients from ecosystems. The loss of nutrients may specifically affect ecosystems with herbaceous vegetation and soils of high permeability. For example, it is possible that only half of the tundra area (gleyic soils) of the FSU will experience an increase in plant productivity (Kolchugina and Vinson, 1993b). Growth of vegetation in dry climatic zones may not change or may even be inhibited by water deficiency. In the present study a 55% increase in plant productivity was assumed for the entire area of the forest-tundra/sparse taiga, taiga, and mixed-deciduous forest, subtropical woodland biomes, and half of the area of tundra ecosystems.

G. Management Options to Conserve and Sequester Carbon in Agricultural Soils and Agroforestry

1. No-Till Management

The precultivation carbon content of arable land was assumed to be equivalent to the corresponding soil type of natural ecosystems. It is generally assumed that over time cultivated soils reach a new equilibrium at a lower level of organic carbon (Haas et al., 1957; Hobbs and Brown, 1965; Unger, 1968; Mann, 1985). Most researchers report that soils cultivated less than 100 years exhibit a continued slow loss of soil carbon to the atmosphere (Haas et al., 1957; Hobbs and Brown, 1965; Unger, 1968). The current cultivation carbon content was estimated for each soil type (Gaston et al., 1993) based on an equation derived by Mann (1986) to estimate an equilibrium carbon content in soils under cultivation for 70-100 years:

$$SOC_c = (SOC_i * 0.76) + 0.88 \quad (R = 0.77, n = 129) \tag{1}$$

in which SOC_c = current soil organic carbon, and SOC_i = initial carbon content of the soil. Equation (1) was applied to each initial soil carbon content and an estimate was made of current soil organic carbon in the agricultural soils of the FSU.

It is difficult to predict the slow rates of soil carbon loss from soils approaching a new equilibrium condition. Since 1913 there has been no significant expansion of agricultural lands in the FSU. Considering this long period of cultivation it is reasonable to assume, notwithstanding ongoing erosion, that an equilibrium condition has been reached. An exception to this assumption is the 33 Mha of so called "virgin lands," which were first converted to agricultural use in the mid 1950s (Medvedev, 1987). At this time it is not possible to quantify the condition of soil carbon in the "virgin lands." However, these soils should continue to lose carbon to the atmosphere through the beginning of the next century.

Conservation tillage is generally defined as a management practice that leaves more than 30% of the crop residue on the soil surface. No-till management leaves virtually 100% of the crop residue undisturbed on the surface of the soil. The cover of crop residue left on the surface of the soil acts as a mulch, reducing soil temperatures and increasing soil moisture. Cool moist soils exhibit significantly slower rates of carbon loss and in some cases act as a long term sink for atmospheric carbon. Conservation tillage reduces the frequency and magnitude of mechanical disturbance. This, in turn, reduces the creation of air filled soil micropores and slows the rate of carbon oxidation. Further, less mixing of crop residue into the soil reduces the rate of biological and chemical interactions (Dick, 1983) slowing the rate of organic matter mineralization.

In the uppermost 8 cm of the soil column the response of soil carbon to no-till management may be predicted by the following equation (Kern and Johnson, 1991)

$$SOC_{nt} = (1.283 * SOC_c) + 0.0510 \quad (R = 0.75, n = 15) \tag{2}$$

in which SOC_{nt} = equilibrium soil carbon content for no-till management. In the 8-30 cm portion of the soil column the rate of carbon accumulation is less under no-till management because the tillage operations no longer transport organic matter from the surface into the lower potions of the soil profile. The equation for this portion of the soil column is (Kern and Johnson, 1991):

$$SOC_{nt} = (1.16 * SOC_c) - 0.018 \quad (R = 0.89, n = 34) \tag{3}$$

Below 30 cm, no significant differences in the rate of carbon accumulation or loss were observed between no-till management and conventional tillage practices.

To use Eqs. (2) and (3) it was necessary to have carbon profiles for each soil type and distribute carbon contents through each profile. Organic carbon profiles for soil types in the FSU reported by Glazovskaya (1972) were used to calculate the percent distribution of carbon in each soil profile. The distributed organic carbon contents and the current soil equilibrium were used with Eqs. (2) and (3) to predict potential carbon sequestration for each agricultural soil type in the FSU.

The area of agricultural soils climatically suitable for no-till management practices was estimated by Gaston et al. (1993). Specifically, areas where crop production is limited by cold soil temperatures, low rates of evapotranspiration, relatively high winter precipitation, and a short growing season are not suitable for no-till management.

2. Agroforestry: Shelter Belt Protection and Afforestation

Shelter belt protection and afforestation enhance microclimate and crop production (Vorobyov, 1985). The main benefit of shelter belt protection is the reduction of soil erosion. The annual rate of water and wind erosion may optimistically be assumed to be 0.11% of the initial carbon content in soils. Shelter belts, which prevent erosion, may be assumed to conserve soil carbon at the same annual rate. Shelter belts are especially important in the steppe and desert-semidesert climatic zones. The average precultivation carbon content of the forest-steppe, steppe, and desert-semidesert biomes was estimated as previously described. Two hundred million hectares of arable lands and pastures may be protected in the FSU by 18 Mha of shelter belts.

In soils under forest vegetation, accumulation of humus increases, and physical mechanical properties are improved (Anonymous, 1991). In a 40 cm soil layer under a 40-yr old oak plantation the humus and nitrogen contents increased by 1.68 and 0.08%, respectively. There is no evidence that soil degradation occurs under forest vegetation. The role of forests in the protection of soil is important in all climatic zones. It is possible that in 100-yr old forest plantations the carbon store in live plant mass and plant detritus (i.e., mortmass and litter) would be 100 Mg C ha^{-1} (Bazilevich, 1986). In this case the average rate of carbon sequestration would be 1.0 Mg C ha^{-1} yr^{-1}. Seventy-one million hectares of forest plantations (including 18 Mha of shelter belts, disfigured landscapes, and heavily polluted lands) may be created in the FSU.

IV. Results and Discussion

The carbon cycle parameters associated with the soil component of the carbon cycle are shown in Table 1. Below-ground phytomass and litter densities were maximum in the forest biomes. Below-ground mortmass density was maximum in the tundra, forest-tundra/sparse taiga, and taiga biomes (subpolar and boreal climatic zones). Soil organic matter density was maximum in the taiga and mixed-deciduous forest biomes due to the fact that peatlands (where carbon contents may be as great as 2,000 Mg ha^{-1}) are widespread within those biomes (Vinson and Kolchugina, 1993).

The NPP density increased from north to south and was maximum in biomes with herbaceous vegetation and deciduous trees (Table 1). The NPP density of the desert-semidesert biome was approximately the same as in the forest-tundra/sparse taiga to mixed-deciduous forest biomes. The density of foliage formation increased from the polar desert to forest-steppe biome, then decreased slightly in the steppe biome, increasing substantially in the subtropical woodland biome. In the desert-semidesert biome the density of foliage formation was the same as in the forest-tundra/sparse taiga biome. The density of below-ground mortmass production increased from the polar desert to the tundra biome, then decreased in the transition to the taiga biome. From the taiga to the steppe biome, the density of below-ground mortmass doubled and tripled in the mixed-deciduous forest and steppe biomes, respectively. The density of below-ground mortmass production was similar in the subtropical woodland and desert-semidesert biomes, and equivalent to the forest-steppe biome. The density of the total soil organic matter formation increased from the polar to temperate climatic zones, and reached a maximum in the subtropical woodland biome. The density of the stable soil organic matter formation generally followed the tendency of the total soil organic matter accumulation.

The vegetation carbon accumulation parameters exhibited the greatest variations in the forest-steppe and subtropical woodland biomes due to the methodology applied in the present study. It was assumed that vegetation in these two biomes was represented equally by forest and herbaceous formations.

The carbon pools which were associated with the soil component of the carbon budget are presented in Figure 1. The soil organic matter was the greatest carbon pool in all nine biomes. The litter, below-ground phytomass, and mortmass carbon pools were approximately equal. The litter carbon pool decreased relative to the below-ground phytomass and mortmass carbon pools from the taiga to desert-semidesert biome. The total soil related carbon store was estimated at 466.7 Pg C (Figure 2). The soil organic matter carbon pool was 87% of the total soil related carbon store. The litter and below-ground mortmass were each 4% and below-ground phytomass was 5%.

Table 2 presents the carbon effluxes from decomposition processes and root respiration under the present climate. The major carbon effluxes resulted from decomposition of below-ground mortmass and litter. They were 20 to 30% greater than the carbon efflux from root respiration. Soil organic matter contributed an order of magnitude less. The sum of carbon effluxes from decomposition of below-ground mortmass, litter, soil organic matter, and root respiration represents the soil surface carbon efflux.

Figure 3 presents a comparison of the soil surface carbon efflux obtained in the present study with estimates based on the results of field experiments reported by Kobak (1988) and Raich and Schlesinger (1992). The soil surface carbon efflux estimated in the present study from data on carbon influxes was 6.8 Pg C yr^{-1}; it was 12% higher than the estimate based on data reported by Raich and Schlesinger, and exceeded the estimate based on Kobak's data by a factor of two.

Raich and Schlesinger included only those data based on most or all of one full year measurements and excluded data obtained with alkali absorption techniques if the surface area of the absorbent was less than 5% of the surface area of the covered ground (this leads to low estimates of soil respiration). Kobak reports the rates of daily CO_2 emissions and the length of the vegetation season. Kobak's CO_2 data were correlated to the growing season and the areal extent of the corresponding soil type in the present study. Soil respiration is not limited by the length of the growing season (Zimov et al., 1991). The disagreement in the

Table 1. Soil related carbon cycle parameters for nine biomes of the FSU (non-arable land)

		Pools (Mg C/ha)					Influxes (MG C/ha/yr)				
Biome	Area (Mha)	Below-ground phytomass	Below-ground mortmass	Litter	Total soil organic matter	Stable soil organic matter	NPP	Foliage formation	Below-ground mortmass production	Total soil organic matter accumulation	Stable soil organic matter accumulation
Polar desert	8.1	0.06±0	0.16±0	0.16±0	19.9±0	13.1±0	0.06±0	0.04±0	0.02±0	0.01±0	0.0±0
Tundra	226.0	5.8±0.07	11.2±0.02	4.8±0	214.3±23.5	142.3±16.9	1.55±0.0	0.63±0.01	0.86±0.01	0.10±0.02	0.02±0
Forest-tundra/sparse taiga	340.0	0.4±0.5	15.6±2.9	17.8±3.6	200±4.4	133.7±2.9	2.0±0.3	1.1±0.21	0.73±0.1	0.10±0.02	0.05±0
Taiga	709.0	20.3±5.0	11.2±3.1	13.1±3.3	270.1±22.3	180.7±15.6	3.1±0.9	1.5±0.36	0.68±0.3	0.12±0.02	0.06±0
Mixed deciduous forest	140.0	23.0±2.7	7.3±0.9	10.0±1.4	287.0±11.4	190.9±7.5	5.6±1.3	1.9±0.4	1.2±0.4	0.14±0.03	0.06±0.01
Forest-steppe	110.0	13.2±3.9	6.6±3.3	7.3±4.8	206.2±62.3	137.5±41.6	6.5±4.5	2.3±1.3	3.1±2.5	0.18±0.05	0.08±0.02
Steppe	117.0	4.9±0.8	5.6±0.5	1.0±0.2	182.9±35.8	121.7±23.8	6.4±0.9	1.7±0.2	4.7±0.7	0.14±0.03	0.06±0.01
Subtropical woodland	0.97	6.0±11	6.2±3.8	5.5±4.5	156.8±99.2	125.5±45.5	7.5±4.8	2.5±1.0	2.7±1.7	0.20±0.01	0.09±0.04
Desert-semidesert	174.0	3.9±0.96	4.5±0.2	0.5±0.1	63.0±11.6	41.9±7.6	4.0±0.52	1.1±0.1	2.8±0.3	0.05±0.01	0.02±0.01

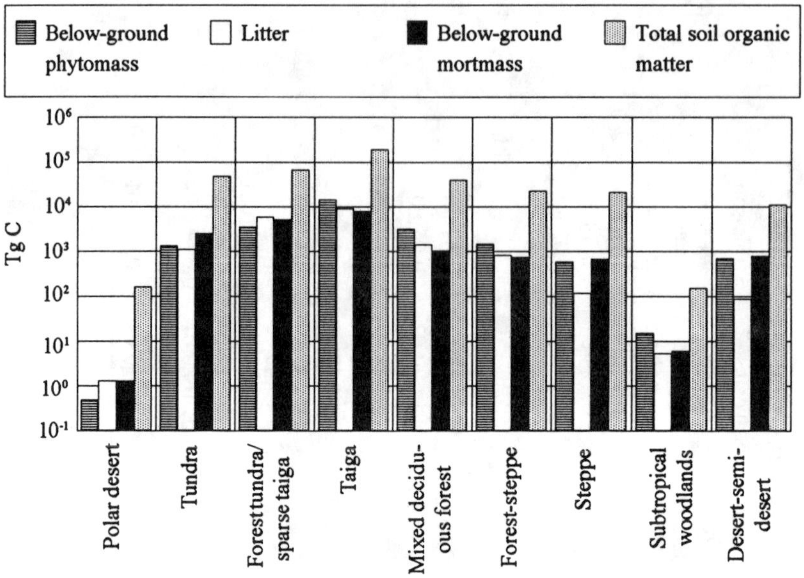

Figure 1. Carbon pools of the FSU biomes.

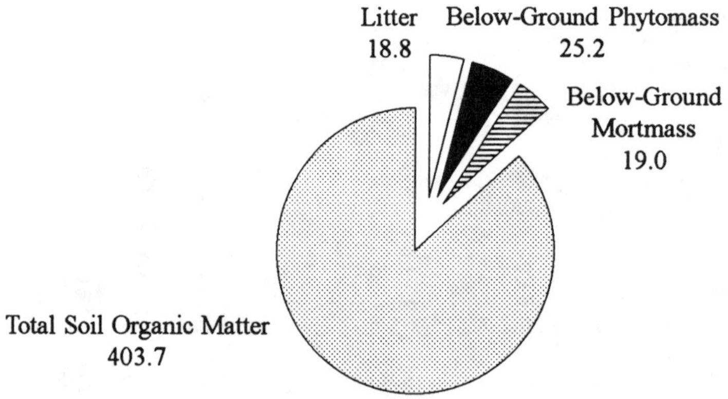

Figure 2. Distribution of carbon in litter, soil, and below-ground plant mass in the FSU (Pg C).

estimates based on carbon influx data and the Raich and Schlesinger data generally occurred in cases where data reported by Raich and Schlesinger were not collected over the entire year.

If climate warms and decomposition rates increase following $Q_{10} = 2.4$, then the total increase in carbon emissions from the decomposition processes may be as great as 1.0 Pg C yr^{-1} (Table 3), which is 16% of the present rate of carbon turnover in the terrestrial ecosystems of the FSU (Vinson and Kolchugina, 1993). The increase in decomposition rates with warming is comparable to 1.02 Pg C yr^{-1} of carbon emissions from the fossil fuel combustion in the FSU (Makarov and Bashmakov, 1990). The increase in plant productivity may be as great as 2.1 Pg C yr^{-1}, assuming (1) productivity would increase by 55% over half of the tundra

Table 2. Carbon effluxes from decomposition and root respiration

	Tg C/yr			
Biome	Below-ground mortmass decomposition	Litter decomposition	Soil organic matter decomposition	Root respiration (R_{AR})
Polar desert	0.2	0.2	0.1	0.2
Tundra	194.4	119.8	18.1	108.6
Forest tundra/ sparse taiga	248.2	340.0	17.0	210.8
Taiga	476.0	966.0	42.0	672.7
Mixed-deciduous forest	168.0	246.4	11.2	243.0
Forest-steppe	341.0	233.2	11.0	221.7
Steppe	549.9	182.5	9.4	232.1
Subtropical woodland	2.6	2.2	0.1	2.3
Desert-semidesert	487.2	182.7	5.2	215.8
Total	2467.4	2273.1	114.0	1907.1

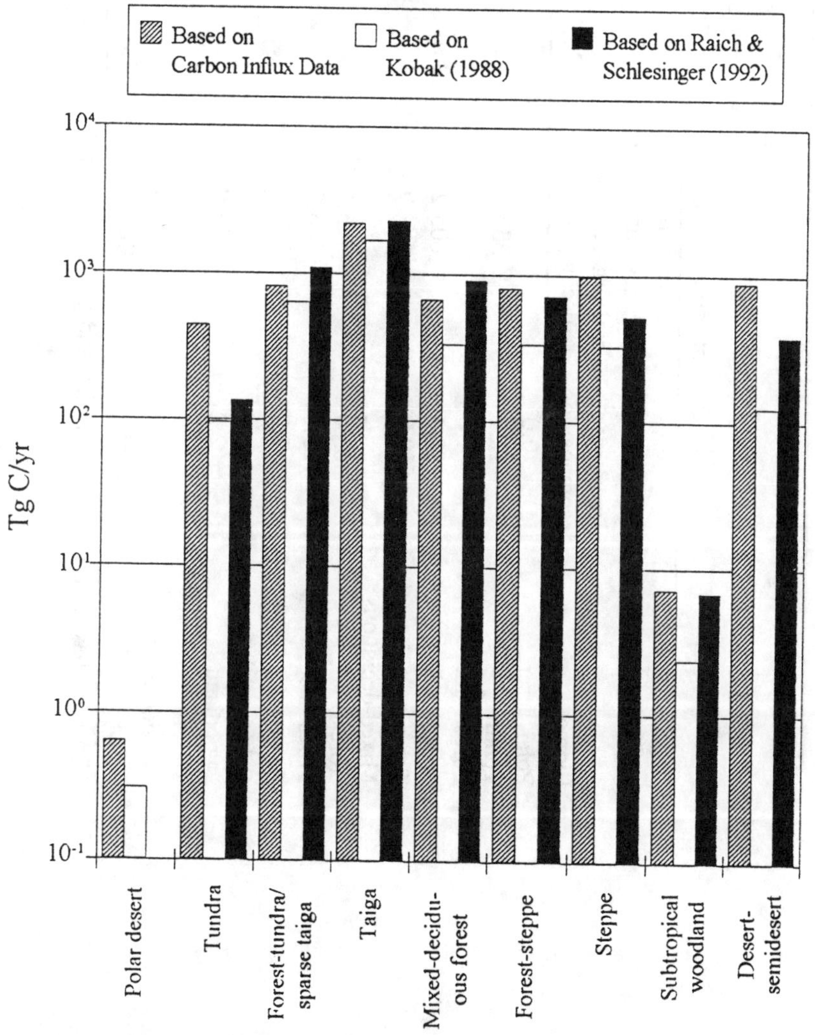

Figure 3. Comparison of soil surface carbon efflux.

Table 3. Change in carbon efflux from decomposition with climate warming (Tg C/yr)

	Present efflux		Change in efflux[a] from above-ground mortmass decomposition			Change in efflux[a] from below-ground mortmass, litter, and soil organic matter decomposition		
Biome	Above-ground mortmass decomposition	Below-ground mortmass litter, and soil organic matter decomposition	4°C	3°C	2°C	2°C	1.5°C	1°C
Polar desert	0.0	0.5	0.0			0.1		
Tundra	12.1	332.2	5.1			63.1		
Forest tundra/ sparse taiga	52.3	605.2	22.0			115.0		
Taiga	612.0	1484.0		183.6			207.8	
Mixed-deciduous forest	363.0	425.6		108.9			59.6	
Forest-steppe	125.0	585.2		37.5			81.9	
Steppe	2.7	741.8			0.5			66.8
Subtropical woodland	2.2	5.0			0.4			0.5
Desert-semidesert	3.3	675.1			0.6			60.8
Total	1172.6	4854.6	27.0	330.0	1.6	178.2	349.3	128.0

[a]Estimate based on Q_{10} = 2.4 (Raich and Schlesinger, 1992) for total *in situ* soil respiration rates.

Table 4. Preculticultivation and current carbon content

Soil type	Arable land (Mha)	Precultivated carbon content (Mg C/ha)	Precultivated carbon pool (Pg C)	Current cultivated carbon content[a] (Mg C/ha)	Current cultivated carbon pool (Pg C)	Area suitable for no-till** (Mha)	Current no-till carbon pool (Pg C)
Gray-brown desert	51.7	120.0	6.2	92.1	4.8	31.4	2.9
Floodplain-meadow	3.9	190.0	0.7	145.3	0.6	1.9	0.3
Gray-forest	44.1	160.0	7.1	122.5	5.4	40.1	4.9
Chernozem	78.0	333.0	26.0	254.0	19.8	76.5	19.4
Chestnut	17.5	110.0	1.9	84.5	1.5	16.9	1.4
Solonetz	6.7	42.0	0.3	32.8	0.2	6.7	0.2
Gray-brown desert	8.3	45.0	0.4	35.1	0.3	7.6	0.3
Total	210.2	---	42.6	---	32.5	181.1	29.4

*Based on equation (1).

**Based on climate factors described in methodology.

Table 5. No-till management option to sequester carbon

Soil type	Current cultivated C content* (Mg C/ha)	Distribution in soil profile (%)**		Current C content in soil horizons (Mg C/ha)		No-till equilibrium (Mg C/ha)		No-till change in C content (Mg C/ha)	Area suitable for no-till (Mha)	No-till C gain (Pg C)
		0-8 cm	8-30 cm	0-8 cm	8-30 cm	0-8 cm***	8-30 cm****			
Podzolic	92.1	39.0	27.0	35.9	24.9	46.1	28.8	14.2	31.4	0.4
Floodplain-meadow	145.3	21.0	39.0	30.5	56.7	39.2	65.7	17.7	1.9	0.0
Gray-forest	122.5	10.0	29.0	12.2	35.5	15.8	41.2	9.2	40.1	0.4
Chernozem	254.0	15.0	45.0	38.1	114.3	48.9	132.5	29.1	76.5	2.2
Chestnut	84.5	28.0	65.0	23.7	54.9	30.4	63.7	15.5	16.9	0.3
Solonetz	32.8	43.0	57.0	14.1	18.7	18.1	21.7	7.0	6.7	0.0
Gray-brown desert	35.1	33.0	67.0	11.6	23.5	14.9	27.2	7.1	7.6	0.1
Total	---	---	---	---	---	213.5	380.9	99.8	181.1	3.4

*Based on equation (1).
**Based on Glazovskaya (1972).
***Based on equation (2).
****Based on equation (3).

biome area and over the entire area of the forest-tundra/sparse taiga, taiga, mixed-deciduous forest, and subtropical woodland biomes, and (2) productivity of the forest steppe, steppe, and desert-semidesert biomes would not change because of the possible water deficiency. Therefore, the atmospheric carbon sink in terrestrial ecosystems of the FSU may increase in the future; the current rate of carbon accumulation in the FSU forests is approximately 0.5 Pg C yr^{-1} (Kolchugina and Vinson, 1993c,d). If productivity increases only by half of the projected amount, the equilibrium of the carbon cycle may be reestablished but at a higher rate of carbon turnover.

No-till management is one of the most promising practices to conserve and sequester carbon in the agricultural soils (Gaston et al., 1993). There are 181.1 Mha of agricultural lands climatically suitable for no-till management (Table 4). Soils of natural ecosystems and arable land of the FSU together stored 436.2 Pg C, or 22% of the global soil carbon pool. There has been a 10.2 Pg C decrease in the soil carbon pool since the onset of agriculture (Gaston et al., 1993). At present arable soils have approximately 24% less carbon compared to the precultivation stage.

The current carbon pool in cultivated soils which may be subjected to no-till management practice was estimated at 29.4 Pg C, or 90% of the total FSU carbon pool of cultivated soils. The no-till management practice would increase the carbon pool in arable lands by 3.4 Pg (Table 5). Carbon sequestration due to no-till management would continue until a new equilibrium of soil carbon is reached in one decade (Gaston et al., 1993). The average annual rate of carbon increase in soils would be 0.3 Pg C yr^{-1}, or one-third of the total increase in plant productivity and one-third to one-sixth of the possible increase in carbon emissions from decomposition processes under climate warming.

Full-scale implementation of no-till management is not possible even though conversion to no-till management requires a lower capital investment than other management practices that have been proposed to sequester carbon in agro-ecosystems. No-till management requires different or specially modified agricultural equipment. It is unlikely that individual farmers or individual collective farms will be able to finance the cost of new equipment, at least in the next decade (Gaston et al., 1993). At present, no-till management has been implemented on 21 Mha, which is 2.5 times more than in 1971-1975 (Ibragimov, 1989).

Shelter belt protection minimizes soil erosion, which may conserve carbon in soils at an average annual rate of 0.11%. The total carbon conserved from the shelter belt protection would be 33.0 Tg C yr^{-1}, assuming an average precultivation carbon content in soil of 150 Mg C ha^{-1} in the forest-steppe, steppe, and desert-semidesert biomes (Table 1). In one century 3.3 Pg C may be conserved.

The annual rate of carbon accumulation in soils and vegetation of forest plantation would be 83.8 Tg C yr^{-1}, or 8.4 Pg C in 100 years, assuming (1) the average rate of carbon accumulation in soils under broadleaf forest plantations is 0.18 Mg C ha^{-1} yr^{-1} (Table 1), (2) the average rate of carbon sequestration in vegetation of forest plantations is 1.0 Mg C yr^{-1}, and (3) 71 Mha of land in the FSU may be afforested. When added to the amount of carbon that may be additionally conserved in soils protected by shelter belts, the total gain in carbon due to both management practices would be approximately 0.12 Pg C yr^{-1}. In one century approximately 12 Pg C may be sequestered and conserved, accounting for approximately 11% of the carbon emitted in the FSU from fossil fuel combustion. Implementation of the no-till management and agroforestry practices would allow soil carbon sequestration and conservation comparable to the present annual net carbon sink in FSU forests (0.5 Pg C yr^{-1}, Kolchugina and Vinson, 1993c,d).

V. Summary and Conclusion

Twenty-one percent (436.2 Pg C) of the global terrestrial soil carbon pool is stored in soils of natural ecosystems and arable lands of the FSU. Soil carbon in natural ecosystems (including peatlands) was estimated at 403.7 Pg C. The current soil carbon in 210.2 Mha of FSU arable land was estimated at 32.4 Pg. There has been a 24% decline in soil carbon in arable land carbon and a total loss of 10.2 Pg C since the onset of cultivation.

Carbon emissions from decomposition of below-ground mortmass and litter were 20 to 30% higher than carbon emissions from root respiration. Decomposition of soil organic matter was an order of magnitude less. The estimated soil surface carbon efflux based on carbon influx data was in good agreement with estimates based on field data collected over an entire year.

Climate warming may increase the rates of decomposition processes in the FSU by 1.0 Pg C yr^{-1}, which is equivalent to current carbon emissions from the fossil fuel combustion in the FSU. It has been suggested that plant productivity may also increase by 55% with climate warming. Under this forecast the increase in plant productivity in the FSU may be as great as 2.1 Pg C yr^{-1}. Therefore, the atmospheric carbon sink in terrestrial ecosystems of the FSU (current rate of carbon accumulation in the FSU forests is 0.5 Pg C yr^{-1}) may increase in the future. If productivity increases only by half of the projected amount, the equilibrium of the carbon cycle may be reestablished but at a higher rate of carbon turnover.

No-till management may be employed to reduce the rate of carbon release from agricultural soils. A number of benefits accrue from this management strategy, including lower rates of carbon oxidation and erosion. The amount of agricultural land climatically suitable for no-till management in the FSU is 181.1 Mha. Implementation of no-till management over this area would result in the sequestration of 3.4 Pg C over a ten year period. This represents a 10% increase in carbon in the agricultural soils of the FSU.

Land-protective forest stands (i.e., shelter belts) in the agricultural sector and afforestation of disfigured landscapes can also be used to conserve and sequester carbon. Full implementation of shelter belt protection and afforestation would result in the conservation of 0.12 Pg C yr^{-1} in soils and plant mass of shelter belts and forest plantations.

Planting shelter belts, afforestation, and implementation of no-till management would result in carbon conservation and sequestration in the agricultural soils of the FSU comparable to the current rate of carbon accumulation in the FSU forests.

Acknowledgements

The work presented herein was funded by the U.S. Environmental Protection Agency (EPA) - Environmental Research Laboratory, Corvallis, Oregon, under Cooperative Agreement CR820239 to Oregon State University. Jeffrey J. Lee is the Project Officer for the project entitled "Carbon Cycling in Terrestrial Ecosystems of the Former Soviet Union." The work presented is a component of the U.S. EPA Global Climate Research Program, Global Mitigation and Adaptation Program, Robert K. Dixon, Program Leader. This paper has not been subjected to the EPA's review and, therefore, does not necessarily reflect the views of the EPA, and no official endorsement should be inferred. Peggy Blair prepared the camera ready copy of the manuscript; Shane Trenary prepared the camera ready copy of the tables and figures.

References

Anonymous. 1991. *Forestry at the Beginning of XXI Century*. X Forestry Congress, Paris. Ecologia Press, Moscow. 188 pp. (in Russian).

Barnwell, T.O., R.B. Jackson, IV, E.T. Elliott, I.C. Burke, C.V. Cole, K. Paustaian, E.A. Paul, A.S. Donigian, A.S. Patwardhan, A. Rowell, and K. Weinrich. 1992. An approach to assessment of management impacts on agricultural soil carbon. *Water, Air, and Soil Poll.* 64:423-435.

Bazilevich, N.I. 1986. Biological productivity of soil-vegetation formats in the U.S.S.R. *Bulletin of Academy of Sciences of the U.S.S.R., Geogr. Ser.*, 2:49-66.

Billings, W.D., J.O. Luken, D.A. Mortensen, and K.M., Peterson. 1982. Arctic tundra: a source or sink for atmospheric carbon dioxide in a changing environment? *Oecol.* 52:7-11.

Bolin, B. 1977. Changes in land biota and their importance in the carbon cycle. *Science* 196:613-615.

Burrough, P.A. 1986. *Principles of Geographical Information Systems for Land Resources Assessment*. Clarendon Press, Oxford, New York. 193 pp.

Chapin, F.S., III. 1984. The impact of increased air temperature on tundra plant communities. p. 43-162. In: *Proceedings of the Conference: The Potential Effects of Carbon Dioxide-induced Climatic Changes in Alaska*. School of Agriculture and Land Resources Management, University of Alaska, Fairbanks.

Chapin, F.S., III, and G.R. Shaver. 1981. Changes in soil properties and vegetation during disturbance of Alaskan Arctic tundra. *J. Appl. Ecol.* 18:605-617.

Cherdantsev, G.N. 1961. Map: *Arable Land in the U.S.S.R. in 1954*. Geography of the U.S.S.R. In: J.P. Cole, F.C. German, *Background to a Planned Economy*. 290 pp.

Dick W.A. 1983. Organic carbon, nitrogen and phosphorus concentrations and pH in soil profiles as affected by tillage intensity. *Soil Sci. Soc. Am. J.* 47:102-107.

Dixon, R.K. and D.P. Turner. 1991. The global carbon cycle and climate change: Responses and feedbacks from below-ground systems. *Env. Poll.* 73:245-262.

Etkin, D. 1990. Greenhouse warming: Consequences for Arctic climate. *J.Cold Reg. Eng.* 4:54-66.

Gaston, G.G., T.P. Kolchugina, and T.S. Vinson. 1993. Potential effect of no-till management on carbon in the agricultural soils of the former Soviet Union. *Agr. Ecos., and Env. J.* 45:295-309.

Glazovskaya M.A. 1972. *Soils of the World. I.* Moscow University Press, Moscow. 214 pp. Translated from Russian. Amerind Publishing Co. Pvt. Ltd., New Delhi, 1983.

Gorshkov, S.P., A.G. Suschevskyi, and G.N. Shenderuk. 1980. Cycle of organic matter. p. 153-181. In: *Cycle of Matter in Nature and its Transformation under Human Activities.* Moscow State University Press, Moscow (in Russian).

Haas H.J., C.E. Evans, and E.F. Miles. 1957. Nitrogen and carbon changes in Great Plains soils as influenced by cropping and soil treatments. *U.S. Dept. Ag. Tech. Bull.* 1164.

Hansen, J., A. Lacis, D. Rind, L. Russel, P. Stone, I. Fung, R. Ruedy, and J. Lerner. 1984. Climate sensitivity and analysis of feedback mechanisms. In: Hansen, J. and R. Thompson (eds.), *Climate Processes and Climate Sensitivity, Geoph. Monogr.*, Am. Geoph. Un., Washington, D.C., 29:13-163.

Hobbs J.A. and P.L. Brown. 1965. Effects of cropping and management on nitrogen and organic carbon contents of a Western Kansas soil. In: *Kansas Agr. Exp. Stat. Tech. Bull.* 144 pp.

Houghton R.A., J.E. Hobbie, J.M. Melillo, B. Moore, B.J. Peterson, G.R. Shaver, and G.M. Woodwell. 1983. Changes in carbon content of terrestrial Biota and soils between 1860 and 1980: A net release of CO_2 to the atmosphere. *Ecol. Monogr.* 53:235-262.

Houghton, R.A. and G.M. Woodwell. 1989. Global climate change. *Sci. Am.* 260:36-44.

Ibragimov, G.G. 1989. For Land - Care and Attention of Foresters. *Lesnoe Khozyaistvo.* 9:11-14 (in Russian).

Intergovernmental Panel on Climate Change (IPCC). 1990. Houghton, J.T., G.J. Jenkins, and J.J. Epgraums (eds.), *Climate Change, The IPCC Scientific Assessment.* Working Group 1 Report, WMD & UNEP Univ. Press, Cambridge. 365 pp.

Intergovernmental Panel on Climate Change (IPCC). 1992. 1992 Supplemental Report of AFOS Working Group. IPCC Secreteriate, Geneva, Switzerland (in press).

Isachenko, A.G. (ed.). 1988. Map: *Landscape Forms of the U.S.S.R.* Institute of Geography, Leningrad State University (in Russian).

Kern J.S. and M.G. Johnson. 1991. Impact of conservation tillage use on soil and atmospheric carbon in the contiguous United States. EPA/600/3-91/056, Env. Res. Lab, Corvallis, OR.

Kobak, K.I. 1988. *Biotical Components of Carbon Cycle.* Hydrometeoizdat, Leningrad. 248 pp. (in Russian).

Kobak, K.I. and N.Yu. Kondrashova. 1987. Anthropogenic impact on soil carbon reservoir and carbon cycle. *Meteor. and Hydrol.* 5:39-46 (in Russian).

Kolchugina, T.P. and T.S. Vinson. 1993a. Equilibrium analysis of carbon pools and fluxes of forest biomes in the former Soviet Union. *Can. J. For. Res.* 23:81-88.

Kolchugina, T.P. and T.S. Vinson. 1993b. Climate warming and the carbon cycle in the permafrost zone of the former Soviet Union. *Perm. and Perigl. Proc.* 4:149-163.

Kolchugina, T.P. and T.S. Vinson. 1993c. Carbon sources and sinks in the forest biomes of the former Soviet Union. *Glob. Biog. Cyc.* 7:291-304.

Kolchugina, T.P. and T.S. Vinson. 1993d. Comparative analysis of carbon budget components for forest biomes in the former Soviet Union. *Water, Air, and Soil Pollution* 70:207-221.

Kononova, M.M. 1984. Organic matter and soil fertility. *Pochvovedenie* 8:6-20 (in Russian).

Makarov, A.A. and I. Bashmakov. 1990. The Soviet Union. p. 35-53. In: W.U. Chandler (ed.), *Carbon Emission Control Strategies: Case Studies in International Cooperation.* World Wildlife Fund and the Conservation Foundation, Washington, D.C.

Makarov, B.N. 1988. *Gaseous Regime.* Agroprom Press, Moscow. 104 pp. (in Russian).

Manabe, S. and R.T. Wetherald. 1987. Large scale changes of soil wetness induced by an increase in atmospheric carbon dioxide. *J. Atm. Sci.* 44:1211-1235.

Mann L.K. 1985. A regional comparison of carbon in cultivated and uncultivated Alfisols and Mollisols in the continental United States. *Geoder.* 36:241-253.

Mann L.K. 1986. Changes in soil carbon storage after cultivation. *Soil Sci.* 142:279-288.

Medvedev, Z.A. 1987. *Soviet Agriculture.* Norton and Co., New York. 464 pp.

Microsoft Corporation. 1991. Microsoft Excel. Redmond, WA.

Miller, P.C., W.A. Stoner, and L.L. Tieszen. 1976. The model of stand photosynthesis for the wet meadow tundra at Barrow, Alaska. *Ecol.* 57:411-430.

Raich, J.W. and W.H. Schlesinger. 1992. The global carbon dioxide flux in soil respiration and its relationship to vegetation and climate. *Tellus* 44B:81-99.

Rastetter, E.B., M.G. Ryan, G.R Shaver, J.M. Melillo, K.J. Nadelhoffer, J.E. Hobbie, and J.D. Aber. 1991. A general biogeochemical model describing the responses of the C and N cycles in terrestrial ecosystems to changes in CO_2, climate, and N deposition. *Tree Phys.* 9:101-126.

Rozanov, B.G., V.O. Targulian, and D.S. Orlov. 1989. Global tendency in soil and soil cover transformation. *Pochvovedenie* 5:5-17 (in Russian).

Ryabchikov, A.M. (ed.). 1988. Map: *Geographical Belts and Zonal Types of Landscapes of the World.* School of Geography, Moscow State University (in Russian).

Schlesinger, W.H. 1977. Carbon balance in terrestrial detritus. *Ann. Rev. Ecol. and Syst.* 8:51-81.

Schlesinger, W.H. 1984. Soil organic matter: a source of atmospheric CO_2. p. 111-127. In: G.M. Woodwell (ed.), *The Role of Terrestrial Vegetation in the Global Carbon Cycle: Measurement by Remote Sensing*, SCOPE, J. Wiley and Sons, NY.

Schlesinger, M.E. and Z.C. Zhao. 1989. Seasonal climate changes induced by doubled CO_2 as simulated by the OSU atmospheric GCM mixed-layer ocean model. *J. Climate* 2:463-499.

Schneider, S.H. 1989. The greenhouse effect: science and policy, *Science* 243:771-781.

Schneider, S.H. 1990. *Global Warming*. Vintage Books, NY. 343 pp.

Shaver, G.R., W.D. Billings, F.S Chapin, III, A.E. Giblin, K.J. Nadelhoffer, W.C. Oechel, and E.B. Rastetter. 1992. Global climate change and the carbon balance of Arctic ecosystems. *BioSci.* 42:433-441.

Unger, P.W. 1968. Soil organic matter and nitrogen changes during 24 years of dryland wheat and cropping practices. *Soil Sci. Soc. Am. Proc.* 32:427-429.

USSR State Committee for the Protection of Nature. 1989. Report on the state of the environment in the USSR in 1988. Moscow. 184 pp.

U.S. Depart. of Agriculture. 1990. *USSR Trade and Agriculture Report: Situation and Outlook*, USDA RS-90-1, May 1990.

Vinson, T.S. and T.P. Kolchugina. 1993. Pools and fluxes of biogenic carbon in the former Soviet Union. *Water, Air, and Soil Pollution* 70:223-237.

Vorobyov, G.I. (ed.). 1985. *Forest Encyclopedia*. Sovetskaya Encyclopedia Press. Moscow. 563 pp.

Washington, W.M. and G.A. Mitchel. 1984. Seasonal cycle experiment on the climate sensitivity due to a doubling of CO_2 with an atmospheric general circulation model coupled to a simple mixed layer ocean model. *J. Geoph. Res.* 89:9475-9503.

Zimov, S.A., S.P. Daviodov, Yu.V. Voropaev, and S.F. Prosiannikov. 1991. Planetary maximum CO_2 and ecosystems of the North. In: T. Kolchugina and T. Vinson (eds.), *Proceedings of Workshop on Carbon Cycling in Boreal Forests and Subarctic Ecosystems*, Corvallis, OR, September 9-14, 1991.

CHAPTER 4

Land Use and Soil Management Effects on Emissions of Radiatively Active Gases from Two Soils in Ohio

R. Lal, N.R. Fausey, and D.J. Eckert

I. Introduction

Terrestrial ecosystems in general and soil resources in particular play a major role in the global C budget and flux of radiatively active gases. Depending on land use and farming systems, soil can be an important source or sink for carbon. Important gases generated within soil include CO_2, N_2O, NO_X, CH_4, C_2H_6, and C_2H_4 under anaerobic conditions, and mostly CO_2 under aerobic conditions. Despite their potential impact, studies of the rate and composition of gaseous emissions from soil to the atmosphere in relation to soil and crop management techniques have not been given the attention they deserve. Important agricultural activities in relation to the emission of radiatively active gases include land use (forest, pasture, or arable), and deforestation. Cropping practices relevant to gaseous flux include crop rotations and combinations, fertility maintenance including use of inorganic fertilizers and organic manures, tillage methods,cover crops, and crop residue management. Important factors affecting gaseous emissions from soil in silviculture and pastoral land use systems may be tree species, tree density, logging and forest management, stocking rate, pasture species and management. In addition, soil and air climates, through their effects on soil and air temperature and soil moisture regime, play an important role in regulating gaseous emissions. Effects of all these factors on the rate and type of gaseous emissions are greatly modified by soil properties, and microclimate.

Because of these interacting factors, gaseous composition and flux from soil are highly variable over time and space (Folorunso and Rolston, 1984; Rochette et al., 1991). There are several mechanisms of origin and distribution of CO_2 in soil (Wood and Petraitis, 1984). CO_2 production in soil is related to biological activity including root respiration (Keller et. al., 1986), and decomposition of soil organic matter by microbial activity (Amundson and Davidson, 1990), and plant growth and irrigation (Robins, 1986). The microbial activity depends on the soil temperature regime. The rate of CO_2 production and flux in soil increases between 1.5 and 3 times for every 10° C increase in temperature between 0 and 50° C (Monteith et. al., 1964; Parada et. al., 1983; Norstadt and Porter, 1984; Amundson and Smith, 1988; Amundson et al., 1989). Soil moisture affects CO_2 concentration through its influence on gaseous diffusion and microbial activity. Increasing soil wetness between the permanent wilting point and 60 to 80% of saturation increases the rate of CO_2 evolution from soil (Kucera and Kirkham, 1971; Alexander, 1977; Solomon and Cerling, 1987). On the basis of a global survey, Brooke et. al, (1983) observed that CO_2 concentration in soil was strongly correlated with evapotranspiration or the rate of biomass production, with higher concentrations in tropical and warm climates than in temperate and cold climates.

The marked seasonal variations in CO_2 concentration in soil are, therefore, partly attributed to changes in soil temperature and soil moisture regimes (Fernandez and Kosian, 1987; Solomon and Cerling, 1987; Amundson and Smith, 1988; Reicosky, 1989; Robertson and Ball, 1989). Concentration of CO_2 also depends on land use, soil management, and other factors that affect soil organic matter content (Moore, 1986) and biomass addition to the soil. The amount of litter added to the soil is an important factor (Andres et. al.,

ISBN 1-56670-117-1/95/$0.00+.50

1983). CO_2 levels are generally higher in notill than in conventional tillage systems (Staley, 1988), which implies that the increase in microbial respiration due to tillage is not a major factor leading to the loss of organic matter in soils under cultivation (Roberts and Chan, 1990). These researchers observed that total C loss due to tillage was 0.0005% to 0.0037%. Yamaguchi et al. (1967) observed CO_2 concentrations as high as 17%. Cleemput and Baert (1983) and Reicosky and Lindstrom (1994) observed that CO_2 concentrations in soil could be as high as 10% during summer. High concentrations were also observed by Lal and Taylor (1969), Buyanovsky and Wagner (1983), and Buyanovsky et al. (1986). Dyer and Brooke (1991) observed that the mean soil CO_2 concentration was 0.207% in an evergreen temperate forest compared with 0.157% in deciduous and mixed forest. The lowest concentration normally occurs in desert regions (Amundson et al., 1989). All other factors remaining the same, forested tropical soils have relatively higher concentrations of CO_2 than soils in other ecosystems (Crowther, 1983; Brooke et al., 1983; Amundson and Davidson, 1990).

Similar to concentrations, CO_2 flux also depends on several factors and is highly variable in time and space. Measured rates of CO_2 flux range from 0 to 4000 mg CO_2 m^{-2} hr^{-1} (Singh and Gupta, 1977; Glinski and Stepniewski, 1985; Reicosky and Lindstrom, 1994).

Soil can be a source or sink of N_2O depending on soil properties, management and environmental factors (Sahrawat and Keeney, 1986; Mooney et al., 1987). N_2O production in soil is related to both nitrification (Goodroad and Keeney, 1984; 1985; Robertson and Tiedje, 1984; Sahrawat et al., 1985) and denitrificatior processes (Duxbury et al., 1982; Goodroad and Keeney, 1984b; Goodroad et al., 1984; Lalisse-Grundmanr and Chalamet, 1987). Factors affecting N_2O production by nitrification include soil pH, organic matte content, O_2 concentration, and availability of nitrifiable N (Grundmann and Rolston, 1987; Grundman e al., 1988). Effects of these factors are greatly influenced by environmental parameters, e.g., temperature rainfall, soil moisture and temperature regimes, etc. Factors affecting N_2O production by denitrificatioı include nitrate and nitrite concentrations in soil, soil O_2 and CO_2 concentrations, pH, organic matter content redox potential, and sulfide content (Keeney et al., 1985; Sahrawat and Keeney, 1986). High levels of N₂(observed in the spring were 5-20 μL L^{-1} and the low level during summer was 0.5 μL L^{-1}. The magnitud of N_2O flux from forest soils may range from 0 to 40 kg N_2O-N $ha^{-1}yr^{-1}$ (Goodroad and Keeney, 1984 Sahrawat and Keeney, 1986). N_2O emissions from arable land are greatly influenced by fertilizer and tillag management practices (Rolston, 1981; Ryden and Rolston, 1983; Goodroad et al., 1984). Form and rat of fertilizer application (Mosier et. al., 1982; 1983; Staley, 1988), soil compaction, tillage and crop cove management (Aulakh et al., 1982; 1983a,b; Smucker and Erickson, 1989; Myburgh and Moolmann, 1991 Simojoki et al., 1991), rainfall and soil moisture regimes have important effects on N_2O emissions from soil Further N_2O emissions from histosols and manured soils are greater than those from mineral or unmanurec soils (Duxbury et al., 1982). Lind (1985) observed that application of organic manure temporarily increasec the concentration of N_2O in soil air. Goodroad et al. (1984) observed that N_2O emissions (Kg N_2O-N ha^{-1}) in two soils of Wisconsin under different treatments were: reduced tillage corn, 3.5 to 6.3; sludge, 1.6; manure, 6.1; alfalfa, 3.2; rye, 1.6; and straw, 3.2. The highest rate of N_2O emissions were often observed at the time of soil thaw (Christenson and Tiedje, 1990). Observed rates of N_2O emissions from arable soils range from 0 to 40 ng N_2O-N $m^{-2}s^{-1}$ with total emission of 0 to 6 kg N_2O-N $ha^{-1}yr^{-1}$ (Goodroad et al., 1984; Sahrawat and Keeney, 1986).

High soil organic matter content and anaerobic conditions favor production of CH_4, C_2H_4, C_2H_6, and C_3H_8 in soil (Tiedje et al., 1984; Neue and Scharpenseel, 1984; Vermoesen et al., 1991). Consequently, CH_4 production is generally higher in rice paddies than in upland soils (Stepniewski and Glimski, 1988; Kimura et. al., 1991). Experiments conducted in West Virginia showed that CH_4 production in surface peat ranged from 0.2 - 18.8 mol mol $(C)^{-1}hr^{-1}$ between February and September (Yavitt et al., 1987). There is a strong seasonal variation in CH_4 evolution from rice soil. Holzapfel-Pschorn and Seiler (1985) observed that CH_4 evolution from rice paddies ranged between 4 and 17 mg CH_4 $m^{-2}hr^{-1}$ resulting in worldwide emission of 33-83 x 10^{12} g hr^{-1}. This flux of CH_4 from rice amounts to 10-20% of the global CH_4 budget From experiments done on peatland soils in Canada, Moore and Knowles (1989) observed that CH evolution decreased logarithmically as the water table was lowered. The highest rate of CH_4 flux (28 mg CH_4 $m^{-2}day^{-1}$) was observed when peat was inundated, and the lowest rate (0.7 mg $m^{-2}day^{-1}$) was when i was well drained. In uplands with moderate concentrations of soil organic matter (<5%) and in well draine conditions, CH_4 production is minimal, and soil is often a sink for CH_4.

Because of an enthusiastic interest in emission of radiatively active gases from soil-related processes there has been proliferation in methods of sampling soil air (Sebacher and Harris, 1982; Bremner an Blackmer, 1982; Kasper, 1984; Klemedtsson et al., 1986; Rolston, 1986; Magnusson, 1989; Moffat et al

1990; Mukhtar et al., 1990), in analytical procedures (Boddy and Lloyd, 1990; Neilson and Pepper, 1990; Freijer and Bouten, 1991; Breitenbeck, 1990), and in developing predictive models at different scales of assessing gaseous emissions (Solomon and Cerling, 1987; Brooke et al., 1983; Naganwa and Kyuma, 1991). There is, however, a strong need to standardize research methods so that results are comparable.

Present experiments were conducted on two sites, one each in central and northwestern Ohio during 1991-92, with the following objectives:

(i) To quantify gaseous emissions from soil as influenced by land use, and soil and crop management factors,

(ii) To evaluate seasonal changes in gaseous emissions under different land use systems, and

(iii) To evaluate the effects of soil and crop management on concentrations of NO_3-N and NH_4-N in the soil solution.

II. Materials and Methods

Sampling to characterize the soil air composition and the emission of radiatively active gasses from the soil was carried out on existing field experiments involving land use and management variables. These field experiments were established at the Agronomy Research Farm of the Ohio State University, Columbus, Ohio (40°N, 83° 02'W), and at the Northwestern Branch of the Ohio Agricultural Research and Development Center (OARDC) at Hoytville, Ohio (41° 13'N, 83° 45'W).

Soil air samples were collected at 15 cm depth using the diffusion chamber technique described by Lal and Taylor (1969) and Taylor and Abrahams (1953). This sampling technique involves excavation of a cavity to 18 cm depth, placement of about 6 cm of fine gravel in the bottom of the cavity, imbedding the lower end of a diffusion tube in this gravel at 15 cm depth, and backfilling around the diffusion tube to establish a permanent sampling location. The upper end of the diffusion tube is fitted with a hypodermic needle which is stoppered until the sampling time. Samples are obtained at specific times by removing the stopper and attaching an evacuated 10-cm^3 vial to the needle. Samples were typically drawn at approximately 10 a.m. and taken to the laboratory for analysis.

Gaseous fluxes were measured using the closed chamber technique developed by Matthias et al. (1980). This sampling technique involves placing an insulated, open-bottom chamber (15 cm diameter and 15 cm height) onto the soil surface for 15 minutes. The chamber is equipped at the top with a hypodermic needle which is stoppered until the sampling time. After 15 minutes, samples are obtained by removing the stopper and attaching an evacuated 10-cm^3 vial to the needle. At the same time, the ambient above ground atmosphere is sampled. Gas flux was identified by the difference between ambient concentration and the concentration in the chamber after 15 minutes. Samples were obtained once every week or once every two weeks and brought to the laboratory for analyses of gaseous composition.

All samples were collected in sterile, 10-cm^3 vials with red rubber septum stoppers. These vials were used directly as supplied by the manufacturer. According to Parkin (1994), the red rubber stoppers have been shown to generate CH_4 being determined for the samples. The vials also contain very small amounts of N_2O in the residual gas that remains in the vials at the evacuation pressure.

All samples were analyzed using a gas chromatograph Varian[1] GC Model 3700 for CO_2 and N_2O and Hewlett Packard[1] Model 5710A for CH_4. Helium was used as the carrier gas at the flow rate of 0.005 L s^{-1}. The temperature of the TCD was 100°C and the column used was Chromosorb 102 with mesh of 80-100. The minimum detectable level with the setup is 30×10^{-10} g ml^{-1} of butane. The FID column used for CH_4 measurement was Pore Pack N with the minimum detectable level of 10 ppb.

A. The Columbus Experiment

The site has a mean annual precipitation of 943 mm, including 714 mm of snowfall. Based on mean monthly temperature, four seasons can be distinguished as: spring (late March-May); summer (June-early September); autumn (September-early November) and winter (November-mid March).

[1]Trade names are presented for the benefit of the reader and do not imply endorsement by the Ohio State University or USDA.

Five land use treatments established were: (i) natural forest, (ii) alfalfa, (iii) corn with plow-till, (iv) corn with ridge-till, and (v) corn with no-till. The forest site was a 25-year old natural regrowth with a mixed stand of predominantly white pine, red pine, and Austrian pine with some stand of oak, beech, maple, and other tree species. The alfalfa treatment was a 5-year old stand which was regularly mowed to simulate harvesting but the hay was left on the field. The plow-till corn involved moldboard plowing to about 20 cm depth across the slope followed by disc harrowing before sowing corn in the spring. No-till corn consisted of sowing directly into the residue of the previous corn after spraying with Paraquat (1, 1 dimethyl-4, 4 bipyridinium ion) at 2.5 L h^{-1}. All corn treatments received N, P, K fertilizer at the rate of 268, 100, and 200 kg ha^{-1}, respectively. The fertilizer was surface broadcast prior to planting and mixed into the soil with disk harrow in the plowtill treatment. In the no-till and ridge-till treatments, fertilizer was broadcast on the surface.

Soils of the Columbus experimental site are mostly well-drained and belong to the Miamian series (a fine, mixed, mesic, Typic Hapludalf). On the lower slopes, under slightly impeded drainage, the soil is classified as a fine, mixed, mesic, Aquic Hapludalf. The surface layer has a pH of 6.9 to 7.3 and soil organic carbon content of 2.3 to 2.5%. Both soils are derived from the glacial till but are different series. All measurements of gaseous composition were made on upper slopes under well-drained conditions.

B. The Hoytville Experiment

Like Columbus, the Hoytville site also has four distinct seasons. Mean annual rainfall at the site ranges from 600 to 1000 mm, with a long-term average of 850 mm. The soil is the Hoytville series (fine, illitic, mesic family of Mollic Ochraqualfs). This soil is derived from a fine-textured, calcareous glacial till. It is characterized by a dark-colored surface horizon of about 20 cm depth, very dark-gray firm clay from 20 to 60 cm depth, and dark grayish-brown but very firm clay and mottled horizon from 60-100 cm depth. The soil has high clay content (about 45%), high swell-shrink capacity, and develops 5 cm wide and 50-80 cm deep cracks on drying. The surface layer has pH of about 6.6 and organic carbon content of about 2.3% to 2.5%.

Determination of the composition of soil air and measurements of gaseous fluxes at Hoytville were made for two experiments:

1. Soil Management Site

This experiment comprised seven treatments involving corn and soybean. Corn was grown following soybeans with four management treatments, e.g., no-input ridge-till, no-input chisel-till, recommended inputs chisel-till, and manure chisel-till. The no-input treatment received neither chemical fertilizers nor organic manure. The manure treatment involved application of 185 Mg ha^{-1} of liquid beef cattle manure. The manure was surface broadcast and then disked in. The manure contains 2.1% total solids. On dry weight basis, manure contained 14.4% total N, 1.0% P, 5.7% K, 2.1% Ca, and 0.6% Mg. The treatment with inputs received 224 kg N/ha, but no P or K, due to high soil test levels. Soybeans were grown with three treatments, e.g., no-input ridgetill, recommended inputs chisel-till, and manure chisel-till. No fertilizer or manure were applied to soybeans. Corn, Pioneer 342.9, was sown in rows 75 cm apart with plant populations of 78,000 plants ha^{-1}. Conrad variety of soybeans was sown in rows 75 cm apart with plant populations of 400,000 plants ha^{-1}.

2. Drainage Site

This experiment involved three treatments, e.g., corn ridge-till, corn plow-till, and soybeans plow-till. Ridge-till is a conservation tillage system whereby semi-permanent ridges were made once and repaired for the following season. Seasonal tillage in ridge treatment involved removing ridge tops at sowing and re-ridging when corn was 30 to 60 cm tall. In ridge-till corn, wheel traffic was restricted to every other furrow. Gaseous flux measurements in ridged treatments were made in the furrow zone without wheel tracks.

Soil solution samples were obtained at the 50-cm depth using 4-5 cm diameter porous cup lysimeters. These samples were analyzed for NO_3-N and NH_4-N by colorimetry using a Lachet auto analyzer.

III. Results

A. Gaseous Composition of Soil Air

1. Columbus Experiment

Gaseous composition of soil air sampled at the 15-cm depth within the plow layer at the Columbus site on Miamian soil is shown in Tables 1, 2, and 3 for CO_2, N_2O, and CH_4 respectively. The data in Table 1 show that CO_2 concentration in soil air was significantly influenced by date of sampling and land use treatment. Seasonal fluctuations showed that CO_2 concentration was in the order of summer > spring > fall > winter. Mean concentration of CO_2 was 1562 ppm, 984 ppm, 462 ppm and 1280 ppm for summer, fall, winter, and spring respectively. In comparison with about 350 ppm of CO_2 concentration in the atmosphere, soil air contained 1.32 to 4.46 times more CO_2 than did the atmosphere. Concentration of CO_2 was also affected by land use treatment. The data in Table 1 show that the mean CO_2 concentration in soil air was in the order of forest > alfalfa > corn. The data averaging all four seasons showed that CO_2 concentration in soil air was 1113 ppm, 1081 ppm, 835 ppm and 867 ppm for forest, alfalfa, corn plow-till and corn ridge-till, respectively. Measurements on corn no-till were not started until winter. Comparison of the data for that period showed that no-till and ridge-till systems were characterized by relatively higher concentrations of CO_2 than plow-till system. Differences among tillage systems were particularly significant during spring and summer. Mean CO_2 concentration in soil air was 529 ppm, 620 ppm, and 638 ppm for plow-till, no-till and ridge-till corn, respectively.

Concentrations of N_2O in soil air for the Miami soil are shown in Table 2. Concentration of N_2O was also measured on all dates on which CO_2 was measured. However, the data presented in Table 2 are only from those samples that contained measurable N_2O concentrations. There were differences in N_2O concentrations with regards to the treatment and date of sampling. In general, N_2O concentration was greater in summer and fall than in winter and early spring. The highest concentration, 101 ppm, was observed in summer. Concentration of N_2O was generally low or unmeasurable during winter. With regards to soil management treatments, N_2O concentration ranged from 0 to 97.2 ppm for forest, 80.4 ppm for alfalfa, 100.5 ppm for corn plow-till, 25.0 ppm for corn notill, and 25.1 ppm for corn ridge-till. Lind (1985) reported N_2O concentrations of 5-20 ppm in a manured treatment in Denmark. For the period when measurements were made in the corn no-till treatment (November 1991 through March 1992), there were no significant differences in N_2O concentration in plow-till (12 ppm), no-till (11 ppm), and ridge-till (13 ppm) corn.

CH_4 concentrations in soil air sampled at 15-cm depth are shown in Table 3. Similar to CO_2, the concentration of CH_4 in soil air was high in summer and low in winter and spring. The highest concentration of about 20 ppm was measured in ridge-till and plow-till corn in August, 1992. There were no definite trends in CH_4 concentrations with regards to land use treatments. With regards to soil management treatments, CH_4 concentration ranged from undetectable to 18.9 ppm for forest, 18.2 ppm for alfalfa, 19.7 ppm for corn plow-till, 16.9 ppm for corn no-till, and 18.0 ppm for corn ridge-till.

2. Hoytville Experiments

a. Soil Management Site

Gaseous composition of soil air data for the soil management site at Hoytville are shown in Tables 4, 5, and 6. The data in Table 4 for CO_2 concentration show differences among treatments and dates of sampling. Grand mean CO_2 concentrations were 653 ppm, 658 ppm, and 700 ppm for soybeans grown with no-input ridge-till, input chisel-till, and manure chisel-till treatments, respectively. Mean CO_2 concentrations for corn were 723 ppm, 540 ppm, 595 ppm, and 600 ppm for no-input ridge-till, no-input chisel-till, input chisel-till, and manure chisel-till treatments, respectively. The data further show that mean concentrations of CO_2 were in the order of summer > fall > early spring > winter. In fact, the CO_2 concentration in soil air was often below 200 ppm during winter. Low concentrations during winter may be due to increase in solubility of CO_2 in water with decrease in temperature (Skopp, 1985; Skopp et al., 1990; Reicosky and Lindstrom, 1994).

Table 1. Seasonal changes in CO_2 concentrations (ppm) in sub-soil air samples in a Miamian soil at Columbus, Ohio

		Land use treatment				
Year	Season	Forest	Alfalfa	Corn plow-till	Corn no-till	Corn ridge-till
1991	Summer	2127±1203	1779±663	1630±657	--	1620±674
	Autumn	820±590	1194±634	813±526	--	805±430
1992	Winter	634±333	612±282	444±167	442±143	500±220
	Spring	665	617±3	558±66	468	446±73
	Summer	1340±215	2025±46	731±97	1160±7	1035±31
LSD			(0.05)	(0.10)		
Treatment			205	171		
Season/dates			593	496		

Table 2. Seasonal changes in N_2 concentrations (ppm) in sub-soil air samples in a Miamian soil at Columbus, Ohio

		Land use treatment				
Year	Season	Forest	Alfalfa	Corn plow-till	Corn no-till	Corn ridge-till
1991	Summer	7.7 (51.4)[a]	14.6 (80.4)	10.5 (100)	--	0.3 (3.6)
	Autumn	16.9 (97.2)	2.9 (16.4)	17.8 (68.7)	--	5.7 (25.1)
1992	Winter	2.1 (25.2)	2.0 (35.3)	2.5 (34.4)	2.1 (25.0)	1.3 (22.7)
LSD			(0.05)	(0.10)		
Treatment			22	18		
Season/dates			32	27		

[a] Values in parentheses are the maximum concentrations measured.

Table 3. Seasonal changes in CH_4 concentrations (ppm) in sub-soil air samples in a Niamian soil at Columbus, Ohio

		Land use treatment				
Year	Season	Forest	Alfalfa	Corn plow-till	Corn no-till	Corn ridge-till
1991	Summer	0.06 (0.76)[a]	0.02 (0.14)	0.05 (0.53)	0.26(0.8)	0.07 (1.02)
	Spring	0.23 (0.45)	0.07 (0.13)	0.06 (0.12)	0.07(0.13)	0.07 (0.13)
1992	Summer	6.75 (18.93)	6.49 (18.2)	7.22 (19.7)	6.3 (25.0)	5.78 (18.0)
LSD			(0.05)	(0.10)		
Treatment			0.46	0.38		
Season/dates			0.77	0.64		

[a] Values in parentheses are the maximum concentrations measured.

Table 4. Soil and crop management effects on CO_2 concentrations (ppm) in subsoil air for the Hoytville series

		Soybean			
Year	Season	NI-RT[a]	NI-CT	I-CT	M-CT
1991	Summer	899±289	--	932±490	670±132
	Autumn	869±192	--	728±303	662±275
1992	Winter	399±109	--	482±218	402±131
		Corn			
1991	Summer	883±265	687±414	706±235	876±325
	Autumn	899±279	632±307	654±192	741±261
1992	Winter	494±290	417±123	509±205	393±177
LSD		(0.05)	(0.10)		
Treatment		184	144		
Season/dates		314	266		

[a] NI = no-input; RT = ridge-till; CT = chisel-till; I = inputs; M = manure.

Table 5. Soil and crop management effects on N_2O concentrations in subsoil air for Hoytville clay

	N_2O concentration (ppm)			
Treatment	08/20/91	09/06/91	09/19/91	10/24/91
Soybean, no-input, ridge-till	0	0	65	0
Soybean, input, chisel-till	0	134	0	0
Soybean, manure, chisel-till	0	57	32	3
Corn, no-input, ridge-till	14	0	0	0
Corn, no-input, chisel-till	8	0	0	2
Corn, input, chisel-till	0	0	0	0
Corn, manure, chisel-till	10	0	11	0
LSD	(0.05)	(0.10)		
Treatment	39	31		
Date	24	20		

Table 6. Soil and crop management effects on CH_4 concentrations in subsoil air for Hoytville clay

	CH_4 concentration (ppm)		
Treatment	12/02/91	02/18/91	03/10/92
Soybean, no-input, ridge-till	29	0	4
Soybean, input, chisel-till	11	0	4
Soybean, manure, chisel-till	8	2	7
Corn, no-input, ridge-till	11	2	2
Corn, no-input, chisel-till	11	2	2
Corn, input, chisel-till	22	0	2
Corn, manure, chisel-till	10	0	11
LSD	(0.10)	(0.10)	
Treatment	31	3	
Date	20	2	

Soil and crop management effects on N_2O concentrations in soil air are shown in Table 5. Only a few samples had measurable levels of N_2O. The maximum concentration recorded was 134 ppm. The data show that manured soybean treatment had a relatively greater concentration of N_2O compared with other treatments. The data in Table 6 show the concentration of CH_4 in sub-soil air sampled at the 15 cm depth. CH_4 concentrations ranged from negligible or detection limit to 24 ppm. In comparison with December and March, concentration was rather low in February.

b. Drainage Site

The data in Tables 7 and 8 show concentrations of CO_2 and CH_4 in soil air sampled at the 15 cm depth. The data in Table 7 show differences in CO_2 concentration due to treatments. Furthermore, concentrations of CO_2 and CH_4 in soil air differed widely depending on the season or month of sampling. The mean CO_2 concentrations were 602 ppm, 558 ppm, 946 ppm, and 527 ppm for alfalfa, corn ridge-till, corn plow-till, and soybeans, respectively. The highest concentration, 2729 ppm, was recorded on 6 September 1991 for plow-till corn. Similar to the data presented in Tables 1 and 4, the CO_2 concentration in soil air in this experiment also was below the atmospheric level on some dates during winter, e.g., 22 November 1991 and 4 February 1992. The analyses of soil air for N_2O concentration showed that all soil air samples contained negligible concentrations of N_2O (data not presented). In comparison, the data in Table 8 show measurable levels of CH_4 for several treatments with concentrations as high as 12 ppm in alfalfa. Plow-till corn and soybeans had comparatively low CH_4 concentrations.

Table 7. Cropping systems and tile drainage effects on CO_2 concentrations in sub-surface soil air samples (ppm) for the Hoytville series

		Treatment			
Year	Season	Alfalfa	Corn ridge-till	Corn plow-till	Soybean
1991	Autumn	740 ± 333	640 ± 349	1269 ± 957	475 ± 188
1992	Winter	463 ± 151	476 ± 201	612 ± 339	695 ± 707
LSD		(0.05)	(0.10)		
Treatment		311	230		
Season/dates		402	335		

Table 8. Cropping systems and tile drainage effects on CH_4 concentration (ppm) in sub-surface soil air sample (ppm) for a clayey soil at Hoytsville

Date		Alfalfa	Corn ridge-till	Corn plow-till	Soybeans
12/2/91		12	10	0	0
2/18/92		0	1	0	0
3.\/10/92		4	7	2	4
LSD	(0.05)		(0.10)		
Treatment	7		5		
Date	3		2		

B. Gaseous Flux

1. Columbus Experiment

The data in Table 9 show calculations of gaseous flux of CO_2 as affected by land use and cropping systems treatments. Sampling date or season had a significant effect on CO_2 flux, being in the order of summer > fall > spring > winter. The maximum flux of CO_2, 1.984 g C m^{-2} hr^{-1}, was observed for the alfalfa treatment in August, 1991. Singh and Gupta (1977) reported a high flux of 4 g C m^{-2} hr^{-1}. The mean

Table 9. Land use and cropping systems effects on CO_2 flux (g C m^{-2} hr^{-1}) in different seasons from the Miamian soil at Columbus, Ohio

		Land use treatment				
Year	Season	Forest	Alfalfa	Corn plow-till	Corn no-till	Corn ridge-till
1991	Summer	0.33±0.24	0.69±0.56	0.35±0.24	--	0.40±0.22
	Autumn	0.07±0.11	0.08±0.09	0.05±0.08	--	0.04±0.12
1992	Winter	0.03±0.05	0.02±0.11	0.03±0.04	0.03±.08	0.02±0.05
	Spring	0.06±0.04	0.09±0.05	0.02±.03	.06±.002	0.04±0.01
	Summer	0.14	0.10	0.10	0.18	0.11

Table 10. Land use and cropping systems effects on N_2O emissions (μg N m^{-2} hr^{-1}) for the Miamian soil at Columbus, Ohio

		Land use treatment				
Year	Season	Forest	Alfalfa	Corn plow-till	Corn no-till	Corn ridge-till
(a) Influx						
1991	Summer	-50 (-550)	-41 (-450)	-48 (-530)	--	-41 (-530)
1992	Winter	-116 (-520)	-116 (-510)	-116 (-540)	151(-970)	-112 (-520)
	Spring	-170 (-510)	-170 (-510)	-160 (-500)	-170 (510)	-160 (-500)
	Summer	-407 (-510)	-402 (-480)	-313 (-550)	-427(-550)	-447 (-480)
(b) Eflux						
1992	summer	3083 (5220)	678 (840)	2604(5800)	2908(4650)	4493 (8720)

Efflux was measured on four separate dates during summer, 1992; the figure in paranthesis refers to the maximum flux.

C flux during August was 0.80, 0.48, 0.41, and 0.37 g C $m^{-2}hr^{-1}$ for alfalfa, corn ridge-till, corn plow-till, and forest respectively. The mean C flux during September was significantly less than during August. During late fall and winter (October through March) CO_2 flux was either low or negative. The negative flux implies that gaseous concentrations of CO_2 in soil air were below those of the atmosphere during the periods of cold climate. There was a high positive flux again during spring and summer of 1992.

Land use and cropping systems effects on N_2O flux were computed for dates in which measurable levels of N_2O were observed. Spot measurements of as high as 0.00024 g N m^{-2} hr^{-1} for corn grown with plow-till, no-till and ridge-till treatments, respectively. These emission rates are higher than those reported by Goodroad and Keeney (1984) at the Wisconsin arboretum site. The absence of data for winter measurements demonstrated the negligible emissions during the cold period when soils were frozen.

Soil is a net sink of CH_4 on well-drained uplands with low to medium levels of soil organic matter contents. The data in Table 10 show that with the exception of a few dates (e.g., 29 May, 1 June, and 10 August), there was a negative flux of CH_4 from the soil. Magnitude of the negative flux ranged from 0 to -0.00055 g C $m^{-2}hr^{-1}$. Whenever the flux was positive, usually during summer, the flux ranged from 0 to 0.006 g C $m^{-2}hr^{-1}$.

2. Hoytville Experiment

Results of both experiments are briefly discussed below.

a. Soil Management Site

The data in Tables 11 and 12 show the flux of CO_2 and CH_4 for different soil management and cropping systems treatments. Similar to results at Columbus, flux of CO_2 was high during summer and early fall, but was generally negative during winter. The maximum flux of 0.5938 g C $m^{-2}hr^{-1}$ was observed for soybeans grown on no-input and ridge-till plots on 19 September 1991. The maximum flux in corn was 0.3648 g C $m^{-2}hr^{-1}$ in the no-input-ridge-till treatment observed on 15 October 1991. The largest negative flux was -0.0778 g C $m^{-2}hr^{-1}$ for the input-chisel-till system in soybeans, and -0.0797 for the input-chisel-till system in corn observed on 22 November 1991. The least negative flux during winter was -0.001 g C $m^{-2}hr^{-1}$ - observed in manured-chisel-till soybeans on 22 January 1992.

Flux of N_2O was computed for those dates which recorded measurable levels of N_2O. Flux of 0.0013 g N $m^{-2}hr^{-1}$ was observed for no-input ridge-till soybean. Flux of CH_4 was mostly negative except for measurements made on 2 December, 1991. The maximum positive flux of CH_4 was 0.0070 g C $m^{-2}hr^{-1}$ for no-input chisel-till corn measured on 2 December, 1991. The highest flux of CH_4 was -0.00055 g C $m^{-2}hr^{-1}$ measured on 18 February 1992.

Table 11. Soil management and cropping systems effects on CO_2 flux (g C m^{-2} hr^{-1}) for a clayey soil in northwest Ohio

		Soybean			
Year	Season	NI-RT[a]	NI-CT	I-CT	M-CT
1991	Summer	0.272 ± 0.068	--	0.118 ± 0.007	0.380 ± 0.020
	Autumn	0.233 ± 0.238	--	0.096 ± 0.061	0.058 ± 0.066
1992	Winter	0.031 ± 0.029	--	0.047 ± 0.093	0.022 ± 0.049
		Corn			
1991	Summer	0.0343	0.124 ± 0.066	0.117 ± 0.068	0.169 ± 0.113
	Autumn	0.179 ± 0.149	0.153 ± 0.113	0.087 ± 0.068	0.125 ± 0.078
1992	Winter	-0.0014 ± 0.1739	0.013 ± 0.050	0.036 ± 0.072	0.024 ± 0.080

[a] NI = no-input; RT = ridge-till; CT = chisel-till; I = inputs; M = manure.

Table 12. Soil management and cropping systems effects on CH_4 flux (μg C m^{-2} hr^{-1}) during winter for a clayey soil in northwest Ohio

	Soybean			
Flux	NI-RT[a]	NI-CT	I-CT	M-CT
(a) Influx	0 ± 600	--	-150 ± 450	-400 ± 200
(b) Eflux	5400	--	3600	4500
	Corn			
(a) Influx	-350 ± 250	-450 ± 150	-250 ± 350	-350 ± 25
(b) Eflux	2100	7000	3300	2200

b. Drainage Site

The data in Tables 13 and 14 show cropping systems' effects on gaseous flux of CO_2, and CH_4, respectively, for clayey soil at Hoytville. The data in Table 13 for CO_2 flux show the highest flux of 0.5854 g C $m^{-2}hr^{-1}$ for alfalfa measured on 6 September 1991. The minimum flux of 0.009 g C $m^{-2}hr^{-1}$ was measured for soybeans on 12 November 1991. Similar to the results of CO_2 flux at Columbus and Hoytville (for the soil management experiment), soil served as an apparent sink during late fall and winter. The maximum negative flux rate of 0.0727 g C $m^{-2}hr^{-1}$ was observed for ridge-till corn on 8 January 1992. Negative CO_2 flux was observed for some treatments until March 10, 1992 when soils were frozen.

Measurements of N_2O flux were negligible for the drainage experiment. With the exception of measurements made on 2 December 1991, there was a negative flux of CH_4 for the remaining observations made in February and March. The maximum flux of CH_4 was 0.00516 g C $m^{-2}hr^{-1}$ and the most negative value was 0.00055 g C $m^{-2}hr^{-1}$ (Table 14).

Table 13. Drainage and cropping systems effects on CO_2 emissions (μg C m^{-2} hr^{-1}) for the Hoytville soil series

Year	Season	Alfalfa	Corn ridge-till	Corn plow-till	Soybean
1991	fall	0.1795 ± 0.175	0.1504 ± 0.132	0.1944 ± 0.2298	0.1300 ± 0.1606
1992	winter	0.0278 ± 0.073	0.07205 ± 0.1338	0.0157 ± 0.0381	0.0285 ± 0.0506

Table 14. Drainage and cropping systems effects on CH_4 emissions (μg C m^{-2} hr^{-1}) for the Hoytville soil series during winter 1991-1992

Year	Season	Alfalfa	Corn ridge-till	Corn plow-till	Soybean
(a) Efflux		5162	3640	4590	4740
(b) Influx		315 ± 265	285 ± 265	245 ± 15	280 ± 270

IV. Discussion

Soil air composition and gaseous flux depend on several factors including: (i) environmental factors e.g., soil and air temperature, rainfall, soil moisture content, etc., (ii) land use, and (iii) cropping systems treatments. Effects of these variables are discussed below for the 3 gases monitored.

A. Environmental Factors

Concentrations of CO_2 in soil air were significantly different among seasons with higher values in summer and fall than in spring and winter. Regression equations relating air temperature with CO_2 concentration in soil air showed positive but weak correlation (Table 15) (Figures 1 and 2). The maximum temporal variability in gaseous composition of soil air in Miamian soil that could be attributed to air temperature was 35% for CO_2, 44% for N_2O, and 18% for CH_4. Similar correlation for the Hoytville soil management experiment showed that the maximum variability that could be attributed to air temperature was 55% for CO_2, 16% for N_2O, and 88% for CH_4 (Table 16).

Soil moisture content is another important factor affecting gaseous concentration in soil (Kucera and Kirkham, 1971; Solomon and Cerling, 1987). Soil moisture content depends on amount and frequency of rain, and evapotranspiration. Regression equations were computed between rainfall received on the previous day and the gaseous composition of soil air. The data from both Columbus and Hoytville showed no correlation between the rainfall amount on the day prior to gas analyses and the composition of soil air. The effect of rainfall depends on its amount, the antecedent moisture content, and the depth of soil saturation. The low correlation coefficient reported herein may be due to the fact that rainfall-related factors were not considered in data analyses.

Table 15. Coefficient of CO_2, N_20, and CH_4 concentrations in soil air with air temperature for a Miamian soil at Columbus, Ohio

	Correlation coefficients (R^2)		
Treatment	CO_2	N_2O	CH_4
Forest	0.12	0.003	0.16
Alfalfa	0.23	0.01	0.14
Corn plow-till	0.30	0.44	0.18
Corn no-till	0.19	----	0.15
Corn ridge-till	0.35	0.02	0.16

Table 16. Correlation coefficient between CO_2, N_2O, and CH_4 concentrations in soil and air temperature for the soil management experiment at Hoytville, Ohio

Treatment	CO_2	N_2O	CH_4
Soybean, no-input, ridge-till	0.41	0.004	0.81
Soybean, inputs, chisel-till	0.32	0.13	0.47
Soybean, manure, chisel-till	0.55	0.16	0.10
Corn, no-input, ridge-till	0.37	----	0.88
Corn, no-input, chisel-till	0.19	0.13	0.76
Corn, input, chisel-till	0.11	0.009	0.88
Corn, manure, chisel-till	0.50	0.08	0.83

B. Land Use

For the Miami soil at Columbus, CO_2 concentration in soil air was in the order for forest > alfalfa > corn with mean average concentration of 1113 ppm for forest, 1081 ppm for alfalfa and 851 ppm for corn. During the months of summer (e.g., August 1991) mean CO_2 concentration in soil air was 2366 ppm for forest, 1956 ppm for alfalfa, and 1756 ppm for plow-till corn (Figure 3). High CO_2 concentrations in forest soils were also observed by Dyer and Brook (1990). Similar trends were observed for the data on CO_2 flux (Figure 4). Soil under forest had a high N_2O flux even though it did not receive any fertilizer.

High N_2O flux from forest soils was also observed by Goodroad and Keeney (1984) and Sahrawat and Keeney (1986). The data in Table 17 show concentrations of NO_3-N in soil solution. Soil solution from forest had low concentration of NO_3-N compared with alfalfa and corn. Mean NO_3-N concentration was 2.8 ppm for forest, 34.3 ppm for alfalfa, 73.6 ppm for plow-till corn, 73.8 ppm for no-till corn, and 52.6 ppm for ridge-till corn (Table 17). Analyses of the rainwater for Columbus site show NO_3-N concentrations of up to 0.52 ppm (Table 18). NO_3-N concentration in soil solution for the drainage experiment shown in Table 19 depicts higher concentrations for leguminous crops than cereals.

C. Soil Management and Cropping System

Manured plots had higher concentration and higher flux of CO_2 than those that received chemical fertilizer. This trend of higher flux from manured treatments was especially true for the summer months of August and September. Mean CO_2 flux for that period was 0.0925 and 0.2431 g C $m^{-2}hr^{-1}$ for soybeans grown with chemical fertilizer and organic manure, respectively. Similarly, mean CO_2 flux for that period was 0.1176 and 0.1702 g C $m^{-2}hr^{-1}$ for corn grown with chemical fertilizer and organic manure, respectively. Similar effects of organic manures were reported by Amundson and Davidson (1990).

The data in Table 11 also show that CO_2 flux was relatively greater from soybeans than from corn. For the summer months of August and September, mean CO_2 flux for no-input ridge-till treatment was 0.4103 g C $m^{-2}hr^{-1}$ for soybean compared with 0.2151 g C $m^{-2}hr^{-1}$ for corn. Similar trends were observed for CO_2 concentration at 15 cm.

The effects of tillage methods on gaseous flux and concentrations can also be assessed from comparison of the data in Tables 7 and 13. The data in Table 7 show comparison of CO_2 concentration in soil air for ridge-till and plow-till corn. Mean CO_2 concentration in soil air was 558 ppm in ridge-till and 945 ppm in

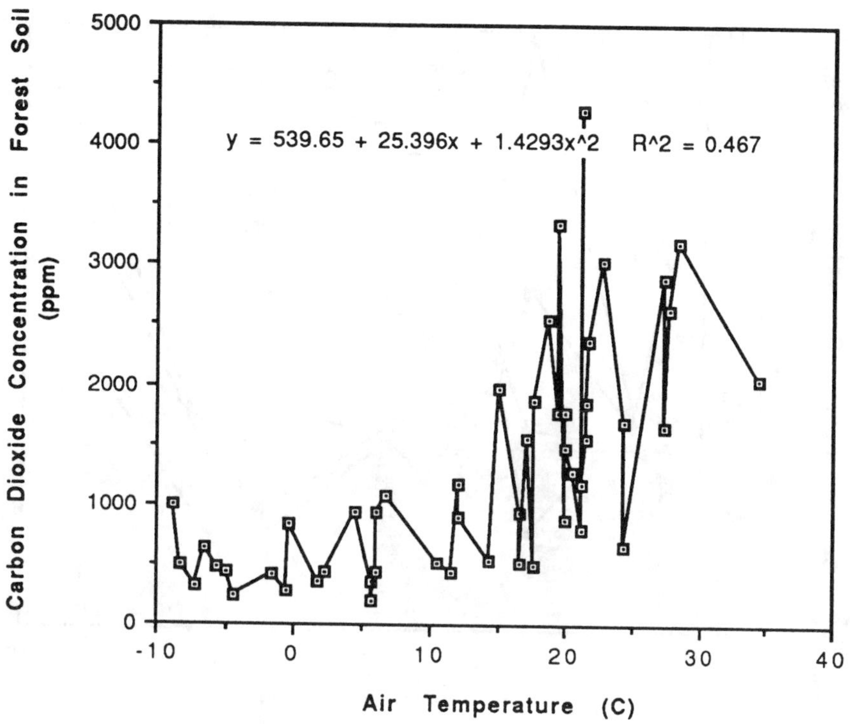

Figure 1. Relation between temperature and carbon dioxide concentration in soil under forest.

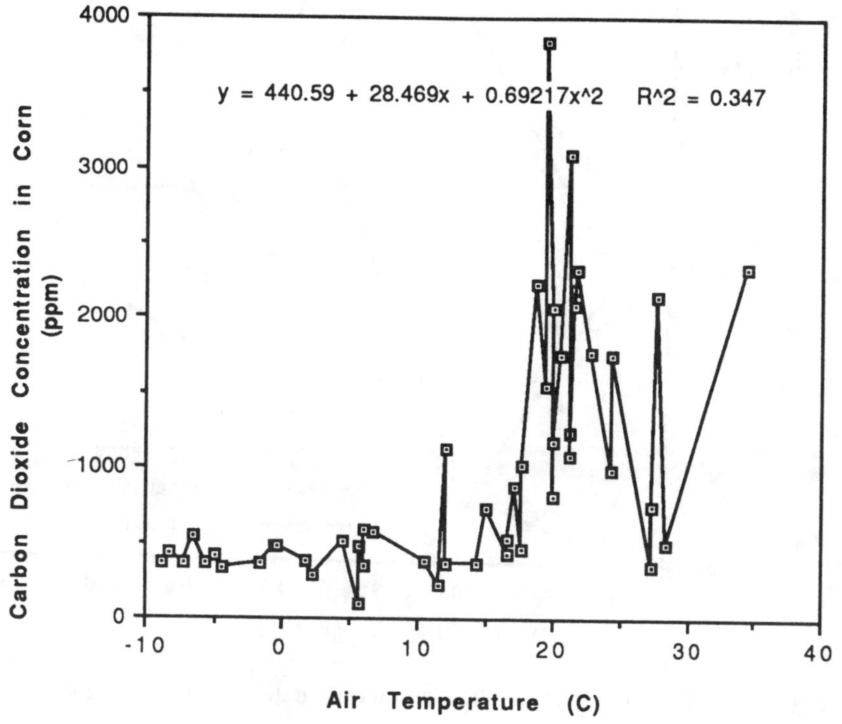

Figure 2. Relation between temperature and carbon dioxide concentration in soil air in plow-till corn.

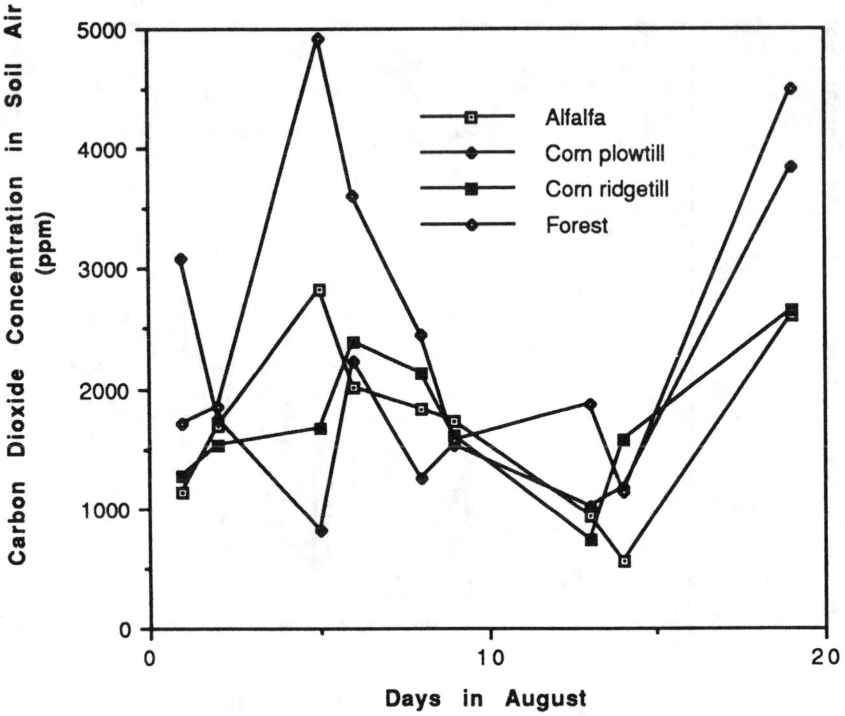

Figure 3. Land use effects on carbon dioxide concentration in soil air.

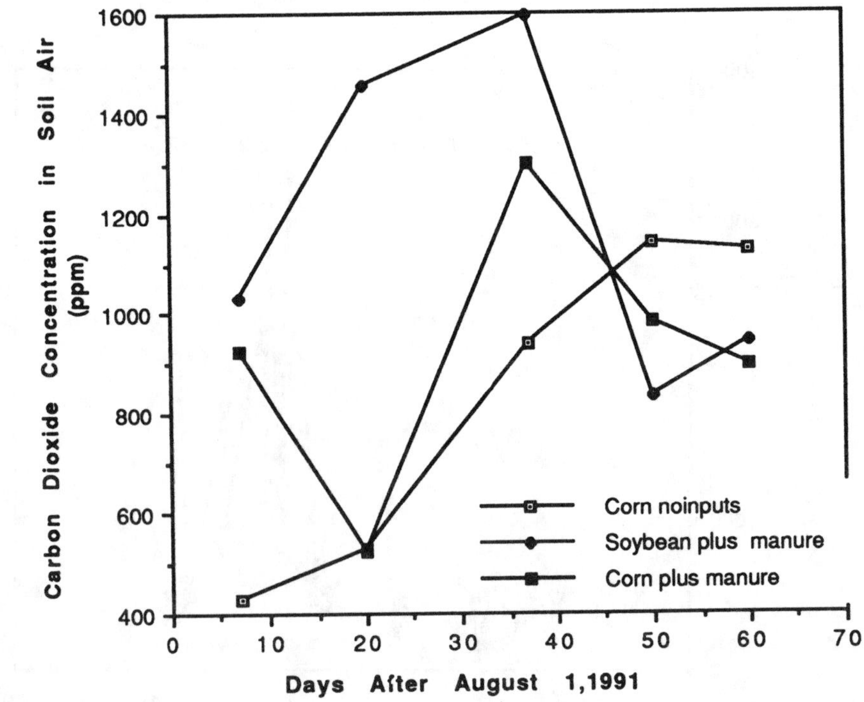

Figure 4. Effects of manure application on carbon dioxide concentration in soil air.

Table 17. Concentration of NO_3-N in soil solution for the Miamian soil (ppm) during winter 1992

Month	Forest	Alfalfa	Corn plow-till	Corn no-till	Corn ridge-till
January	3.6 ± 2.1	31.9	72.1 ± 7.5	75.0 ± 14.6	54.1 ± 16.8
February	1.8 ± 0.5	36.7	75.7 ± 5.6	72.7 ± 1.4	51.1 ± 9.2

Table 18. Chemical analysis of rainfall received at the Columbus, Ohio site during winter 1992

Month	NO_3-N (ppm)	NH_4-N (ppm)
February	0.16 ± 0.15	0.29 ± 0.19
March	0.52 ± 0.38	0.33 ± 0.22

Table 19. Drainage and cropping systems effects on concentration of NO_3-N (ppm) and NH_4-N (ppm) in soil solution for the Hoytville experiment

Treatment	2-4-92		2-18-92	
	NO_3-N	NH_4-N	NO_3-N	NH_4-N
Alfalfa	21.8	<0.1	21.8	0.85
Corn plow-till	8.4	<0.1	3.9	<0.1
Corn ridge-till	5.3	<0.1	4.9	<0.1
Plow-till	44.8	<0.1	11.9	<0.1
Rain (2-1-92)	1.99		0.54	
Snow (2-17-92)	1.21		<0.1	

plow-till treatments, respectively. CO_2 concentration in soil air for the plow-till treatment was especially high during the summer, e.g., September 1991. The mean CO_2 concentration during September was 856 ppm for ridge-till, compared with 2100 ppm for plow-till treatments. Similar trends were observed for the flux measurements shown in Table 13. Difference in CO_2 concentration due to tillage methods may be attributed to differences in soil temperature (Singh and Gupta, 1977).

Comparison between no-till and plow-till systems of seedbed preparation on gaseous composition and flux can be made for the data of Miamian soil shown in Tables 1 and 4, respectively. Because measurements in no-till treatment were started late, realistic comparisons can be made for the data of late spring and early summer of 1992. Two measurements made in May and June 1992 show that the no-till treatment had higher CO_2 concentration and flux than plow-till treatment. However, no conclusive trends were evident from the data of 1991.

V. Conclusions

The data reported in this paper highlight the need for systematic studies on quantification of gaseous composition and flux in relation to predominant land use and management systems, e.g., forest, pasture, arable land, cropping system, tillage methods, and manuring. All these factors have important impacts on gaseous composition and flux.

The data show marked seasonal fluctuations in gaseous composition and flux. With regards to CO_2, concentrations in soil air and flux were in the order of summer > spring > fall > winter. Higher concentrations in summer were only weakly related to air temperature. There is a need to establish the cause-effect relationships between soil temperature and soil moisture content on the one hand, and gaseous composition and flux on the other. Effects of soil temperature and wetness may be altered by differences in soil organic matter content, bulk density and macroporosity, and crust characteristics of the surface layer.

Lower than ambient concentrations of CO_2 and negative flux observed in the winter months warrant further studies. It seems that cold or frozen soil may be an apparent or transient sink for CO_2 during winter because CO_2 solubility in water increases as temperature decreases. Soil can also be a sink for CH_4 and N_2O. Identifying the environmental factors, land uses, and soil and crop management systems that make soil a sink for radiatively active gases is a researchable priority.

Acknowledgements

This project was financially supported by the U.S. EPA. Mr. J. McLaughlin was involved during the initial phases of the project and assisted in getting gas chambers fabricated and placing orders for laboratory equipment. The help received from Mr. Keith Serafy in field work and laboratory analyses is gratefully acknowledged.

References

Alexander, M. 1977. *Introduction to soil microbiology. 2nd ed.* J. Wiley and Sons, NY, 467 pp.

Amundson, R.G. O.A. Chadwick, and J.M. Sowers. 1989. A comparison of soil climate and biological activity along an elevation gradient in the eastern Mojave Desert. *Oecologia* 80: 395-400.

Amundson, R.G. and E.A. Davidson. 1990. Carbon dioxide and nitrogenous oxide gases in the soil atmosphere. *Journal Geochemical Exploration* 38(1/2):13-41.

Amundson, R.G., and V.S. Smith. 1988. Annual cycles of physical and biological properties in an uncultivated and irrigated soil in the San Joaquin Valley of California. *Agricultural Ecosystem Environ.* 20:195-208.

Andres E., K.W., Becker and B. Meyer. 1983. CO_2 evolution from soil as a measurement of carbon transformations in a brown loam rendzina under beech forest. Comparison of belljar methods and particle pressure gradient. *Mitteilunger der Deutschen Bodenkundlichen Gesellschaft* 38:189-194.

Aulakh. M.S., D.A. Rennie, and E.A. Paul. 1982. Gaseous N losses from cropped and summer fallowed soils. Canadian *Journal Soil Science* 62:187-195.

Aulakh, M.S., D.A. Rennie, and E.A. Paul. 1983a. The effect of various clover management practices on gaseous N losses and mineral N accumulation *Canadian Journal Soil Science* 63:593-605.

Aulakh, M.S., D.A. Rennie, and E.A. Paul. 1983b. Field studies on gaseous N losses from soils under continuous wheat-fallow rotation. *Plant Soil* 75:15-28.

Boddy, L. and D. Lloyd. 1990. Portable mass spectrometry: a potentially useful ecological tool for simultaneous, continuous measurement of gases in situ in soils and sediments. p. 139-152. In: A.F. Ineson and O.W. Heal (eds.), *Nutrient Cycling in Terrestrial Ecosystems: Field Methods, Application and Interpretation* Elsevier Applied Science Publisher.

Breitenbeck, G.A. 1990. Sampling the atmospheres of small vessels. *Soil Sci. Soc. Am. J.* 54(6): 1794-1797.

Bremner, J.M. and A.M. Blackmer. 1982. Composition of soil atmospheres In: *Methods of Soil Analyses, Part 2: Chemical and Microbiological Properties* A.L. Page (ed). ASA, Madison, WI. 873-901.

Brook, G.A., M.E. Folkoff, and E.O. Box. 1983. A world model of soil CO_2. Earth Surface Processes Landforms 8:79-88.

Buyanovsky, G.A., and G.H. Wagner, 1983. Annual cycles of CO_2 level in soil air. *Soil Sci. Soc. Am. J.* 47:1139-1145.

Buyanovsky, G.A., G.H. Wagner and C.J. Gantzer, 1986. Soil respiration in a winter wheat ecosystem. *Soil Sci. Soc. Am. J.* 50:338-344.

Christensen, S. and J.M. Tiedje. 1990. Brief and vigorous N_2O production by soil at spring thaw. *Journal Soil Science.* 41:1-4.

Cleemput, O. Van and L. Baert. 1983. Soil aeration data of sandy and sandy loam profiles in Belgium. *Pedologie* 33:105-115.

Crowther, J. 1983. CO_2 concentrations in some tropical karst soils, *West Malaysia. Catena* 10: 29-39.

Duxbury, J.M., D.R. Bouldin, R.E. Terry, and R.L. Tate III. 1982. Emissions of N_2O from soils. *Nature* 198: 462-464.

Dyer, J M and G A. Brook. 1991. Spatial and temporal variations in temperate forest soil carbon dioxide during the non-growing season. *Earth Surface Processes and Landforms* 16:411.

Fernandez, I.J. and P.A. Kosian. 1987. Soil air CO_2 concentrations in a new England spruce-fir forest. *Soil Sci. Soc. Am. J.* 51:261-263.

Folorunso, O.A. and D.E. Rolston. 1984. Spatial variability of field-measured denitrification gas fluxes. *Soil Sci. Soc. Am. J.* 48:1214-1219.

Freijer, J.I. and W. Bouten. 1991. A comparison of field methods of measuring CO_2 evolution: experiments and simulation. *Plant Soil.*135:133-142.

Glinski, J. and W. Stepniewski. 1985. *Soil aeration and the role for plants.* CRC Press. Boca Raton, FL.

Goodroad, L.L. and D.R. Keeney. 1984a. N_2O production in aerobic soils under varying pH, temperature, and water content. *Soil Biol Biochem.* 16:39-43.

Goodroad, L.L. and D.R. Keeney. 1984b. N_2O emissions from forest, marsh and prairie ecosystems. *J. Environmental Qual.* 13:448-452.

Goodroad, L.L. and D.R. Keeney. 1985. Site of N_2O production in successional and old-growth Michigan forests. *Soil Sci. Soc. Am. J.* 48:383-389.

Goodroad, L.L., D.R. Keeney, and L.A. Peterson. 1984. N_2O from agricultural soils in Wisconsin. *J. Environmental Qual.* 13:557-561.

Grundmann, G.L. and D.E. Rolston. 1987. A water function approximation to degree of anaerobiosis with denitrification. *Soil Sci.* 144:437-441.

Grundmann, G.L., D.E. Rolston, and R. Kachanoski. 1988. Field soil properties influencing the variability of denitrification gas fluxes. *Soil Sci. Soc. Am. J.* 52:13511355.

Holzapfel-Pschorn, A. and W. Seiler. 1985. Contribution of CH_4 produced in rice paddies to the global CH_4 budget. Biometeorology 9:53-61.

Kasper, H.F. 1984. A simple method for the measurement of N_2O and CO_2 flux rates across undisturbed soil surfaces. *New Zealand J. Sci.* 27:243-246.

Keeney, D.R., K.L. Sahrawat, and S.S. Adams. 1985. CO_2 concentration in soil: effects on nitrification, denitrification, and associated N_2O production. *Soil Biol. Biochem.* 17:541-577.

Keller, M., W.A. Kaplan, and S.C. Wofsy. 1986. Emissions of N_2O, CH_4 and CO_2 from tropical forest soils. *J. Geophysics Research* 91 (D 11):11, 791-11,802.

Kimura, M., H. Murakami and H. Wads. 1991. CO_2, H2 and CH_4 production in rice rhizosphere. *Soil Sci. and Plant Nutrition* 37:55-60.

Klemedtsson, L. S. Simkins, and B.H. Svensson. 1986. Tandem thermal-conductivity and electron-capture detectors and non-linear calibration curves in quantitative N_2O analysis. *J. Chromatography* 361: 107-116.

Kucera, C.L. and D.L. Kirkham. 1971. Soil respiration studies in tallgrass prairie in Missouri. *Ecology* 52: 912-915.

Lal, R. and G.S. Taylor. 1969. Drainage and nutrient effects in a field lysimeter study I: Corn yield and soil conditions. *Soil Sci. Soc. Am. J.*. 33:937-941.

Lalisse-Grundmann, G. and A. Chalamet. 1987. Diffusion of C2H2 and N_2O in soil in relation to measurement of denitrification. *Agronomie* 7:297-301.

Lind, A.M. 1985. Soil air concentration of N_2O over 3 years of field experiments with animal manure and inorganic N-fertilizer. *Tidsskrift Planeaul* 89:331 -340.

Magnusson, T. 1989. A method of equilibration chamber sampling and gas chromatographic analysis of the soil atmosphere. *Plant Soil* 120:39-47.

Matthias, A.D., A.M. Blackmer, and J.M. Bremner. 1980. A simple chamber technique for field measurement of emissions of nitrous oxide from soils. *J. Environ. Qual.* 9:251-256.

Moffat, A.J., M. Johnson, and J.S. Wright. 1990. An improved probe for sampling soil atmospheres. *Plant Soil* 121:145-147.

Monteith, J.L., G. Sceica, and K. Yabuky. 1964. Crop photosynthesis and the flux of CO_2 below the canopy. *J. Applied Ecology* 1: 321-327.

Mooney, H.A., P.M. Vitousek, and P.A. Matson. 1987. Exchange of material between terrestrial ecosystems and atmosphere. *Science* 238:926-938.

Moore, T.R. and R. Knowles. 1989. The influence of water table levels on CH_4 and CO_2 emissions from peatland soils. *Canadian J. Soil Sci.* 69:33-38.

Moore, T.R. 1986. CO_2 evolution from subarctic peatlands in eastern Canada. *Arctic and Alpine Research* 18:189-193.

Mosier, A.R., F.L. Hutchinson, B.R. Sabey, and J. Baxter. 1982. N_2O emissions from barley plots treated with NH4NO3 or sewage sludge. *J. Environmental Qual.* 11:78-81.

Mosier, A.R., W.J. Parton, and G.L. Hutchinson. 1983. Modeling N_2O evolution from cropped and native soils. *Environmental Biogeochemistry Ecol. Bulletin* (Stockholm) 35:229-241.

Mukhtar, S., J.L. Baker, and R.S. Kanwar. 1990. Soil atmosphere access chamber and analytical assembly to monitor soil aeration. *Soil Sci. Soc. Am. J.* 54:167-172.

Myburgh, P.A. and J.M. Moolman. 1991. Ridging a soil preparation practice to improve aeration of vineyard soils. *South African J. Plant Sci.* 8:189-193.

Neilson, J.W. and I.L. Pepper. 1990. Soil respiration as an index of soil aeration. *Soil Sci. Soc. Am. J.* 54: 528-432.

Naganwa, T. and K. Kyuma. 1991. Concentration dependence of CO_2 evolution from soil in chamber with low CO_2 concentration (<200 ppm), and CO_2 diffusion/sorption model in soil. *Soil Sci. and Plant Nutrition* 37:381-386.

Neue, H.U. and H.W. Scharpenseel. 1984. Gaseous products of the decomposition of organic matter in submerged soils. p. 311-327. In: *Organic Matter and Rice*, IRRI, Los Banos, Philippines.

Norstadt, F.A. and L.K. Porter. 1984. Soil gases and temperature: a beef cattle feedlot compared to alfalfa. *Soil Sci. Soc. Am. J.* 48:783-789.

Pareda, C.S. A. Long and S.N. Davis. 1983. Stable isotopic composition of soil CO_2 in the Tucson Basin, Arizona, USA. ISA. *Geoscience* 1:219-236.

Reicosky, D.C. 1989. Diurnal and seasonal trends in CO_2 concentrations in corn and soybean canopies as affected by tillage and irrigation. *Agriculture Forest Met.* 48: 285-303.

Reicosky, D.C. and M.J. Lindstrom, 1995. Impact of fall tillage on short-term carbon dioxide flux. In: Lal et al. (eds.), *Soil Management and Greenhouse Effect*, Lewis Publishers, Chelsea, MI.

Roberts, W.P. and K.Y. Chan. 1990. Tillage-induced increases in carbon dioxide loss from soil. *Soil Tillage Research.* 17:143-151.

Robertson, G.P. and J.M. Tiedje. 1984. Denitrification and N_2O production in successional and old-growth Michigan forests. *Soil Sci. Soc. Am. J.* 48:383-389.

Robins, C.W. 1986. Carbon dioxide partial pressure in lysimeter soils. *Agronomy J.* 78:151-158.

Robertson, E.A.G. and B.C. Ball. 1989. Soil atmosphere composition and straw breakdown rate in a straw incorporation experiment, 1985-88. Departmental Note - Scottish Center of Agriculture. England N. 23, 18 pp.

Rochette, P., R.L. Desjardins, and E. Pattey, 1991. Spatial and temporal variability of soil respiration in agricultural fields. *Can. J. Soil Sci.* 71:189-196.

Rolston, D.E. 1981. N_2O and N_2 gas production in fertilizer loss. p. 127-149. In: C.C. Delwicke (ed.), *Denitrification, nitrification, and atmospheric N_2O.* J. Wiley and Sons, NY.

Rolston, D.E. 1986. Gas flux. p. 1103-1119. In: A.Klute (ed.) *Methods of soil Analysis: Part 1. Physical and Mineralogical Methods.* Amer. Soc. Agron., Madison, WI.

Ryden, J.C. and D.E. Rolston. 1983. The measurement of denitrification. p. 91-132. In: J.R. Freney and J.F. Simpson (eds.), *Gaseous loss of N from plant-soil systems.* Developments in Plant Soil Science. 9. Martinus Nijhoff, The Hague, Netherlands.

Sahrawat, K.L. and D.R. Keeney. 1986. Nitrous oxide emissions from soil. *Adv. Soil Science.* 4:103-148.

Sahrawat, K.L., D.R.. Keeney and S.S. Adams. 1985. Rate of aerobic N transformations in six acid climax forest soils and the effect of phosphorous and $CaCO_3$. *Forest Science.* 31:680-684.

Sebacher, D.I. and R.C. Hariss. 1982. A system for measuring CH_4 fluxes from inland and coastal wetland environments. *J. Environmental Qual.* 11: 34-37.

Singh, J.S. and S.R. Gupta, 1977. Plant decomposition and soil respiration in terrestrial ecosystems. *Bot. Rev.* 43:449-527.

Simojoki, A., A. Jaakkole, and L. Alakukku. 1991. Effect of compaction on soil air in a pot experiment and in the field. Soil Tillage Research. 19:175-186.

Skopp, J., 1985. Oxygen uptake and transport in soils: analysis of the air-water interfacial area. *Soil Sci. Soc. Am. J.* 49:1327-1331.

Skopp, J., M.D. Jawson, and J.W. Doran, 1990. Steady-state aerobic microbial activity as a function of soil-water content. *Soil Sci. Soc. Am. J.* 54:1619-1625.

Smucker, J.A.M. and A.E. Erickson. 1989. Tillage and compactive modifications of gaseous flow and soil aeration. p. 205-221. In: W.E. Larson et. al. (eds.), *Mechanics and Related Processes in Structured Agricultural Soils.* NATO ASI Series. Series E, Applied Sciences (1989) No 172:

Solomon, D.K. and T.E. Cerling. 1987. The annual CO_2 cycle in a Montana soil: observations, modeling and implications for weathering. *Water Resources Res.* 23:2257-2265.

Staley, T.E. 1988. Carbon, nitrogen and gaseous profiles in a humid, temperate region, maize field soil under no-tillage. *Comm. Soil Sci. Plant Analysis* 19:625-642.

Stepniewski, W. and J. Glimski. 1988. Gas exchange and atmospheric properties of flooded soils. p. 269-278. In: D.D. Hook et al. (eds.), *Ecology and Management of Wetlands*, Beckenham, Kent, U.K.

Taylor, G.S. and J.H. Abrahams. 1953. A diffusion equilibrium method for obtaining soil gases under field conditions. *Soil Sci. Soc. Am. J.* 17:201-206.

Tiedje, J.M., A.J. Sexstone, T.B. Parkin, N.P. Reusbech, and D.R. Shelton. 1984. Anaerobic process in soil. *Plant Soil* 76:197-212.

Vermoesen,j A., H. Ramon, and O.Van Cleemput 1991. Composition of the soil gas phase. Permanent gases and hydrocarbons. *Pedologie* 41:119-132.

Wood, W.W. and M.J. Petraitis. 1984. Origin and distribution of CO_2 in unsaturated zone of the Southern High Plains of Texas. *Water Resources Res.* 20:1993-1208.

Yamaguchi, M., W.J. Flocker, and F.D. Howard. 1967. Soil atmosphere as influenced by temperature and moisture. *Soil Sci. Soc. Am. Proc.* 31:164-167.

Yavitt, J.B., G.E. Land, and R.K. Wider. 1987. Control of C mineralization to CH_4 and CO_2 in anaerobic, sphagnum-derived peat from Big Run Bog, West Virginia. *Biogeochemistry* 4:141-157.

CHAPTER 5

Carbon Sequestration in Corn-Soybean Agroecosystems

D.R. Huggins, C.E. Clapp, R.R. Allmaras, and J.A. Lamb

I. Introduction

Soil organic matter (SOM) has been reduced by cultivation in many agroecosystems, irrespective of climatic and edaphic characteristics (Voroney et al., 1981; Tiessen and Stewart, 1983; Dalal and Mayer, 1986). Losses of soil carbon (C) have contributed significantly to global CO_2 levels (Post et al., 1990). Globally, soil C is three times that in plant biomass and two times that in the atmosphere (Post et al., 1990; Johnson and Kerns, 1991). Levels of SOM can be managed through crop rotation, tillage regime, fertilizer practices, and other cropping-system components (Campbell et al., 1990; Janzen et al., 1992) to increase the terrestrial C pool and provide a sink for atmospheric CO_2 (Barnwell et al., 1992). Quantifying SOM dynamics as related to agroecosystem management is fundamental to identifying pathways for soil C sequestration.

Chemical characterization and dynamics of SOM have not been successfully linked (Jenkinson, 1971; Oades and Ladd, 1977; Oades et al., 1988, Capriel et al., 1992). Isolation of mineral-organic soil fractions by physical methods has provided a more functional characterization of SOM that is related to biological turnover (Christensen, 1987). Physical separation of soil structural and textural components (i.e. macro-aggregates, micro-aggregates, primary particles) using water stable aggregation or other resistance to physical dispersion in water has linked cropping influences and SOM dynamics (Tisdall and Oades, 1982; Tiessen and Stewart, 1983; Christensen, 1987; Balesdent et al., 1988; Cambardella and Elliott, 1992). Elliott (1986) reported that SOM associated with macro-aggregates was more labile than SOM associated with micro-aggregates. Enriched labile fractions have been isolated from inside macro-aggregates (Cambardella and Elliott, 1993). Large particulate organic matter (POM) has been reported to be labile (Ladd et al., 1977; Turchenek and Oades, 1979; Elliott, 1986) or to contain organic fractions that are more stable or slowly decomposable (Cambardella and Elliott, 1992). The most stable SOM is often associated with silt and clay fractions (Tiessen and Stewart, 1983; Balesdent et al., 1987; Balesdent et al., 1988).

Natural ^{13}C abundance can complement physical fractionation analyses and further access the fate of recent organic inputs into SOM. Its principle is based on variations of stable ^{13}C natural abundance induced by plants with a high $\delta^{13}C$ value (C_4 photosynthetic pathway). *In situ* labeling of the organic matter occurs as a result of differential isotopic enrichment. This method can be used to characterize soil organic fractions and soil C turnover rates where C_4 and C_3 crops are used (Balesdent et al., 1987; Balesdent et al., 1988; Balesdent et al., 1990). This technique can be applied in corn (*Zea mays* L.; C_4 plant)-soybean [*Glycine max* (L.) Merr.; C_3 plant] agroecosystems to explore and quantify SOM dynamics under different management regimes.

A long-term study (10 years) with different corn-soybean sequence combinations provided a unique investigative opportunity. Our objective was to examine the effect of crop sequence on soil organic C concentration, source (corn vs. soybean), and C dynamics.

ISBN 1-56670-117-1/95/$0.00+.50

II. Materials and Methods

A field study with 14 crop sequence treatments was initiated in 1981 at the University of Minnesota Southwest Experiment Station near Lamberton, MN (Crookston et al., 1991). The treatments consisted of various lengths of corn and soybean crop sequences including continuous corn and continuous soybean. Only eight of these treatments were used for analysis (Table 1). Prior to 1981, a 2-year corn-soybean rotation had been primarily used since 1960. The experimental design was a randomized complete block with four replications. Primary tillage consisted of fall moldboard plowing to a depth of 25 to 30 cm followed by spring disking in a diagonal direction without traffic control. All crop residues were annually incorporated by the moldboard tillage, which generally incorporated >80% of the crop residue below 15 cm (Staricka et al., 1991). Planting, post-plant cultivation, and harvest were all done with controlled traffic on the 12-row plots. Soil within the study area is classified as a Webster clay loam (Typic Haplaquolls).

Composite soil samples (0 to 15 cm) were collected (6 cores/plot x 18 mm diam.) from each of the 8 treatments in mid-July 1991 and stored in a field-moist condition at 4°C. The samples were partially air-dried, fragmented, and passed through an 8 mm sieve to provide representative sub-samples. Soil samples were also collected from an alleyway adjacent to each replication block, where fallow (no crop) conditions had existed since the start of the experiment. Tillage of this area was the same as that in the plots with a crop sequence.

Soil samples from continuous corn and continuous soybean crop sequence treatments (11 and 12) were physically separated into primary size fractions. Sub-samples were gently crushed by hand and passed through a 2 mm sieve. Fractionation into particle separates was then accomplished using a combination of sonication, sieving for 5 min, sedimentation, and centrifugation (Figure 1). Low energy (1500 J) ultrasonic treatment of a 70 g soil sample in water allowed floatable, particulate organic matter (POM) to be removed (Edwards and Bremner, 1967). Following POM removal, a high energy sonication (12500 J) was applied for 15 min to complete aggregate breakdown and dispersion. The disrupted soil suspension was passed through a 50 μm (300 mesh) sieve to separate the sand fraction (>50 μm). Coarse silt (20-50 μm) and fine silt (2-20 μm) were separated using sedimentation. A SORVALL[1] (RC-5B) centrifuge with GSA head was used to separate the remaining suspended materials into coarse clay (0.2-2 μm) and fine clay (<0.2 μm) fractions.

Simultaneous analyses of total C and ^{13}C:^{12}C ratios were made on whole soil samples from treatments 1, 5, 6, 10, 11, 12, 13 and 14, and the fallow areas (treatment 0) and on each of the physical separates from treatments 11 and 12. The analyses were made with a CARLO ERBA[1] (NA1500) carbon-nitrogen analyzer interfaced with a VG-ISOGAS[1] (OPTIMA) isotope ratio mass spectrometer. The C isotope ratios were expressed as $\delta^{13}C$ values:

$$\delta^{13}C(‰) = [(R_{sam}/R_{std}) - 1] \times 10^3 \qquad [1]$$

where R_{sam} and R_{std} are the ^{13}C:^{12}C ratio of sample, and working standard, respectively. To determine the proportion (f) of organic C originating from source A in a mixture A+B whose $\delta^{13}C$ is equal to δ, the following equation is applied:

$$\delta = f\delta_A + (1 - f)\delta_B \qquad [2]$$

where δ_A and δ_B are the $\delta^{13}C$ of sources A and B, respectively. If the total organic carbon (C) in the sample is known, then the amount of C from source A (C_A) and source B (C_B) are as follows:

$$C_A = fC \qquad [3]$$

$$C_B = (1 - f)C. \qquad [4]$$

Sources A and B are the C originating from C_4 and C_3 plants, respectively.

[1]Mention of trade names is for reader convenience only and does not imply endorsement by the USDA-ARS or the University of Minnesota over similar products of companies not mentioned.

Table 1. Crop sequence treatments with corn (C), soybean (S), and fallow (F)

Treatment number	Year (19XX) 81	82	83	84	85	86	87	88	89	90
0	F	F	F	F	F	F	F	F	F	F
1	C	C	C	C	C	S	S	S	S	S
5	S	S	S	S	C	C	C	C	C	S
6	S	S	S	S	S	C	C	C	C	C
10	C	C	C	C	S	S	S	S	S	C
11	C	C	C	C	C	C	C	C	C	C
12	S	S	S	S	S	S	S	S	S	S
13	C	S	C	S	C	S	C	S	C	S
14	S	C	S	C	S	C	S	C	S	C

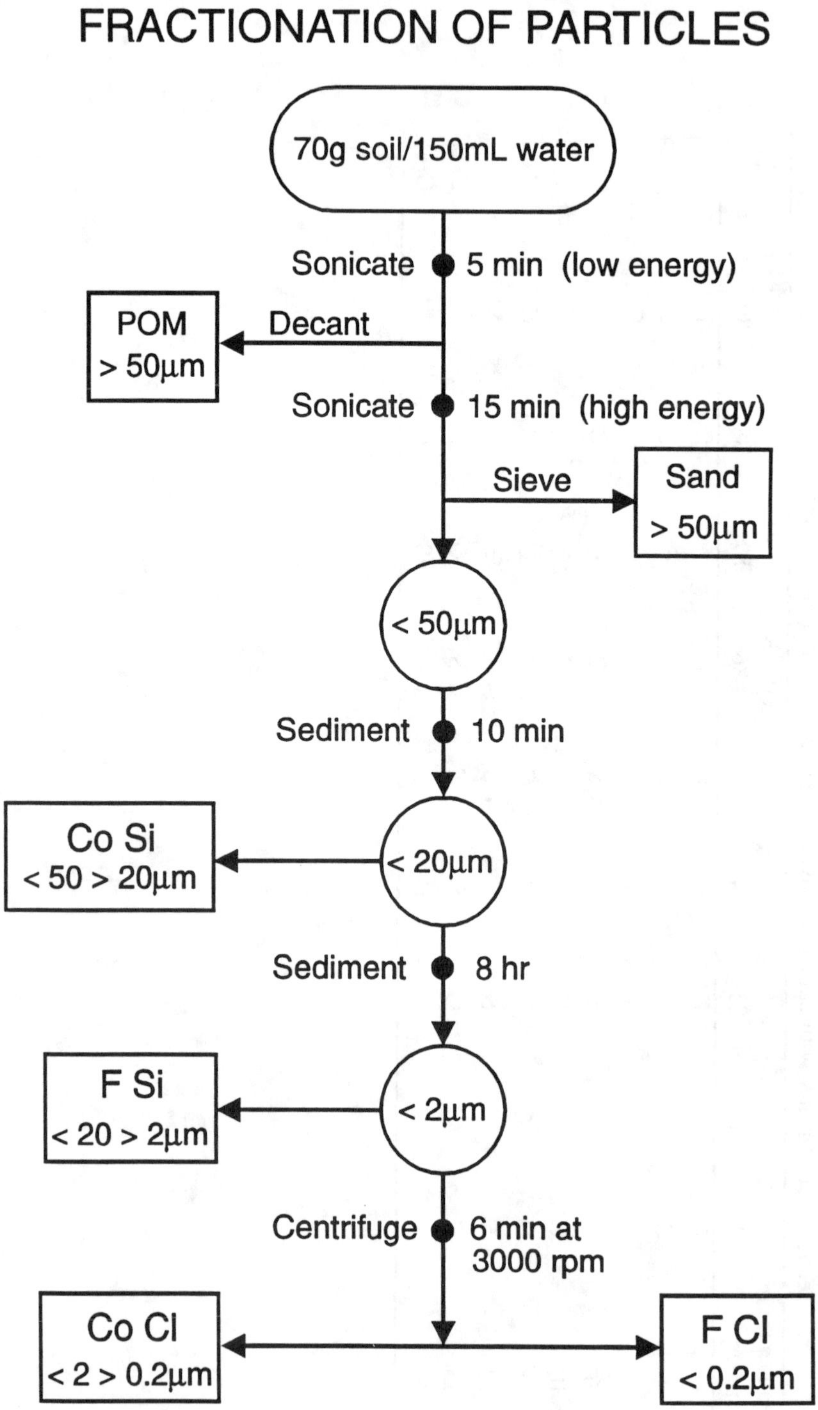

Figure 1. Physical fractionation procedure.

Table 2. Carbon concentrations of whole soils and partitioning of C into C_4 and C_3 sources

SEQUENCE[a]				C_4	C_3
NO.	CROP	C ± S.D.	$\delta^{13}C$ ± S.D.	SOURCE	SOURCE
		mg/g	‰	------fraction[b]------	
0	Fallow	21.8 ± 1.5	-18.0 ± 0.2	0.58	0.42
11	Cont. corn	22.7 ± 0.5	-17.3 ± 0.1	0.63	0.37
10	4 yr C / 5 yr S / 1 yr C	22.8 ± 1.4	-17.9 ± 0.1	0.59	0.41
6	5 yr S / 5 yr C	23.5 ± 1.4	-17.8 ± 0.4	0.61	0.39
14	1 yr S / 1 yr C	22.8 ± 1.8	-17.9 ± 0.1	0.59	0.41
12	Cont. soybean	21.8 ± 2.0	-18.4 ± 0.2	0.56	0.44
5	4 yr S / 5 yr C / 1 yr S	21.4 ± 1.2	-17.8 ± 0.2	0.60	0.40
1	5 yr C / 5 yr S	22.3 ± 0.7	-17.6 ± 0.2	0.61	0.39
13	1 yr C / 1 yr S	23.2 ± 1.3	-17.6 ± 0.1	0.61	0.39

[a]C = Corn; S = Soybean; [b]C_3 and C_4 fractions have S.D. ranging from ± 0.01 to ± 0.02.

Table 3. Carbon concentrations of soil fractions and partitioning of C into C_4 and C_3 sources

SEQUENCE	FRACTION	SAMPLE	C ± S.D.	δ ^{13}C ± S.D.	C_4 SOURCE	C_3 SOURCE	± S.D.
		%	mg/g	‰	fraction		
Cont. corn	Plant residue	--	404 ± 6	-12.0 ± 0.2	--	--	--
(11)	POM	0.20	212 ± 36	-14.6 ± 0.5	0.82	0.18	± 0.04
	Sand	37.0	1.2 ± 0.3	-19.4 ± 0.6	0.48	0.52	± 0.01
	Co Si	14.2	4.3 ± 2.2	-19.0 ± 1.4	0.51	0.49	± 0.12
	F Si	23.9	29.6 ± 10.3	-17.6 ± 0.3	0.61	0.39	± 0.03
	Co Cl	6.6	59.2 ± 2.8	-17.6 ± 0.1	0.61	0.39	± 0.01
	F Cl	18.3	41.0 ± 6.6	-16.4 ± 0.2	0.69	0.31	± 0.02
Cont. soybean	Plant residue	--	423 ± 5	-26.4 ± 0.1	--	--	--
(12)	POM	0.13	240 ± 52	-22.4 ± 1.2	0.28	0.72	± 0.09
	Sand	39.2	0.9 ± 0.2	-21.7 ± 0.1	0.33	0.67	± 0.01
	Co Si	16.0	6.0 ± 5.2	-19.9 ± 1.0	0.45	0.54	± 0.08
	F Si	20.7	39.7 ± 11.9	-18.4 ± 0.3	0.56	0.44	± 0.02
	Co Cl	6.9	45.9 ± 5.2	-17.8 ± 0.4	0.59	0.41	± 0.04
	F Cl	17.1	44.4 ± 10.2	-17.4 ± 0.3	0.62	0.38	± 0.02

III. Results and Discussion

Whole soil analyses indicated that the proportion of C_4- and C_3-derived C was influenced by the residual crop sequence treatments. Carbon concentrations in whole soils ranged from 21.4 to 23.5 mg/g and were unaffected by crop sequence ($p < 0.05$; Table 2). The $\delta^{13}C$ values for corn residue averaged -12.0‰ and soybean residue -26.4‰ (Table 3). All $\delta^{13}C$ values of whole soils were intermediate to C_4 and C_3 crop residue values indicating that a mixture of these two C sources was present (Table 2). The proportion of soil C was 60% from C_4 sources and 40% from C_3 sources, averaged across all crop sequences (Table 2). Whole soil in the continuous corn sequence had the least negative $\delta^{13}C$ value (-17.3‰) which was greater ($p < 0.05$) than $\delta^{13}C$ values for continuous soybean (-18.4‰) and fallow (-18.0‰). The similar soil C values but greater proportion of C_4-derived C in the continuous corn sequence as compared to the continuous soybean sequence and fallow is an indication of labile C turnover. No initial $\delta^{13}C$ soil values are available, however, if the alternate corn/soybean sequences are assumed to reflect initial values, then comparisons of continuous corn and soybean to these values can be used to estimate C_4- or C_3-derived C enrichment over 10 years. This comparison indicates a C_4-derived C enrichment of 4% in the continuous corn sequence and a C_3-derived C enrichment of 5% in the continuous soybean sequence (Table 2).

Corn vs. soybean produced a different distribution of C among the soil fractions. The POM fraction in soils from the continuous corn and soybean sequences had the greatest C concentration (Table 3). The proportion of C_4 and C_3 carbon in the POM fraction was very sensitive to crop sequence with C_4-derived C declining from 82% in the continuous corn sequence to 28% in the continuous soybean sequence. The sand and coarse silt fractions had relatively low C concentrations (1 to 6 mg/g) while C concentrations in fine silt, coarse clay, and fine clay were 6 to 10 times greater than the coarser fractions (Table 3). With the exception of the POM fraction in the continuous corn sequence, the proportion of C from C_4 sources is inversely related to particle size, whereas the proportion from C_3 sources is positively related to particle size (Table 3). Accompanying the greater proportion of C_4-derived C in the clay fraction was a 29% greater C concentration of the coarse clay fraction of the continuous corn sequence as compared to continuous soybean. These data suggest fundamental differences in the dynamics of C derived from corn versus soybean residue that, in turn, may influence commonly observed differences in soil aggregate size and stability, crusting, susceptibility to erosion, and other physical characteristics (McCracken et al., 1985).

IV. Summary

Analysis of $\delta^{13}C$ values was useful for accessing crop sequence effects on C sequestration in corn-soybean agroecosystems. Partitioning C from whole soils into C_4 and C_3 sources of origin indicated a 4% enrichment of C_4- and C_3-derived sources of C with either continuous corn or soybean, respectively. Partitioning of C in particle fractions showed an increasing proportion of C_4-derived C in finer textured fractions with continuous corn, whereas the proportion of C_3-derived C decreased with finer textured fractions with continuous soybean. These data suggest fundamental differences in the dynamics of C derived from corn versus soybean residue that may be related to commonly observed differences in soil physical properties.

Acknowledgement

The authors recognize and thank R.K. Crookston, W.W. Nelson, and R.H. Dowdy. R.K. Crookston, Head and Professor of Agronomy and W.W. Nelson, Emeritus Professor of Soil Science, University of Minnesota conducted the long-term crop sequence experiments; R.H. Dowdy, USDA, Agricultural Research Service secured USDA funds for purchase and installation of the elemental analyzer-mass spectometer.

References

Balesdent, J., A. Mariotti, and D. Boisgontier. 1990. Effect of tillage on soil organic carbon mineralization estimated from ^{13}C abundance in maize fields. *J. Soil Sci.* 41:587-596.

Balesdent, J., A. Mariotti, and B. Guillet. 1987. Natural ^{13}C abundance as a tracer for studies of soil organic matter dynamics. *Soil Biol. Biochem.* 19:25-30.

Balesdent, J., G.H. Wagner, and A. Mariotti. 1988. Soil organic matter turnover in long-term field experiments as revealed by carbon-13 natural abundance. *Soil Sci. Soc. Am. J.* 52:118-124.

Barnwell, T.O. Jr., R.B. Jackson IV, E.T. Elliott, I.C. Burke, C.V. Cole, K. Paustian, A.S. Donigian, A.S. Patwardhan, A. Rowell, and K. Weinrich. 1992. An approach to assessment of management impacts on agricultural soil carbon. *Water, Air, and Soil Pollution.* 64:423-435.

Cambardella, C.A. and E.T. Elliott. 1992. Particulate soil organic-matter changes across a grassland cultivation sequence. *Soil Sci. Soc. Am. J.* 56:777-783.

Cambardella, C.A. and E.T. Elliott. 1993. Methods for physical separation and characterization of soil organic matter fractions. *Geoderma* 56:449-457.

Campbell, C.A., R.P. Zentner, H.H. Janzen, and K.E. Bowren. 1990. Crop rotation studies on the Canadian prairies. Res. Branch, Agric. Can. Publ. 1841/E. Supply and Services Canada, Ottawa.

Capriel, P., P. Harter, and D. Stephenson. 1992. Influence of management on the organic matter of a mineral soil. *Soil Sci.* 153:122-128.

Christensen, B.T. 1987. Decomposability of organic matter in particle size fractions from field soils with straw incorporation. *Soil Biol. Biochem.* 19:429-435.

Crookston, R.K., J.E. Kurle, P.J. Copeland, H.J. Ford, and W.E. Lueschen. 1991. Rotational cropping sequence affects yield of corn and soybean. *Agron. J.* 83:108-113.

Dalal, R.C. and R.J. Mayer. 1986. Long-term trends in fertility of soils under continuous cultivation and cereal cropping in southern Queensland. I. Overall changes in soil properties and trends in winter cereal yields. *Aust. J. Soil Res.* 24:265-279.

Edwards, A.P. and J.M. Bremner. 1967. Dispersion of soil particles by sonic vibration. *J. Soil Sci.* 18:47-63.

Elliott, E.T. 1986. Aggregate structure and carbon, nitrogen, and phosphorus in native and cultivated soils. *Soil Sci. Soc. Am. J.* 50:627-633.

Janzen, H.H., C.A. Campbell, S.A. Brandt, G.P. Lafond, and L. Townley-Smith. 1992. Light-fraction organic matter in soils from long-term crop rotations. *Soil Sci. Soc. Am. J.* 56:1799-1806.

Jenkinson, D.S. 1971. Studies on the decomposition of ^{14}C-labelled organic matter in soil. *Soil Sci.* 111:64-70.

Johnson, M.C. and J.S. Kerns. 1991. Sequestering carbon in soils: A workshop to explore the potential for mitigating global climate change. USEPA Rep. 600/3-91/031. Environ. Res. Lab., Corvallis, OR.

Ladd, J.N., J.W. Parton, and M. Amato. 1977. Studies of nitrogen immobilization and mineralization in calcareous soils. I. Distribution of immobilized nitrogen amongst soil fractions of different particle size and density. *Soil Biol. Biochem.* 9:309-318.

McCracken, D.V., W.C. Moldenhauer, and J.M. Laflen. 1985. The impact of soybeans on soil physical properties and soil erodibility. p. 988-994. In: R. Shibles (ed.) World Soybean Res. Conf. III Proceedings. Westview Press, Boulder, CO.

Oades, J.M. and J.N. Ladd. 1977. Biochemical properties: carbon and nitrogen metabolism. In: J.S. Russel and E.L. Greacen (eds.) Soil factors in crop production in a semiarid environment. p. 127-162. Univ. of Queensland Press, St. Lucia, Queensland.

Oades, J.M., A.G. Waters, A.M. Vassallo, M.A. Wilson, and G.P. Jones. 1988. Influence of management on the composition of organic matter in a red-brown earth shown by ^{13}C nuclear magnetic resonance. *Aust. J. Soil Res.* 26:289-299.

Post, W.M., T.H. Peng, W.R. Emanuel, A.W. King, V.H. Dale, and D.L. DeAngelis. 1990. The global carbon cycle. *Am. Sci.* 78:310-326.

Staricka, J.A., R.R. Allmaras, and W.W. Nelson. 1991. Spatial variation of crop residue incorporated by tillage. *Soil Sci. Soc. Am. J.* 55:1668-1674.

Tiessen, H. and J.W.B. Stewart. 1983. Particle size fractions and their use in studies of soil organic matter. II. Cultivation effects on organic matter composition in size fractions. *Soil Sci. Soc. Am. J.* 47:509-514.

Tisdall, J.M. and J.M. Oades. 1982. Organic matter and water-stable aggregates in soils. *J. Soil Sci.* 33:141-163.

Turchenek, L.W. and J.M. Oades. 1979. Fractionation of organo-mineral complexes by sedimentation and density techniques. *Geoderma* 21:311-343.

Voroney, R.P., J.A. van Veen, and E.A. Paul. 1981. Organic carbon dynamics in grassland soils. 2. Model validation and simulation of long-term effects of cultivation and rainfall erosion. *Can. J. Soil Sci.* 61:211-224.

CHAPTER 6

Management Impacts on Carbon Storage and Gas Fluxes (CO_2, CH_4) in Mid-Latitude Cropland

Keith Paustian, G. Philip Robertson, and Edward T. Elliott

I. Introduction

Agricultural ecosystems comprise an estimated 11% of the land surface of the earth (Houghton et al., 1983) and include some of the most productive and carbon-rich soils. As a result they play a significant role in the storage and release of C within the terrestrial carbon cycle. At the same time these ecosystems are highly impacted through human activities and thus processes determining net carbon fluxes to the atmosphere are to a large degree influenced by land management practices.

In the context of global change, three major issues dealing with soil C balance and the emission of greenhouse gases from soil have come to the fore. These are (i) the potential for increased CO_2 emissions from soil under conditions of global warming, giving rise to a positive feedback on the greenhouse effect (Jenkinson et al., 1991), (ii) increased emission of other radiatively-active trace gases from soil as a consequence of land management practices [(e.g., greater N_2O emission and less CH_4 consumption with increased N fertilizer use (Bronson and Mosier, 1993; Robertson, 1993)], and (iii) the potential for increasing C storage in soils to explain what appears to be a contemporary mid-latitude CO_2 sink of 1.3-2.4 Pg yr^{-1} C (Tans et al., 1990) and to help ameliorate future increases of CO_2 in the atmosphere (Barnwell et al., 1992).

In this paper we will examine processes determining the carbon balance in agricultural soils, focusing in particular on the role of land management practices in influencing these processes. Our objectives are to provide a perspective on which practices are most likely to be important for shifting the balance towards greater storage of carbon in soils and reducing the net emission of CO_2 and maximizing the net capture of CH_4 in soil, and to examine the interactions between management and environmental controls on soil C under conditions of global climate change.

II. Historical Perspective on Temperate Zone Agriculture and Soil C

Throughout most of the ~8,000 year history of agriculture, human use of the land for farming and grazing purposes probably had little effect on the global C cycle. This was due to low population densities, and consequently the relatively small area of land devoted to agriculture, and the extensive agricultural techniques used. However, locally and even regionally, some human-induced changes in amounts and distribution of soil carbon were likely significant, such as with deforestation and soil erosion occurring in the Mediterranean region during the Greek and Roman eras.

In temperate regions, early agriculture was primarily based on slash-and-burn methods. Carbon losses incurred by land clearing and cropping were largely restored upon abandonment of fields during a fallow period. As agriculture became more settled, with a permanent landbase, and as cultivation practices intensified, soil fertility and soil organic matter became more severely depleted. The 2- and 3-field farming

ISBN 1-56670-117-1/95/$0.00+.50

systems, where one field was fallowed annually in rotation, typified European agriculture through the Middle Ages and into the 16th-17th centuries. However, replenishment of soil organic matter and plant nutrients through the addition of animal manure and with a brief fallow period in the rotation was generally insufficient, resulting in chronically low productivity and periodic cycles of famine and land abandonment (Seebomn, 1976).

By the late 17th and 18th centuries, land reform, the introduction of new crop rotations, including root and hay/legume crops, improved stock breeding and animal husbandry (hence greater manure production), and more efficient tillage methods gradually improved agricultural productivity in Europe and probably increased soil C stocks on historically tilled areas. By the mid 1850's these practices, typified by the famous Norfolk four-course rotation, were well established in western Europe and may have represented a near optimum situation for maintaining organic matter in agricultural soils (Steen, 1989).

Ironically, this same period also marks the beginning of a massive global loss of soil C associated with the rapid expansion of agriculture onto grassland and forest soils in North America, Australia, southern Africa and Eastern Europe (Wilson, 1978; Haas et al., 1957). While the intensive European-style of tillage based on the moldboard plow was widely adopted, the high native fertility of these newly cultivated soils as well as various economic and ecological constraints generally discouraged the use of crop rotations, manuring, ley cropping, and other practices beneficial to soil C maintenance (Cochrane, 1979). The net effect of this expansion of "exploitative" agriculture may have released as much as 110 Pg C from soils (Wilson, 1978), contributing to perhaps 25% of the ppm increase in atmospheric CO_2 prior to 1900.

The technological revolution in temperate zone agriculture during the 20th century, including farm mechanization and the use of industrially-produced fertilizers and pesticides, had major impacts on soil C. In the older, established agricultural areas of Europe and eastern North America these changes were mainly detrimental to soil C maintenance, with shifts towards more cereal-based monocultures relying on chemical fertilizers and away from animal-based production systems employing perennial hay crops, legumes, and manuring (Steen, 1989). In the more recently exploited lands, however, such as the former prairie soils of North America, the use of chemical fertilizers and introduction of higher yielding crop varieties greatly increased productivity and appears to have helped stabilize soil C levels (Cole et al., 1989). The current interest in sustainable agriculture, land use and global change issues, and the wider range of alternative tillage and cropping systems now available, together offer the potential for significant enhancements of soil C storage in temperate zone agroecosytems.

III. Management Links to Controls on Soil C Balance

The assimilation of CO_2 through photosynthesis and the release of CO_2 to the atmosphere through plant and heterotroph respiration ultimately determine the C balance of terrestrial ecosystems. In principle, a direct accounting of the various CO_2 fluxes involved in these processes (i.e., gross photosynthesis, photorespiration and dark respiration by plants, respiration by soil organisms) could be used to quantify fluxes and predict long-term changes in soil and vegetation C storage. In practice, such an approach is problematic due to the high spatial and temporal variability of these gas fluxes as well as methodological difficulties in performing large-scale gas exchange measurements. Furthermore, in most agroecosystems there is often little or no change in the long-term C storage in vegetation (e.g., annual crops) due to harvesting, and therefore the principle factors of interest are those affecting the amount of organic matter entering the soil and its rate of decomposition. Because of the long residence time of C in most soils, data from the long-term field experiments are invaluable in accessing how management practices, climate and edaphic factors interact in determining the C balance. Data from such experiments will be used to examine how management might be directed towards increasing C sequestration in soil.

Considering soil C change, in the simplest terms, as the net difference between C added to soil and the C mineralization from soil organic matter (Figure 1), there are three potential ways to increase C storage: (i) by increasing C inputs, (ii) by decreasing decomposition rates, and (iii) by reducing the amount of CO_2 produced per unit of organic matter decomposed. Each of these processes will be evaluated in turn as to how they may be influenced by management.

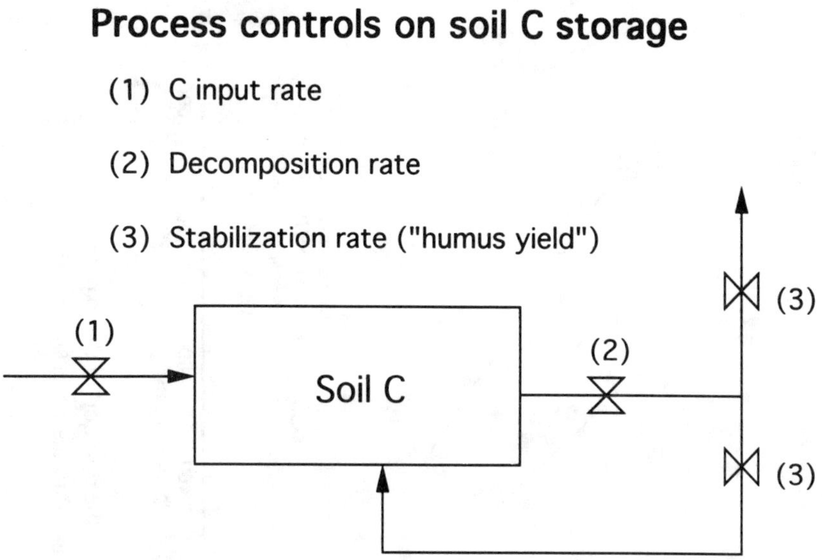

Figure 1. Characterization of control points at which management practices can influence soil C storage.

A. Controls on C Inputs

The main objective of agroecosystem management is to produce dry matter (as food and fiber) which, in turn, affects the amount of non-harvested organic matter returned to soil. Thus, C inputs to soil are influenced by nearly every facet of agricultural management. In cropland and pasture these management factors include the crop type (i.e., production potential), frequency of fallow, fertilization, and residue management. In addition, the use of organic amendments such as manure and sewage sludge constitute a direct management control on C supply to soil.

Residue amounts do not necessarily vary in direct proportion to crop yields. For instance, yield increases from N application, may not necessarily result in greater soil C input. Production responses to management are likely to be different for the harvested vs. non-harvested portions of the plant, and until recently relatively little information has been available for relating total crop productivity (including root production) to management. Therefore, it is difficult to extrapolate knowledge of how management affects crop yield to the management effects on C inputs and soil organic matter dynamics.

Nevertheless, several long-term field experiments in which C inputs have been directly controlled (through crop removal and subsequent readdition of known amounts of organic matter) shed considerable light on the response of soil to C inputs. Such experiments generally impose the same cropping treatments across C addition treatments such that differences in root C inputs and abiotic conditions as they affect decomposition rates are minimized. These experiments have generally found a close linear relationship between C addition rates and soil C amounts (Figure 2). The relationship between annual C addition rates and the average annual change in soil C varies depending on climatic and edaphic factors affecting decomposition rates at a particular site, as well as on the duration of the experiment, e.g., mean annual rates of change are higher for the short-term experiments at Lind, WA (17 years) and Culbertson, MT (7 years) (Figure 2). Note also that while the slopes for the wheat-fallow and the continuous wheat treatments at Pullman are roughly similar, the intercept for the wheat-fallow system is lower, suggesting a more favorable decomposition regime associated with the higher soil moisture in the fallowed system.

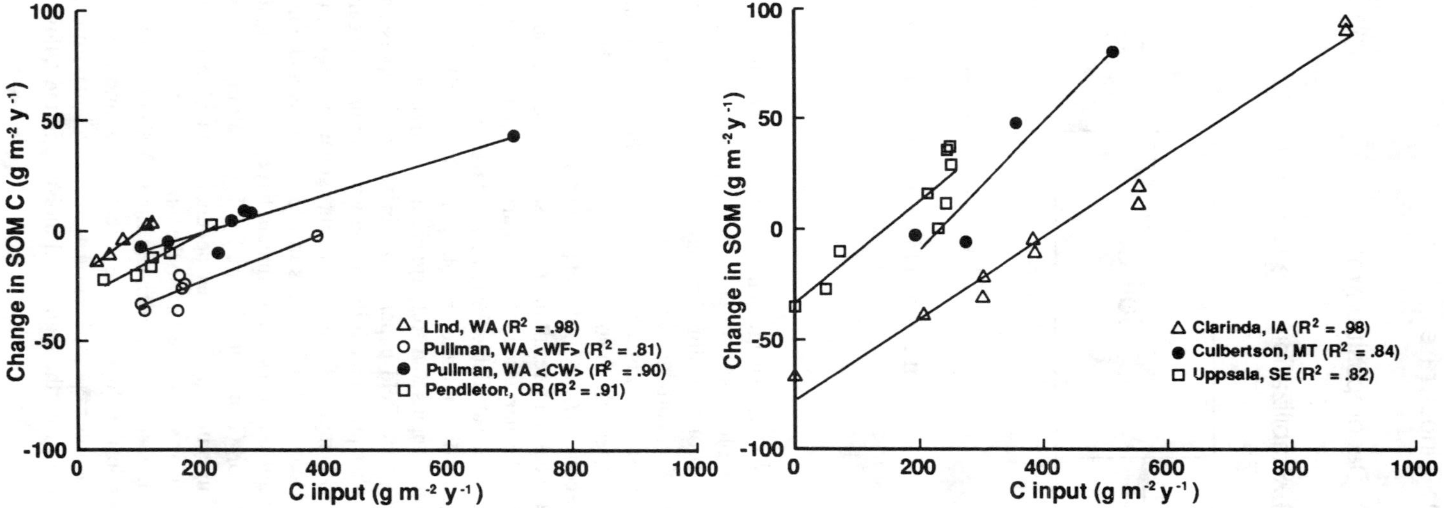

Figure 2. Relationship between annual C input rates and net changes in soil C, averaged over the duration of the experiment. Data are from Black (1973), Horner et al. (1960), Larson et al. (1972), Paustian et al. (1992) and Rasmussen et al. (1980).

The linear relationship between C inputs and soil C storage agrees with what would be predicted theoretically if the decomposition of the total soil C is assumed to follow first-order kinetics, i.e.,

$$dC_t/dt = I - kC_t,$$

where the soil C (C_t^*) at steady-state, ($dC_t/dt = 0$), is directly proportional to C inputs (I), i.e.,

$$1 = kC_t^*.$$

The same relationship holds if multiple soil C pools having different characteristic turnover times (e.g., Parton et al., 1987; Jenkinson et al., 1987) are included, provided they follow first-order decomposition, i.e.,

$$dC_t/dt = d(C_1 + C_2 + \ldots + C_n)/dt = I - k_1C_1 - k_2C_2 - \ldots - k_nC_n.$$

Since at steady-state each individual pool would make up a constant fraction of total C (i.e., $C_1 = f_1C_t^*$), then,

$$\begin{aligned} I &= k_1(f_1C_t^*) + k_2(f_2C_t^*) + \ldots + k_n(f_nC_t^*) \\ &= (k_1f_1 + k_2f_2 + \ldots + k_nf_n)\, C_t^* \\ &= k_tC_t^*. \end{aligned}$$

In summary, both empirical and theoretical evidence suggests that C storage in soils can be increased in direct proportion to increases in C inputs, provided there is no change in specific decomposition (more specifically, C mineralization) rates. Management controls on decomposition rates are discussed in the following section.

B. Controls on Decomposition Rate

The decomposition rates of soil C and crop residues are controlled by a variety of factors including soil abiotic conditions, residue composition, and soil disturbance, all of which can be affected, directly and indirectly, by management. However, designing management systems to reduce decomposition rates may not always be commensurate with production goals, particularly with respect to modifying the abiotic environment. In most temperate soils (excluding wetlands) decomposition rates could be reduced by decreasing temperature, decreasing soil water, and decreasing soil oxygen, all of which would be likely to reduce crop growth. One exception would be the reduction in soil moisture occurring in conjunction with reduced fallow frequency. In semiarid regions a major consequence of conversion from wheat-fallow to more continuous cropping are drier soils and reduced potential decomposition rates.

Other strategies could include manipulating chemical and/or physical factors of crop residues to restrict decomposition activity in soil. It is well recognized that different chemical compounds in plant materials vary widely in their decomposition rate. In particular, the lignin content of plant tissue has often been found to correlate well with litter decomposition rates (Aber and Melillo, 1982). The complex structure of lignin and the high energetic costs of lignin degradation restrict the number and activity of organisms adapted to decompose lignin and lignin-associated products.

The influence of residue quality on soil organic matter formation has been examined in a few long-term field experiments. Figure 3 shows results from two experiments where similar amounts of different residues were added to small field plots. The same cropping regime was imposed across treatments (within an experiment) with similar amounts of N inputs between treatments. In the Swedish experiment, the highest soil C values were found in manure and sawdust-amended plots, having lignin contents estimated to be ~30%, and the lowest where grass litter (~6% lignin) was added (Paustian et al., 1992). Wheat straw (15% lignin) yielded intermediate soil C increases. Similar results were found by Sowden and Atkinson (1968) in Canada, with the highest gains of soil C in peat and manure-amended soil and the lowest with alfalfa additions. However, Larson et al. (1972) found no significant effect of residue type on soil C accumulation in an 11 year experiment.

Physical protection of soil organic matter is widely believed to play an important role in restricting the access to substrates by microorganisms and extracellular enzymes, thereby reducing decomposition rates.

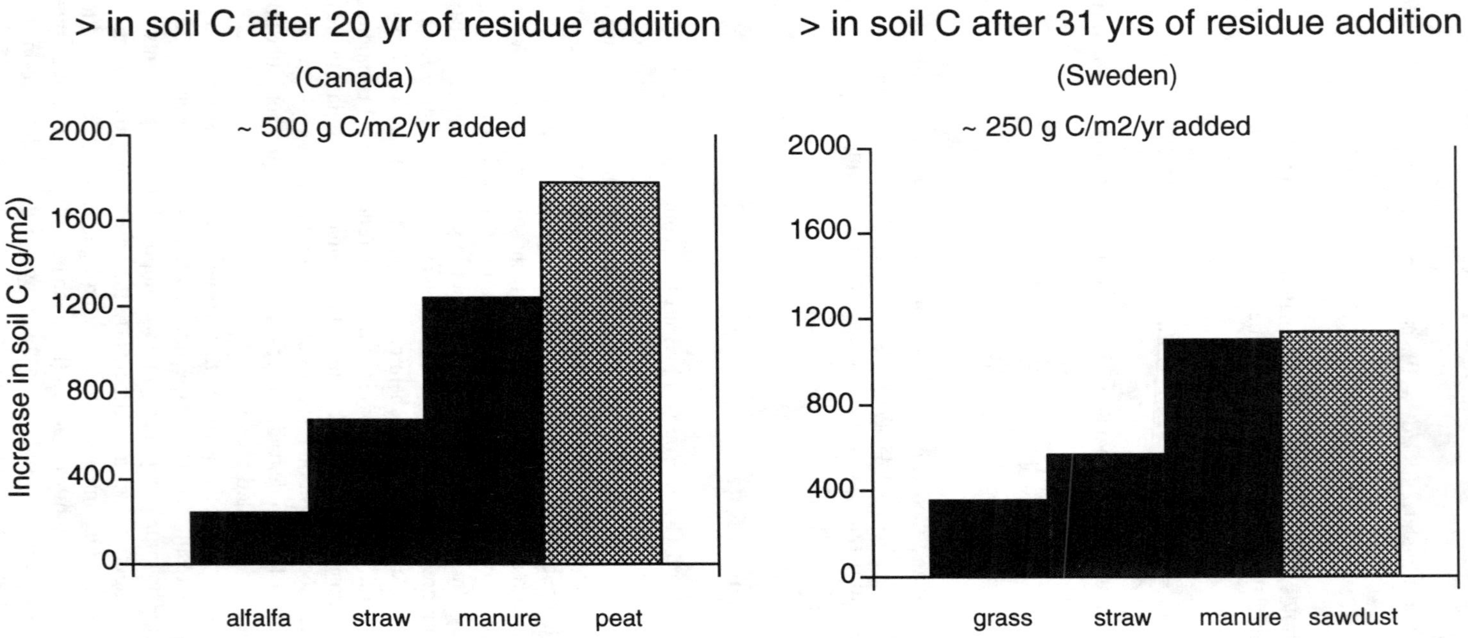

Figure 3. Net increments in soil C with different residue types added in equal amounts to a sandy soil in Canada (data from Sowden and Atkinson, 1968) and a sandy clay loam in Sweden (data from Paustian et al., 1992).

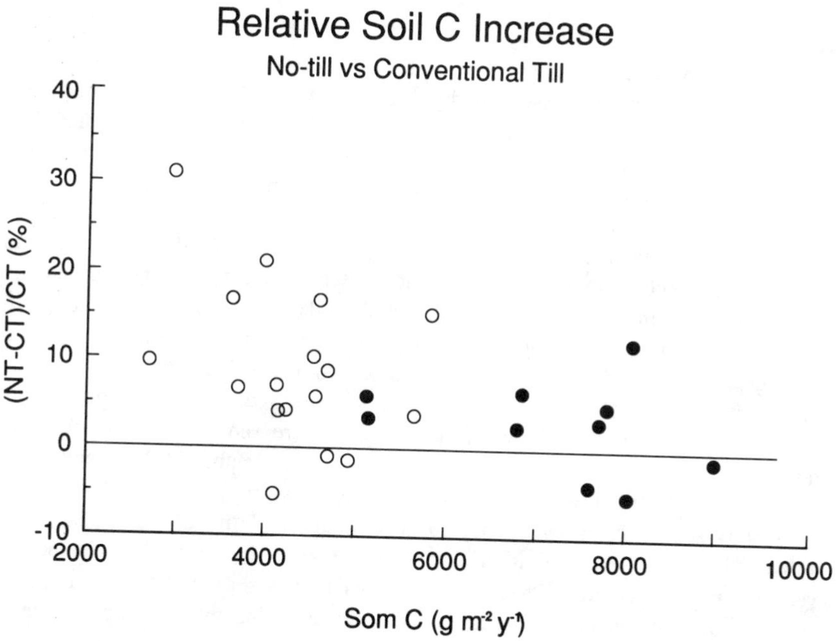

Figure 4. Differences in C in surface soils between no-tilled and moldboard plowed treatments in paired field plots. Values are for total C to depths at or below the depth of plowing, adjusted for differences in bulk density. Data are from Balesdent (1990), Blevins et al. (1983), Dalal (1989), Dick (1983), Dick et al. (1986a,1986b), Doran (1987), Havlin et al. (1990), Groffman (1985) and Powlson and Jenkinson (1981).

In simple terms this physical protection has been conceptualized as an "encapsulation" of organic matter by clay particles and soil aggregates (Tisdall and Oades, 1982; Elliott, 1986). Tillage and other mechanical disturbance of soil has been found to decrease aggregate stability which may result in increased susceptibility to decomposition of physically-protected organic matter. While tillage influences a number of other factors affecting decomposition rates, including soil moisture, temperature, and aeration, the degradation of soil structure and loss of physical protection has been postulated as a major cause of soil C losses when undisturbed soils are cultivated (Cambardella and Elliott, 1992). Thus, greater use of reduced and no-till cultivation could help to decrease organic matter decomposition in cropland soils.

Results from a number of field experiments comparing no-till with conventional (moldboard plow) tillage show in most cases higher C storage under no-till (Figure 4). Only data for paired tillage plots were included, in which soil C values were available to or below depth of plowing (to ~20-30 cm depth in most cases). Differences between no-till and conventional till systems were calculated on an absolute (per m^2) basis, corrected for differences in bulk density. The duration of the experiments varied from 5 to 30 years.

For the experiments surveyed, no-till had up to 30% more C in the upper profile, although in most cases increases were below 20%. On a relative basis, the difference between no-till and plow tillage decreased with increased soil C content which may indicate a greater impact of changes in tillage for low C and/or degraded soils. Fine textured soils (filled circles) tended to have the highest C amounts, irrespective of tillage, and showed less response to no-till. No-till may affect other factors which can act differentially on decomposition rates, such as reducing soil temperature (<decomposition) and increasing soil water (> decomposition). Also decreased soil erosion under no-till might result in substantial differences in soil C as a function of tillage at sites with a high erosion potential. Therefore, the high variability in the apparent effects of no-till across sites is not surprising. However, the available empirical evidence does support the potential for tillage reductions to help sequester C.

C. Controls on Humus Yield

The main products of aerobic decomposition of organic carbon compounds are CO_2 and microbial biomass and metabolic products, including non-assimilated decomposition products. Soil organic matter is formed from the amalgamation and stabilization of dead microbial material, metabolites and residual decomposition products. Conceptually, one way to increase soil C storage is to decrease the relative proportion of C mineralized directly to CO_2 and increase the formation of more stable C forms in SOM. This stabilization efficiency is sometimes referred to as the "humus yield".

At the microbial level, it is well know that the growth yield (biomass produced/C assimilated) can vary widely depending on growth conditions as well as species differences. As yet there is little knowledge about microbial energetics and growth yield applicable at the whole soil or ecosystem level although it's reasonable to suppose that these factors might be influenced through management practices. For instance, no-till may favor a greater degree of fungal (vs. bacterial) dominance compared with conventionally tilled systems (Hendrix et al., 1986; Beare et al., 1992). Fungi are important agents of soil aggregation (Tisdall and Oades, 1982; Gupta and Germida, 1988), which may lead to increased C stabilization through protection of soil aggregates in fungal-dominated no-till systems (M. Beare, unpubl.). It has been further suggested (Holland and Coleman, 1987) that the product yield (i.e., biomass and metabolite production per unit CO_2 respired) of fungi may be higher, possibly due to accumulation of fungal cell walls (C. Cambardella, unpubl.), than that of bacteria, which would effectively increase the humus yield under no-till. While data is limited at present, a more in-depth understanding of how decomposer communities and their metabolism respond to environmental and management influences may give insight into appropriate strategies for increasing soil C sequestration.

Other chemical and physical factors can affect soil C stabilization. It is widely accepted that lignin degradation products and proteinaceous compounds (peptides, amino acids) are major constituents of recalcitrant humic substances in soils. Thus formation of stabile organic matter in soil should be enhanced where these humic precursors are in abundant supply. Data from the Swedish field experiment cited earlier indicate a potential strong interaction between the lignin content of residues and nitrogen availability (Figure 5). Where either straw or sawdust were added there was a strong additional affect of nitrogen addition on the accumulation of soil organic matter. Some additional C inputs (i.e., more root production) and, potentially, drier soils in the fertilized treatments (due to higher evapotranspiration) could have contributed to these differences in C buildup. However, simulation analyses of these data (Paustian et al., 1992) indicated that these factors could not fully explain the magnitude of the differences, suggesting the importance of a direct stabilization mechanism associated with N availability in the system. Thus, the combination of high lignin residues and N addition may be a management option conducive to increasing soil C storage.

IV. Agricultural Soils as Attenuated Sinks for CH_4

That forest and grassland soil can act as sinks for atmospheric methane, and that these sinks can be effectively switched off by nitrogen addition, is by now well established (Steudler et al., 1989; Mosier et al., 1991). The importance of this sink is globally significant: ca. 30 Tg CH_4 yr^{-1} of a total sink of 500 Tg CH_4 yr^{-1} is thought to be captured by upland soils (IPCC, 1992). About 37 Tg CH_4 yr^{-1} is now accumulating in the atmosphere. Thus, in the absence of such uptake, atmospheric methane concentrations would likely be significantly greater than current concentrations of ca. 2.1 ppmv, leading to more atmospheric heat trapping and to important changes in tropospheric $[OH^-]$ scavenging by CH_4 molecules.

That nitrogen additions to soil can reduce methane uptake potentials (e.g., Steudler et al., 1989; Bronson and Mosier, 1993) suggests a potential for substantial historical changes in the soil CH_4 sink strength due to agricultural activity, and in particular an attenuation of this sink as tillage and fertilizer practices reduce CH_4 uptake in former grassland and forest soils. Available evidence, while limited, suggests that this change is in fact occurring: in one U.S. corn belt experiment (Robertson et al., 1993) uptake in conventionally-tilled soybeans (*Glycine max* L.) was 7-8 times less than uptake in a never-tilled soil under grassland vegetation (Table 1). Uptake under other cropping systems was equally attenuated, although less so in the organic-based no-fertilizer systems.

Such results suggest yet another impact of agriculture on a carbon-cycle greenhouse gas. And again agriculture acts as a net source of the gas, although in this case by attenuating an existing sink rather than

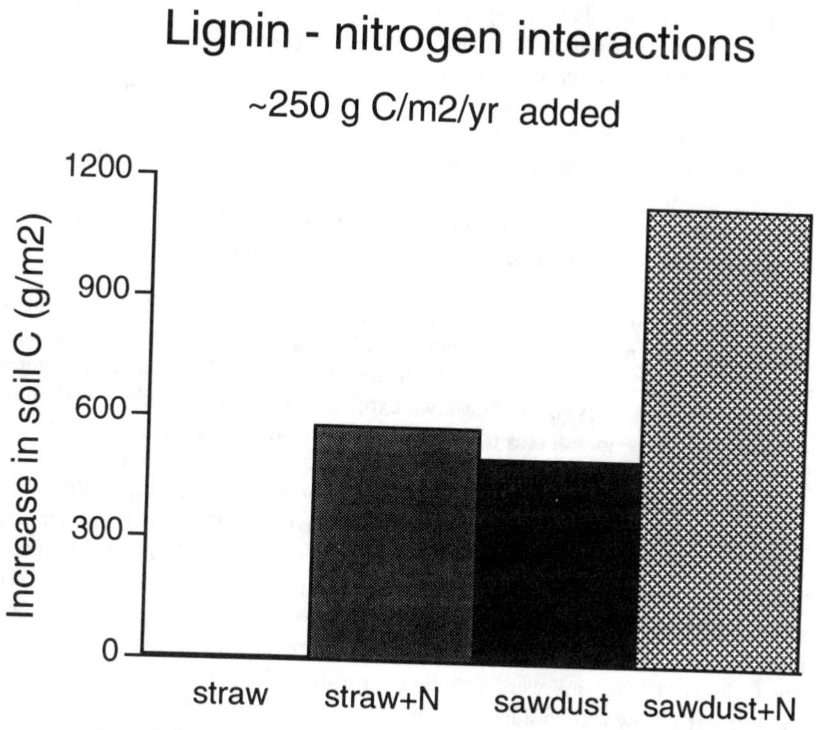

Figure 5. Effect of residue quality and N addition rate on net soil C increase after 31 years. Treatments received the same amounts of residue added to the soil after removal of all aboveground plant material. C inputs from roots are included in the input estimate. Fertilizer nitrogen was added to two of the treatments at a rate equivalent to 80 kg N ha^{-1} yr^{-1} as $Ca(NO_3)_2$, the other two treatments received no N fertilizer. Data from Paustian et al. (1992).

Table 1. Daily mean CH_4 uptake in soil under conventional till soybeans (*Glycine max* L.) vs. a never-tilled soil under native vegetation in southern Michigan. Values are means of 15 or 16 dates in 1992 for each of 4 replicate cropping/native systems

System	Daily Mean Uptake g CH_4-C ha^{-1} d^{-1}	Standard Error
CT Soybeans	0.176	.078 (n=16)
Never-tilled native	1.440	.408 (n=15)

by an absolute increase in source (as for CO_2). Early results suggesting that the effect of nitrogen on CH_4 uptake may be closely related to the organic matter status of soil (Robertson, unpubl.) suggest in turn that future management for soil C sequestration may also lead to the re-establishment of significant CH_4 uptake in these soils.

V. Management and Climate Change Interactions

Forecasting changes in the carbon balance of agricultural soils over the next 50 to 100 years is complicated by the anticipated rapid rate of change of climate conditions and land-use practices. Current projections are for mean global temperature to increase by about 1-4 °C by the middle of the next century with somewhat greater increases predicted for mid- and high-latitude regions (IPCC, 1990). There is greater uncertainty about changes in precipitation patterns but the potential for major regional shifts in weather patterns associated with global warming is apparent. In addition, agricultural land use is changing rapidly in response to a variety of social, economic, and environmental considerations. However, our current understanding of soil C dynamics as embodied in simulation models (e.g., Jenkinson et al., 1987; Parton et al., 1987) and experience from long-term field experiments provide a powerful tool for assessing changes in soil C (see Metherell et al., this volume; Patwardhan et al., this volume).

To evaluate changes in potential soil C storage as a consequence of global warming and changes in agricultural productivity and management, we performed an equilibrium analysis of the CENTURY model (Parton et al. 1987). At equilibrium, the model can be solved analytically and the interactions between different factors affecting soil C levels can be shown explicitly (Paustian, unpubl.). It should be emphasized that this kind of analysis does not address transient changes in soil C but rather provides a measure of the potential soil C level under a given set of conditions.

We analyzed the model for two long-term field sites, the High Plains Agricultural Laboratory near Sidney, Nebraska, and the Long-Term Ecological Research site at Kellogg Biological Station (KBS) in Michigan, which are included in a network of long-term sites being used to evaluate management effects on soil C (Elliott et al., 1994). The cropping treatments at Sidney include wheat-fallow (*Triticum aestivum* L.) rotations under different tillage regimes and at KBS the dominant row crops are corn (*Zea mays* L.) and soybeans with different tillage and with and without winter cover crops.

Equilibrium C levels (0-20 cm) were calculated for five different scenarios (Figure 6). "Present" management was wheat-fallow with moldboard plowing at Sidney and corn-soybean with moldboard plowing at KBS. "Best" management at Sidney represented no-till with a chemical fallow and, at KBS, corn-soybean under no-till with a winter cover crop. Climate change was represented by a 2°C increase in mean annual temperature by 2050, based on the best estimate for temperature increase in the central U.S. (IPCC, 1992). Three scenarios also include an increase in crop residue inputs of 40% (assumed to accompany projected productivity increase) above current levels by 2050. Over the past 40 years, agricultural productivity has increased by an average of 1.8% per year and is projected to increase at an annual rate of 1.4% over the next 20 years (Crosson, 1993). Finally, the potential effect of breeding crop varieties to have higher lignin contents was represented by assuming increases in residue lignin contents to 20% of total dry matter.

The model was executed using rate parameters determined by Parton et al. (1987), except for reduction in the yield efficiency of litter decomposition from 0.45 to 0.35, based on model applications to other agricultural treatments (Paustian et al. 1992). This parameter change has little effect on the relative differences between scenarios and results in slightly lower C levels than if the higher yield efficiency is assumed. Long-term (30 year) averages of temperature and precipitation measurements at the sites were used to calculate the effect of present temperature and soil moisture on decomposition rates. The 2°C temperature increment assumed in the global warming scenarios was distributed over the year such that the increase in winter temperature would be 50% higher and summer temperatures 50% lower than the mean annual increase, reflecting seasonal differences projected by the climate models. Precipitation rates for the warming scenarios were assumed to be unchanged from the 30 year averages. Present C inputs levels were estimated from field measurements of aboveground productivity at Sidney (Power et al., 1986) and KBS (S. Halstead, pers. commun.) and estimates of relative belowground allocation were based on Buyanovsky and Wagner (1986).

Potential (equilibrium) C levels predicted under present management, 3.4 kg C m^{-2} at Sidney and 4.6 kg C m^{-2} at KBS (Figure 6), are roughly in line with current measured C levels (about 3 kg C m^{-2} for 0-20 cm at both sites; Elliott et al., 1994; S. Halsted, pers. commun.). Under the global warming scenario, assuming no changes in management or C input levels, the model predicts a slight decrease in potential C storage at KBS due to increased decomposition rates. At the drier site in Sidney, increased temperature increases evapotranspiration and thus decreases soil moisture, such that the net effect of warming on decomposition rates is negative, resulting in increased C storage potential. The interaction between temperature and C inputs as they affect equilibrium C levels for the two sites are shown in more detail in Figure 7.

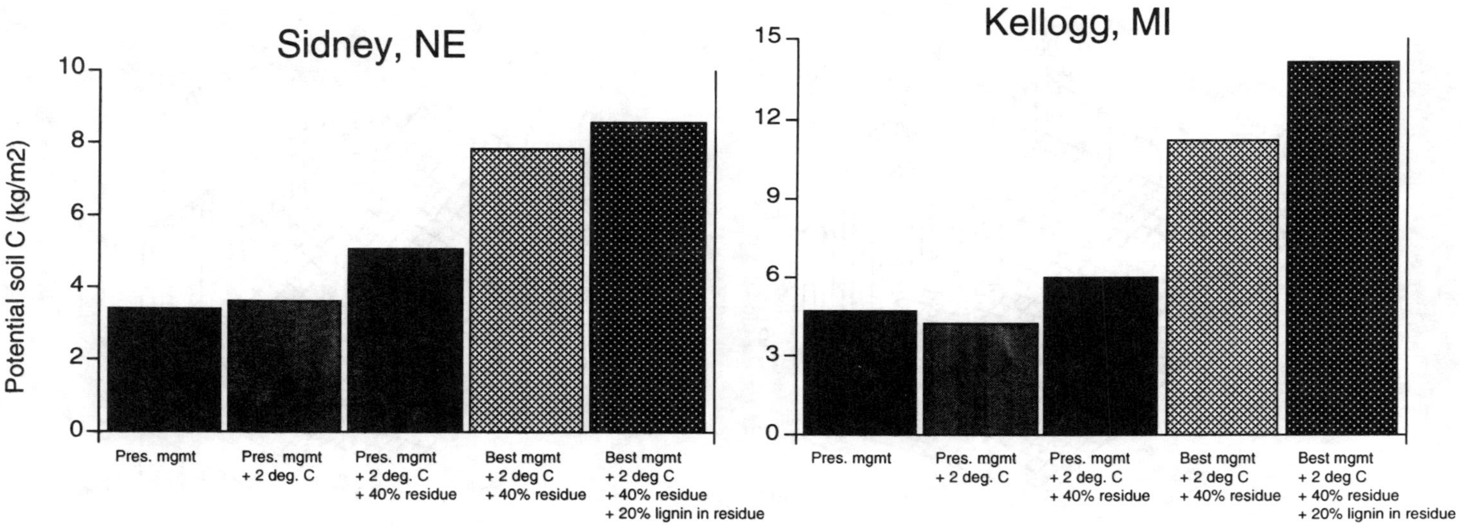

Figure 6. Steady-state solutions of the CENTURY model, showing potential soil C levels in surface soil for 5 management-climate change scenarios, for a semiarid wheat-fallow at Sidney, NE and a corn-soybean based rotation in southwest Michigan (KBS). See text for explanation of model scenarios.

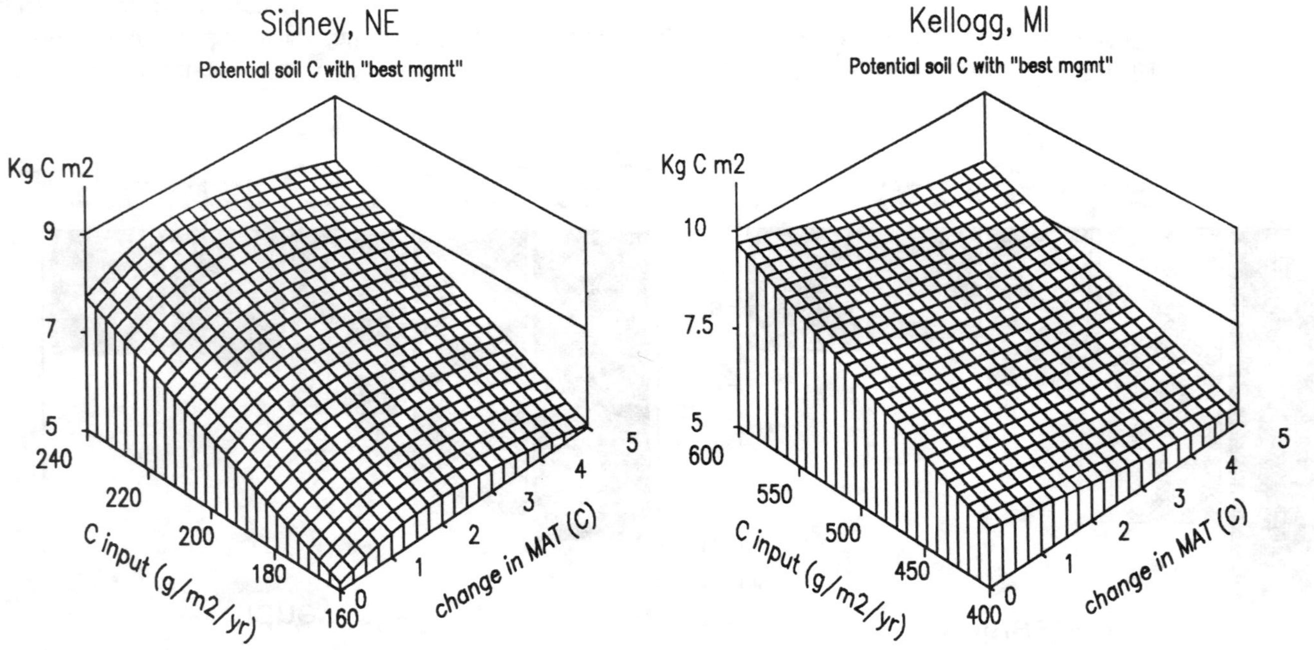

Figure 7. Steady-state values of soil C as a function of C inputs at the NE and MI sites, ranging from current levels up to 50% higher annual C inputs, and change in mean annual temperature, ranging from present climate (0) conditions to a 5 degree rise in average temperature. Other model parameters were the same as used in the "present" management scenario.

In scenario 3, with an increase in C inputs from crop residues of 40%, C storage potential increases and at KBS, C levels under the increased input scenario more than compensate for the increase in decomposition rates due to warming. Using "best" management practices, i.e., no-till at both sites and a winter cover crop (adding an additional 200 g C m^{-2} yr^{-1}) at KBS, substantially increases C storage potential. Finally, the effect of increased residue lignin is higher at KBS due to the greater relative increase for corn and soybean stover (~10 to 20%) than for wheat (15 to 20%).

VI. Summary and Conclusions

Data from long-term field experiments and model analyses clearly illustrate the key role of C inputs, and to a lesser extent reduced tillage intensity, in altering soil C sequestration. The potential for modifying crop residue quality and the stabilization efficiency of C provide additional means for increasing C levels in agricultural soils. Thus, both theory and empirical evidence from long-term field experiments suggest that gains in C storage in proportion to increases in C additions to soil are feasible, particularly in C-depleted agricultural soils. Considering only climate effects on decomposition, model analyses suggest that relatively modest gains in C inputs would be sufficient to compensate for global warming influences. However, climate change, including increased drought risk, and other factors may constrain increases in productivity, particularly in semiarid croplands. The challenge to agricultural management will thus be to design systems capable of maintaining economically viable harvests while increasing the allocation of NPP to the soil and minimizing the soil disturbance.

Acknowledgments

Support for the research reported here was provided by grants from NSF (BSR 87-02332) to Michigan State University, EPA (AERL9101) to Colorado State University and Michigan State University and DOE (Midwestern Regional NIGEC) to Michigan State University.

References

Aber, J.D. and J.M. Melillo. 1982. Nitrogen immobilization in decaying hardwood leaf litter as a function of initial nitrogen and lignin content. *Can. J. Bot.* 60:2263-2269.

Balesdent, J., A. Mariotti, and D. Boisgontier. 1990. Effect of tillage on soil organic carbon mineralization estimated from ^{13}C abundance in maize fields. *J. Soil Sci.* 41:587-596.

Barnwell, T.O., R.B. Jackson, E.T. Elliott, I.C. Burke, C.V. Cole, K. Paustian, E.A. Paul, A.S. Donigian, A.S. Patwardhan, A. Rowell, and K. Weinrich. 1992. An approach to assessment of management impacts on agricultural soil carbon. *Water, Air, Soil Pollut.* 64:423-435.

Beare, M.H., R.W. Parmelee, P.F. Hendrix, W. Cheng, D.C. Coleman, and D.A. Crossley, Jr. 1992. Microbial and faunal interactions and effects on litter nitrogen and decomposition in agroecosystems. *Ecol. Monogr.* 62:569-591.

Black, A.L. 1973. Soil property changes associated with crop residue management in a wheat-fallow rotation. *Soil Sci. Soc. Am. Proc.* 37:943-946.

Blevins, R.L., G.W. Thomas, M.S. Smith, W.W. Frye, and P.L. Cornelius. 1983. Changes in soil properties after 10 years continuous no-tilled and conventionally tilled corn. *Soil Till. Res.* 3:135-146.

Bronson, K.F. and A.R. Mosier. 1993. Nitrous oxide emissions and methane consumption in wheat and corn-cropped systems in northeastern Colorado. In: *Agricultural Ecosystem Effects on Trace Gases and Global Climate Change*. ASA Spec. Publ. No. 55. pp. 133-144, ASA-CSA-SSSA, Madison, WI.

Buyanovsky, G.A. and G.H. Wagner. 1986. Post-harvest residue input to cropland. *Plant Soil* 93:57-65.

Cambardella, C.A. and E.T. Elliott. 1992. Particulate organic matter loss and turnover under different cultivation practices. *Soil Sci. Soc. Am. J.* 56:777-783.

Cochrane, W.W. 1979. *The development of American agriculture. A historical analysis*. Univ. Minn. Press, Minneapolis.

Cole, C.V., J.W.B. Stewart, D.S. Ojima, W.J. Parton, and D.S. Schimel. 1989. Modelling land use effects of soil organic matter dynamics in the North American Great Plains. In: M. Clarholm and L. Bergström (eds.), *Ecology of Arable Land - Perspectives and Challenges*. pp. 89-98. Kluwer, Dordrecht.

Crosson, P.R. 1993. United States agriculture and the environment: Perspectives in the next twenty years. *U.S. EPA* (in press).

Dalal, R.C. 1989. Long-term effects of no-tillage, crop residue, and nitrogen application on properties of a Vertisol. *Soil Sci. Soc. Am. J.* 53:1511-1515.

Dick, W.A. 1983. Organic carbon, nitrogen, and phosphorus concentrations and pH in soil profiles as affected by tillage intensity. *Soil Sci. Soc. Am. J.* 47:402-407.

Dick, W.A., D.M. van Doren Jr., G.B. Triplett Jr., and J.E. Henry. 1986a. Influence of long-term tillage and rotation combinations on crop yields and selected soil parameters. I. Results obtained for a Mollic Ochraqualf soil. *Res. Bull. 1180*, The Ohio State University, Ohio Agricultural Research and Development Center. 30 p.

Dick, W.A., D.M. van Doren Jr., G.B. Triplett Jr., and J.E. Henry. 1986b. Influence of long-term tillage and rotation combinations on crop yields and selected soil parameters. I. Results obtained for a Typic Fragiudalf soil. *Res. Bull. 1181,* The Ohio State University, Ohio Agricultural Research and Development Center. 34 p.

Doran, J.W. 1987. Microbial biomass and mineralizable nitrogen distributions in no-tillage and plowed soils. *Biol. Fert. Soils* 5:68-75.

Elliott, E.T. 1986. Aggregate structure and carbon, nitrogen, and phosphorus in native and cultivated soils. *Soil Sci. Soc. Am. J.* 50:627-633.

Elliott, E.T., K. Paustian, H.P. Collins, E.A. Paul, C.V. Cole, I.C. Burke, R.L. Blevins, D.J. Lyon, W.W. Frye, A.D. Halvorson, D.R. Huggins, R.F. Turco, M. Hickman, C.A. Monz, and S.D. Frey. 1994. Terrestrial carbon pools in grasslands and agricultural soils: Preliminary data from the Corn Belt and Great Plains Regions. In: J.W. Doran, D.C. Coleman, D.F. Bezdicek, and B.A. Stewart (eds.), *Defining Soil Quality for a Sustainable Environment.* ASA Spec. Publ. (in press).

Groffman, P.M. 1985. Nitrification and denitrification in conventional and no-tillage soils. *Soil Sci. Soc. Am. J.* 49:329-334.

Gupta, V.V.S.R. and J.J. Germida. 1988. Distribution of microbial biomass and its activity in different soil aggregate size classes as affected by cultivation. *Soil Biol. Biochem.* 2:777-786.

Haas, H.J, C.E. Evans and E.F. Miles. 1957. Nitrogen and carbon changes in Great Plains soils as influenced by cropping and soil treatments. *USDA Tech. Bull. 1164*.

Havlin, J.L., D.E. Kissel, L.D. Maddux, M.M. Claassen, and J.H. Long. 1990. Crop rotation and tillage effects on soil organic carbon and nitrogen. *Soil Sci. Soc. Am. J.* 54:448-452.

Hendrix, P.F., R.W. Parmelee, D.A. Crossley, Jr., D.C. Coleman, E.P. Odum, and P.M. Groffman. 1986. Detritus food webs in conventional and no-tillage agroecosystems. *Bioscience* 36:374-380.

Holland, E.A. and D.C. Coleman 1987. Litter placement effects on microbial and organic matter dynamics in an agroecosystem. *Ecology* 68:425-433.

Horner, G.M., M.M. Oveson, G.O. Baker, and W.W. Pawson. 1960. Effect of cropping practices on yield, soil organic matter and erosion in the Pacific Northwest wheat region. *Washington, Idaho and Oregon Agric. Exp. Stn. and ARS-USDA Coop. Bull. No. 1.* 25 p.

Houghton, R.A., J.E. Hobie, J.M. Melillo, B. Moore, B.J. Peterson, G.R. Shaver, and G.M. Woodwell. 1983. Changes in the carbon content of terrestrial biota and soils between 1860: and 1980: A net release of CO_2 to the atmosphere. *Ecol. Monogr.* 53:235-262.

Intergovernmental Panel on Climate Change. 1992. *Climate Change 1992: The Supplementary Report to the IPCC Scientific Assessment*. Univ. Press, Cambridge, England.

Jenkinson, D.S., P.B.S. Hart, J.H. Rayner, and L.C. Parry. 1987. Modelling the turnover of organic matter in long-term experiments at Rothamsted. *Intecol Bull.* 1987-15:1-8.

Jenkinson, D.S., D.E. Adams, and A. Wild. 1991. Model estimates of CO_2 emissions from soil in response to global warming. *Nature* 351:304-306.

Larson, W.E., C.E. Clapp, W.H. Pierre, and Y.B. Morachan. 1972. Effects of increasing amounts of organic residues on continuous corn: II. Organic carbon, nitrogen, phosphorus and sulfur. *Agron. J.* 64:204-208.

Mosier, A., D. Schimel, D. Valentine, K. Bronson, and W. Parton. 1991. Methane and nitrous oxide fluxes in native, fertilized, and cultivated grasslands. *Nature* 350:330-332.

Parton, W.J., D.S. Schimel, C.V. Cole, and D.S. Ojima. 1987. Analysis of factors controlling soil organic matter levels in Great Plains grasslands. *Soil Sci. Soc. Am. J.* 51:1173-1179.

Paustian, K., W.J. Parton, and J. Persson. 1992. Modeling soil organic matter in organic-amended and N-fertilized long-term plots. *Soil Sci. Soc. Am. J.* 56:476-488.

Power, J.F., W.W. Wilhelm, and J.W. Doran. 1986. Recovery of fertilizer nitrogen by wheat as affected by fallow method. *Soil Sci. Soc. Am. J.* 50:1499-1503.

Powlson D.S. and D.S. Jenkinson. 1981. A comparison of the organic matter, biomass, adenosine triphosphate and mineralizable nitrogen contents of ploughed and direct-drilled soils. *J. Agric. Sci.(Camb.)* 97:713-721.

Rasmussen, P.E., R.R. Allmaras, C.R. Rohde, and N.C. Roager, Jr. 1980. Crop residue influences on soil carbon and nitrogen in a wheat-fallow system. *Soil Sci. Soc. Am. J.* 44:596-600.

Robertson, G.P. 1993. Fluxes of nitrous oxide and other nitrogen trace gases from intensively managed landscapes: A global perspective. In: *Agricultural Ecosystem Effects on Trace Gases and Global Climate Change*. ASA Spec. Publ. No. 55. pp. 95-108, ASA-CSA-SSSA, Madison, WI.

Robertson, G.P., W.A. Reed, and S.J. Halstead. 1993. Methane uptake in agricultural ecosystems of the U.S. Midwest. *Global Atmospheric Biospheric Chemistry Conference Abstracts*, p. 27, OHOLO Conference Series, Eilat, Israel.

Seebomn, M.E. 1976. *The evolution of the English farm.* E.P. Publishing Ltd, Ilkeley, West Yorkshire, England.

Sowden, F.J. and H.J. Atkinson. 1968. Effect of long-term annual additions of various organic amendments on the organic matter of a clay and a sand. *Can. J. Soil Sci.* 48:323-330.

Steen, E. 1989. Agricultural Outlook. In: O. Andrén, T. Lindberg, K. Paustian, and T. Rosswall (eds.), Ecology of Arable Land - Organisms, Carbon and Nitrogen Cycling. *Ecol. Bull. (Copenhagen)* 40:181-192.

Steudler, P.A., R.D. Bowden, J.M. Melillo, and J.D. Aber. 1989. Influence of nitrogen fertilization on methane uptake in temperate forest soils. *Nature* 341:314-316.

Tans, P.P., I.Y. Fung, and T. Takahashi. 1990. Observational constraints on the global atmospheric CO_2 budget. *Science* 247:1431-1438.

Tisdall, J.M. and J.M. Oades. 1982. Organic matter and water-stable aggregates in soil. *J. Soil Sci.* 33:141-163.

Wilson, A.T. 1978. Pioneer agriculture explosion and CO_2 levels in the atmosphere. *Nature* 273:40-41.

CHAPTER 7

Estimated Soil Organic Carbon Losses from Long-Term Crop-Fallow in the Northern Great Plains of the USA

L.J. Cihacek and M.G. Ulmer

I. Introduction

In the semiarid, cultivated lands of the U.S. Great Plains west of the 100° meridian, crop-fallow rotation has been practiced since the early 1900's to conserve water, reduce the risk of crop failure, and provide mineral nitrogen (N) for subsequent crops. Typically, in a crop-fallow system, wheat (*Triticum aestivium* L.) or other small grains are planted only in alternate years. During the intervening fallow period, vegetative growth is restricted by shallow cultivation and/or herbicide applications during the summer months. Cool temperatures and irregular rainfall generally deficient for optimum plant growth characterize this region and encourage an agriculture based on hard red spring wheat production (Norum et al., 1957).

In the crop-fallow cycle, carbon is sequestered in the crop residue during the period when the crop is growing on the soil. After the crop is harvested, tillage incorporates the residue back into the soil where the carbon is released by microbial activity. The carbon released from residue then enters either the soil organic matter pool or the atmospheric pool as CO_2 (Alexander, 1961).

This paper reports an estimate of carbon released to the atmosphere from a spring wheat-fallow cropping system in a Major Land Resource Area (MLRA) from the central part of the northern Great Plains and compares this estimate with an estimate of C released from cultivated soils during the period that these soils have been taken out of grassland production.

II. Crop Fallow and Carbon Loss

Declines in organic carbon (OC) of grassland soils after cultivation have been previously documented (See Table 1). Organic carbon loss is variable but generally follows a trend where an increase in the amount of OC loss is related to the length of time that soils have been cultivated.

Tillage and residue management can influence the OC content of soil. Examples of studies are reported in Table 2. Increasing the amount of residue remaining on the soil surface due to reduced tillage intensity or a lower frequency of fallowing soil decreases OC loss.

Many reports of OC decline due to cultivation have also examined continuous cropping systems. But, cropping systems other than an alternating crop fallow system based on spring wheat or small grains are subject to greater risk due to a lack of moisture at critical periods during the crop growth cycle. Crop-fallowing allows for additional soil moisture storage to minimize drought risk.

Crop-fallow produces residue (sequestered C) only once in a two-year cycle and the residue is subject to decomposition and C release to the atmosphere during the extended fallow period. This, then, represents a worst case scenario of C release (as CO_2) to the atmosphere and minimization of longer term C sequestration.

ISBN 1-56670-117-1/95/$0.00+.50

III. Study Area

The Northern Great Plains Spring Wheat Region (USDA, 1981) is made up of 9 MLRA's which have climates with average annual temperature ranges from 1-9 °C and average annual precipitation ranges from 250-550 mm (Figure 1). The frost free period varies from 100 to 155 days. In most years, precipitation is inadequate for maximum production potential.

Inadequate moisture is responsible for wide application of crop-fallow agriculture and the short frost free growing period is the impetus for the wide production spring wheat and spring small grains. Crop-fallow is the predominant cropping system. Crop fallow is practiced on 46 to 74% (average, 66%) of the land area seeded to spring wheat in the study area (N.D. Agricultural Statistics Service, 1992).

Major Land Resource Area 54 was selected for this discussion because of size, central location, and predominance of dry-land crop-fallow agriculture (see Figure 1). Soils in this MLRA are Borolls which are characterized as moderately deep to deep, well drained and moderately well drained, and loamy or clayey textures with high potential productivity. Most cropland soils in this MLRA were brought under cultivation by 1910.

IV. Methods

The surface horizon of seventeen paired cultivated and native rangeland sites were sampled for SOC throughout MLRA 54 in North Dakota. The paired sites were within 75m of one another and had similar slope and aspect parameters. Soils from all sites were classified as fine-loamy, mixed, Typic Haploborolls and were located on Pleistocene age terrace deposits. Surface textures included sandy loams, sandy clay loams, loams, and silt loams. Soils from the Amor, Bowdle, Shambo, and Stady series represent the major cultivated cropland soils of the region. Cropland sites have been under cultivation over 40 years.

Three 6-cm diameter cores were collected from an area representative of the cultivated or rangeland soils at each site. The cores were composited by horizon.

Organic carbon was determined by the method of Yeomans and Bremner (1988) and bulk density was determined on the cores as described by Blake and Hartge (1986). Carbon loss was calculated by calculating the difference contained in the surface 15 cm depth of grassland soils minus the C in a 15 cm depth of cultivated cropland soils as adjusted for bulk density.

Spring wheat straw yields were calculated from forty-two years of yield and harvested area statistics for MLRA 54 (N.D. Crop and Livestock Reporting Service, 1960, 1962, 1964, 1966, 1968, 1970, 1972, 1978, 1981, 1985; N.D. Agricultural Statistics Service, 1990, 1992). Straw yields were calculated at 1.58 kg straw for each 1.0 kg grain produced (Bauer and Zubriski, 1978; Bauer et al., 1987) with a carbon content of straw of 42% (Rasmussen et al., 1980).

V. Carbon Losses in a Spring Wheat-Fallow Cropping System

Soil carbon loss due to cultivation averaged 14344 kg ha^{-1} compared to the carbon in the native grassland soil on a volumetric basis (Table 3). Bulk density increased by 0.28 g cm^{-3} due to cultivation which reduced the apparent carbon loss from 44% on a soil weight basis to 28% on a volumetric basis. The lower volumetric loss indicates a shift in less stable OC in native soil to more stable OC in cultivated soil but is comparable with previously reported losses (Table 1). Using average land area of 358,095 ha under wheat-fallow cropping in MLRA 54, a total of 5,136,514 Mg C was lost due to cultivation. Regression analysis showed no significant relationships between SOC content and sand, silt, or clay content for either cropland or rangeland soils.

Annual wheat yield, area seeded, and estimated C loss from the residue are shown in Figure 2. Although yields were low during the first 15 years of the study period, they remained relatively constant during the remaining 27 years. The low yields during the early years were due to erratic weather conditions, lack of disease resistant varieties, and low fertilizer usage. Low yields in 1980 and 1988 were mainly due to drought conditions.

Average annual straw production on spring wheat-fallow in MLRA 54 was 795,655 Mg and its C content was 334,175 Mg. Over the 42-year period examined here, total straw production was 34,213,150 Mg and the C lost through residue decomposition to the atmosphere was estimated 14,369,523 Mg. Based on a 42

Table 1. Studies on soil organic carbon (SOC) losses in the Great Plains

Reference	Study area	Crops or rotations	Cultivation	Loss of SOC
			-years-	--%--
Alway (1909)	Nebraska	Various	20-40	10-51
Russel (1929)	Nebraska	Grain	3-7	7
			8-15	12
			17-30	27
			32-44	26
			45-60	28
Sewell and Gainey (1932)	Kansas	Wheat-fallow	13	13-32
Hide and Metzger (1939)	Kansas	Unknown	30+	37-44
Hill (1954)	Alberta	Wheat-fallow	33	11
Haas et al. (1957)	Kansas	Wheat-fallow	30	23
	North Dakota		30	28
	Montana		31	53
	Wyoming		30-34	36-41
Hobbs and Brown (1957)	Kansas	Wheat-fallow	30	25-33
Hobbs and Brown (1965)	Kansas	Wheat-fallow	40	31
Tiessen et al. (1982)	Saskatchewan	Various	65-90	19-51

Table 2. Effects of tillage and residue management on SOC in the Great Plains

Reference	Study area	Crops or rotation	Time -years-	Tillage[†] or residue management	Change in SOC
Unger (1968)	Texas	Wheat-fallow	24	One way	-35
				Stubble-mulch	-32
				Delayed stubble mulch	-17
Ridley and Hedlin (1968)	Manitoba	Wheat-fallow	37	F-W[‡]	
				F-W-W	
				F-W-W-W	
Black (1973)	Montana	Wheat-fallow	6[§]	Straw rates[¶]	
				0	
				1680	
				3360	
				6730	
Bauer and Black (1981)	North Dakota	Wheat-fallow	25	Conventional	-38
				Stubble mulch	-27
Tiessen et al. (1982)	Saskatchewan	Wheat-fallow	65-90	Conventional	-19 to -51

[†]One way tillage - conventional large disk plow tillage; conventional tillage - tillage where <30% residue remains after all tillage operations; stubble mulch - reduced tillage where >30% residue remains after all tillage operations; delayed stubble mulch - initiation of tillage delayed until later in fallowing cycle.

[‡]F - fallow; W - wheat; Frequency of fallow in cropping cycle is noted in this study. Original SOC values unknown. Diferences based on changes compared with F-W management.

[§]Length of study was 6 years; field in a standard wheat-fallow system for 50 years perior to that study.

[¶]Rates in Kg ha^{-1}. SOC changes based on SOC measurements at beginning and end of study.

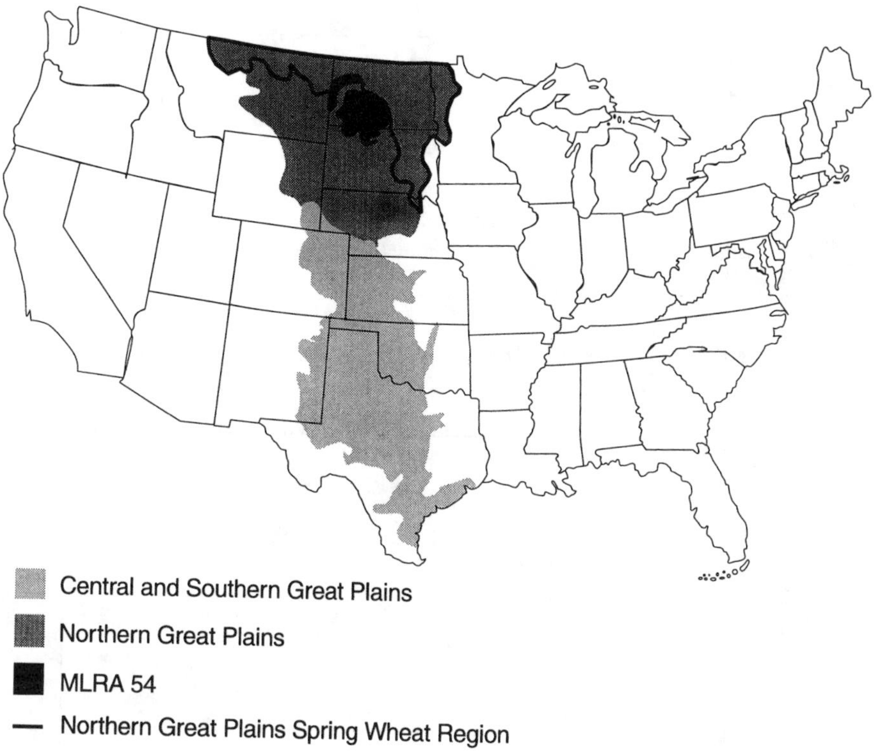

Figure 1. Locations of the Great Plains, the Northern Great Plains Spring Wheat Region and Major Land Resource Area (MLRA) 54.

Table 3. Changes in SOC at 17 sites in MLRA 54

Management status	SOC minimum value	SOC maximum value	SOC average value	Bulk density	Total SOC[a]
		%		g cm^{-3}	Kg ha^{-1}
Cultivated soil (C)	0.76	2.42	1.51 ± 0.41	1.23 ± 0.11	37146
Native soil (N)	2.05	4.52	2.71 ± 0.58	0.95 ± 0.16	51490
Change (N-C)	1.29	2.10	1.20	- 0.28	14344

[a] Adjusted for bulk density

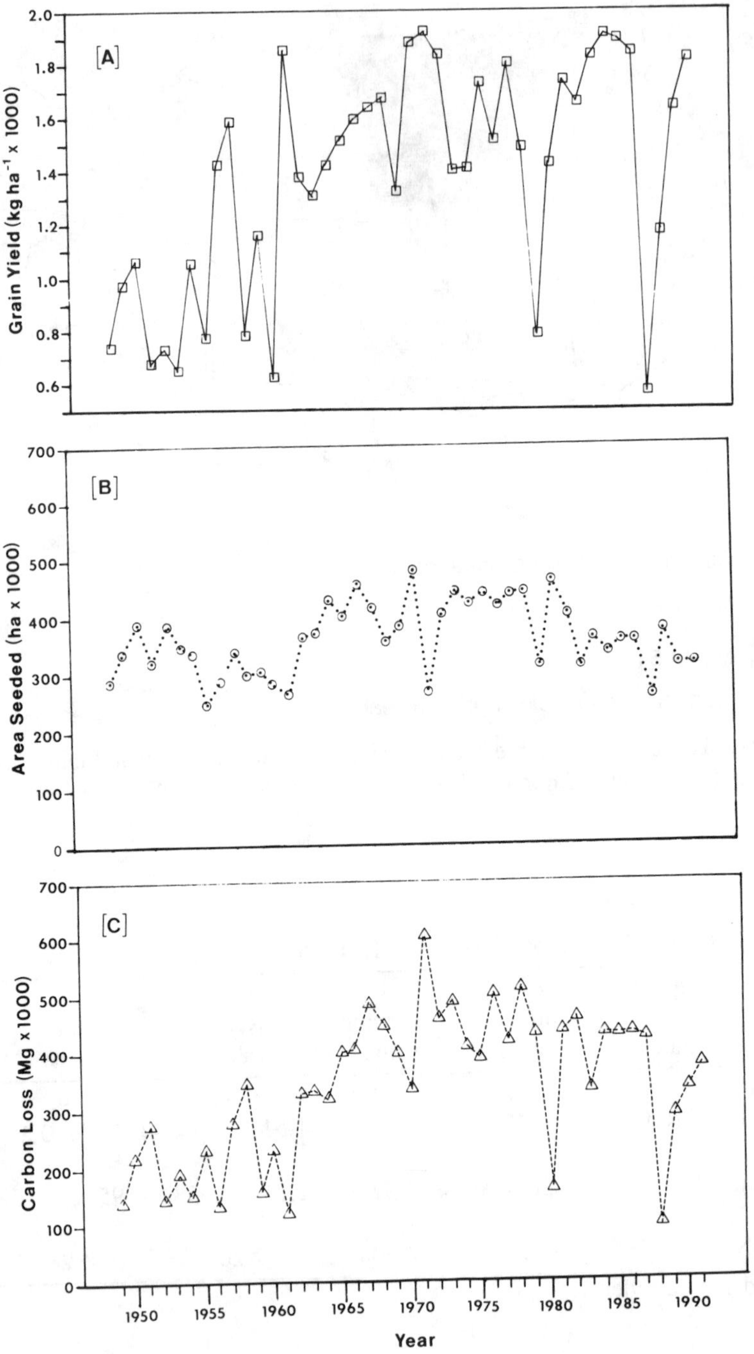

Figure 2. Crop yield (A), area seeded (B) and carbon loss (C) for MLRA 54, 1949-1991.

year average of 358,095 ha in crop-fallow wheat, annual C loss was 0.93 Mg C ha^{-1} and long term loss was 40.1 Mg C ha^{-1}. Relatively stable or lower C losses from 1980 on were influenced by a stabilization in the area seeded to crop-fallow wheat due to governmental price support and soil conservation programs and removal of land area from production by the Conservation Reserve Program (CRP).

VI. Discussion

Carbon loss from tillage incorporated crop residue was a greater contributor to atmospheric C losses than was due to cultivation when cultivation loss is prorated over 80+ years of cultivation. It should be noted that C loss from crop residue is only a portion of the total C sequestered by the crop and that much of the C lost will be sequestered by the crop in the following year. However, management practices can increase the amount of C sequestered by the soil as shown in Table 2. Soil organic C can be maintained by (i) reducing tillage intensity during the fallow period, thereby maintaining more C in undecomposed crop residue, or (ii) lengthening the interval between the fallow periods, thereby maintaining a greater frequency of crop on the soil and increasing total residue production.

Recent research (Peterson et al., 1992) indicates the latter practice may not only improve C retention in the soil and on the land but also provides greater sequestration due to higher total yields.

Although reduced tillage intensity has been practiced in spring wheat-fallow culture for a long time, recent legislation in the Food Security Acts of 1985 and 1990 has encouraged the reduction of tillage on lands subject to erosion. Less tillage will influence the maintenance of C in undecomposed residue and increase sequestered C in the soil. Maintaining C sequestered in crop residue and soil organic matter is the most practical way to reduce agricultural CO_2 emissions to the atmosphere.

Since Montana (1990), North Dakota (1987), and South Dakota (1987) had recent totals of 1.98, 2.56, and 0.74 million ha, respectively, in fallow (Montana Agricultural Statistics Service, 1991; USDA Soil Conservation Service, 1989) even small increases in C sequestration would significantly reduce CO_2 emissions from cropland in the northern Great Plains.

References

Alexander, M. 1961. *Introduction to soil microbiology*. John Wiley & Sons, Inc. New York.

Alway, F.J. 1909. Changes in the composition of the loess soils of Nebraska caused by cultivation. Nebr. Agric. Exp. Sta. Bull. 111.

Bauer, A. and A.L. Black. 1981. Soil carbon, nitrogen, and bulk density comparisons in two cropland tillage systems after 25 years and in virgin grassland. *Soil Sci. Soc. Am. J.* 45:1166-1170.

Bauer, A. and J.C. Zubriski. 1978. Hard red spring wheat straw yields in relation to grain yields. *Soil Sci. Soc. Am. J.* 42:777-781.

Bauer, A., A.B. Frank, and A.L. Black. 1987. Aerial parts of hard red spring wheat. I. Dry matter distribution by plant development stage. *Agron. J.* 79:845-852.

Black, A.L. 1973. Soil property changes associated with crop residue management in a wheat-fallow rotation. *Soil Sci. Soc. Am. Proc.* 37:943-946.

Blake, G.R and K.H. Hartge. 1986. Bulk density. In: A. Klute (ed.) Methods of soil analyses, Part 1. 2nd ed. Monogr. of ASA and SSSA, Madison, WI.

Haas, H.J., C.E. Evans, and E.F. Miles. 1957. Nitrogen and carbon changes in Great Plains soils as influenced by cropping and soil treatments. USDA Tech. Bull. 1164. U.S. Government Printing Office. Washington, D.C.

Hide, J.C. and W.H. Metzger. 1939. The effect of cultivation and erosion on the nitrogen and carbon of some Kansas soils. *Agron. J.* 31:625-632.

Hill, K.W. 1954. Wheat yields and soil fertility on the Canadian prairies after a half century of farming. *Soil Sci. Soc. Am. Proc.* 18:182-184.

Hobbs, J.A. and P.L. Brown. 1957. Nitrogen and organic carbon changes in cultivated western Kansas soils. Kansas Agric. Exp. Sta. Tech. Bul. 89.

Hobbs, J.A. and P.A. Brown. 1965. Effects of cropping and management on nitrogen and organic carbon contents of a western Kansas soil. Kansas Agric. Exp. Sta. Tech. Bull. 144.

Montana Agricultural Statistics Service. 1991. Montana Agricultural Statistics - 1991. Vol. 28.

North Dakota Crop and Livestock Reporting Service. 1960. North Dakota wheat. Agric. Statistics No. 3.

North Dakota Crop and Livestock Reporting Service. 1962. North Dakota crop and livestock statistics: Annual summary for 1961 with revisions for 1956-60. Agric. Statistics No. 8.

North Dakota Crop and Livestock Reporting Service. 1964. North Dakota crop and livestock statistics. 1963. Agric. Statistics No. 12.

North Dakota Crop and Livestock Reporting Service. 1966. North Dakota crop and livestock statistics. 1965. Agric. Statistics No. 15.

North Dakota Crop and Livestock Reporting Service. 1968. North Dakota crop and livestock statistics. Agric. Statistics No. 18.

North Dakota Crop and Livestock Reporting Service. 1970. North Dakota crop and livestock statistics. Agric. Statistics No. 21.

North Dakota Crop and Livestock Reporting Service. 1972. North Dakota crop and livestock statistics. Agric. Statistics No. 26.

North Dakota Crop and Livestock Reporting Service. 1978. North Dakota crop and livestock statistics: Historic estimates 1971-1975. Agric. Statistics No. 43.

North Dakota Crop and Livestock Reporting Service. 1981. North Dakota agricultural statistics: 1976-1979 final and 1978 Agric. census data. Agric. Statistics No. 49.

North Dakota Crop and Livestock Reporting Service. 1985. North Dakota agricultural statistics 1985. Agric. Statistics No. 54.

North Dakota Agricultural Statistics Service. 1990. North Dakota Agricultural Statistics 1990. Agric. Statistics No. 59.

North Dakota Agricultural Statistics Service. 1992. North Dakota Agricultural Statistics 1992. Agric. Statistics No. 61.

Norum, E.B., B.A. Kranz, and H.J. Haas. 1957. The northern Great Plains. p. 494-505. In: A. Stefferud (ed.). *Soil.* 1957 Yearbook of Agriculture. USDA, Washington, D.C.

Peterson, G.A., D.G. Westfall, L. Sherrod, E. McGee, and R. Kolberg. 1992. Crop and soil management in dryland agroecosystems. Colorado State Univ. Tech. Bull. TB92-2.

Rasmussen, P.E., R.R. Allmaras, C.R. Rohde, and N.C. Roager, Jr. 1980. Crop residue influenced on soil carbon and nitrogen in a wheat fallow system. *Soil Sci. Soc. Am. J.* 44:596-600.

Ridley, A.O. and R.A. Hedlin. 1968. Soil organic matter and crop yields as influenced by the frequency of summerfallowing. *Can. J. Soil Sci.* 48:315-322.

Russel, J.C. 1929. Organic matter problems under dry farming conditions. *Agron. J.* 21:960-969.

Sewell, M.C. and P.L. Gainey. 1932. Organic matter changes in dry-farming regions. *Agron. J.* 24:275-283.

Tiessen, J., J.W.B. Stewart, and J.R. Bettany. 1982. Cultivation effects on the amounts and concentration of carbon, nitrogen and phosphorus in grassland soils. *Agron. J.* 74:831-835.

Unger, P.W. 1968. Soil organic matter and nitrogen changes during 24 years of dryland wheat tillage and cropping practices. *Soil Sci. Soc. Am. J.* 32:426-429.

USDA Soil Conservation Service. 1981. Land resource regions and major land resource areas of the United States. USDA Agric. Hndbk 296. Washington, D.C.

USDA Soil Conservation Service. 1989. Summary report 1987 National Resources Inventory. USDA-SCS Stat. Bul. 790.

Yeomans, J.C. and J.M. Bremner. 1988. A rapid and precise method for routine determination of organic carbon in soil. *Commun. Soil Sci. Plant Anal.* 19:1467-1476.

CHAPTER 8

Fertilizer N, Crop Residue, and Tillage Alter Soil C and N Content in a Decade

M. Nyborg, E.D. Solberg, S.S. Malhi, and R.C. Izaurralde

I. Introduction

The maintenance of organic matter in soils has long been recognized as a strategy to prevent soil degradation. In addition, there is increased recognition today that soils, as open systems, can either regulate or contribute to atmospheric gas pools. A current hypothesis is that soils can function as net sinks of atmospheric C and therefore attenuate the increase in atmospheric CO_2. On a global scale, however, there is considerable variety in production and soils systems and, therefore, not all of them would serve that purpose. Consequently, there is a need to identify, locally, agronomic systems able to attenuate the increase of atmospheric CO_2 by increasing the mass of C stored in soil.

In Alberta, the cultivation of land over the last 100 years has decreased, in most cases, both concentration and mass of soil organic C (McGill et al., 1988). In a recent survey of the status of soil organic matter conducted in 72 Alberta farms, across major soil zones, Reinl (1984) found that all Ap horizons had less concentration and mass of organic C than their native Ah counterparts. On average, greater carbon losses from A horizons occurred when the results were expressed as concentration (48%) than when calculated as mass (24%).

Inclusion of an alfalfa (*Medicago sativa* L.)/brome (*Bromus inermis* Leyss.) mixture in a 5-year rotation, practiced for 50 years on a Cryoboralf, resulted in a greater concentration of soil organic C than that in a wheat (*Triticum aestivum* L.) - fallow rotation (McGill et al., 1986). Grain production, however, predominates across major soil zones of Alberta. In these systems, fertilizer N is the preferred method used by farmers for achieving yield goals and maintaining soil fertility. In regions where water is limiting, producers often practice summerfallow to accumulate water and mineralize nutrients. Janzen (1987), however, found that a non-fertilized wheat monoculture succeeded more at maintaining soil organic C than cereal rotations that included summerfallow. Current conservation efforts in Alberta favor the use of reduced tillage methods for the prevention of erosion and organic matter decline, especially on land used predominantly for grain production.

In 1979 we initiated an experiment at two sites in north central Alberta with original objectives of studying the interactions among tillage method, straw-disposal method, and fertilizer N on barley (*Hordeum vulgare* L.) productivity and on soil quality. We examined how, after more than a decade of continuous treatment; methods of tillage, straw-disposal, and fertilizer N influenced soil organic C.

II. Materials and Methods

The experimental sites selected for this study represented major and distinctly different soil types commonly found in north central Alberta. A Typic Cryoboralf of the Breton loam series dominates the rolling landscape at site 1 near the town of Breton, Alberta (53° 07' N, 114° 28' W). Its concentration of organic

ISBN 1-56670-117-1/95/$0.00+.50

C in the 0-15 cm depth ranged from 18 to 24 g kg^{-1}. A Typic Cryoboroll of the Malmo loam series is common to the flat lacustrine landscape at site 2 near the hamlet of Ellerslie, Alberta (53° 25' N, 113° 33' W). The concentration range of organic C (58-70 g kg^{-1}) of the Malmo loam was three times greater than that of the Breton loam.

The experiment, established in the fall of 1979 as a randomized complete block with four replicates, served to examine the interactive effect of tillage (zero and conventional), straw (removed from or left on the plot after harvest), and N fertility (0 and 56 kg N ha^{-1} yr^{-1}) on barley yield and plant-N uptake. The sizes of the experimental units were 2.8 m x 6.9 m. Plots under conventional tillage (CT) received two passes, one in the fall and one in the spring, with a small sweep-chisel cultivator followed by a coil packer. Zero tillage (ZT) plots underwent no other disturbance but that of the drill itself. The N-fertilizer source used was urea. The method of placement was broadcasting during the first half of the experiment but was banding during the second half. From 1980 to 1990, we used a hoe-opener drill with a 0.23-m spacing between rows to sow barley at a rate of 100 kg ha^{-1}. The procedure followed at maturity consisted of collecting a biomass sample from a 2.3 m^2 area and determining, after proper drying at 65 °C, the values of grain, straw, and total biomass. We then removed the rest of the plant biomass from each plot and individually threshed and returned straw and chaff to the plots requiring it.

We took soil samples from all plots after the termination of the 11th growing season (October 1990). The sampling scheme consisted of taking ten soil cores from each treatment in each replicate in 5-cm increments from 0 to 15 cm soil depth. We proceeded to bulk the soil cores from each plot into a composite sample. These then underwent the application of standard procedures for air-drying and sieving through a 2-mm mesh. Bulk densities were determined by the soil core method. We utilized a LECO Carbon Determinator (Model CR12; St. Joseph, MI) to measure concentration of total C in all soil samples. None of the samples analyzed contained significant amounts of free carbonates. For calculation purposes, we assumed the concentration of C in grain and straw samples of barley to be 44%. We determined total soil N with a colorimetric analysis (Technicon Industrial Systems, 1977) of a H_2SO_4-H_2O_2 digested sample.

III. Results and Discussion

The concentrations of soil C in a 15-cm thick layer at the onset of the experiment at Breton and Ellerslie were 13.75 and 56.45 g kg^{-1}, respectively. The corresponding concentrations of soil N were 1.2 and 4.9 g kg^{-1}.

At both sites, tillage, crop residue, and fertilizer-N induced significant changes both in concentration and mass of C and N in the surface 0.15 m after 11 years of continuous treatment (Table 1). The Typic Cryoboralf at Breton, however, exhibited greater changes of C and N than the Typic Cryoboroll at Ellerslie. Returning the straw produced to the land after harvest tended to result in a greater concentration and mass of C and N in the soil than that found when straw was not returned. Fertilizer N had a similar effect. The comparison between zero tillage and conventional tillage, however, was less clear. Overall, ZT and CT had similar concentration of soil organic C. Surface soil under ZT, however, had greater mass of organic C than soil under CT. Results of soil N were similar to those of organic C. The experimental variables studied did not alter C/N ratios within this soil depth.

Zero tillage, however, may affect the distribution of C and N inputs to soil within small depth increments. This physical process may, in turn, affect turnover rates and pool distribution of these systems. We therefore proceeded to analyze the concentration and mass of these nutrients in 0.05-m depth increments from 0 to 0.15 m. At Breton (Table 2), contrast analyses of the first soil layer dataset indicated that ZT, straw left on the surface, and use of fertilizer N all individually increased concentration and mass of soil C and of soil N (probabilities reported herein for all contrasts are less than 0.05). Under ZT, both concentration and mass of soil C and soil N were greater when the straw was left than when it was removed from the plots after harvest. The same comparisons, however, did not hold true under conventional tillage (CT). In the second soil layer, the use of fertilizer N provided the most significant contribution towards increasing concentration and mass of soil C and N. Under ZT, leaving crop residues on the soil increased concentration and mass of soil N only. The same fertilizer effect observed in the second soil layer extended down to the third soil layer. Leaving the straw on the soil surface increased concentration and mass of soil C on ZT plots but not on CT plots. Ratios of C/N decreased with the use of fertilizer N or increased by leaving straw on the soil surface under ZT. We observed this effect in the first soil layer only.

Table 1. Total soil C and total soil N in the surface 15 cm found after 11 years of continuous treatment at Breton and Ellerslie

Treatment				Carbon		Nitrogen		
No.	Tillage	Straw	N	Conc.	Mass	Conc.	Mass	C/N
			kg ha^{-1}	g kg^{-1}	Mg ha^{-1}	g kg^{-1}	Mg ha^{-1}	
				Breton				
1	No	No	0	13.51	30.52	1.19	2.70	11.3
4	No	No	0	14.64	32.07	1.24	2.77	11.8
3	No	No	56	18.15	42.03	1.61	3.74	11.3
4	Yes	Yes	0	13.61	28.71	1.17	2.51	11.6
8	Yes	Yes	56	16.26	34.53	1.40	3.01	11.6
2	Yes	Yes	0	14.22	30.26	1.19	2.58	12.0
10	Yes	Yes	56	15.87	34.41	1.41	3.08	11.3
LSD $_{0.05}$				2.07	5.05	0.15	0.37	0.7
				Ellerslie				
1	No	No	0	53.96	84.41	4.71	7.38	11.4
4	Yes	Yes	0	57.31	89.70	4.99	7.83	11.5
3	Yes	Yes	56	56.37	88.25	5.02	7.86	11.2
5	Yes	Yes	0	56.06	87.82	4.82	7.56	11.6
8	Yes	Yes	56	57.56	90.00	5.10	7.95	11.3
2	No	No	0	53.36	83.59	4.74	7.43	11.3
10	No	No	56	53.26	83.35	4.72	7.38	11.3
LSD $_{0.05}$				3.39	5.50	0.22	0.36	0.3

The first soil layer of the Typic Cryoboroll at Ellerslie (Table 3) showed a greater mass of soil C under ZT than under CT. The removal of straw from the plots decreased both concentration and mass of soil C. Soil N, however, was the variable that responded more dramatically to the treatment effects. Overall in the first layer, ZT plots had more soil N than CT plots. Straw and fertilizer N contributed both towards increasing soil N. Results of the second soil layer showed similar trends. In the third soil layer, however, leaving crop residues consistently increased concentration and mass of soil C and soil N. Our results also showed a decrease in C/N ratios with inputs of fertilizer N.

At Breton, our results indicate that the poorest treatments (Nos. 1 and 2, Table 1) did not reduce soil-C concentration in the first 0.00-0.15 m layer but rather, the best treatments (Nos. 3 and 8, Table 1) increased it by approximately 30%. At Ellerslie, we observed a slight decrease of 5% in the average concentration of soil C of treatments 1 and 2 when compared to the original concentration measured. It appears then that more potential exists to store atmospheric C in soils low in C such as the Alfisol studied at Breton than in the Mollisol rich in C examined at Ellerslie.

We then calculated the change in C mass of the surface 0.15 m of all treatments with respect to treatment No. 2. In addition, we calculated the total mass of straw C that each of the plots had received, according to the planned treatments, from 1980 to 1990. We found a significant relationship between these two variables (Figure 1). In this analysis we did not include the value of treatment 3 at Breton since it would have increased the slope of the regression line considerably. Therefore the adjusted regression line of Figure 1 is a conservative estimate of the increase in soil C content with addition of straw C. In Saskatchewan, however, Campbell et al. (1991b) found no relationship between the organic matter of a Mollisol (at Melfort, Saskatchewan) and either residue C or N added after a 31-year period. Similar to our results of the Typic Cryoboralf, these authors reported a significant relationship between these variables on another

Table 2. Total soil Ca and total soil N by 5-cm depth increments found after eleven years of continuous treatment at Breton

--------------Treatment--------------				-----Carbon-----		----------Nitrogen---------		
No.	Tillage	Straw	N kg ha^{-1}	Conc. g kg^{-1}	Mass Mg ha^{-1}	Conc. g kg^{-1}	Mass Mg ha^{-1}	C/N
			0.00-0.05 m depth					
1	No	No	0	15.82	11.35	1.42	1.02	11.1
4	No	Yes	0	17.87	12.82	1.50	1.08	11.9
3	No	Yes	56	21.42	15.37	1.87	1.34	11.4
5	Yes	Yes	0	15.81	10.21	1.32	0.86	11.9
8	Yes	Yes	56	17.90	11.56	1.58	1.02	11.4
2	Yes	No	0	15.48	10.00	1.28	0.82	12.2
10	Yes	No	56	16.94	10.94	1.50	0.97	11.3
LSD $_{0.05}$				2.04	1.40	0.16	0.11	0.5
			0.05-0.10 m depth					
1	No	No	0	14.24	11.59	1.20	0.97	11.8
4	No	Yes	0	14.54	11.84	1.25	1.01	11.6
3	No	Yes	56	17.42	14.17	1.54	1.25	11.3
5	Yes	Yes	0	14.45	10.82	1.25	0.94	11.6
8	Yes	Yes	56	18.00	13.48	1.48	1.10	12.2
2	Yes	No	0	15.90	11.90	1.30	0.97	12.3
10	Yes	No	56	17.06	12.77	1.48	1.10	11.6
LSD $_{0.05}$				2.57	2.01	0.17	0.15	1.2
			0.10-0.15 m depth					
1	No	No	0	8.92	7.58	0.82	0.70	10.8
4	No	Yes	0	8.72	7.41	0.80	0.68	10.8
3	No	Yes	56	14.69	12.48	1.35	1.15	10.8
5	Yes	Yes	0	9.32	7.68	0.88	0.72	10.6
8	Yes	Yes	56	11.52	9.49	1.08	1.88	10.6
2	Yes	No	0	10.15	8.35	0.95	0.78	10.6
10	Yes	No	56	13.00	10.70	1.22	1.01	10.6
LSD $_{0.05}$				3.08	2.56	0.23	0.19	1.2

Table 3. Total soil C and total soil N by 5-cm depth increments found after eleven years of continuous treatment at Ellerslie

--------------Treatment--------------				-----Carbon-----		----------Nitrogen---------		
No.	Tillage	Straw	N	Conc.	Mass	Conc.	Mass	C/N
			kg ha^{-1}	g kg^{-1}	Mg ha^{-1}	g kg^{-1}	Mg ha^{-1}	
			0.00-0.05 m depth					
1	No	No	0	57.09	27.72	4.95	2.40	11.5
4	No	Yes	0	61.49	29.85	5.28	2.56	11.7
3	No	Yes	56	59.12	28.70	5.42	2.63	10.9
5	Yes	Yes	0	58.46	27.77	4.92	2.34	11.9
8	Yes	Yes	56	61.40	29.17	5.40	2.56	11.4
2	Yes	No	0	55.40	26.32	4.82	2.29	11.5
10	Yes	No	56	55.72	26.47	4.82	2.29	11.6
$LSD_{0.05}$				4.43	2.14	0.20	0.10	0.6
			0.05-0.10 m depth					
1	No	No	0	54.64	29.71	4.70	2.56	11.6
4	No	Yes	0	56.98	30.98	4.85	2.64	11.8
3	No	Yes	56	56.62	30.78	4.85	2.64	11.7
5	Yes	Yes	0	57.29	30.02	4.88	2.55	11.8
8	Yes	Yes	56	58.78	30.80	5.35	2.80	11.0
2	Yes	No	0	54.90	28.77	4.82	2.53	11.4
10	Yes	No	56	55.04	28.85	4.98	2.61	11.1
$LSD_{0.05}$				2.46	1.32	0.24	0.13	0.4
			0.10-0.15 m depth					
1	No	No	0	49.78	26.98	4.48	2.43	11.1
4	No	Yes	0	53.25	28.87	4.85	2.63	11.0
3	No	Yes	56	53.08	28.77	4.78	2.59	11.1
5	Yes	Yes	0	52.58	30.03	4.68	2.67	11.2
8	Yes	Yes	56	52.55	30.02	4.52	2.58	11.6
2	Yes	No	0	49.90	28.51	4.58	2.61	10.9
10	Yes	No	56	49.08	28.03	4.35	2.48	11.3
$LSD_{0.05}$				4.83	2.64	0.37	0.20	0.3

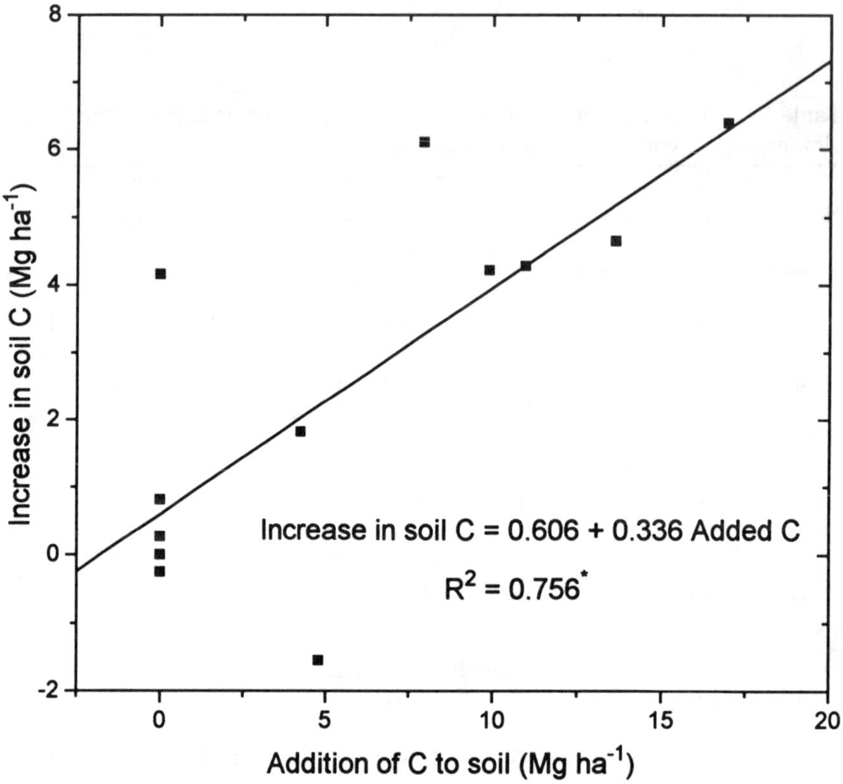

Figure 1. The increase of soil organic C of a Typic Cryoboralf and a Typic Cryoboroll of north central Alberta as a function of straw C additions over an 11-year period.

Mollisol at Indian Head, Saskatchewan, with considerably less organic matter than the soil at Melfort. They attributed the lack of response of organic matter to C and N additions of the soil at Melfort to its inherent richness.

We did not measure root mass in this experiment. However, in another experiment placed near our two sites, Izaurralde et al. (1993) measured root mass of barley from 0 to 0.4 m depth. They found similar production of root mass at both sites and calculated a root/shoot ratio of 0.12 at harvest. Using this ratio and also assuming a harvest index of 0.5 for barley, an estimate of the static input of root C into soil would be 0.24 Mg for each Mg of straw C. Dividing the slope of the regression line by 1.24 then gives the rate of soil C increase per unit of root C plus straw C added over an 11-year period, namely 0.271. Our data suggest then that soils such as those present in north-central Alberta, under improved management, can sequester every 11 years from 0.27 to 0.34 Mg ha^{-1} of atmospheric CO_2-C for every Mg ha^{-1} of C added. These values are approximately two times greater than that estimated by Campbell et al. (1991a) from a 30-year old experiment at Indian Head. It is possible that as our experiment becomes older the value of the slope line will decrease and eventually approximate those reported by Campbell et al. (1991a).

IV. Summary

Soils as open systems can either regulate or contribute to atmospheric gas pools. We conducted two experiments on a Typic Cryoboralf and on a Typic Cryoboroll in north central Alberta for 11 years to study the interactions among tillage (zero and conventional), straw disposal (removed or left on the plot after harvest), and fertilizer N (0 and 56 kg N ha^{-1} yr^{-1}) on barley productivity and soil quality. We examined how, after more than a decade, these treatments influenced soil organic C and total soil N in the top 15 cm. At both sites, tillage, crop residue, and fertilizer-N induced significant changes both in concentration and mass of C and N in the surface 0.15 m after 11 years of continuous treatment. The concentration of organic C in non-fertilized soil under conventional tillage without return of straw changed little the original concentration measured 11 years earlier. The concentration of organic C increased over that of the original soil only on the low-organic matter Typic Cryoboralf at Breton. The Typic Cryoboralf, however, exhibited greater changes of C and N than did the Typic Cryoboroll. Returning the straw produced to the land after harvest resulted in greater concentration and mass of C and N in soil than that found not returning it. Fertilizer N had a similar effect. Overall, zero-tillage and conventional tillage produced soils with similar concentrations of soil organic C. Surface soil under zero tillage, however, had greater mass of organic C than soil under conventional tillage. The results of soil N were similar to those of organic C. The increase in mass of soil C observed was a linear function of the total residue added over 11 years ($y = 0.606 + 0.336\,x$, $r = 0.756^{**}$). Under the environmental conditions of north-central Alberta, soils under minimal disturbance receiving annual inputs of C and N can increase soil organic matter and function, therefore, as net sinks of atmospheric CO_2.

V. Conclusions

We conclude that (i) concentration of organic C in non-fertilized soil under conventional tillage without return of straw changed little the original concentration measured 11 years earlier, (ii) concentration of organic C increased over that of the original soil only on the low-organic matter Typic Cryoboralf at Breton, (iii) the elimination of tillage, the return of straw to soil, and use of fertilizer favored accumulation of soil organic C and total N, and (iv) the increase in soil organic C observed after 11 years was a linear function of the total residue C added to soil during that period.

Acknowledgements

We thank M. Molina-Ayala, Z. Zhang, and J. Thurston for their invaluable technical assistance. We also thank the following agencies and programs for financial support: Farming For the Future, Alberta Agricultural Research Institute, and Canada-Alberta Soil Conservation Initiative.

References

Campbell, C.A., V.O Biederbeck, R.P. Zentner, and G.P. Lafond. 1991a. Effect of crop rotations and cultural practices on soil organic matter, microbial biomass and respiration in a thin Black Chernozem. *Can. J. Soil Sci.* 71:363-376.

Campbell, C.A., K.E. Bowren, M. Schnitzer, R.P. Zentner, and L. Townley-Smith. 1991b. Effect of crop rotations and fertilization on soil organic matter and some biological properties of a thick Black Chernozem. *Can. J. Soil Sci.* 71:377-387.

Izaurralde, R.C., N.G. Juma, W.B. McGill, D.S. Chanasyk, S. Pawluk, and M.J. Dudas. 1993. Performance of alternative cropping systems in cryoboreal-subhumid central Alberta. *J. Agric. Sci.* 120: 33-41.

Janzen, H.H. 1987. Soil organic matter characteristics after long-term cropping to various spring wheat rotations. *Can. J. Soil Sci.* 67:845-856.

McGill, W.B., K.R. Cannon, J.A. Robertson, and F.D. Cook. 1986. Dynamics of soil microbial biomass and water soluble C in Breton L after 50 years of cropping to two rotations. *Can. J. Soil Sci.* 66:1-19.

McGill, W.B., J.F. Dormaar, and E. Reinl-Dwyer. 1988. New prespectives on soil organic matter quality, quantity, and dynamics on the Canadian Prairies. Canadian Society of Soil Science and Canadian Society of Extension Joint Symposium Land Degradation: Assessment and Insight into a Western Canadian Problem. August 23, 1988. Agriculture Institute of Canada. Calgary, Alberta.

Technicon Industrial Systems. 1977. Individual/simultaneous determinations of nitrogen and/or phosphorus in Bd acid digests. Industrial method 334-74 w/B+. Technicon Industrial Systems, Tarrytown, New York.

Reinl, E. 1984. Changes in soil organic carbon due to agricultural land use in Alberta. M.Sc. thesis. Department of Soil Science, University of Alberta. Edmonton, Alberta.

CHAPTER 9

Modeling Impact of Agricultural Practices on Soil C and N_2O Emissions

Changsheng Li

I. Introduction

Soil organic carbon (SOC) generally decreases with cultivation, and the carbon lost from soil transfers into atmospheric carbon dioxide (CO_2), a greenhouse gas. Meanwhile agricultural activities also enhance nitrous oxide (N_2O), another greenhouse gas, emissions from soils (Eichner, 1990; Mosier et al., 1991; Aulakh et al., 1984). There is potential for restoring organic carbon levels in and reducing N_2O emissions from soils through changing agricultural practices including tillage, fertilization, manure amendment, crop rotation, and others. Since any change in agricultural practice could simultaneously alter the SOC storage in and N_2O flux from the soils, the net benefit produced changing agricultural practice should be considered. There is a substantial literature on the impacts of tillage, fertilization, manure application, and crop rotation on soil organic matter or N_2O emissions. Based on reviewing the literature, a process-oriented simulation model, DNDC, was recently developed for predicting effects of agricultural practices on carbon (C) and nitrogen (N) dynamics and N_2O evolution in soils (Li et al., 1992a,b). For reviews of the extensive literature on the effects of agricultural practices on soil organic matter or N_2O emissions, the reader can consult Sahrawat and Keeney (1986), Eichner (1990), Dalal and Mayer (1986a,b,c), Doran and Power (1983), Jenkinson (1990), Li et al. (1992a,b), and Li et al. (1993). DNDC has been calibrated and validated with quite a number of field studies including CO_2 emissions, residue decomposition, long-term soil C storage dynamics, and daily N_2O emissions (Li et al., 1992b, 1993, and 1994). This study was to use the model to estimate the net benefits of changing agricultural practices regarding mitigating the greenhouse effect.

II. Materials and Methods

A. The DNDC Model

Soil C and N dynamics are mainly controlled by little inputs, root exudates, root uptake, decomposition, ammonification, ammonia volatilization, nitrification, and denitrification rates in soils. Organic C and N are converted into inorganic forms (e.g. CO_2, NH_4^+, or NO_3^-) during decomposition and other relevant processes. Nitrate is reduced into N_2O during denitrification. The denitrification and decomposition (DNDC) model contains three interacting submodels. The thermal-hydraulic submodel uses soil texture, air temperature, and precipitation data to calculate soil temperature and moisture profiles and soil water fluxes over time. This information is fed into either a denitrification or a decomposition submodel. The denitrification submodel calculates hourly denitrification processes and N_2 (dinitrogen) and N_2O production during wet periods (water-filled porosity exceeds 40 percent of total). The decomposition submodel calculates daily decomposition, nitrification, and ammonium volatilization processes, and production. Effects

ISBN 1-56670-117-1/95/$0.00+.50

Table 1. Site characteristics

Location	Crop	Annual prec. (cm)	Annual temp. (°C)	Soil texture	SOC (kg C/kg soil)
Hamilton, IA	Soybean (*Glycine max*)	82.5	9.1	Udolls (clay loam)	0.036
Montgomery, IL	Soybean (*Glycine max*)	96.6	11.3	Udalfs (silt loam)	0.013
Butler, KS	Winter wheat (*Triticum vulgare*)	69.5	12.7	Ustolls (silty clay)	0.019
Dundy, NE	Soybean (*Glycine max*)	57.9	9.6	Ustolls (sandy loam)	0.006
Merced, CA	*Cotton (Gossypium barbadense)*	55.9	15.1	Xeralfs (loam)	0.018
Glades, FL	*Barley (Hordeum vulgare)*	137.6	21.5	Histo-sols	0.10

(Adapted from USDA, 1991; USDA, 1989; and USDA/SCS, 1962-1976.)

of anthropogenic activities (fertilization, tillage, application of manure, and other agricultural practices) are incorporated into the model. Climate scenario (temperature and precipitation), soil properties (SOC, pH, and density), and cropping practices (tillage, fertilization, manure application, crop rotation, and irrigation) are needed as input parameters to run DNDC. DNDC output includes CO_2, N_2O, and NH_3 emissions, CH_4 uptake by soil, SOC and N contents in soil and other pools, NH_4^+ and NO_3^- concentrations in soil profile, and grain and residue yields. The time-step in the DNDC model is daily and hourly in decomposition and denitrification submodels, respectively.

B. Site Selection

For estimating effects of changing agricultural practices on SOC storage and N_2O emission under different climatic conditions, six sites were selected from various climatic zones in the United States. Sites were located in Hamilton County in Iowa (IA), Montgomery County in Illinois (IL), Butler County in Kansas (KS), Dundy County in Nebraska (NE), Merced County in California (CA), and Glades County in Florida (FL). Data of land use and soil properties were obtained from USDA (1989b, 1991) and USDA/SCS (1962-1976). Major site characteristics (i.e. location, annual average temperature and precipitation, soil properties, and crop) are listed in Table 1.

C. Climate Scenarios

DNDC model is driven by climate scenarios. Annual climate scenarios used in the study were adopted from the 39-year (1950-88) average monthly temperature and precipitation data published by the United States Department of Agriculture (1989a). Errors could be introduced by using monthly averages, especially for daily N_2O fluxes. But the simulated annual N_2O emission rates have been shown consistent with the field data at 23 sites worldwide (Li, 1993).

D. Agricultural Practice Scenarios

For estimating effects of changing cropping practices on SOC storage and N_2O emissions at the test sites, a baseline scenario and 8 projected scenarios were designed to represent conventional practices and future options, respectively.

Under the baseline scenario (BASE), conventional practices included: (1) plowing with moldboard and chisel at crop seeding and harvest time, respectively, (2) application of N-fertilizer at a high rate (200 kg N of ammonium per ha), (3) no manure amendment, and (4) no planting of winter-cover crop.

Two reduced tillage scenarios were designed. Under the scenario of conservation tillage (CT) the field was disked once before seeding. Under the scenario of no-till (NT) the field was mulched after harvest. Two reduced fertilization scenarios were designed to reduce the ammonium application rate to 100 (F1) and 50 (F2) kg N/ha. Two manure amendment scenarios were designed to add manure of 1,000 (M1) and 2,000 (M2) kg C/ha into the field. A winter-cover scenario (WC) was designed to plant clover after harvest of the major crops. Under each of the 7 single-practice change scenarios, only one practice was changed keeping others same as in the baseline scenario.

A comprehensive scenario was designed to allow several practices to be changed at one site. This comprehensive scenario (CM) included no-till, low rate (50 kg N/ha) of fertilization, manure application (2,000 kg C/ha), and sowed winter-cover clover.

Nominal scenarios for fertilizer practices were developed from statistical data by TVA (1990) and USDA (1989b). Telephone interviews and written materials provided by local agricultural extension agents provided information of tillage and manure application.

E. Simulation

A 5-year run was conducted with the DNDC model for each scenario at each site. While operating, model reads climate data, soil properties, and cropping practices from the pre-set input files. The output data are recorded in the electronic files or printed out as hard copies.

F. Net Benefit Calculations

Carbon dioxide, N_2O, and CH_4 have different warming effect because of their different impacts on radiative forcing and residence time in the atmosphere. The change in SOC is regarded as net emission of CO_2 from soil in the study. Since any change in agricultural practice alters N_2O emission, CH_4 uptake and SOC storage simultaneously, the net benefit of SOC, CH_4 and N_2O changes should be calculated based on their own contributions to the greenhouse effect. The estimates of Global Warming Potentials (GWP) for 100 year time horizon are 1, 270, and 11 for CO_2, N_2O, and CH_4, respectively (IPCC, 1992). The Global Warming Potential (GWP) of each site was calculated based on the annual SOC loss, N_2O flux, and CH_4 uptake at the site. Since the contributions to global warming of CH_4 uptake were much smaller than that of CO_2 and N_2O at all sites in this study, CH_4 uptake was not included in net benefit calculations. The net benefit of both SOC and N_2O changes was estimated by equation 1.

$$NB = [N2O_b \times 270 + SOC_b] - [N2O_c \times 270 + SOC_c] \quad \text{(Eq. 1)}$$

where NB is the net benefit (kg CO_2 equivalent/ha/yr); $N2O_b$ and $N2O_c$ are the annual N_2O fluxes (kg N_2O/ha/yr) under baseline and projected scenarios, respectively; SOC_b and SOC_c is annual loss of SOC (kg CO_2/ha/yr) under baseline and projected scenarios, respectively.

A positive value of NB imply net benefit by converting the baseline scenario into the projected scenario and reducing the greenhouse effect.

Table 2. Change in SOC and N_2O flux and their global warming potential (kg CO_2 /ha/yr) at six sites in the USA

	Cropping	----N_2O flux----		---SOC change---		Total	Net
Site	scenario	kg N/ha	GWP	kg C/ha	GWP	GWP	benefit
IA	Base	2.51	1067	-449	1720	2787	0
	CT	2.27	963	-197	722	1685	-1103
	NT	2.40	1019	-90	330	1349	-1438
	F1	2.43	1029	-620	2277	3306	519
	F2	2.40	1017	-701	2571	3588	801
	M1	2.87	1217	110	-403	814	-1973
	M2	3.09	1309	1042	-3823	-2513	-5300
	WC	2.52	1071	424	-1558	-487	-3274
	CM	2.86	1213	2205	-8091	-6877	-9665
IL	Base	2.13	904	1022	-3751	-2848	0
	CT	1.96	831	789	-2895	-2064	784
	NT	1.84	781	753	-2765	-1984	863
	F1	1.68	713	618	-2270	-1557	1291
	F2	1.58	670	406	-1491	-822	2026
	M1	2.18	927	1865	-6845	-5918	-3070
	M2	2.35	995	2690	-9874	-8878	-6031
	WC	2.17	921	1914	-7026	-6105	-3257
	CM	1.76	745	2865	-10516	-9771	-6923
NE	Base	1.65	699	1166	-4278	-3579	0
	CT	1.45	614	1080	-3962	-3348	231
	NT	1.36	578	1093	-4012	-3434	145
	F1	1.42	604	1127	-4135	-3530	49
	F2	1.37	581	1020	-3745	-3164	416
	M1	1.83	775	1943	-7131	-6356	-2777
	M2	1.99	843	2717	-9973	-9130	-5551
	WC	1.76	748	2188	-8030	-7282	-3703
	CM	1.44	610	3306	-12132	-11521	-7942
KS	Base	3.16	1342	-942	3456	4798	0
	CT	2.34	993	-540	1980	2973	-1825
	NT	2.29	972	-506	1857	2829	-1969
	F1	3.07	1301	-973	3570	4871	73
	F2	3.17	1346	-986	3619	4965	167
	M1	3.59	1523	-50	184	1707	-3091
	M2	3.69	1566	726	-2665	-1099	-5897
	CM	2.71	1148	998	-3663	-2515	-7313

Table 2. --continued

Site	Cropping scenario	----N_2O flux---- kg N/ha	GWP	---SOC change--- kg C/ha	GWP	Total GWP	Net benefit
CA	Base	2.51	1067	-449	1720	2787	0
	CT	1.37	581	-720	2642	3224	-339
	NT	1.43	606	-700	2570	3176	-386
	F1	1.21	514	-881	3235	3749	186
	F2	1.14	485	-928	3406	3890	328
	M1	1.62	686	182	-669	17	-3546
	M2	1.60	678	1171	-4297	-3619	-7181
	WC	1.59	673	500	-1835	-1162	-4725
	CM	1.07	455	1940	-7121	-6666	-10228
FL	Base	22.84	9689	-6581	24151	33840	0
	CT	24.42	10361	-6093	22363	32724	-1116
	NT	23.15	9824	-5989	21979	31803	-2037
	F1	17.56	7452	-6543	24014	31467	-2373
	F2	20.42	8662	-6577	24139	32801	-1039
	M1	24.12	10234	-5904	21668	31902	-1938
	M2	24.75	10501	-5234	19207	29708	-4131
	WC	23.61	10016	484	-1776	8240	-25600
	CM	34.22	14518	-210	770	15288	-18552

III. Results and Discussion

A. Contribution of Baseline Scenario to Global Warming

Conventional cropping practices were simulated with the baseline scenario for each site. Under the scenario, SOC storage decreased at the sites IA, KS, CA, and FL, but increased at IL and NE (Table 2). The average annual loss of SOC was about 6,600, 940, 790, 470, -1,000, and -1,200 kg C/ha at FL, KS, CA, IA, IL, and NE, respectively. The magnitude of change in SOC storage was related to the initial SOC content and climate at each site. Generally, higher initial SOC and higher temperature favored SOC loss through enhanced decomposition. The N_2O emission rates appeared to be related to SOC content and annual precipitation. The highest N_2O flux, 22.84 kg N/ha/yr, was obtained in Florida where SOC and precipitation were the highest. On the contrary, lower N_2O fluxes (1.65 and 1.57 kg N/ha/yr) were obtained in Nebraska and Kansas where SOC and precipitation were relatively low (Table 2). Under the baseline scenario, the contributions to global warming of IA, IL, NE, KS, CA, and FL sites were equivalent flux of 2,787, −2,848, −3,579, 4,798, 3,562, and 33,840 kg CO_2/ha/year, respectively (Figure 1). The contribution of FL site was about 10 times higher than that of other sites. The contributions of N_2O emission were accounted for 62, 24, 16, 39, 23, and 40 percent of GWP for IA, IL, NE, KS, CA, and FL sites, respectively (Table 2).

B. Reduced Tillage

Impact of reduced tillage on CO_2 and N_2O emissions and SOC storage change was computed for conservation tillage (CT) and no-till (NT). Reducing tillage may decrease decomposition rate, and reduce CO_2 emission and the production of inorganic dissolved nitrogen (i.e. nitrate and ammonium) in soil. The data showed that when the conventional tillage was converted to conservation tillage or no-till, both CO_2 emissions and N-uptake by crops were reduced at all sites (Figure 2A and 2B). Reduction in CO_2 emission favors SOC conservation, but reduction in N-uptake decreases residue yield and, hence, organic C storage in soils. The dynamics of SOC storage depends on the balance between the CO_2 emitted from and the

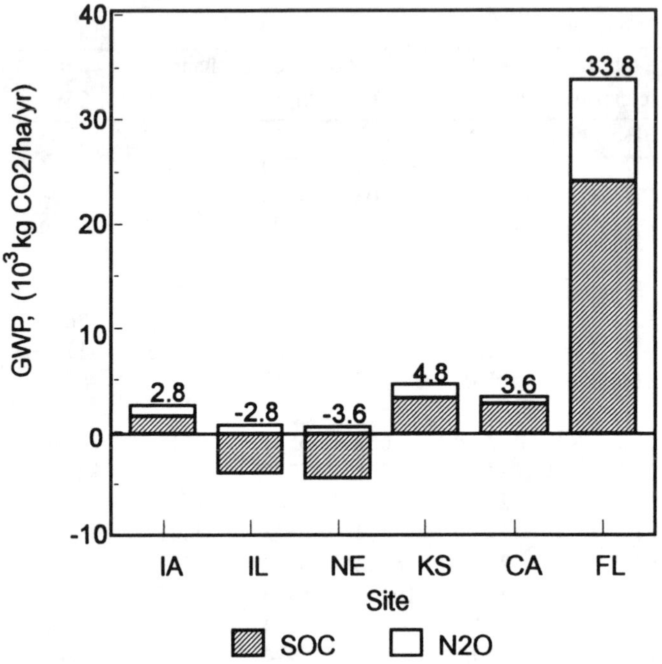

Figure 1. Under the baseline scenario, there were negative contributions to global warming of IL and NE sites, and positive of IA, KS, CA and FL sites. The contributions of N_2O emission were accounted for 16-62 percent of GWP for the test sites.

residue returned to the soil. Reducing tillage significantly decreased SOC loss at the SOC-rich sites (e.g. FL, IA, KS, and CA), but not at IL and NE sites with lower SOC contents. Converting conventional tillage to conservation tillage or no-till slightly decreased N_2O emissions at most sites except FL with organic soil where N_2O fluxes were higher under conservation tillage than under conventional tillage (Table 2). The net benefit of converting conventional tillage to conservation tillage was positive at IA (1,100-1,400 kg CO_2/ha/yr equivalent), KS (1,800 to 1,970 kg CO_2/ha/yr equivalent), CA (340 to 390 kg CO_2/ha/yr equivalent), and FL (1,100 to 2,000 kg CO_2/ha/yr equivalent), but negative at IL (-780 to -860 kg CO_2/ha/yr equivalent) and NE (-230 to -150 kg CO_2/ha/yr equivalent) (Figure 3).

Reducing tillage favors SOC conservation in soils with high SOC content, but may increase N_2O emissions. For the SOC-rich sites in this study, the negative effect of high N_2O emission was offset by the positive effect of increase in SOC when conventional tillage was converted to conservation tillage or no-till.

B. Reduced Fertilizer Use

The effect of reducing fertilizer use was tested with two scenarios, F1 and F2, with lower rates (100 and 50 kg ammonium-N/ha) applied. N_2O flux decreased with reduced fertilizer use for all 6 sites (Table 2). The magnitude of the change in annual N_2O flux at each site was related to SOC content. The lower the SOC content, the less the change in N_2O flux caused by reducing fertilizer rate. Reduced fertilizer use decreased SOC storage because the crop residue produced decreased. In case of reduced fertilizer scenarios, the positive effect of N_2O decrease was offset by the negative effect of SOC decrease at sites with mineral soils. In case of organic soil at FL, reduction in N_2O flux by reducing fertilizer use was significant enough to mitigate negative effect decrease in SOC (Figure 4). Reducing fertilizer use didn't produce net benefit regarding mitigating the greenhouse effect except in some organic soils.

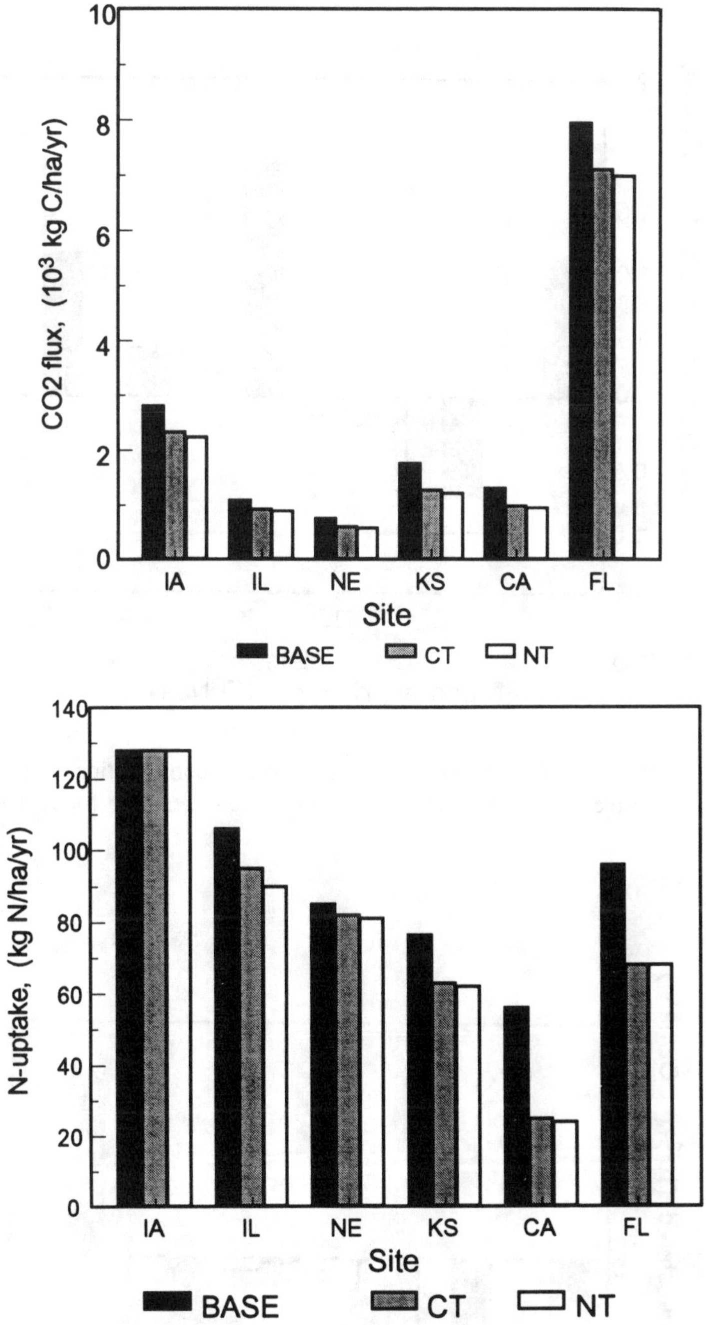

Figure 2. A) In comparison with conventional tillage (baseline scenario, BASE), conservation tillage (CT) and no-till (NT) reduced CO_2 emissions at all sites; B) In comparison with conventional tillage (baseline scenario, BASE), conservation tillage (CT) and no-till (NT) reduced availability of soil N for crops at all sites except IA.

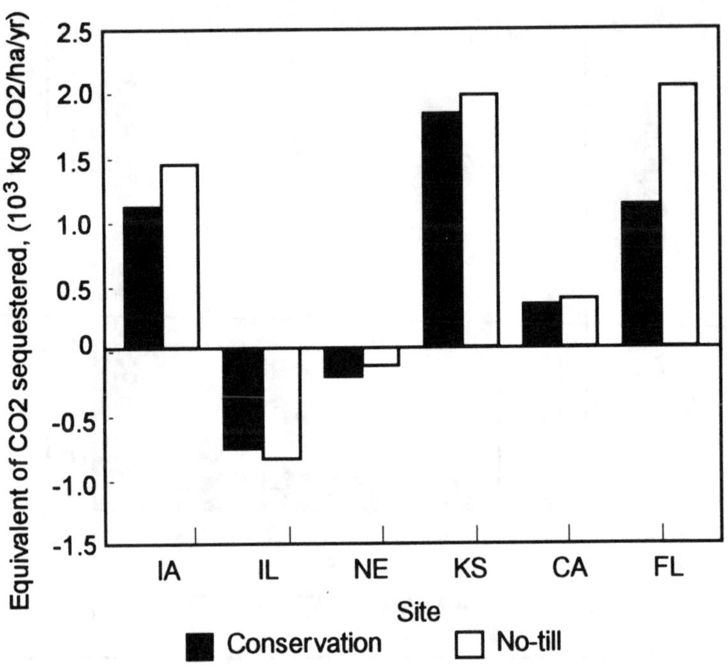

Figure 3. Net benefits of converting conventional to conservation tillage or no-till were positive for IA, KS, CA, and FL sites, but negative for IL and NE.

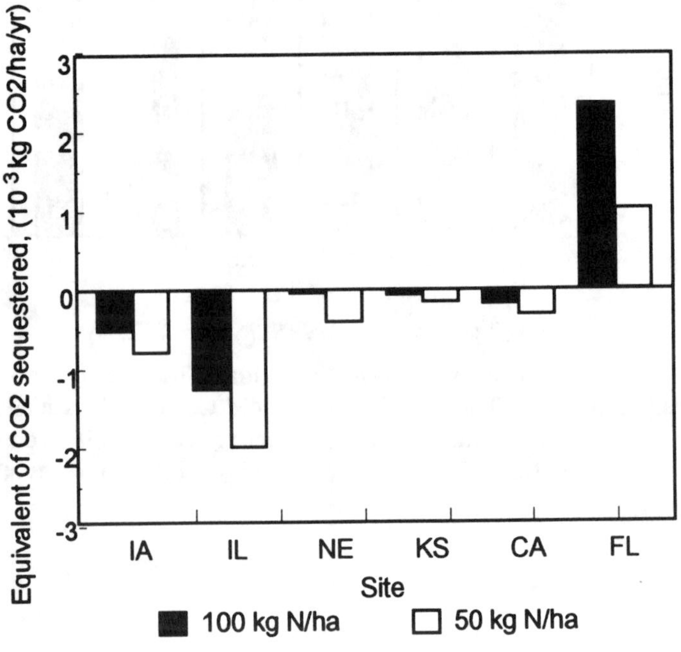

Figure 4. Net benefits of reduced fertilizer use were negative for all sites except FL.

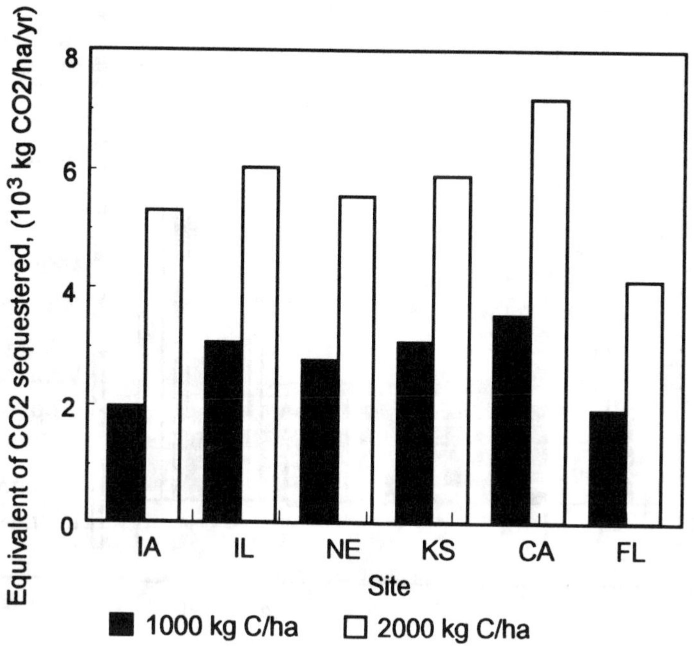

Figure 5. Net benefits of manure application were positive for all sites.

D. Manure Application

Two rates (1,000 and 2,000 kg C/ha) of manure use simulated revealed that manure amendment increased both N_2O emissions and C storage in all 6 sites (Table 2). Manure application directly added organic matter into the SOC pool and increased N_2O emission rate through elevating nitrate and soluble C concentrations in soils. The net benefit of manure use was always positive because the negative effect of increase in N_2O emission was offset by the positive effect of increase in SOC at each site. The magnitude of net benefit by manuring was related to SOC content and the amount of manure applied (Figure 5). Generally, higher net benefit was gained at the SOC-poor sites (e.g. NE, IL, CA, and KS) than at the SOC-rich sites (e.g. FL and IA). The net benefit at 1,000 kg C/ha of manure application was equivalent to 1,900-3,500 kg CO_2/ha/yr sequestered, and the net benefit at 2,000 kg C/ha was 4,100-7,200 kg CO_2/ha/yr equivalent.

E. Cover Crops

Planting winter-cover crop increased the amount of residue produced and, hence, increased SOC input (Table 2). The yield of winter-cover crop was related to soil fertility and climate. In Florida, the winter-cover clover produced highest yield because of fertile soil and warm winter. However, planting winter-cover crop had little effect on N_2O emissions (Table 2). The net benefit of planting winter-cover crop ranged from 3,200 to 5,900 kg CO_2/ha/yr equivalent at all sites with mineral soils. At FL with organic soil, the net benefit was as high as 25,600 kg CO_2/ha/yr equivalent (Figure 6).

F. Alternative Cropping Practices

A comprehensive scenario of multi-practice change was designed to simultaneously change several practices at each site. The changes included reduced tillage from conventional to no-till, reduced fertilizer rate from 200 to 50 kg N/ha, manure application at 2,000 kg C/ha, and planting winter-cover crop. The simulated results revealed that the multi-practice change had the most significant effect on SOC dynamics in

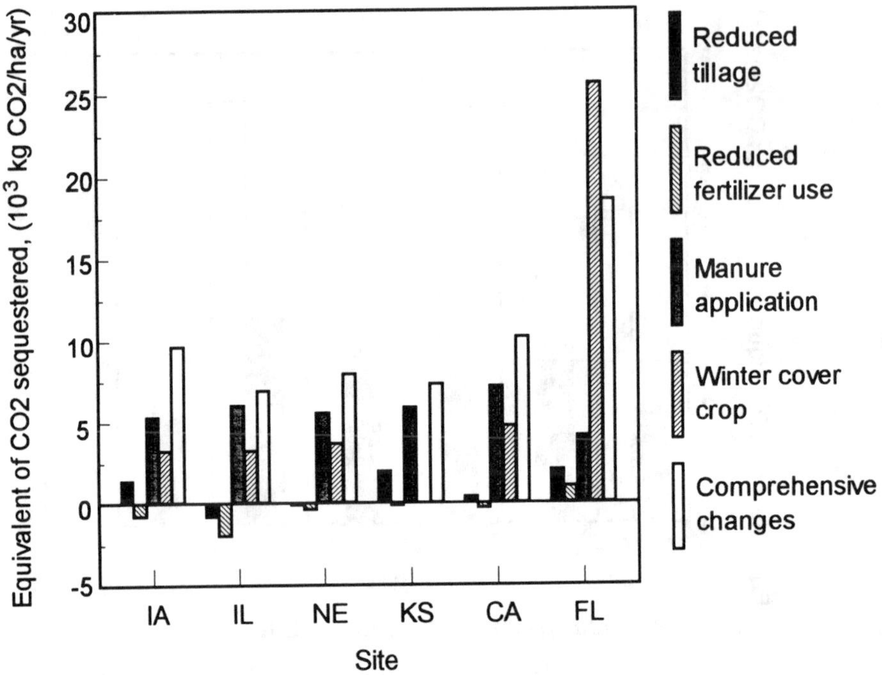

Figure 6. Priority scenarios of changing agricultural practices. The net benefit gained by single-practice changes depended on soil and climate. Multi-practice changes are recommended for all climate zones and soil types. Where applicable, planting winter-cover crop and using manure produce net benefit. Reduced tillage is encouraged for SOC-rich farmlands, and reduced fertilizer use in organic soils.

comparison with the single-practice change scenarios. The multi-practice change increased SOC accumulation rates by about two times at IL and NE sites, converted SOC loss into gain at IA, KS, and CA and greatly decreased SOC loss from 6,600 to 210 kg C/ha/yr in the organic soil at FL (Table 2). The multi-practice change increased N_2O emissions at FL and IA, but decreased N_2O emissions at IL, NE, KS, and CA. The net benefit of applying the comprehensive scenario was 7,000-10,000 kg CO_2/ha/yr equivalent at sites with mineral soils, and more than 18,000 kg CO_2/ha/yr equivalent at FL with organic soil (Figure 6).

G. Toward Mitigating Greenhouse Gases

Net benefits of changing agricultural practices at 6 sites are compared in Figure 6. The net benefit gained by single-practice changes depended on soil and climate. For example, the net benefits of manure application ranged from 5,300 to 7,200 kg CO_2/ha/yr equivalent, with higher gains for SOC-poor than SOC-rich sites. The net benefits from planting winter-cover crop ranged from 3,200 to 25,600 kg CO_2/ha/yr equivalent, with higher benefits for fertile soil and warm weather than poor soil and cold weather. Manure application and planting winter-cover had always net benefits at all sites. Reduced tillage had net benefit at IA, KS, CA, and FL where SOC was relatively high. At SOC-poor sites, reduced tillage did not produce significant net benefit. Reduced fertilizer use decreased N_2O fluxes at all 6 sites, but did not have net benefit except at Florida. All 6 sites sequestered net benefits when the baseline scenario was shifted into the comprehensive scenario. Therefore, multi-practice changes are highly recommended for all climate zones and soil types; where applicable, planting winter-cover crop and manure produce net benefit; reduced tillage is encouraged for SOC-rich (>1.5%) farmlands; and reduced fertilizer use in organic soils (SOC > 10%).

IV. Conclusions

A. Carbon-Rich Soils the Priority Target

Contributions of farmland to global warming depend on soil and climate. Farmlands with rich soils (organic C > 1.5%) and warm weather are loosing carbon and generating higher N_2O emissions under the conventional cropping practices. Farmlands under conventional practices contribute an equivalent of 3,000-5,000 kg CO_2/ha/yr. N_2O emission accounts for 20-40 percent contribution to greenhouse emission. Cultivating organic soils significantly leads to high emission of CO_2, and of N_2O at the rate of 20-30 kg N/ha/yr. In contrast, carbon-poor soils are sequestering atmospheric carbon under the conventional cropping practices. Carbon-rich soils should be set as a priority target for mitigating CO_2 and N_2O emissions from farmland.

B. Alternative Agricultural Practices

1. Planting winter-cover crop or using manure at 1,000 kg C/ha can sequester an equivalent of 2,000-5,000 kg CO_2/ha/yr. These estimates are similar to the current contributions (3,000-5,000 kg CO_2/ha/yr) of farmlands in Iowa, Kansas, and California.

2. The effect of reduced tillage on greenhouse gases depends on soil and climate. Reducing tillage is beneficial in fertile soils. The net benefit by converting conventional to conservation or no-till ranges from 350 to 2,000 kg CO_2/ha/yr equivalent. These benefits are equivalent to 10-50 percent of current emissions from farmlands.

3. Reduced fertilizer use decreases N_2O emissions, but may not be beneficial to mineral soils. Decrease in biomass production leads to low residue returned to soils. The benefits are, however, substantial in all organic soils.

4. Multi-practice changes are highly recommended. Benefits are 2-3 times higher than the current contributions of greenhouse gases from mineral soils.

C. Research Needs

Further studies are needed at: (1) more sites representing major climatic zones and soil types and design more agricultural scenarios to produce a nation-wide data base of change in SOC and N_2O flux with regards to alternative agricultural practices, and (2) link the DNDC model with geographic information system (GIS).

Acknowledgements

The author thanks R. Harriss (University of New Hampshire), K. Andrasko and S. Winnett (US EPA, Washington, D.C.) for the discussions regarding this work. Thank is also given to D. Blaha (University of New Hampshire) for providing the agricultural data. The work was funded by the Climate Change Division, Office of Policy, Planning and Evaluation, United States Environmental Protection Agency under grant 68-W8-0113.

References

Aulakh, M.S., D.A. Rennie, and E.A. Paul. 1984. Gaseous nitrogen losses from soils under zero-till as compared with conventional-till management systems. *J. Environ. Qual.* 13:130-136.

Dalal, R.C. and R.J. Mayer. 1986a. Long-term trends in fertility of soils under continuous cultivation and cereal cropping in southern Queensland. I. Overall changes in soil properties and trends in winter cereal yields. *Aust. J. Soil Res.* 24:265-279.

Dalal, R.C. and R.J. Mayer. 1986b. Long-term trends in fertility of soils under continuous cultivation and cereal cropping in southern Queensland. II. Total organic carbon and its rate of loss from the soil profile. *Aust. J. Soil Res.* 24:281-292.

Dalal, R.C. and R.J. Mayer. 1986c. Long-term trends in fertility of soils under continuous cultivation and cereal cropping in southern Queensland. III. Distribution and kinetics of soil organic carbon in particle-size fractions. *Aust. J. Soil Res.* 24:293-300.

Doran, J.W. and J.F. Power. 1983. The effects of tillage on the nitrogen cycle in corn and wheat production. In: R.R. Lawrence et al. (eds.), *Nutrient Cycling in Agricultural Ecosystems.* Univ. Georgia Agri. Expt. Sta. Special Pub. 23, Athens, GA.

Eichner, M.J. 1990. Nitrous oxide emissions from fertilized soils: Summary of available data. *J. Environ. Qual.* 19:272-280.

Intergovernmental Panel on Climate Change (IPCC). 1992. 1992 IPCC Supplement: Scientific Assessment of Climate Change. Submission from Working Group 1. WMO and UNEP.

Jenkinson, D.S. 1990. The turnover of organic carbon and nitrogen in soil. *Phil. Trans. R. Soc. Lond.* B 329:361-368.

Li, C., S. Frolking, and T.A. Frolking. 1992a. A model of nitrous oxide evolution from soil driven by rainfall events: 1. Model development and sensitivity. *J. Geophys. Res.* 97:9759-9776.

Li, C., S. Frolking, and T.A. Frolking. 1992b. A model of nitrous oxide evolution from soil driven by rainfall events: 2. Model application. *J. Geophys. Res.* 97:9777-9783.

Li, C., S. Frolking, and R.C. Harriss. 1993. Modeling soil C biogeochemistry in agricultural ecosystems. Global Biogeochemical Cycles (in review)

Li, C. 1993. Soil N_2O simulation with generalized climatic scenarios. (draft)

Li, C., S. Frolking, R.C. Harriss, and R.E. Terry. 1994. Modeling nitrous oxide emissions from agriculture: A Florida case study. *Chemosphere* (in press)

Mosier, A., D. Schimel, D. Valentine, K. Bronson, and W. Parton. 1991. Methane and nitrous oxide fluxes in native, fertilized and cultivated grasslands. *Nature.* 350:330-332.

Sahrawat, K.L. and D. Keeney. 1986. Nitrous oxide emissions from soils. *Adv. Soil Sci.* 4:103-148.

Tennessee Valley Authority. 1990. Consumption of primary plant nutrients and total fertilizer. National Fertilizer and Environmental Research Center, Huntsville, AL.

United States Department of Agriculture. 1991. Agricultural Statistics, 1991. United States Government Printing Office, Washington.

United States Department of Agriculture, Economic Research Service. 1989a. Weather in U.S. Agriculture: Monthly Temperature and Precipitation by State and Farm Production Region, 1950-88. L.D. Teigen and F. Singer (eds.). Statistical Bulletin No. 789. Washington, D.C. 20005-4788.

United States Department of Agriculture. 1989b. Census of Agriculture. Bureau of the Census, Washington, D.C.

United States Department of Agriculture, Soil Conservation Service. 1962-1976. Soil survey laboratory data and description for some soils.

CHAPTER **10**

Analysis of the Short-Term Effects of Management on Soil Organic Matter Using the CENTURY Model

R. Paul Voroney and Denis A. Angers

I. Introduction

In agricultural systems many factors altering soil organic matter (SOM) levels are controlled by management decisions. These include choices of cropping systems, management of crop residues and organic waste application, and methods and intensity of soil tillage. The adoption by farmers of those agricultural management practices which promote conservation of SOM requires that changes be evident within their planning horizons, typically 1 to 5 years (Kay, 1990). However, such short-term effects of management on SOM levels are usually not easy to detect in the field because the content of SOM in surface field soils can be highly variable. Furthermore, the changes in SOM levels are difficult to predict quantitatively owing to the large number of possible combinations of crops and management practices, coupled with the complexity of SOM constituents and their relative importance in varying soils, climates and crops. Recent developments in simulation models of SOM dynamics suggest that models may be useful tools for analyzing management effects on the turnover of the various constituents of SOM (Paustian et al., 1992).

While there have been many reports of the effects of different farming systems on SOM content showing long-term effects, few studies have investigated changes occurring in the short term. Previous research in Québec has shown that 5 years of alfalfa increased organic C content by 15% whereas silage corn and fallow had apparently no effect (Angers, 1992). In silage corn production for 11 years, tillage practices had no effect on soil organic C content (Angers et al., 1993), whereas applications of cattle manure resulted in increases ranging from 15 to 45% depending on the rate of manure application (Angers and N'dayegamiye, 1991).

Several simulation models have been developed to describe the dynamics of organic C in soil over the years-to-centuries time scale (eg. Jenkinson and Rayner, 1977; van Veen and Paul, 1981; Juma and McGill, 1986; Parton et al., 1987). These models were largely intended to be used to assess the long-term impacts of soils, climate a,nd management on SOM, or to predict the implications of regional management on nutrient cycling (Burke et al., 1989). There have been few studies of the short-term effects compatible with the time period for management decision making.

The CENTURY model developed by Parton et al. (1987) has been used to analyze management practices as they affect changes to SOM in the grassland soils of North America. A recent version of the model allows inclusion of crop rotations and manure additions, thereby making it possible to describe organic matter dynamics in the agricultural soils of Eastern Canada. The objective of this study was to examine the usefulness of the CENTURY model for predicting short-term changes in SOM due to management. Data from three field experiments at two locations in Québec, Canada, were analyzed using the model to simulate management practices, crop production and SOM dynamics at each field site.

ISBN 1-56670-117-1/95/\$0.00+.50

II. Materials and Methods

A. Field Sites and Management Practices

The research at the field sites has been described previously and only a brief description will be given here. Unless stated otherwise, management practices were those recommended for each crop by the Québec Council for Crop Production.

1. Cropping Effects:

This study was located at the Agriculture Canada Experimental Farm in La Pocatière, Québec. The site had been used for research plots (small grain production) with conventional tillage (moldboard ploughing in fall and rototilling in spring) for more than 15 years prior to initiating this study in 1986. The soil at the site is a Kamouraska clay (Typic Humic Gleysol) containing 16% sand, 37% silt, and 47% clay. At the beginning of the experiment, plot soil to 20-cm depth had an organic C content of 2.5%, pH of 5.5, and a soil bulk density of 1.2 Mg m^{-3}.

Over 5 years the cropping systems were in continuous production of barley (*Hordeum vulgare* L.) with straw removal, silage corn (*Zea mays* L.), and alfalfa (*Medicago sativa* L.) with two cuttings taken each season. Average annual production for barley grain, silage corn dry matter, and alfalfa dry matter was 5, 13, and 4.7 Mg ha^{-1}, respectively. A fallow (bare-soil) plot was included as a control treatment. Tillage depth and intensity for annual crop production were reduced to a minimum using only shallow cultivation (<5 cm depth) to prepare the seedbed. Other than standing stubble, little above-ground residue was incorporated into the soil. Previous research results of cropping effects included changes in soil aggregation (Angers and Mehuys, 1989; Angers, 1992) and in SOM content and composition (Angers and Mehuys, 1989; 1990; Angers, 1992).

2. Tillage Effects:

The research was located at the Québec Ministry of Agriculture Experimental Farm in St-Lambert, Québec, on a Neubois silt loam (fine-loamy, mixed, frigid Aeric Haplaquept). Prior to this experiment, which began in 1978 and continued until 1989, the site had been under permanent grass pasture for more than 20 years. The texture of the surface soil (0-24 cm) was 27% sand, 50% silt, and 23% clay; the organic C content was 2.8%, pH was 5.3, and the soil bulk density was 1.3 Mg m^{-3}.

The tillage treatments were: (i) minimum tilling consisting of shallow cultivation twice in spring, (ii) ridge tilling which involved reforming of the ridges each spring before planting, and (iii) moldboard ploughing (15- to 18-cm ploughing depth) in fall followed by harrowing in the spring. Silage corn (*Zea mays* L.) was grown continuously. No yield differences were found due to tillage and the average annual dry-matter harvest over 11 years was 9.5 Mg ha^{-1}. Effects of tillage on SOM and soil microbial biomass have been reported by Angers et al. (1993).

3. Manure Additions

The manure experiments were located near the tillage plots and initiated at about the same time. Surface soil (0-15 cm) at the site had a pH of 5.3 and contained 35% sand, 53% silt, and 12% clay; the initial organic C content was 2.7% and the soil bulk density was 1.2 Mg m^{-3}.

Applications of solid beef cattle manure were made every second year in fall over 10 years, starting in 1978, at rates up to 100 Mg ha^{-1} (wet-weight basis). Manure applications were followed by moldboard ploughing (15- to 18-cm ploughing depth), and in spring plots were harrowed prior to seeding. The research plots were continuously cropped to silage corn (*Zea mays* L.) without addition of mineral fertilizers. Effects of manure applications on crop yields and soil chemical properties have been reported by N'dayegamiye (1990); soil aggregation and microbial activity by N'dayegamiye and Angers (1990); and organic C, N, and carbohydrate contents by Angers and N'dayegamiye (1991).

B. CENTURY Model

The CENTURY model is a general model that has been used extensively to describe the SOM dynamics for different ecosystems. The model was originally developed for simulating SOM dynamics in native grassland ecosystems, and it was later extended to have the capability of simulating agricultural ecosystems typical of the semi-arid Great Plains. A detailed description of the CENTURY model is given by Parton et al. (1987; 1988); more recently Paustian et al. (1992) have described the use of the model to study SOM dynamics in Swedish soils.

In order to be able to simulate a wide range of cropping systems and different geographical regions, recent versions of CENTURY have included a pre-processor to allow scheduling an annual sequence of management events such as cultivation, planting, fertilization, harvest and organic matter addition over a number of years, and to control model execution. CENTURY also contains a crop production sub-model for simulating growth and nutrient uptake of a variety of crops. The PC version that we used was interfaced with Time-Zero™, which is an integrated modelling environment for running the model and viewing output.

1. Model Parameters

Permanent meteorological stations were located at or near the research sites and provided long-term climatic records. A 30-year average (1951-1980) of the mean monthly precipitation and temperature for the study sites is given in Table 1.

Potential plant production in CENTURY is defined for each crop and is a function of a genetic maximum and an optimal growing temperature. These two parameters were adjusted to reflect the local crop varieties and seasonal growing conditions, and to match the site-specific crop yields. The fraction of the total plant production allocated below-ground was based on crop-specific values obtained from the literature for root production: 0.25 for barley (Hansson et al., 1987), 0.085 for corn (Foth, 1965), and 0.60 for alfalfa (Paustian et al., 1990).

In the model, parameters associated with describing each method of tillage affect both the formation of SOM and its turnover rate. They control the fraction of above-ground residue transferred to the topsoil layer, and modify the decomposition rates of each organic matter constituent. These factors increase with increasing tillage intensity and burial of above-ground residues. Relative to no-tillage, they were set 1.5 times higher for the moldboard plough and cultivator tillages.

Soil properties, including texture, bulk density, pH, and organic C content, were initial field-site values for a defined sampling depth. In the preliminary runs, the model was initialized using the proportions of total SOM in each of the three SOM pools and their specific decomposition rate constants as proposed by Parton et al. (1987) (Table 2). However, utilization of these parameters, which were derived from studies of the native prairies and wheat-fallow agriculture in the Great Plains (Cole et al., 1989; Parton et al. 1987; 1988), consistently over-estimated decomposition rates and resulted in the predicted SOM levels being lower than the measured values. In contrast to the grassland soils which have developed on calcareous parent material under arid conditions, the soils in Québec have originated from boreal forest vegetation and acidic parent material and experience a cool, humid climate.

Paustian et al. (1992) chose to alter the parameter t_{opt} in the decomposition-temperature response function, to fit the model to Swedish soils. Our approach to fitting the model to Québec soils was to modify individually the parameters governing SOM dynamics. Adjustments were made to the proportions of the total organic matter in each of the SOM pools and to their specific decomposition rate constants, to obtain the best fit for the data obtained in the 5-year study of cropping effects on SOM. In addition, the active organic matter pool was substituted for the soil microbial biomass (Paustian et al., 1992). Data obtained from the tillage experiments were used to further adjust the microbial biomass pool size and its specific decomposition rate constant.

In these short-term studies, the passive organic matter pool would not be expected to play a significant role. However, selection of the parameters for describing the microbial biomass and the slow organic matter pool essentially fixed the values for the passive pool (Paustian et al. 1992), providing the SOM constituents are assumed to be at steady-state. Once the values for the initial SOM distribution and their decay rate constants were chosen, the same values were used in all subsequent simulations of management effects at both sites (Table 2).

Table 1. 30-year mean precipitation and temperature for the field sites in Québec

	La Pocatière			St. Lambert		
	Prec.	T_{min}	T_{max}	Prec.	T_{min}	T_{max}
Month	(cm)			(°C)		
January	7.9	-15.5	-7.0	7.8	-15.0	-7.6
February	7.1	-14.7	-5.8	7.3	-13.9	-6.2
March	6.7	-8.6	-0.2	6.9	-7.9	-0.2
April	6.3	-1.5	7.1	6.9	-0.4	7.6
May	6.9	4.4	15.4	8.0	6.2	16.0
June	9.0	10.2	21.2	11.3	11.8	21.9
July	9.5	13.3	24.1	13.6	14.3	24.4
August	9.8	12.1	22.5	11.5	12.9	22.6
September	9.5	7.6	17.5	11.5	8.2	17.4
October	7.1	2.7	10.9	8.7	3.3	10.6
November	7.8	-2.9	3.6	8.4	-2.7	2.8
December	9.0	-11.9	-4.4	10.8	-11.5	-5.3

Table 2. Model parameters for SOM constituents

	Distribution of soil C (% of total SOM)		Specific decay constant (yr^{-1})	
SOM constituent	Parton et al., 1987	This study	Parton et al., 1987	This study
Microbial biomass	7	3	7.3	4[a]
Slow	50	65	0.2	0.05
Passive	43	32	0.0068	0.0013

[a] Conventional tillage.

III. Results and Discussion

A. Cropping

The different cropping treatments resulted in changes to SOM content which were predicted using the CENTURY model (Figure1). Angers (1992) reported that while no detectable differences in SOM content existed between the cropping treatments after two growing seasons, thereafter significant differences were evident. In the third year organic C content followed a sigmoid shape under alfalfa, increasing from 6240 up to 6884 g m^{-2} and then remained relatively constant. The increase observed under barley cropping was smaller but more constant, about 72 g C m^{-2} yr^{-1}. SOM decreased by 60 g C m^{-2} yr^{-1} in both the fallow and silage corn cropping treatments during the 5 years. The measured changes in the quantity of SOM ranged from decreases of 4.6 and 2.0% in the bare fallow and corn-cropped soils, respectively, to increases of 8.3% in barley and 15.1% in alfalfa.

The magnitude of the changes in the quantity of SOM was directly related to the quantity of organic matter added to the soil by each cropping treatment. Inputs of organic matter to soil by the cropping systems are the single most important factor determining SOM levels (Paustian et al., 1992). In all three of the cropped treatments, above-ground plant residues were almost entirely removed leaving below-ground inputs to maintain SOM levels. Contributions from standing stubble would be minimal. Estimates of the annual organic C inputs for each of the cropping treatments and their predicted affect on SOM are reported in Table 3.

Previous reports of this research indicated that although differences in total SOM content between cropping treatments were not detected until the third growing season, significant changes in the constituents of SOM could be observed earlier (Angers, 1992; Angers and Mehuys, 1989; 1990). Minor changes in total SOM content can be difficult to measure due to the relatively large quantity of organic matter already present in the soil, to its spatial variability in surface soils, and to the variability in soil bulk density

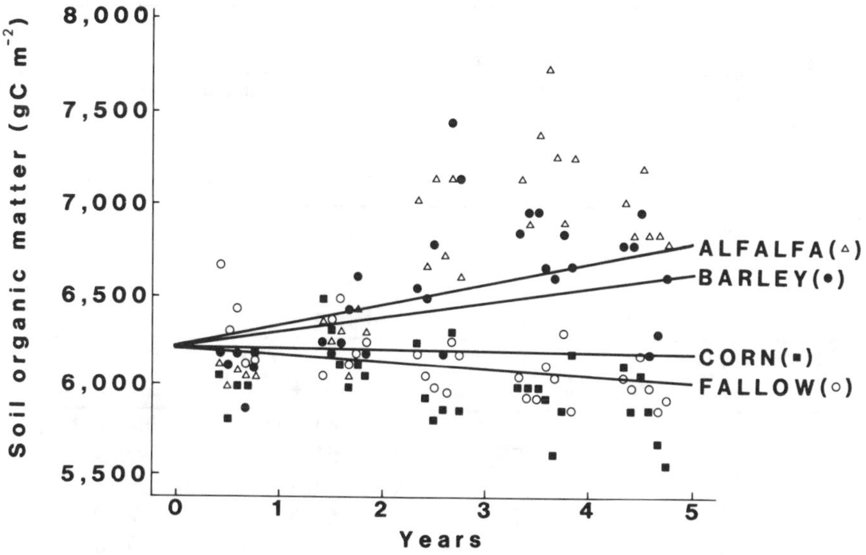

Figure 1. Cropping effects on SOM.

Table 3. Annual input of crop residues at La Pocatière and its affect on SOM content

	Crop residue inputs	SOM change
	$g\ C\ m^{-2}\ yr^{-1}$	
Alfalfa (*Medicago sativa* L.)	288	62
Barley (*Hordeum vulgare* L.)[a]	285	47
Silage corn (*Zea mays* L.)[a]	95	-14
Fallow	0	-48

[a] Includes an estimate of the contribution from stubble.

Table 4. Predicted cropping effects on soil microbial biomass after two growing seasons.

Crop	Microbial biomass[a]
	($g\ C\ m^{-2}$)
Fallow	65
Corn (*Zea mays* L.)	90
Barley (*Hordeum vulgare* L.)	163
Alfalfa (*Medicago sativa* L.)	216

[a] Initial microbial biomass content, 125 $g\ C\ m^{-2}$.

measurements. For these reasons, measurements of constituents of SOM such as microbial biomass can be more useful for assessing the impacts of management (Carter, 1986; McGill et al., 1986). The model predicted that changes in microbial biomass within two years would range from 30% and 73% increases in barley and alfalfa, respectively, to decreases of 28% in corn cropped soils and 48% in the bare fallow (Table 4). Such changes to soil microbial biomass can be easily measured (Voroney et al. 1993), and they can help also to explain the observed cropping effects both on the soil carbohydrate content and on the mean-weight diameter of water-stable aggregates (Angers and Mehuys, 1989).

Table 5. Effects of tillage practices on soil microbial biomass

Tillage	Microbial biomass ($g\ C\ m^{-2}$)		Microbial biomass/SOM (%)	
	Predicted	Measured[a]	Predicted	Measured[a]
Moldboard plow	95	96	1.3	1.3
Ridge tillage	225	214	2.9	3.0
Minimum tillage	344	306	4.5	4.2

[a] Angers et al., 1993.

Table 6. Effects of manure application on SOM content

Manure application[a]	Crop inputs	SOM content	
----------($g\ C\ m^{-2}\ yr^{-1}$)----------		----------($g\ C\ m^{-2}$)----------	
		Predicted	Measured[b]
0	35	6135	4969
87	38	6402	6078
174	43	6668	6459
261	48	6936	7080
348	53	7203	7505
435	60	7413	7708

[a] Manure applications of 0, 20, 40, 60, 80, and 100 Mg ha^{-1} every two years for 10 years; [b] estimated from data reported by N'dayegamiye, 1990.

B. Tillage

No significant differences in total SOM (0-24 cm) due to the different tillage practices were detectable (Angers et al. 1993), and this was predicted by the model. Tillage had no affect on crop yields, and annual C inputs with silage corn production, estimated as 68 g C m^{-2}, would largely be derived from below-ground production.

The effects of specific cultivation treatments on the processes controlling SOM dynamics are not well-known. Balesdent et al. (1990), using tracer techniques, reported that the production of SOM from crop residues was much lower with no tillage compared with conventional tillage. This was attributed to a lack of protection of surface residues from microbial attack, and was afforded by soil particles if the residues were incorporated. Furthermore, the absence of tillage slowed the overall turnover of SOM. Others have reported an accumulation of the more labile components of SOM in the surface soil of reduced tillage systems including microbial biomass (Carter, 1991; Doran, 1987) and particulate organic matter (Angers et al., 1993; Cambardella and Elliot, 1992). However, reducing tillage intensity does not necessarily result in increased total SOM levels if crop inputs are maintained and erosion is controlled (Angers et al., 1993; Haynes and Knight, 1989).

Although the scheduler in the model was set to reflect the three tillage practices, the model predicted that there would not be any changes to microbial biomass. To simulate tillage effects on the microbial biomass, its specific decay rate constant was reduced from 4 yr^{-1} in the moldboard plough treatment to 1.7 yr^{-1} and 1 yr^{-1} in the ridge tillage and minimum tillage treatments, respectively. These adjustments improved the predictions of the effects of tillage on the microbial biomass and its proportion of the total SOM (Table 5).

C. Organic Matter Addition

Applications of solid beef cattle manure significantly affected SOM contents, and measured trends were correctly described by the model (Table 6). The model, however, over-estimated SOM levels at low manure application rates and under-estimated levels at high rates. Besides the organic C contained in the manure which would promote microbial activity and SOM formation, nutrients limiting crop growth were supplied with the manure applications and these increased crop productivity (N'dayegamiye, 1990). Over a 10-year period, corn yields on plots receiving manure were about double those of the unmanured treatment. Adjustments were made to the plant production sub-model to reflect actual crop production for each manure treatment. However, other factors which were not considered could have affected SOM levels. For example, the efficiency of manure C incorporation into SOM may be very different from that of crop residues.

Manure applications of at least 40 Mg ha^{-1} every two years, equivalent to an input of 217 kg C m^{-2} yr^{-1}, were required to maintain SOM levels at this site. These soils had been in long-term pasture prior to this research and presumably the levels of SOM were relatively high and approaching steady-state.

IV. Conclusion

Although not specifically designed to be used as a tool for examining management effects, this study has shown that the basic structure of the CENTURY model is sufficiently robust to accurately simulate short-term changes in SOM content due to management, i.e., SOM can be adequately described by three different pools: an active pool or microbial biomass, a slow pool and a passive pool. Critical elements controlling the dynamics of SOM constituents which are affected by management, such as the quantity and quality of organic matter inputs, and the effects of tillage on rates of SOM decomposition were easily evaluated using simulation techniques. The model can be useful in developing management guidelines for identifying those practices promoting the sequestration of organic matter in soil.

Acknowledgements

This research was conducted while the senior author was on sabbatical leave at Agriculture Canada, Ste-Foy, Québec. The research assistance provided by M. Bolinder is gratefully acknowledged. Financial support was provided by grants from Agriculture Canada and the Natural Sciences and Engineering Research Council of Canada.

References

Angers, D.A. 1992. Changes in soil aggregation and organic carbon under corn and alfalfa. *Soil Sci. Soc. Am. J.* 56:1244-1249.

Angers, D.A., A. N'dayegamiye, and D. Côté. 1993. Tillage-induced differences in organic matter of particle size fractions and microbial biomass. *Soil Sci. Soc. Am. J.* 57:512-516.

Angers, D.A. and A. N'dayegamiye. 1991. Effects of manure application on carbon, nitrogen, and carbohydrate contents of a silt loam and its particle-size fractions. *Biol. Fertil. Soils* 11:79-82.

Angers, D.A. and G.R. Mehuys. 1989. Effects of cropping on carbohydrate content and water-stable aggregation. *Can. J. Soil Sci.* 69:373-380.

Angers, D.A. and G.R. Mehuys. 1990. Barley and alfalfa cropping effects on carbohydrate contents of a clay soil and its size fractions. *Soil Biol. Biochem.* 22:285-288.

Balesdent, J., A. Mariotti, and D. Boisgontier. 1990. Effect of tillage on soil carbon mineralization estimated from ^{13}C abundance in maize fields. *J. Soil Sci.* 41:587-596.

Burke, I.C., C.M. Yonker, W.J. Parton, C.V. Cole, K. Flach, and D.S. Schimel. 1989. Texture, climate and cultivation effects on soil organic matter content in U.S. grassland soils. *Soil Sci. Soc. Am. J.* 53:800-805.

Cambardella, C.A. and E.T. Elliot. 1992. Particulate soil organic-matter changes across a grassland cultivation sequence. *Soil Sci. Soc. Am. J.* 56:777-783.

Carter, M.R. 1986. Microbial biomass as an index for tillage-induced changes in soil biological properties. *Soil Tillage Res.* 7:29-40.

Carter, M.R. 1991. The influence of tillage on the proportions of organic carbon and nitrogen in the microbial biomass of medium textured soils in a humid climate. *Biol. Fert. Soils* 11:135-139.

Cole, C.V., I.C. Burke, W.J. Parton, D.S. Schimel, D.S. Ojima, and J.W.B. Stewart. 1989. Analysis of historical changes in soil fertility and organic matter levels of the North American Great Plains. In: P.W. Unger (ed.) *Proceedings of the International Conference on Dryland Farming.* Amarillo, TX. 15-19 August. 1988. Texas A & M Univ., College Station, TX.

Doran, J.W. 1987. Microbial biomass and mineralizable nitrogen distributions in no-tillage and plowed soils. *Biol. Fert. Soils* 5:68-75.

Foth, H.D. 1965. Root and top growth of corn. *Agron. J.* 54:49-52.

Jenkinson, D.S., and J.H. Rayner. 1977. The turnover of soil organic matter in some of the Rothamsted classical experiments. *Soil Sci.* 123:298-305.

Juma, N.G. and W.M. McGill. 1986. Decomposition and nutrient cycling in agro-ecosystems. In: M.J. Mitchell and J.P. Nakas (eds.), *Microfloral and faunal interactions in natural and agro-ecosystems.*

Hansson, A.-C., R. Petterson, and K. Paustian. 1987. Shoot and root production and nitrogen uptake in barley, with and without nitrogen fertilizer. *J. Agron. Crop Sci.* 158:163-171.

Haynes, R.J. and T.L. Knight. 1989. Comparison of soil chemical properties, enzyme activities, levels of biomass N and aggregate stability in the soil profile under conventional and no-tillage in Canterbury, New Zealand. *Soil Tillage Res.* 14:197-208.

Kay, B.D. 1990. Rates of change of soil structure under different cropping systems. p. 1-52. In: B.A. Stewart (ed.), *Advances in Soil Science.* Vol.12, Springer-Verlag, New York, Inc.

McGill, W.B., K.R. Cannon, J.A. Robertson, and F.D. Cook. 1986. Dynamics of soil microbial biomass and water-soluble organic C in Breton L after 50 years of cropping to two rotations. *Can. J. Soil Sci.* 66:1-19.

N'dayegamiye, A. 1990. Effets à long terme d'apports de fumier solide de bovins sur l'évolution des charactéristiques chimiques du sol et de la production de maïs-ensilage. *Can. J. Plant Sci.* 70:767-775.

N'dayegamiye, A. and D.A. Angers. 1990. Effets de l'apport prolongé de fumier de bovins sur quelques propriétés physiques et biologiques d'un loam limoneux Neubois sous culture de maïs. *Can. J. Soil Sci.* 70:259-262.

Parton, W.J., J.W.B. Stewart, and C.V. Cole. 1988. Dynamics of C,N,P and S in grassland soils: a model. *Biogeochem.* 5:109-131.

Parton, W.J., D.S. Schimel, C.V. Cole, and D.S. Ojima. 1987. Analysis of factors controlling soil organic matter levels in Great Plains grasslands. *Soil Sci. Soc. Am. J.* 51:1173-1179.

Paustian, K., O. Andrén, M. Clarholm, A.-C. Hansson, G. Johansson, J. Lagerlöf, T. Lindberg, R. Pettersson, and B. Sohlenius. 1990. Carbon and nitrogen budgets of four agroecosystems with annual and perennial crops, with and without N fertilization. *J. Appl. Ecol.* 27:60-84.

Paustian, K., W.J. Parton, and J. Persson. 1992. Modeling soil organic matter in organic-amended and nitrogen-fertilized long-term plots. *Soil Sci. Soc. Am. J.* 56:476-488.

van Veen, J.A. and E.A. Paul. 1981. Organic C dynamics in grassland soils. 1. Background information and computer simulation. *Can. J. Soil Sci.* 61:185-201.

Voroney, R.P., J.P. Winter, and R.P. Beyaert. 1993. Soil microbial biomass C and N. In: M.R. Carter (ed.), *Soil sampling and methods of analysis.* Lewis Publishers.

CHAPTER 11

Modeling the Impacts of Agricultural Management Practices on Soil Carbon in the Central U.S.

A.S. Donigian, Jr., A.S. Patwardhan, R.B. Jackson, IV., T.O. Barnwell, Jr., K.B. Weinrich, and A.L. Rowell

I. Introduction

As part of the U.S. EPA Climate Change Program, a research effort was sponsored to evaluate the potential to sequester additional carbon, and thereby reduce atmospheric CO_2 emissions in agricultural systems through selected management practices and polices. This paper, in coordination with other papers in the symposium (submitted for presentation), describes the soil carbon modeling procedures used to evaluate the impacts of cropping practices, crop rotations, tillage practices, soils, and climate on soil carbon levels in the Central U.S.

The study region extends from the Great Lakes to Eastern Colorado, and from the Canadian border to northern Arkansas, thereby including portions of the agricultural regions of the Cornbelt, Great Plains, and Great Lakes (Figure 1). The impacts of projections reflecting increased adoption of conservation tillage and increased use of cover crops are compared to "status quo" conditions based on current practices and policies. Comparisons are made in terms of both the "unit area" impact of crop rotations and practices throughout the study region, and the cumulative impacts for the entire study region to assess the potential role of agricultural management, within a national and global perspective, to sequester soil carbon for mitigation of global warming.

A. Overview and Methodology Components

Barnwell et al (1993) presented the overall operational strategy for the study in terms of the high-level integration of models, databases, and their linkage for assessment of policy impacts on agriculture and resulting soil carbon levels. The objective is to estimate greenhouse gas emissions and C sequestration potential of agricultural production systems, and the impact of various policy alternatives designed to reduce emissions by increasing soil C levels.

In order to effectively use models to assess the soil carbon sequestration potential of agricultural regions of the U.S., a *framework, or methodology*, was needed to address issues of model integration, data needs and availability, temporal and spatial scales of analysis, detail of representation of agricultural practices, and associated technical model application questions. The integration of these concerns is implemented through the overall project methodology that was followed to assess both current conditions of soil carbon, and soil organic matter, in agricultural systems, and projected changes. These changes were evaluated under a continuation of current conditions (i.e. status quo) and selected alternative futures derived from policy changes.

ISBN 1-56670-117-1/95/$0.00+.50

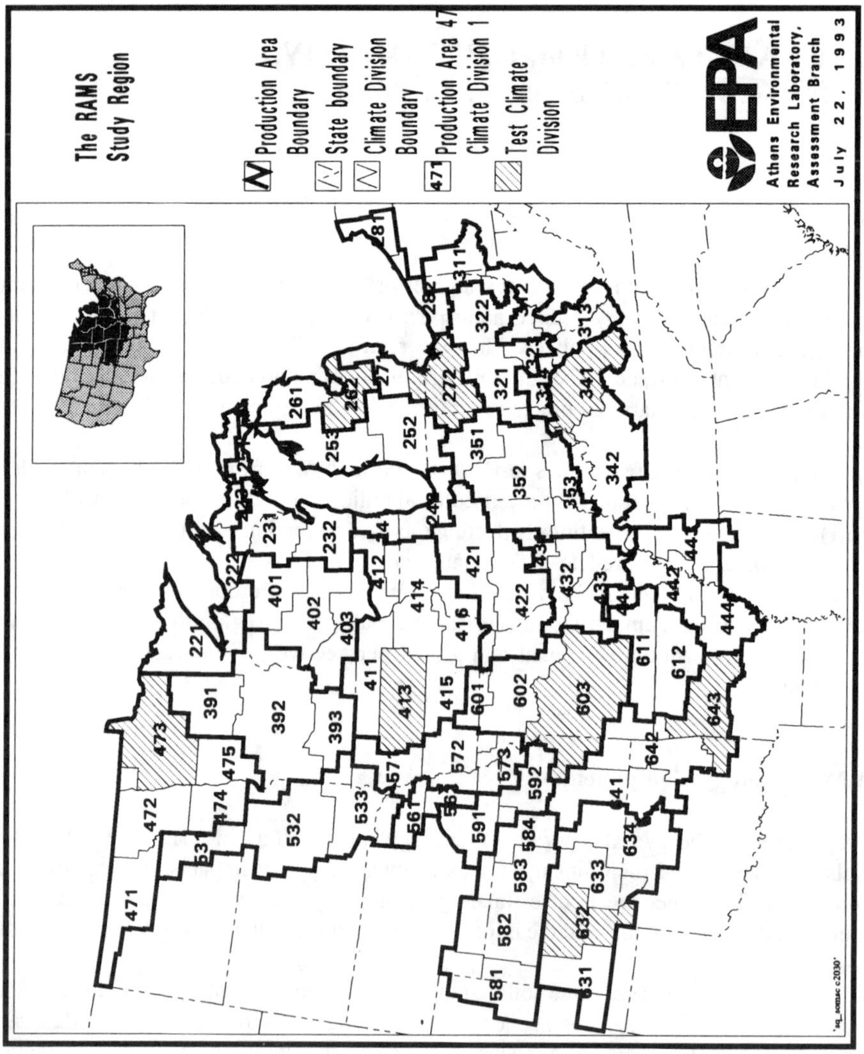

Figure 1. Study region, production areas, and climate divisions.

The primary components of the modeling system include: the Resource Adjustment Modeling System (RAMS) (Bouzaher et al., 1992) to provide baseline production systems and cropping practices, and changes due to policy alternatives; the CENTURY soil carbon model (Parton et al, 1988) to provide unit-area (i.e. per hectare) changes in soil carbon and greenhouse gas emissions for each designated agricultural production system; the EPA Agroecosystem Carbon Pools database (Eliott, 1993, personal communications) to provide field site data for model testing and validation; and the EPA Athens GIS capabilities to integrate the unit-area soil C and emission values with cropping acreages for display and analysis of impacts of baseline and policy alternatives. These components are discussed by Barnwell et al (1993) and other papers in this symposium.

A review of soil carbon models was performed and led to the selection of the CENTURY model for use in the modeling component of the study. A model was needed for this study due to the lack of a sufficiently comprehensive SOM data base for the entire study region, including all crops, soils, rotations, and impacts of tillage practices. Although soil carbon/soil organic matter models are still largely "research-type" tools, CENTURY has been applied and tested on a variety of sites throughout the world, and has shown an acceptable degree of success in reflecting observed soil data. However, further refinement was needed to represent the range of crops, cropping/tillage practices, and crop rotations common to U.S. agriculture. (see Metherell, 1992).

B. CENTURY and RAMS Model Overviews

Parton et al. (1988) developed the CENTURY model to simulate the dynamics of C, N, P, and S in cultivated and uncultivated grassland soils. The model uses monthly time steps for simulation, and model runs can be performed for time periods ranging from 10 to 10,000 years. CENTURY has submodels for estimating plant growth and soil water balance. These submodels provide information on plant litter inputs and climatic variables for soil organic matter computations. Options in the model allow the user to either specifically define fertilizer and irrigation applications, or direct the model to calculate (and satisfy) the plant nutrient and moisture needs based on user-defined criteria e.g. fraction of optimum production or moisture deficits at which irrigation is initiated. These options were used in this study to avoid the need to develop extensive data on actual fertilizer and irrigation applications.

The Resource Adjustment Modeling System (RAMS) was developed in 1990 by the Center for Agricultural and Rural Development (CARD) at Iowa State University. RAMS is a regional, short-term, static, profit maximizing, linear programming model of agricultural production defined at the producing area level. RAMS provides the CENTURY model information about production patterns and production practices (i.e. crops, rotations, tillage, and production inputs) throughout the study region. Such linking of environmental and economic realms generates a more comprehensive evaluation of policies designed to sequester carbon in agricultural soils.

C. Initial Study Region and Production Areas

This initial assessment was restricted to the RAMS study region shown in Figure 1; this region encompasses the traditional agricultural regions of the Corn Belt, the Lakes States, and portions of the Northern Plains. Approximately 216 million acres of cropland are included in the RAMS region, representing 60-70% of the total U.S. cropland, depending on specific definitions of harvested versus total cropland acres (USDA, 1992). In addition, the RAMS region is *25% of the continental U.S.*, and the *216 million cropland acres represents 44% of the entire study region.* Also shown in Figure 1 are the RAMS 'Production Areas' (PAs) representing the basic spatial unit of analysis for the economic and policy analysis. PAs are hydrological areas, adjusted to county boundaries, representing aggregated subareas defined by the U.S. Water Resources Council (1970). The areas are small enough so that the assumption of homogenous production across the area can be reasonably made.

II. Project Methodology

The project modeling methodology was developed to integrate the issues of spatial detail of the required data, modeling of agricultural practices, estimation of current or initial soil carbon conditions, and procedures for assessment of alternative future conditions resulting from policy scenarios. An overriding concern throughout the work was to strike a balance between the number of combinations of soils, climate, crop rotations, tillage practices, and associated model runs needed to adequately represent the heterogeneities of the study region, within the time and resource constraints of the assessment.

Figure 2 shows the basic steps in the methodology and its application to the RAMS study region; these steps are discussed in more detail below:

1. Divide the study region into "Climate Divisions" (CD) that can be represented with individual climate timeseries of the monthly precipitation and max/min air temperature needed by CENTURY. The CD boundaries were defined to coincide with both Production Area (PA) and county boundaries to facilitate the transformations (i.e aggregation/disaggregation) of croppingpractice and soils data. Figure 1 shows the climate divisions along with PA and state boundaries. *The "Climate Division" is the basic spatial unit of analysis for this study*. Jackson et al (1993) discuss the CD boundaries and associated meteorologic data.

2. For each CD, identify the distribution of soil textures on the agricultural cropland. We used the NRI/SOILS5 database (Goebel, 1987) and the EPA DBAPE system (Imhoff et al., 1990) to determine the surface soil textural distributions for all counties in each CD (see Jackson et al., 1993).

3. For each CD, identify the primary crops/rotation/tillage (C/R/T) scenarios as defined by the RAMS output for the "status quo" condition and alternative policy conditions. The RAMS output identified 80 different rotations used in the study region; these were grouped into 35 rotations for modeling purposes, based on similarity of crops, cropping sequence, and modeling capabilities. The RAMS output for each PA (and each policy alternative) is transformed to a CD based on the ratio of "county cropland" to "PA cropland" and then aggregated for the specific counties in each CD.

4. Establish Initial Conditions (Current, i.e. 1989-90) for SOC (and other needed model variables) to be used for each C/R/T for each soil texture within each CD. This was based on calibrating model results to historical crop yields from 1907 through 1989 for the major rotations and practices in each CD. (See Project Report by Donigian et al., 1993).

5. For each CD, extend the historical CENTURY model runs (from Step 4, above) through the 40-year projection period of 1990-2030 for each C/R/T and each soil texture group, and output SOC (year-end values, gC/m^2) and other model output for analysis.

6. Weight the output from each soil-C/R/T (CENTURY) run by the soil texture distribution for that CD to get one time series curve of SOC for each C/R/T within each CD.

7. For each CD, we then multiply the weighted curve (e.g. for SOC, gC/m^2) by the area for each C/R/T combination; these areas are provided by the RAMS output for each PA, which is transformed to a CD basis. The resulting values of SOC for each C/R/T are then analyzed to determine the total SOC changes for each CD. Since the projections are available for each yearof the projection period, the analysis can be done on an annual basis, in 10-year increments (i.e. 1990–2000, 2000–2010, 2010–2020, 2020–2030), or for the entire 40-year period.The totals for each CD are then summed to provide totals for the entire study region for the entire projection period or any subset thereof.

8. The evaluation of "Status Quo" or baseline conditions, and comparison with each alternative scenario, is based on the relative impacts on SOC changes and N emissions. Alternatives that simply change the area of the rotations (i.e. each C/R/T combination) do not require additional model runs;

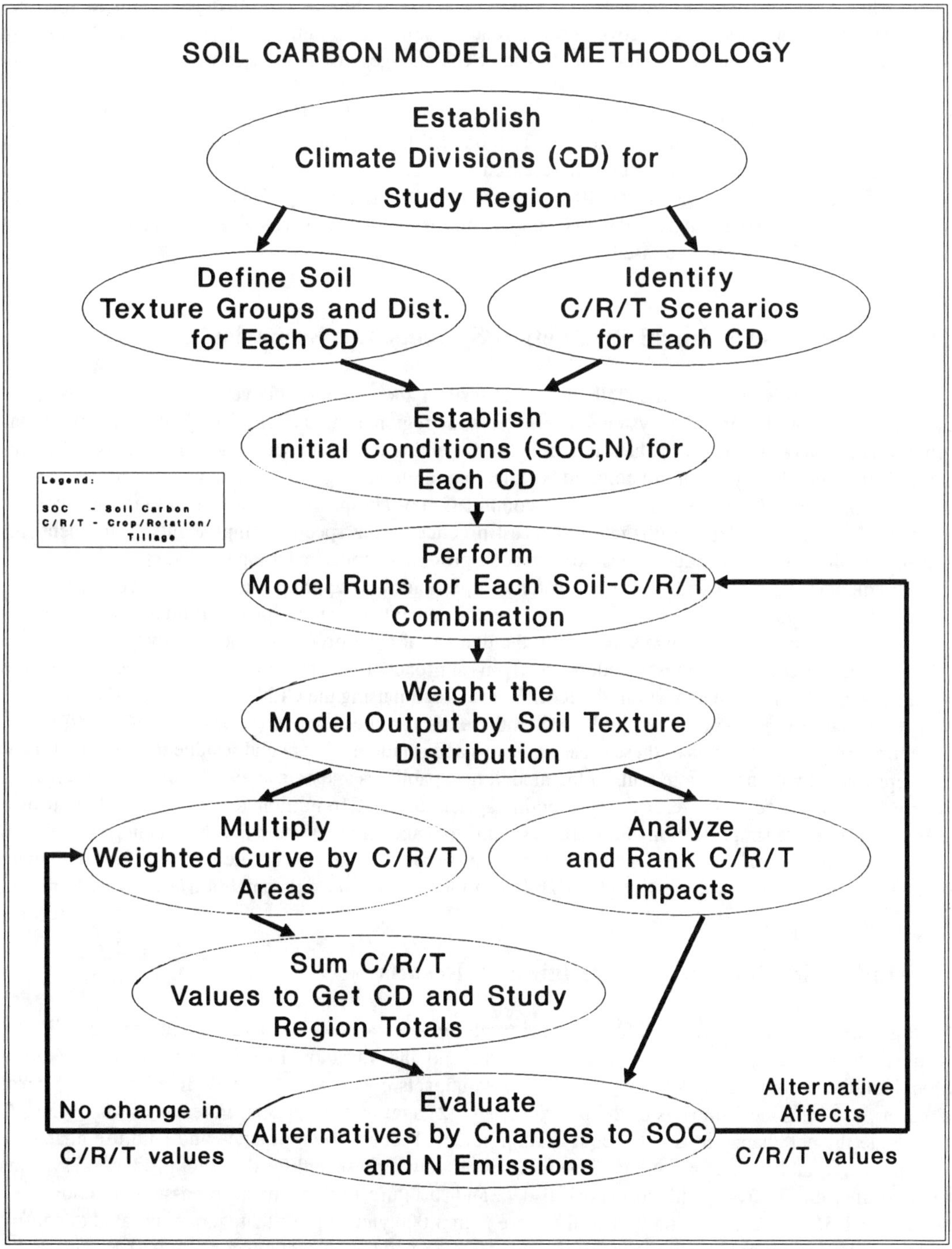

Figure 2. Soil carbon modeling methodology.

Step 7 is repeated for each new set of rotation acreages, in a simple spreadsheet-type calculation. Alternatives that lead to changes in the individual C/R/T curves (e.g. changes in tillage practices, new rotations, addition of cover crops) require new model runs, so Steps 5 through 7 are all repeated, along with calculation of the total SOC changes and N emissions.

A key factor in the modeling of SOC during the projection period of 1990 to 2030 is the assumption of changes, primarily increases, in crop yields. Based on U.S.D.A. estimates derived from the deliberations of a group of agronomic and agricultural research experts, *an annual increase of 1.5% in yields was assumed* for each crop throughout the 40-year period (U.S.D.A., 1990). This assumption has a critical impact on projected changes in SOC due to its impact on carbon inputs to the soil; higher yields will provide increased carbon inputs and increased SOC.

III. Modeling Agricultural Production Systems and Scenarios

The objectives of this study and the methodology described above required development of methods to (1) represent agricultural production systems within the study region using the CENTURY model, (2) estimate initial soil carbon (and nitrogen) values as a basis for projecting future changes, and (3) define both baseline (i.e. current) and alternative future conditions reflecting the modeling scenarios developed by RAMS. An "*agricultural production system*" is a complex combination of (1) specific crops grown in defined rotation patterns; (2) seasonally defined tillage and harvest practices using specified implements (equipment) that control the soil, crop, and residue disposition; and (3) agronomic inputs including fertilizers, manure, and/or organic amendments along with irrigation water and pesticides, as appropriate. All these components are integrated within the local climate and soil resource environment, and are ultimately influenced by the policy and market (economic) mechanisms that drive the demand for the products produced by the system.

The Project Report describes the details of our approach for representing the physical aspects of agricultural production systems within the RAMS study region using the CENTURY model; the policy and market mechanisms are represented by RAMS and are discussed elsewhere (Bouzaher et al., 1993). Each of the three major components of these systems - crops and rotations, tillage and residue management inputs - are represented within the limitations of the modeling capabilities and data available for this investigation. Clearly, as in any modeling exercise, representing agriculture on 216 million acres with 100,000 farms or more, requires some simplification of the "real system" to make the assessment feasible. Thus, our approach was to focus on the dominant crops, rotations, and tillage practices, and to represent the soil carbon impacts of these combinations within the spatial variability of the climate and soils conditions of the study region.

IV. Model Simulations and Preliminary Results

For this initial assessment we have focussed primarily on soil carbon (SOC) values, and changes in SOC as impacted by the agricultural production systems and the variation in soils and climate conditions throughout the study region. We have analyzed these impacts in two ways: (1) unit area impacts, in terms of gC/m^2, to assess local impacts of the practices, and (2) aggregated impacts, in terms of total grams C, to evaluate the net change in C (i.e. sequestration potential) due to both practice and land use changes.

In order to establish that carbon inputs were appropriately represented for different production systems, we calibrated model crop yields to historical data and to future projections at the assumed annual yield increase of 1.5%. Figure 3 shows four of the more than 3,000 yield curves that were generated to confirm model predictions for both the historical yields for 1924-88 and the projections for 1990-2030. The graphs in Figure 3 are for the major crops and rotations, and associated tillage practices, for CDs 272, 413, 473, and 611; these rotations occupied the largest percent of cropland within the respective CDs. The agreement between historical and simulated yields shown in Figure 3 is typical of the results obtained for all other C/R/T combinations in the study region, and confirms the models representation of the crop yield portion of the methodology.

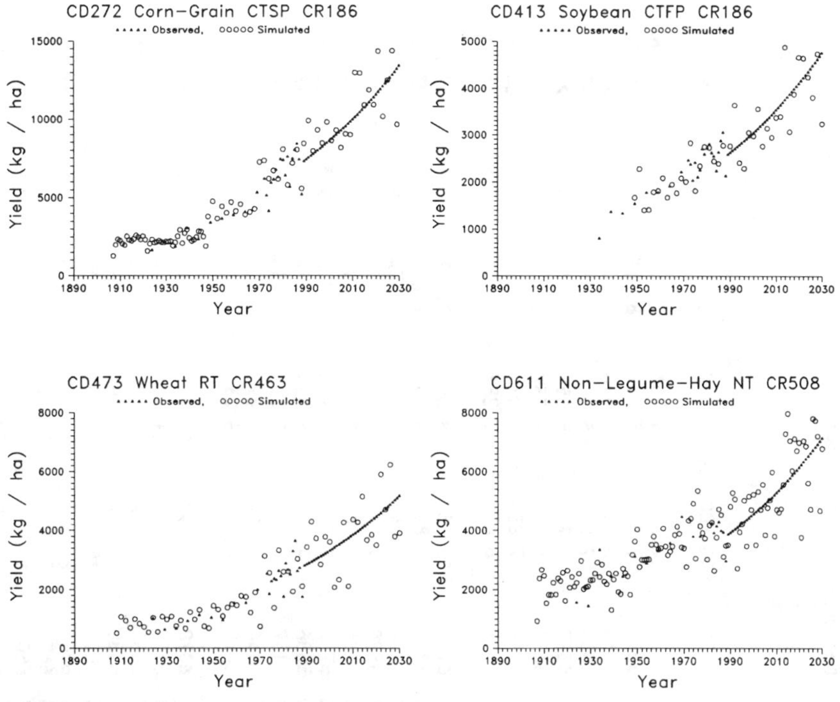

Figure 3. Simulated vs actual crop yields for corn, soybean, wheat, and non-legume hay for selected CD's, rotations, and tillage practices.

A. Impacts of Crops, Rotations, and Tillage Practices

Table 1 shows model predictions of SOC for rotations and tillage combinations for selected CDs distributed throughout the study region; the locations of these CDs are shown in Figure 1. The values in Table 1 are year-end conditions at ten-year intervals from 1980 to 2030, and the percent change is calculated for each ten-year period starting in 1990 and for the entire 40-year projection period, 1990-2030. The SOC values pertain to the *top 20 cm of the soil profile* as this is the soil layer depth represented by CENTURY. The Project Report includes results for all CDs within the study region.

Figure 4 is an example of the CENTURY model predictions for SOC for the entire simulation period of 1907-2030 for CD 413 in central Iowa. These are presented to demonstrate the historical and future trends in SOC predicted by CENTURY under the model application assumptions of the study. Analysis of the model results provided the following observations:

1. Values in Table 1 are generally representative of those for all CDs and C/R/T combinations; the percent change for the 40-year projection period is typically about *20 to 50%*. However, selected C/R/T combinations have much larger and smaller changes, including some with negative changes i.e. loss of SOC during the projection period.

   The greatest percent changes, ranging up to 170%, usually occur when a dominant hay or grain rotation, simulated during the historic period, is followed by a corn or corn-silage based rotation during the 40-year projection period. This is due to the large biomass production of corn, and the resulting high level of carbon inputs to the soil for these rotations. Conversely, the smallest increases, and some decreases, in SOC are for the reverse situation.

2. Historical practices, as represented by the model assumptions used in this study, have led to decreases in SOC until about the 1940's and 1950's. Since that time period, the model predictions

Table 1. Soil cargon (gC/M^2) and percent change in soil carbon for selected climatic divisions, predicted by CENTURY

				Soil Carbon (gC/m2)						Percent Change in Soil Carbon for Selected Time Periods				
CD	CR #	ROTATION SEQUENCE	TILL	1980	1990	2000	2010	2020	2030	1990/2000	2000/2010	2010/2020	2020/2030	1990/2030
272	100	CRN	NT	4137	4421	4965	5426	5858	6219	12.30%	9.28%	7.96%	6.16%	40.66%
272	138	CRN,CRN,SOY,WWT,HLH	CTSP	4137	4399	4743	5016	5443	5696	7.81%	5.75%	8.51%	4.64%	29.48%
272	186	CRN,SOY	CTSP	4137	4399	4699	4930	5218	5461	6.81%	4.91%	5.84%	4.65%	24.14%
272	186	CRN,SOY	RT	4137	4401	4710	4956	5266	5534	7.02%	5.22%	6.25%	5.08%	25.74%
272	186	CRN,SOY	NT	4137	4420	4673	4909	5155	5372	5.72%	5.05%	5.01%	4.20%	21.53%
272	201	CRN,SOY,WWT	RT	4137	4403	4706	4910	5141	5451	6.88%	4.33%	4.70%	6.02%	23.80%
272	243	CSL,CSL,OTS,HLH,HLH	CTSP	4137	4391	5174	5747	6645	7396	17.83%	11.07%	15.62%	11.30%	68.43%
272	366	OTS,NLH,NLH,NLH	CTSP	4137	4417	4684	5025	5355	5972	6.04%	7.28%	6.56%	11.52%	35.20%
413	100	CRN	CTSP	4486	4779	5428	5895	6326	6749	13.58%	8.60%	7.31%	6.68%	41.22%
413	138	CRN,CRN,SOY,WWT,HLH	CTSP	4486	4779	5059	5263	5664	5899	5.85%	4.03%	7.61%	4.14%	23.43%
413	138	CRN,CRN,SOY,WWT,HLH	RT	4486	4782	5052	5264	5660	5900	5.64%	4.19%	7.52%	4.24%	23.37%
413	186	CRN,SOY	CTFP	4486	4782	4980	5138	5331	5588	4.14%	3.17%	3.75%	4.82%	16.85%
413	186	CRN,SOY	CTSP	4486	4779	5025	5210	5439	5734	5.14%	3.68%	4.39%	5.42%	19.98%
413	186	CRN,SOY	RT	4486	4781	5040	5240	5470	5765	5.41%	3.96%	4.38%	5.39%	20.58%
413	186	CRN,SOY	NT	4486	4800	5025	5229	5450	5686	4.68%	4.05%	4.22%	4.33%	18.45%
413	243	CSL,CSL,OTS,HLH,HLH	CTSP	4486	4779	5593	6227	7207	8157	17.03%	11.33%	15.73%	13.18%	70.68%
413	503	HLH,HLH,HLH,HLH	CTSP	4486	4788	4803	5115	5413	6134	0.31%	6.49%	5.82%	13.31%	28.11%
413	508	NLH,NLH,NLH,NLH	CTFP	4486	4790	4905	5309	5714	6489	2.40%	8.23%	7.62%	13.56%	35.46%
473	002	BAR,BAR,SOY	CTSP	3803	3921	4607	5094	5555	6078	17.49%	10.57%	9.04%	9.41%	55.01%
473	002	BAR,BAR,SOY	RT	3803	3915	4621	5135	5618	6159	18.03%	11.12%	9.40%	9.62%	57.31%
473	100	CRN	NT	3803	3898	4398	4816	5278	5665	12.82%	9.50%	9.59%	7.33%	45.33%
473	125	CRN,CRN,OTS,NLH,NLH	CTSP	3803	3915	4672	5219	5833	6380	19.33%	11.70%	11.76%	9.37%	62.96%
473	145	CRN,CRN,WWT,HLH,HLH,HLH	CTFP	3803	3916	4525	5068	5555	6027	15.55%	12.00%	9.60%	8.49%	53.90%
473	215	CRN,SWT,SWT	CTSP	3803	3915	4761	5435	6031	6641	21.60%	14.15%	10.96%	10.11%	69.62%
473	215	CRN,SWT,SWT	RT	3803	3912	4742	5422	6036	6687	21.21%	14.33%	11.32%	10.78%	70.93%
473	215	CRN,SWT,SWT	NT	3803	3898	4591	5251	5883	6506	17.77%	14.37%	12.03%	10.58%	66.90%
473	262	CSL,OTS,HLH,HLH,HLH	CTFP	3803	3921	4712	5282	6111	6892	20.17%	12.09%	15.69%	12.78%	75.77%
473	463	SMF,SWT	CTSP	3803	3880	4080	4181	4369	4588	5.15%	2.47%	4.49%	5.01%	18.24%
473	463	SMF,SWT	RT	3803	3880	4100	4204	4414	4637	5.67%	2.53%	4.99%	5.05%	19.51%
473	508	NLH,NLH,NLH,NLH	CTFP	3803	3919	4416	4989	5462	6519	12.68%	12.97%	9.48%	19.35%	66.34%

Table 1 continued--

643	100	CRN	NT	1814	1954	2462	2832	3181	3408	25.99%	15.02%	12.32%	7.13%	74.41%
643	131	CRN,CRN,SOY	CTSP	1814	1940	2314	2495	2687	2892	19.27%	7.82%	7.69%	7.62%	49.07%
643	131	CRN,CRN,SOY	RT	1814	1943	2328	2526	2736	2935	19.81%	8.50%	8.31%	7.27%	51.05%
643	218	CRN,WWT	CTFP	1814	1958	2354	2551	2814	3029	20.22%	8.36%	10.30%	7.64%	54.69%
643	366	OTS,NLH,NLH,NLH	CTSP	1814	1942	1988	2112	2298	2544	2.36%	6.23%	8.80%	10.70%	30.99%
643	490	WWT	RT	1814	1888	2166	2301	2586	2805	14.72%	6.23%	12.38%	8.46%	48.56%
643	503	HLH,HLH,HLH,HLH	CTSP	1814	1907	1820	1848	1986	2153	-4.56%	1.53%	7.46%	8.40%	12.89%

CD	-	Climate Division	CR	-	Crop Rotation
CTSP	-	Conventaion Till Fall Plow	CTSP	-	Conventional Till Spring Plow
RT	-	Reduced Till	NT	-	No Till
BAR	-	Barley	OTS	-	Oats
CRN	-	Corn	SMF	-	Summer Month Fallow
CSL	-	Corn Silage	SOY	-	Soybean
HLH	-	Legume Hay	SWT	-	Spring Wheat
NLH	-	Non-Legume Hay	WWT	-	Winter Wheat

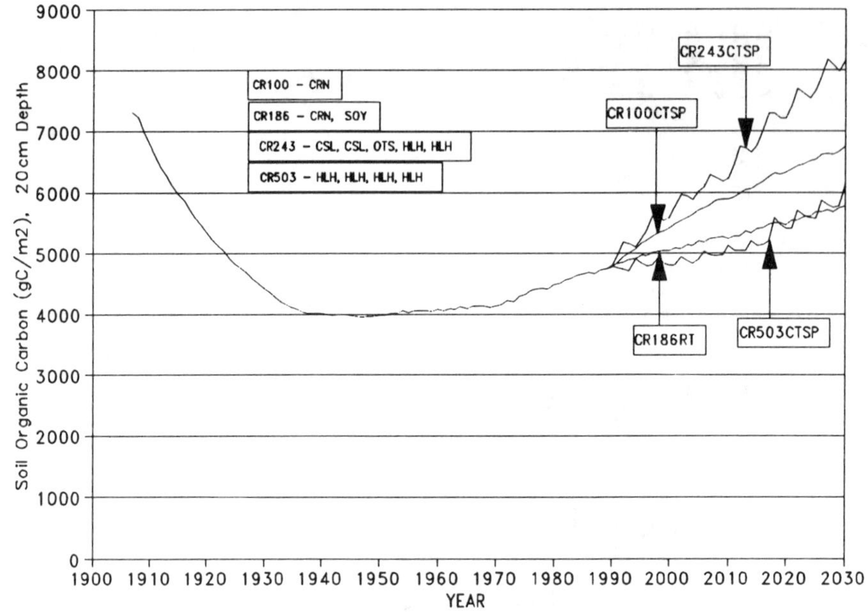

Figure 4. SOC changes and impacts of agricultural production systems on SOC in CD413, 1907-2030, predicted by the CENTURY model.

of SOC are increasing for most crop production systems that leave significant amounts of crop residue on the field. At this time, these predicted trends appear to be reasonable and consistent with the application assumptions; however, further direct confirmation of these predictions is needed.

3. Except for a few hay or grain-based rotations, in CDs located in the northern or western portions of the study region, the model results show *consistent and significant increases in SOC* throughout the projection period. This is due primarily to the assumption of a 1.5% annual increase in crop yields. Sensitivity runs with CENTURY (Donigian et al., 1993) indicate that if the annual yield increase is only 0.5%, the increase in SOC would be reduced by a factor of two or more, and for some C/R/Ts there would be no significant change during the projection period.

4. The impacts of the *tillage* practices - conventional-till, reduced-till, no-till - are much less than generally assumed by conventional wisdom. For dominant rotations simulated during both the historic and projection periods, differences are usually only a few (3 to 5) percentage points for tillage alternatives. This amount of change is probably not significant. For other rotations, differences are larger, but could be due to a mismatch or inconsistency between the dominant and projection period rotation. These preliminary results are being re-evaluated and revised (Donigian et al., 1993).

5. Usually the largest increase in SOC occurred for RT, and in many cases SOC under NT was less than both RT and conventional-till. The reasons for this are two-fold: (1) although the methodology was designed to maintain essentially the same crop yields (and therefore the C inputs to the soil) for all tillage alternatives, the NT yields were often slightly less than the others; and (2) the parameter changes to distinguish RT and NT are uncertain and require further investigation. As noted above, these initial conclusions are being re-evaluated (Donigian et al., 1993).

6. Differences in the 1980 SOC conditions within a PA, i.e. among CDs with the same first two digits (e.g. 411, 412, 413) are due entirely to soil and climate differences since the historical management practices and dominant rotations were the same within the PA.

B. Impacts of Cover Crops

Implementation of cover crops in appropriate CDs and C/R/T combinations identified by the CARD analysis resulted in dramatic increases in SOC in many cases. The Project Report (Donigian et al., 1993) includes complete results for all CDs and rotations with cover crops which were too extensive to present here. Only 45 CDs out of the total number of 80 were considered conducive to cover crops due to climate, crops, and length of the growing season. As a result, these CDs were located in the central and southern portions of the study region.

The observations derived from the cover crop simulations include:

1. Use of cover crops can significantly increase SOC for many C/R/T combinations, especially for CDs located in the southern portions of the study region. For example, in CDs 603 and 612 the increase during the 40-year projection period is on the order of 50% greater with cover crops. For some CDs and C/R/T combinations, the increase is up to twice as much with cover crops.

2. For the CDs located in the central portion of the study region, the SOC increase is much less, and in some cases SOC decreases under the cover crop scenario due to yield reductions in the crops following the cover crop. In these areas the benefits of cover crops would be "marginal" at best.

3. Even if climate conditions are favorable for cover crops, the extent of the impact on SOC depends on the specific crop rotation sequence. Thus, crop rotation 350 (CSL,CSL,SOY) can support cover crops in each winter (inter-crop) period, whereas rotation 186 (CRN,SOY) includes a cover crop only after the soybean crop, i.e. only 50% of the time. Clearly, the more often a cover crop is grown, the greater the potential impact on SOC.

C. Aggregate SOC Impacts for CDs and the Study Region

To evaluate the aggregate impact on SOC within each CD and the entire study region, the unit SOC soil storage values discussed above were converted to total SOC, i.e. gC, by multiplying times the appropriate areas associated with each C/R/T combination and summing the values for the CD. The results of this are shown in Figures 5 and 6, spatial displays of SOC by CD for 1990 and 2030, respectively, for the Status Quo scenario prepared by the EPA Athens GIS group. Below we have summarized the net change in SOC for the Status Quo; Low, Medium, and High Conservation; and Cover Crop scenarios simulated in the study. Note that the level of usage of conservation tillage increased from 30% under the Status Quo to 33%, 37%, and 56% under the three tillage levels, respectively.

Modeled	Gain in SOC, 1990-2030	
Scenarios	Gigatonnes C	Percent of 1990 Value
Status Quo/Baseline	1.13	31.7%
Low Conservation	1.13	31.7%
Medium Conservation	1.13	31.7%
High Conservation	1.08	30.5%
Cover Crops	1.21	34.0%

From our modeling results presented here, with greater detail in the Project Report, the following observations are provided:

1. Under the Status Quo or Baseline scenario, *1.13 Gt C* will be retained or sequestered within the study region from 1990 to 2030 under the Status Quo assumptions described earlier. This represents a *32% increase* over the current 1990 level of SOC of *3.56 Gt C in the study region* based on the model calculations.

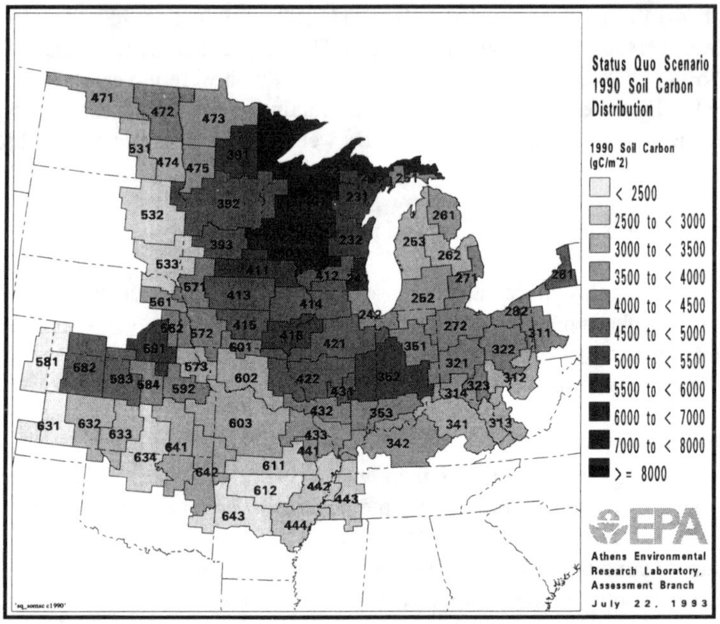

Figure 5. CENTURY simulation of 1990 soil carbon (gC/m^2) distribution within the study region.

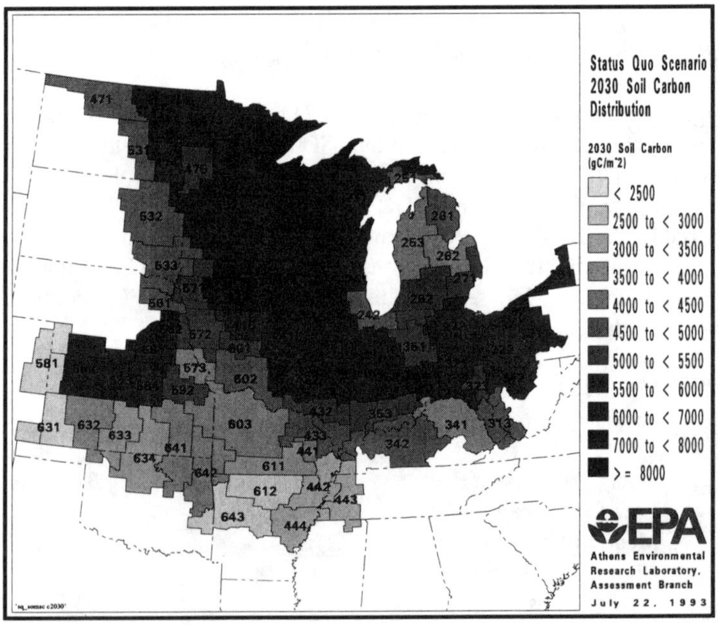

Figure 6. CENTURY simulation of 1990 soil carbon (gC/m^2) distribution within the study region.

2. The range in increase in SOC among the CDs is from 12% to 70% during the 40-year projection period.

3. The three conservation tillage alternatives had no significant impact on SOC or carbon sequestration in the study region as a whole. The differences shown above, especially for the High Conservation alternative, are not considered significant within the accuracy of the application methodology, and are currently being re-evaluated (Donigian et al., 1993). For the Medium Conservation scenario, the SOC change relative to the Status Quo ranges from +8% to -11%; thus the impacts vary among the CDs. Clearly, conservation tillage has other benefits (e.g. water quality) not considered in this study.

4. *Cover crops have a significant impact on SOC* in the study region leading to a 34% increase from 1990, compared to 32% for Status Quo, and an *additional 0.08 Gt C sequestered* in the study region. Although these numbers are not large, they are the result of cover crops being *implemented on only 10%* of the cropland in the study region. Among the CDs with cover crops, the average SOC increase over the Status Quo was 26% with a range from -3% to over +150%.

5. The pattern of higher values of SOC in the northern and eastern portions of the study region, and lower values in the southern and western portions, is consistent with other soil carbon mapping efforts (e.g. Kern and Johnson, 1991). The 1990 and 2030 maps of SOC show the same generally spatial pattern, although with higher values for 2030.

6. The regions with the greatest change and percent change in SOC appear to be those CDs with dominant corn-based rotations and/or those located in the northern portions of the study region. The smallest absolute increases are generally in the south.

V. Summary Study Conclusions and Recommendations

The primary conclusions and recommendations from these initial study results are as follows:

1. Reasonable extrapolation of current agricultural practices and trends will lead to an increase (sequestration) of *about 1 Gt C* within the study region by the year 2030. This represents a *32% increase* over current 1990 levels. Nationwide the increase could be 50% greater since our study region includes 60-70% of total U.S. cropland.

2. The key assumption underlying this prediction is an *annual crop yield increase of 1.5% per year.* The validity of this assumption needs to be re-assessed or confirmed, and if valid, policies and research need to be promoted to support the chances of agriculture attaining this level of yield increase.

3. Although *conservation tillage practices* can increase soil carbon in some combinations of crops, soils, and climate, the overall impact of increased reduced-till and no-till practices was *not significant for the study region as a whole*, as reflected in the approach for this preliminary effort and our limited knowledge of the quantitative impacts of tillage practices. More research and development is needed in this area, and these initial study conclusions are being revised (Donigian et al., 1993).

4. *Cover crops* can lead to significant increases in soil carbon in crop, soil and climate regimes where they are feasible and appropriate. Although only 10% of the study region cropland included cover crops (under the Cover Crop scenario), this increased soil carbon by *80 Mt* through 2030. Since southern and eastern portions of our study region were most appropriate for cover crops, this may be an attractive alternative for promoting carbon gains in the south and southeastern U.S.

In addition, the study indicated a number of recommendations related to refinements in the modeling procedures and application to the study region. The key areas for further investigation included: further model testing and validation, especially related to historical trends and current SOC levels, corn and soybean based rotations, and impacts of tillage practices; sensitivity of predictions to alternative yield assumptions;

impacts of erosion which was not considered in this study; and improved modeling procedures for the integrated assessment of SOC with the nitrogen cycle, and associated N_2O emissions, as impacted by agricultural management alternatives.

The study results provide a strong indication, even with the uncertainty associated with model-based projections, that agricultural trends are leading to generally improved soil fertility and increased SOC sequestration even without specific policies designed to promote these objectives. The modeling procedures and methodology described herein, with further refinements, can be used to help quantify both spatially and temporally the carbon sequestration potential of alternate practices and policies. In this way, perhaps we can better identify those policies that will support and nourish current agricultural trends for continued improvements in soil fertility and productivity, since these goals are coincident with our efforts to sequester carbon and mitigate global warming and climate change impacts.

Acknowledgements

This study was funded by the Environmental Research Laboratory in Athens, GA, under EPA Contract No. 68-CO-0019 with AQUA TERRA Consultants, Mountain View, CA. Mr. Tom Barnwell was the Project Officer from the EPA Athens ERL. His support and assistance was critical to the success of the project. In addition, we would like to acknowledge the support and participation of the Natural Resources Ecology Laboratory at Colorado State University, the Center for Agricultural and Rural Development at Iowa State University, and the Kellogg Biological Center at Michigan State University.

References

Barnwell, T.O., Jr. R.B. Jackson, IV, and L.A. Mulkey. 1993. An assessment of alternative agricultural management practice impacts on soil carbon in the corn belt. (This symposium). Environmental Research Laboratory, U.S. EPA, Athens, GA.

Bouzaher, A.B., D.J. Holtkamp, and J.F. Shogren. 1992. The Resource Adjustment Modeling System (RAMS). Research Report, Center for Agricultural and Rural Development, Iowa State University, Ames, Iowa.

Bouzaher, A.B., D.J. Holtkamp, R. Reese, and J.F. Shogren. 1993. Economic and resource impacts of policies to increase organic carbon in agricultural soils. (this symposium). Center of Agricultural and Rural Development.

Donigian, A.S. Jr., T.O. Barnwell, Jr., R.B. Jackson, IV, A.S. Patwardhan, K.B. Weinrich, A.L. Powell, R.V. Chinnaswamy, and C.V. Cole. 1993. Assessment of alternative management practices and policies affecting soil carbon in agroecosystems of the central United States. Final Report under EPA Contract No. 68-CO-0019, W.A. #13. Prepared for U.S. EPA, Environmental Research Laboratory, Athens, GA.

Goebel, J.J. 1987. National resources inventory. U.S. Soil Conservation Service. Resource Inventory Division. USDA Soil Conservation Service. P.O. Box 2890. Washington, D.C. 20013.

Imhoff, J.C., R.F. Carsel, J.L. Kittle Jr., and P.R. Hummel. 1990. Data Base Analyzer and Parameter Estimator (DBAPE) Interactive Program User's Manual. EPA/600/3-89/083. Environmental Research Laboratory, U.S. Environmental Protection Agency, Athens, GA.

Jackson, R.B., IV., A.L. Rowell, and K.B. Weinrich. 1993. Spatial modeling using partially spatial data. (this symposium). Environmental Research Laboratory, U.S. EPA, Athens, GA, and Computer Science Corporation, Athens, GA.

Kern, J.S. and M.G. Johnson. 1991. The impact of conservation tillage use on soil and atmospheric carbon in the contiguous United States. EPA/600/3-91/056. Environmental Research Laboratory, Corvallis, OR.

Parton, W.J., J.W.B. Stewart, and C.V. Cole. 1988. Dynamics of C, N, P and S in grassland soils: A Model. *Biogeochemistry* 5:109-131.

Metherell, A.K. 1992. Simulation of soil organic matter dynamics and nutrient cycling in agroecosystems. Ph.D. dissertation. Colorado State University, Fort Collins, Colorado.

U.S. Department of Agriculture. 1990. The second RCA appraisal. USDA Miscellaneous Publication No. 1482, Washington, D.C.

U.S. Department of Agriculture. 1992. Agricultural Resources: cropland, water, and conservation. Situation and outlook report. Economic Research Service. AR-27.

U.S. Water Resources Council. 1970. Water resources: regions and sub-regions for the national assessment of water and related land resources. U.S. Government Printing Office, Washington, D.C.

CHAPTER 12

Effects of Forest Management and Elevated Carbon Dioxide on Soil Carbon Storage

Dale W. Johnson and Phyllis Henderson

1. Introduction

On a global scale, soil organic matter (SOM) contains more C (approximately 1500-1600 x 10^{15} g) than either vegetation (500-800 x 10^{15} g) or the atmosphere (750 x 10^{15} g) (Post et al., 1990). Thus, there is justifiable concern over whether soils are a source or a sink for carbon. Of particular interest is the "missing sink" of C, which arises from the difference in CO_2 release by fossil fuels (approximately 6 x 10^{15} g) and the annual CO_2 increase in the atmosphere (approximately 3.4 x 10^{15} g) (Lugo 1992). It will be difficult, if not impossible, to definitively identify this missing sink because the background fluxes and pools are very large by comparison. For example, CO^2 fixation by photosynthesis is estimated at 120 x 10^{15} g, and autotrophic and heterotrophic respiration are each estimated at 60 x 10^{15} g (Post et al., 1990). Nonetheless, it is important to try to at least identify the direction, if not the exact magnitude, of C flux into and out of major global C reservoirs such as soils.

The role of soils as a source or sink for carbon on a global scale is somewhat controversial. Schlesinger (1990) estimated that soils can, at most, sequester about 0.4 x 10^{15} g yr^{-1} of carbon based upon a review of chronosequence studies. He also stated that, "if the terrestrial biosphere is indeed to act as a carbon sink under future elevated levels of carbon dioxide, this would be more likely to be a result of changes in the distribution and biomass of terrestrial vegetation than of changes in the accumulation of soil organic matter". Jenkinson et al. (1991) estimated that soils will release about 61 x 10^{15} g of C over the next 60 years (or about 1 x 10^{15} g yr^{-1}) due to climate warming based upon the Rothamsted model and a prediction of mean annual temperature rise of 0.03° C yr^{-1}. On the other hand, there is a substantial potential for sequestering C in soils by converting former agricultural land to forest (Mann, 1986; Brown et al., 1992) and by increasing the abundance of nitrogen fixers (Johnson 1992b).

This review will confine itself to two aspects of soil C that are currently under study at our institute: the effects of forest management (harvesting and nitrogen fixers) and the effects of elevated CO_2. These topics have received less attention than changes in land use and climate, but represent important feedbacks between atmosphere and the soil in the global C budget. For example, forest management may cause changes in soil C storage, affecting the atmospheric CO_2 levels, which in turn may affect both primary productivity and soil C storage. An assessment of the net effects of these processes requires modeling which is well beyond the scope of this paper; our purpose here is to present a review of the literature, updating previous reviews (Johnson and Ball, 1991; Johnson, 1992b) with recent results from literature and our own studies.

II. Effects of Forest Management

Soil C data from a variety of studies of forest harvesting, site preparation, fertilization, and N-fixers were recently summarized (Johnson, 1992b). While there were problems in this assessment arising from

ISBN 1-56670-117-1/95/$0.00+.50

differences in sampling protocols, the overall trends were quite clear: little change in soil C with harvesting, on average, regardless of sampling intensity or time since harvest, and large increases in soil C with the presence of N-fixers. There were exceptions to the general rule with regard to harvesting - some studies documented a large decrease in soil C and others documented a large increase - but if the results are scaled up as a crude average, there would be no net effect on global C budgets.

A. Harvesting

Since the last literature review (Johnson, 1992b), two additional studies of the effects of harvesting and site preparation have been completed. Alban and Perala (1992) studied ecosystem carbon contents of a chronosequence of aspen (*Populus tremuloides*) stands in Minnesota ranging in age from 0 to 80 years. They found that soil C stayed relatively constant throughout the stand's development over this time period, and that harvesting had no effect. Johnson (1993) found that the treatment of slash during site preparation had a major effect on the soil C in forest ecosystems of the eastern Sierra Nevada Mountains. At sites where the slash was piled and burned, soil C levels ranged from 49% lower to 27% higher than in control sites. On the other hand, harvested sites where the slash was masticated and spread had 45 to 120% more soil C than adjacent unharvested sites.

The data sets on harvesting used in the original review (Johnson, 1992b) have been updated with the Minnesota and Sierra Nevada results and are summarized in Figure 1. The results are presented separately for harvesting only, harvesting + site preparation (where residues where incorporated into the soil but not burned), and harvesting + residue burning. Percentage changes due to treatment are shown regardless of whether they were statistically significant at the given site or not. This approach was taken with the view that the individual data sets should be considered as analogous to points on a regression, whereas plotting of statistically significant differences only would over-emphasize the "no effects" results. The number of stastically significant differences in each category are shown as numbers above each bar.

The data summarized in Figure 1, top support the current view that forest harvesting alone has, on average, little effect on soil C (Detwiler, 1986; Brown et al., 1992) and not the earlier assumption that large (35-50%) losses of soil C normally occur following forest harvesting (Houghton et al., 1983). However, it is clear that harvesting can have a major effect on soil C (either positively or negatively) at specific sites, and that the direction and magnitude of this effect is highly dependent upon the amount of logging residue that is left on site and how it is treated during site preparation. Harvesting followed by site preparation which involves residue incorporation into the soil without burning can lead to large increases in soil C (Figure 1, middle), whereas harvesting followed by residue burning can lead to large decreases (Figure 1, bottom). As noted previously, harvesting followed by cultivation causes large losses of soil C in nearly all cases (Mann, 1986).

B. Nitrogen Fixers

The recent literature review of forest management (Johnson, 1992b) reconfirmed what has long been known: presence of nitrogen-fixing trees nearly always results in increased soil C and N (Crocker and Major, 1955; Tarrant and Miller, 1963; Zavitkovski and Newton, 1968; Binkley 1983; Boring et al., 1988). Our studies of *Ceanothus velutinus* in the Sierra Nevada showed that soils beneath Ceanothus sites had 20-50% more C (and N) than adjacent pine (*Pinus jeffreyi*) sites (Johnson, 1983). The Sierra Nevada results combined with other results from the literature show a nearly uniform increase in soil C with the presence of N fixers (Figure 2) (as compared to non-fixers).

A recent study of interest in terms of N-fixers is that of Wood et al., (1992). They investigated the effects of tree (mixed deciduous species) and herbaceous species on soil C and N accumulation in loblolly pine (*Pinus taeda*) stands in Alabama. They noted significant accretions of N in the stands with herbaceous species, and speculated that this was due to symbiotic N-fixation. Their results have significant implications to management practices involving weed control.

While the mechanism by which N fixers cause increased soil N accumulation seems self-evident (i.e., increased N input), the mechanism(s) by which they cause increased SOM have received little attention. In some cases, N fixers may cause increased SOM because of increased primary productivity (i.e., in disturbed or recently deglaciated sites; Crocker and Major, 1955) or reduced decomposition rates. However, this is

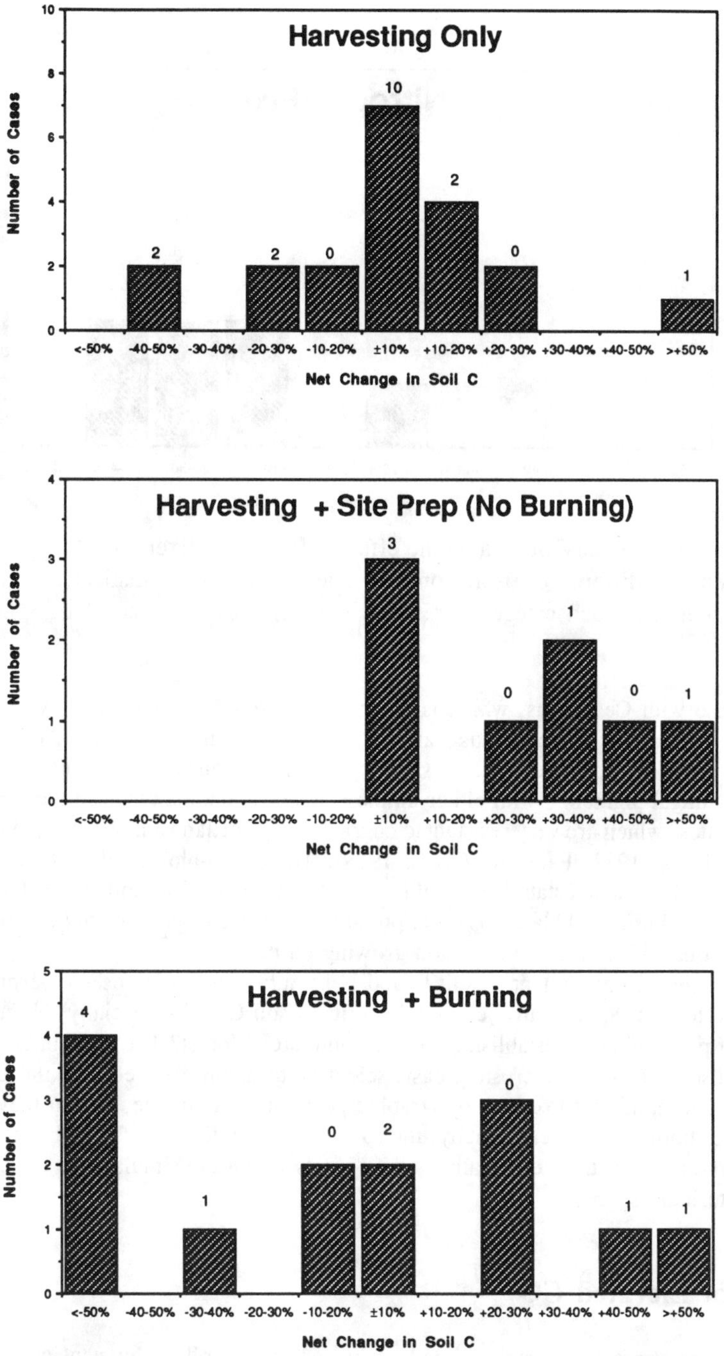

Figure 1. Summary of data on the effects of harvesting (top), harvesting + site preparation (middle), and harvesting + residue burning (bottom) on soil C expressed as percentage difference from control. The number of statistically significant differences in each category are shown as numbers over each bar.

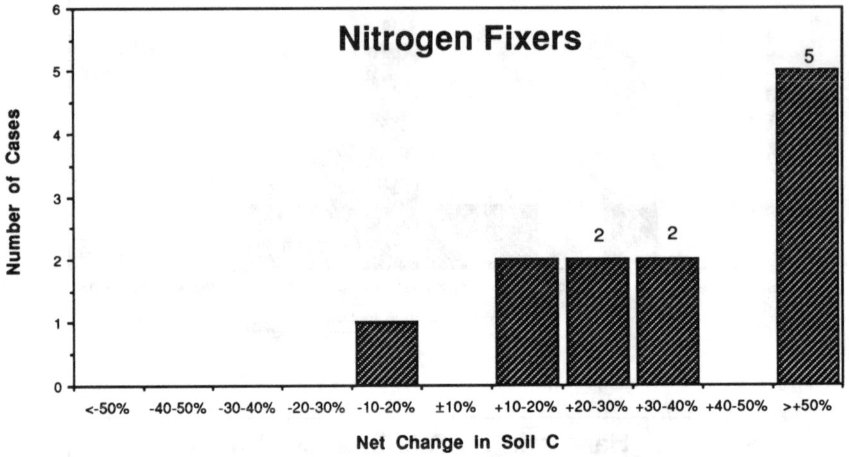

Figure 2. Summary of data on the effects of nitrogen-fixers soil C expressed as percentage difference from control. The number of statistically significant differences in each category are shown as numbers over each bar.

not likely the case with Ceanothus, which is a shrub of 1 to 1.5 m in stature. We hypothesize that the accumulation of SOM beneath Ceanothus stands is due to the stabilization of organic matter through the incorporation of nitrogen into humus. Condensation reactions of phenols (originating from partially degraded lignin and some fungal pigments) with either amino acids or ammonia result in the formation of "brown, nitrogenous humates" which are very resistant to degradation (Mortland and Wolcott, 1965; Nommik, 1965; Nommik and Vahtras, 1982; Paul and Clark, 1989). These non-biological, autocatalytic reactions are important in the production and stabilization of humus (Mortland and Wolcott, 1965; Paul and Clark, 1989; see also review by Johnson, 1992a). It is quite possible that these organic matter stabilizing processes are important in all soils with N-fixing vegetation growing on them.

The well-documented enrichment of soil C and N by nitrogen fixers poses a perplexing management problem in some forests. Specifically, can the benefits of soil C and (especially) N enrichment offset the disadvantages of postponing the establishment of a commercial forest? The enrichment of soil C and N by nitrogen fixers also poses some interesting basic scientific questions; namely, are the C and N contents of soils ultimately determined by texture, topographic position, and climate, or can they be "permanently" changed by the addition of nitrogen? Finally, the role of nitrogen fixers in the global carbon balance needs further review and investigation, especially as the models attempt to simulate the effects of fire and other catastrophic disturbances.

III. Effects of Elevated CO_2

Elevated CO_2 may affect (and might already have affected) soil C by causing changes in primary productivity, litter quality, decomposition rates, root exudation, and turnover. Elevated CO_2 might also affect soil C indirectly by causing changes in soil CO_2 partial pressure, which may in turn affect root growth and decomposition.

A. Changes in Litter and Soil Organic Matter Decomposition

It has been hypothesized that elevated CO_2 will indirectly cause changes in litter and soil C pools via changes in litter quality and decomposition (Strain, 1985; Norby et al., 1986b; Couteaux et al., 1991). This hypothesis arises from the often-noted decrease in foliar N concentration with elevated CO_2 and the assumption that these changes in foliar N will be manifested in lower litterfall N concentrations. Whether the latter is true depends upon the nature of N translocation from live foliage during senescence. Turner

(1977) found that imposing changes in soil available N status caused large changes live foliage N concentration and litterfall amount but very minor changes in litterfall N concentration in Douglas-fir (*Pseudotsuga meziesii*) plantations. He noted that foliar N concentrations were reduced to an almost constant concentration (approximately 0.6%) by translocation during senescence. Reductions in soil N availability caused increases in the amount of N translocation from old foliage necessary to meet current growth requirements. However, the increases in translocation were manifested as changes in litter quantity rather than quality, as old foliage was shed when it reached a threshold N concentration.

In contrast, Nambiar and Fife (1991) found that litter fall N concentration was closely tied to live foliage N concentration in radiata pine (*Pinus radiata*), and they disputed the conceptual model posed by Turner (1977). The authors also showed that translocation from roots prior to senescence in radiata pine was virtually nil, so that changes in root litter quality would be directly manifested in root decomposition rates. Given the observation that root N concentration decreases and root biomass increases with elevated CO_2 (Norby et al., 1986a and b; Luxmoore et al., 1986; Higginbotham et al., 1985; Rogers et al., 1992), one would hypothesize an increase in soil C via root turnover under an elevated CO_2 environment.

It is well-established that litter decomposition is more closely related to lignin:N than to C:N ratios (Melillo et al., 1982). Norby et al. (1986b) found that leaves from CO_2-fumigated white oak (*Quercus alba*) seedlings had lower lignin and higher tannin and sugars than leaves from unfumigated trees. However, Norby et al. (1986a) concluded that rates of litter decomposition will not be greatly affected by CO_2 based upon lignin:N and lignin:P ratios. Our initial studies of the effects of CO_2 on the decomposition of aspen leaves have also shown little lasting effect of CO_2 (Henderson and Johnson, unpubl. data).

The study of Couteaux et al. (1991) demonstrated that initial indices of litter quality such as C:N or lignin:N ratio may not be accurate predictors of decomposition rates over the long term. They found that CO_2 enrichment caused increased C:N ratios (from 40 to 75) in chestnut litter, and that microbial respiration and N mineralization rates reflected this initial difference for the first 3 weeks of decomposition. Subsequently, however, respiration in the CO_2-enriched litter was greater than in ambient CO_2-litter. This reversal during the second phase of decomposition was due to the greater populations of white-rot fungi (which are better able to use lignin and aromatic polymer-protein as sources of N) in the CO_2 - enriched litter. The better "quality" of the ambient CO_2-litter had prevented colonization by these fungi because of competition from other decomposer organisms. Thus, it is clear that decomposition is not merely a function of traditional, initial parameters of litter quality such as C:N and lignin:N ratios, but also of the dynamics of microbial populations as they develop on the litter substrate.

To the best of our knowledge, only two studies have measured actual changes in soil C due to elevated CO_2. Körner and Arnone. (1992) found a reduction in soil C in an artificial tropical ecosystem fumigated with elevated CO_2. The authors attributed the soil C loss to root exudation which stimulated decomposition in the rhizosphere. Both fine root biomass and soil respiration were greater in the elevated CO_2 treatments. Some significant questions concerning these results remain, however. The soil used was artificially constructed (a mixture of vermiculite and sand covered with mulch), and thus contained no stable soil organic matter. The longer-term effects of litter quality changes and stabilization of humus by chemical combination with N and polyvalent cations (Oades, 1988; Anderson, 1992) remain unknown.

A recent study be Zak et al. (1993) also suggested that elevated CO_2 may cause changes in soil C by affecting root exudation. The authors grew *Populus grandidentata* seedlings in twice ambient and ambient CO_2 and measured various growth, tree physiological, and soil parameters. They found that the seedlings allocated more C to roots with elevated CO_2, as is often the case, and that this in turn caused increases in labile C and N in rhizosphere soil under elevated CO_2. The authors pose a conceptual model whereby elevated CO_2 creates a positive feedback on soil C and N dynamics and tree growth because of increased carbohydrate allocation and, consequently, increased N availability in the rhizosphere.

B. Changes in Soil pCO_2

Soil C accumulation rates may also be indirectly affected by plant and microbial physiological responses to elevated soil pCO_2. With the often-noted increase in fine root biomass with elevated CO_2 (Norby et al., 1986a; Rogers et al., 1992), an increase in root respiration and, consequently, in soil pCO_2 can be expected and has been observed (Körner and Arnone, 1992; Vose et al. this volume). Increases in soil pCO_2 may have significant effects upon root growth and decomposer activity. Several studies have shown that soil pCO_2 can significantly affect root growth, either positively or negatively (Stolwijk and Thimann, 1957;

Baron and Gorski, 1986; Mauney and Hendrix, 1988). $^{14}CO_2$ labeling studies have shown that some plants, at least, can take up CO_2 through their roots, thus explaining the positive response. The negative response, which tends to occur at high pCO_2 levels (e.g., 1.5 to 15%) may be in part due to lack of oxygen. Stolwijk and Thimann (1957) noted a slightly positive response of root growth in peas (*Pisum sativum* var Alaska) to increased pCO_2 from ambient atmosphere up to 1.5%, but a negative growth response from 1.5% to 6%. Baron and Gorski (1986) found variable growth responses in eggplant (*Solanum melonaena* L.) to soil pCO_2, depending upon environmental conditions, and demonstrated CO_2 uptake by roots and translocation to shoots using $^{14}CO_2$. Thus, there are interesting possible feedbacks between atmospheric CO_2, root growth, and soil pCO_2 that have yet to be explored.

Increases in soil pCO_2 May also have significant effects upon microbial communities. Researchers have recognized for years the direct effects Of CO_2 concentration on microbial growth and biochemistry. The anoxygenic phototrophic bacteria like the *Chromatiaceae* (Staley et. al., 1989) grow in a CO_2-saturated environment. Most species of the genus, *Clostridium*, a common soil bacteria, require a 100% CO_2 environment or a 90% N_2 to 10% CO_2 environment for growth. Bacteria in the *Nitrobacteraceae*, obligate chemolithotrophs that are important soil nitrifiers, have shown altered biochemistry due to CO_2 concentration. The activity of carbonic anhydrase in *N. europea* is stimulated by decreased CO_2 concentration. Members of the family *Rhodospirillaceae*, which include photoautotrophs dependent on CO_2 as a carbon source, are widely distributed in aquatic and moist soil environments. They have also shown altered metabolism at various CO_2 concentrations. *Rhodospirillum rubrum* (Sarles et al 1983) demonstrated significant increases in ribulose 1,5-bisphosphate carboxylase production at decreased CO_2 concentrations. At CO_2 concentration of 1.5 to 2.0% carboxylase made up as much as 50% of the total cellular protein as compared to the 4% concentration at a CO_2 level of 6.4%. The chemoorganotrophic *Actinomycetes* are common soil bacteria that are considered to have an obligate CO_2 requirement (Howell and Pine, 1956). Aerobic and anaerobic growth is greatly enhanced by increased CO_2. Buchanan and Pine (1965) found that CO_2 was essential for the fermentation of glucose to succinate in *Actinomyces naeslundii*. They also demonstrated that CO_2 was essential for the synthesis of aspartic acid. Metabolism in the chemolithothautotrophic, aerobic hydrogen-oxidizing bacteria, *Alcaligenes eutrophus*, is altered by CO_2 concentration. Friedrich (1982) demonstrated increased ribulosebisphosphate carboxylase production at decreased carbon dioxide levels and slight increases in hydrogenase activity with increased CO_2 in A. *eutrophus*.

In addition to the well documented effects of carbon dioxide concentration on bacterial growth, CO_2 is known to influence fungal growth. Alexander (1961) suggests that qualitative changes in fungal populations at different horizons within the soil may be due to increases in CO_2 at deeper levels. Fungal populations at higher levels may be inhibited by CO_2 and therefore grow at the surface where less carbon dioxide is available. Those at lower levels are adapted to higher CO_2 concentration and are stimulated by its presence.

IV. Summary and Conclusions

This two-part review addressed the effects of forest management and elevated CO_2 on soil C. Forest harvesting can result in either increases or decreases in soil C, depending upon how the residues are treated during site preparation. On average, however, it appears that harvesting has little net effect on soil C unless followed by cultivation. Nitrogen-fixers cause almost uniform increases in soil N and C. Reasons for the greater C accumulation are not explicitly known; we hypothesize that increased N inputs cause greater SOM stabilization by N incorporation into humus.

Elevated atmospheric CO_2 may cause changes in soil C by altering primary productivity, litter decomposition rates, and indirectly by changing root exudation and soil pCO_2 levels. Potential effects of elevated atmospheric CO_2 on soil pCO_2, and subsequent effects on root and microbial activity are real but, as yet, unexplored. The mechanisms and potential feedbacks by which increased atmospheric CO_2 could affect soil C accumulation are very complex, making predictions very difficult. Initial indices of litter quality such as C:N and lignin:N ratios may not accurately reflect actual long-term trends in decomposition.

V. Synthesis and Speculation

Given the very large uncertainties in the global C budget and the processes affecting it, we are left largely with speculation as to how it might affect or be affected by soil C storage. The results of this review

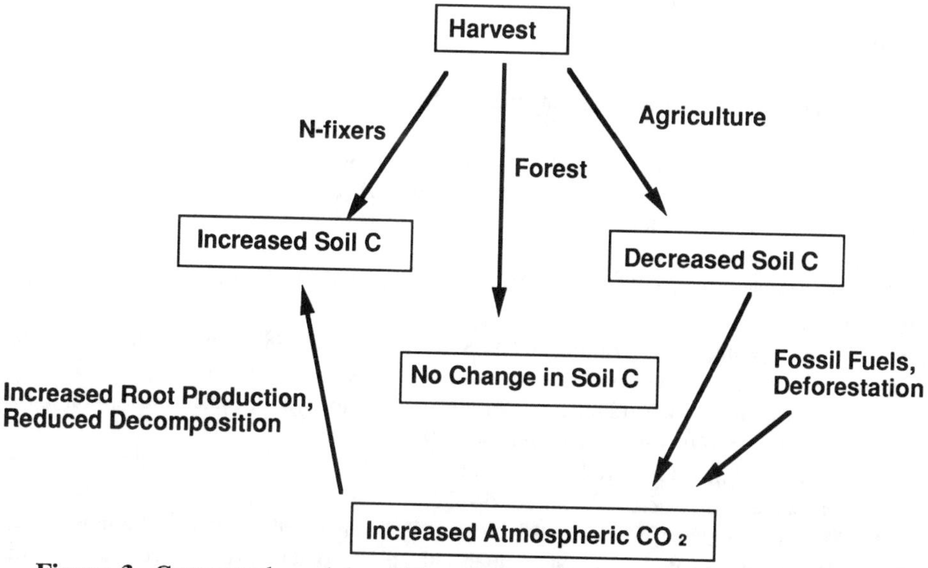

Figure 3. Conceptual module of literature results and speculations as to the feedbacks between soil C and atmospheric CO_2.

combined with the speculations as to what they mean to global soil C storage are depicted in Figure 3. Both this and previous reviews (Mann, 1986; Johnson, 1992b) suggest that soil C will change in certain ways following forest harvesting, depending upon the land use after the harvest. If harvesting is followed by reforestation, there will, in most instances and on average, be little or no net change in soil C. Exceptions to this may occur at individual sites, as noted above. If the site is allowed to convert to N-fixers, there will nearly always be an increase in soil C. If the site is converted to agriculture, the preponderance of evidence suggests that soil C levels will decrease, leading to CO_2 release to the atmosphere. This, combined with other sources of CO_2, may have caused and may continue to cause increased atmospheric CO_2 levels. We speculate that increased atmospheric CO_2 will cause (and perhaps has already caused) a tendency toward increased soil C because of increased root production and turnover, even though there may be some offsetting effect of root exudation over the short term.

For several reasons, the set of results and speculations embodied in the conceptual model in Figure 3 cannot be converted into a quantitative model of the feedbacks between soil C and atmospheric CO_2. First, feedbacks between soil C and CO_2 via nutrient cycles on an ecosystem level are very poorly understood. A decrease in litter (foliar or root) quality may create or exacerbate nitrogen deficiency, perhaps causing further deterioration of litter quality and creating a feedback loop that ultimately must result in a decline in primary productivity. On the other hand, slowed decomposition rates may cause litter-soil C pools to increase. Thus, the net ecosystem C balance with elevated CO_2 is very difficult to predict with the current paucity of information. Second, the available data for carbon pool sizes and fluxes on a global scale lack the quality assurance and statistical rigor needed to quantify the model in Figure 3. Third, and perhaps most important, the model in Figure 3 represents only part of the picture; climate change will be a major (and perhaps overriding) variable. We offer Figure 3 merely by way of synthesizing the results of this review and perhaps as a module (rather than model) for incorporation into larger, more comprehensive models.

Acknowledgments

Research supported by the Electric Power Research Institute (RP3041-02), the National Council of the Paper Industry for Air and Stream Improvement, Inc., and the Nevada Agricultural Experiment Station, University of Nevada, Reno.

References

Alban, D.H. and D.A. Perala. 1992. Carbon storage in Lake States aspen ecosystems. *Can. J. For. Res.* 22: 1107-1110.

Alexander, Martin. 1961. *Soil Microbiology*. John Wiley and Sons, New York.

Anderson, J.M. 1992. Responses of soils to climate change. *Adv. Ecol. Res.* 22: 163-210.

Baron, J.J. and S.F. Gorski. 1986. Response of eggplant to a root environment enriched with CO_2. *HortScience* 21: 495-498.

Binkley, D. 1983. Ecosystem production in Douglas-fir plantations: Interactions of red alder and site fertility. *For. Ecol. Manage.* 5:215-227

Boring, L.R., W.T. Swank, J.B. Waide, and G.S. Henderson. 1988. Sources, fates, and impacts of nitrogen inputs to terrestrial ecosystems: review and synthesis. *Biogeochemistry* 6:119-159

Brown, S., A.E. Lugo, and L.R. Iverson. 1992. Processes and lands for sequestering carbon in the tropical forest landscape. *Water, Air, and Soil Pollut.* 64:139-156

Buchanan, B. B. and L. Pine. 1965. Relationship of carbon dioxide to aspartic acid and glutamic acid in *Actinomyces naeslundii*. *J. Bacteriol.* 89(3):729-733.

Couteaux, M.-M., M. Mousseau, M.-L. Celkerier, and P. Bottner. 1991. Increased atmospheric CO_2 and litter quality decomposition of sweet chestnut litter with animal food webs of different complexities. *Oikos* 61: 54-64.

Crocker, R.L. and J. Major. 1955. Soil development in relation to vegetation and surface age at Glacier Bay, Alaska. *J. Ecol.* 43:427-448.

Detwiler, R.P. 1986. Land use change and the global carbon cycle: the role of tropical soils. *Biogeochemistry* 2:67-93.

Friedrich, Cornelius G. 1982. Derepression of hydrogenase during limitation of electron donors and derepression of rubulosebisphosphate carboxylase during carbon limitation of *Alcaligenes eutrophus*. *J. Bacteriol.* 149(l):203-210

Higginbotham, K.O., J.M. Mayo, S.L'Hirondelle, and D.K. Krystofiak. 1985. Physiological ecology of lodgepole pine *(Pinus contorta)* in an enriched CO_2 environment. *Can. J. For. Res.* 15:417-421.

Houghton, R.A., J.E. Hobbie, J.M. Melillo, B. Moore, B.J. Peterson, G.R. Shaver, and G.M. Woodwell. 1983. Changes in the carbon content of terrestrial biota and soils between 1860 and 1980: a net release of CO_2 to the atmosphere. *Ecol. Monogr.* 53:235-262.

Howell, A. and L. Pine. 1956. Studies on the growth of the species *Actinomyces*. *J. Bacteriol.* 71:47-53.

Jenkinson, D.S., D.E. Adams, and A. Wild. 1991. Model estimates of CO_2 emissions from soil in response to global warming. *Nature* 351:304-306.

Johnson, D.W. 1992a. Nitrogen retention in forest soils. *J. Environ. Qual.* 21:1-12

Johnson, D.W. 1992b. The effects of forest management on soil carbon storage. *Water, Air, and Soil Pollut.* 64:83-120.

Johnson, D.W. 1993. Effects of forest management on soil C storage in eastern Sierra Nevada forests. Report to the National Council of the Paper Industry for Air and Stream Improvement, Inc.

Johnson, D.W. and J.T. Ball. 1991. Environmental pollution and impacts on soils and forests of North America. *Water, Air, and Soil Pollut.* 54:3-20.

Körner, C. and J.A. Arnone. 1992. Responses to elevated carbon dioxide in artificial tropical ecosystems. *Science* 257:1672-1675

Lugo, A.E. 1992. The search for carbon sinks in the tropics. *Water, Air, and Soil Pollut.* 64: 3-9.

Luxmoore, R.J., E.G. O'Neill, J.M. Ells, and H.H. Rogers. 1986. Nutrient uptake and growth responses of Virginia pine to elevated atmospheric carbon dioxide. *J. Environ. Qual.* 15:244-251.

Mann, L.K. 1986. Changes in soil carbon storage after cultivation. *Soil Sci.* 142:279-288.

Mauney, J.R. and D.L. Hendrix. 1988. Responses of glasshouse grown cotton to irrigation with carbon dioxide-saturated water. *Crop Sci.* 28:835-838.

Melillo, J.M., J. Aber, and J.F. Muratore. 1982. Nitrogen and lignin control of hardwood lead litter decomposition dynamics. *Ecology* 63:621-626.

Mortland, M.M. and A.R. Wolcott. 1965. Sorption of inorganic nitrogen compunds by soil materials. pp. 150-197. In: W.V. Bartholomew and F.E. Clark (eds.), *Soil Nitrogen*. Agronomy 10. Amer. Soc. Agron., Madison, Wisconsin.

Nambiar, E.K.S. and D.N. Fife. 1991. Nutrient retranslocation in temperate conifers. *Tree Physiology* 9: 185-207.

Nommik, H. 1965. Ammonium fixation and other reactions involving a nonenzymatic immobilization of mineral nitrogen in soil. p. 198-257. In: W.V. Bartholomew and F.E. Clark (eds.), *Soil Nitrogen*. Agronomy 10. Amer. Soc. Agron., Madison, Wisconsin.

Nommik, H. and K. Vahtras. 1982. Retention and fixation of ammonium and ammonia in soils. pp. 123-172. In: F.J. Stevensen, J.M. Breamer, R.D. Hauck, and D.R. Keeney (eds.), *Nitrogen in Agricultural Soils*. Agronomy, Vol 22. Amer. Soc. Agron., Madison, Wisconsin.

Norby, R.J., E.G. O'Neill, and R.J. Luxmoore. 1986a. Effects of atmospheric CO_2 enrichment on the growth and mineral nutrition of *Quercus alba* seedlings in nutrient-poor soil. *Plant Physiol.* 82:83-89.

Norby, R.J., J. Pastor, and J.M. Melillo. 1986b. Carbon-nitrogen interactions in CO_2-enriched white oak: physiological and long-term perspectives. *Tree Physiol* 2:233-241.

Oades, J.M. 1988. The retention of organic matter in soils. *Biogeochemistry* 5:35-70.

Paul, E.A. and F.E. Clark. 1989. *Soil Microbiology and Biochemistry*. Academic Press, San Diego, New York.

Post, W.M., T-H Peng, W.R. Emmanuel, A.W. King, V.H. Dale, and D. L. DeAngelis. 1990. The global carbon cycle. *American Scientist* 78:310-326

Rogers, H.H., C.M. Peterson, J.N. McCrimmon, and J.D. Cure. 1992. Response of plant roots to elevated atmospheric carbon dioxide. *Plant, Cell, and Environment* 15:749-752.

Sarles, L.S. and F.R. Tabita. 1983. Derepression of the synthesis of D-ribulose 1,5-bisphophosphate carboxylase/oxygenase from *Rhodospirillum ruburm*. *J. Bacteriol.* 153(l):458-464.

Schlesinger, W.H. 1990. Evidence from chronosequence studies for a low carbon-storage potential of soils. *Nature* 348: 232-234.

Staley, J.T., M.P. Dryant, N. Pfenning, and J.G. Holt. 1989. *Bergey's Manual of Systematic Bacterology*. Willliams and Wilkins, London.

Strain, B.R. 1985. Physiological and ecological controls on carbon sequestering in terrestrial ecosystems. *Biogeochemistry* 1:219-232.

Stolwijk, J.A.J. and K.V. Thimann. 1957. On the uptake of carbon dioxide and bicarbonate by roots, and its influence on growth. *Plant Physiol.* 32:513-520.

Tarrant, R.F. and R.E. Miller. 1963. Accumulation of organic matter and soil nitrogen beneath a plantation of red alder and Douglas-fir. *Soil Sci. Soc. Amer. Proc.* 27:231-234.

Turner, J. 1977. Effect of nitrogen availability on nitrogen cycling in a Douglas-fir stand. *For. Sci.* 23:307-316

Vose, J.M., K.J. Elliot, and D.W. Johnson. Diurnal and daily patterns in soil CO_2 flux in response to elevated atmospheric CO_2 and nitrogen fertilization. This volume.

Wood, C.W., R.J. Mitchell, B.R. Zufter, and C.L. Lin. 1992. Loblolly pine plant community effects on soil carbon and nitrogen. *Soil Sci.* 154:410-419.

Zak, D.R., K.S. Pregitzer, P.S. Curtis, J.A. Teeri, R. Fogel, and D.L. Randlett. 1993. Elevated atmospheric CO_2 and feedback between carbon and nitrogen cycles. *Plant Soil* 51:105-117.

Zavitkovski, J. and M. Newton. 1968. Ecological importance of snowbrush Ceanothus velutinus in the Oregon Cascades. *Ecology* 49:1134-1145

CHAPTER 13

Carbon Pools and Trace Gas Fluxes in Urban Forest Soils

Peter M. Groffman, Richard V. Pouyat, Mark J. McDonnell, Steward T.A. Pickett, and Wayne C. Zipperer

I. Introduction

Urbanization is a process that affects large areas of the earth's surface. Urban environments are distinct from more natural environments in several factors that influence biogeochemical processes such as carbon (C) storage and trace gas fluxes. Key environmental changes of biogeochemical significance in urban areas include increases in temperature, enhanced deposition of nitrogen (N), metals and organic compounds, increased concentrations of atmospheric pollutants, especially ozone, introductions of non-native species and increased physical disturbance.

Study of C pools and trace gas fluxes in urban areas is important due to the large spatial extent of these areas, and perhaps more importantly, it may contribute significantly to understanding global climate change. Comparing biogeochemical processes in ecosystems within urban areas with comparable ecosystems outside of urban areas may be a useful approach for evaluating how changes in climate and atmospheric chemistry affect biogeochemistry (McDonnell and Pickett, 1990). In addition to evaluation of relatively straightforward temperature effects, urban-rural comparisons may be useful for evaluating more complex long-term effects of climate change, such as how changes in species composition and litter quality affect the flow of C between plant and microbial communities and how these changes influence trace gas fluxes.

In this paper we describe the unique physical, chemical, and biological aspects of ecosystems within urban areas and describe how urbanization can serve as an analog for global climate change. We then present a conceptual model of C flow and trace gas fluxes in forest ecosystems and discuss how urbanization and/or global change can affect key pools and processes within this model. Finally, we present data from a comparison of biogeochemical processes in urban and rural forest soils in the New York City Metropolitan Area.

II. Characterizing Urban Environments

Urbanization is a dramatic land alteration process that eliminated 10 million ha of agricultural and forested land in the United States between 1960 and 1980 alone (Alig and Healy, 1987; Frey, 1984). The proportion of the U.S. population that lives in urban areas is expected to increase from 74% in 1986 to over 80% by the year 2025 (Alig and Healy, 1987; Haub and Kent, 1989). Urbanization is also a vigorous process in developing countries (Kellogg and Schware, 1981; French, 1990).

Although urban areas are characterized by human-constructed features, ecosystems which are not intensively managed or disturbed by humans exist in most urbanized landscapes (Dickinson, 1966; Stearns and Montag, 1974; McDonnell and Pickett, 1990). Although these ecosystems may look like non-urban

ISBN 1-56670-117-1/95/$0.00+.50

ecosystems, they may be ecologically altered due to the complex set of physical, chemical, and biological factors that are affected by urbanization (Bornkamm et al., 1982; Baker, 1986; Seinfeld, 1989; Graedel and Crutzen, 1989; Pouyat and Zipperer, 1992). Changes associated with urbanization include alterations in climate, air quality, and invasions by non-native species (Landsberg, 1981; Oke, 1990; McDonnell and Pickett, 1990).

Perhaps the most well documented environmental effect of urbanization is a localized increase in temperature or the "heat island" effect (Landsberg, 1981; Oke, 1990). Decreases in evapotranspiration and increases in heat absorbing surfaces (e.g. pavement) associated with urbanization lead to increases in maximum, minimum, and average temperatures. Urban-rural temperature differences are usually greatest, 3 to 8° C, in early evening near the urban core. Concentrations of urban O_3 are enhanced by increases in ambient temperature caused by the heat-island effect (Candelino and Chameides, 1990).

Urbanization has been observed to have significant but variable effects on precipitation. Urban activities produce large quantities of condensation nuclei that can increase precipitation downwind of urban sources (Oke, 1990). The interaction of extra condensation nuclei and local climate conditions can lead to increased variability in precipitation in urban areas, especially precipitation associated with local heavy rainfall events (Changnon, 1980).

Urbanization results in marked changes in atmospheric chemistry and deposition. Ecosystems in urban areas receive relatively high amounts of N, heavy metals, organic compounds, ozone, sulfur, and acid in wet and dry deposition from the atmosphere (Seinfeld, 1989). Depending on local weather conditions, air quality can change markedly with distance from the urban core for some parameters (metals, N), while other urban air pollutants, especially ozone and acid, spread over larger regional areas (Schwartz, 1989; Sillman et al., 1990; Chang et al., 1992).

Ecosystems in urban areas usually contain a higher proportion of exotic and naturalized plant and animal species than ecosystems in rural areas (Bagnall, 1979; Airola and Bucholz, 1984; Hobbs, 1988; Rudnicky and McDonnell, 1989; McDonnell and Pickett, 1990). The effects of the presence of non-native species are variable and complex. Changes in tree species composition can change water and nutrient fluxes in the ecosystem through differences in uptake and more subtly, through changes in litter quality and organic matter dynamics. Changes in soil fauna, such as earthworms, many of which are non-native, can have important effects on nutrient cycling and organic matter dynamics in forests in urban areas and can buffer the effects of pollutants on soil processes (Scheu and Wolters, 1991; Pouyat and Parmelee, submitted).

The wide variety of environmental changes associated with urbanization complicates the use of ecosystems within urban landscapes for study of the effects of global change on biogeochemical processes. While differences in temperature and precipitation between similar ecosystems in urban and rural areas provide an ideal opportunity for study of global climate change, differences in air quality and species composition associated with urbanization may not be suitable analogues of global scale changes. Urban versus rural comparisons are most valuable in that urbanization acts as a long-term whole ecosystem manipulation experiment. Urbanization exposes entire ecosystems to altered climate and allows for replication of sites within an urban area and for comparison with rural "control" sites. Experimental use of "patches" in which the various urbanization factors are combined in different ways, coupled with manipulation experiments, can isolate the diverse environmental and ecological effects associated with urbanization.

III. Conceptual Approach

We have used conceptual models of ecosystem and soil processes to guide our studies of biogeochemical processes in forest ecosystems in urban areas. By linking models that depict the influence of climate and site factors on ecosystem structure and function with detailed models of soil C flow and microbial processes, we have been able to develop hypotheses about C storage and trace gas fluxes in urban forest soils. Our approach allows us to separate (conceptually at least) the wide range of environmental and ecological factors that influence biogeochemical processes in ecosystems in urban areas.

The conceptual model presented by Pastor (1986) depicts the influence of climate and site factors, primarily water and N availability, on net primary production (NPP), litter C chemistry, decomposition and N availability in forest ecosystems (Figure 1). This model emphasizes the reciprocal links between N and C cycling in ecosystems, which are expressed through the influence of N availability on litter quality and quantity. Microbial processes and soil C dynamics fall within the decomposition compartment of this model, which is driven by the nature and amount of litter input.

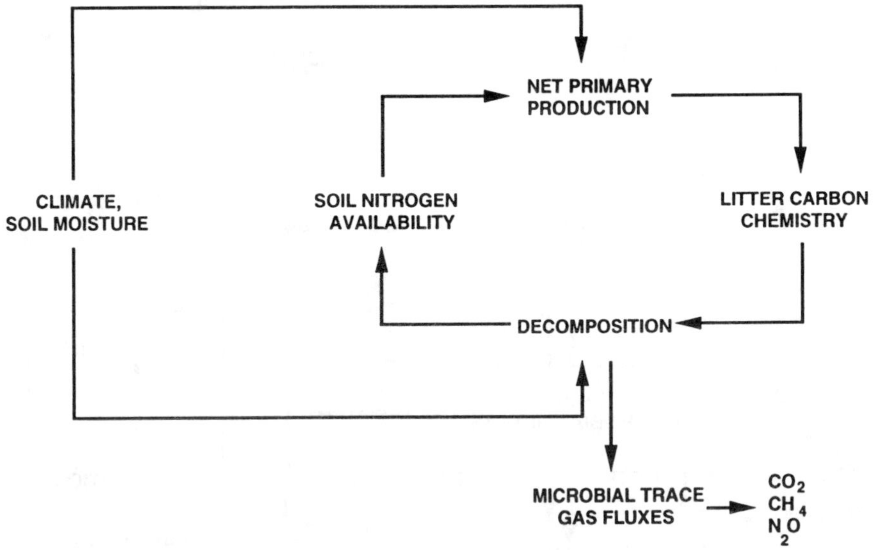

Figure 1. Conceptual model of links between net primary production, litter C chemistry, decomposition, microbial trace gas fluxes and soil N availability, and how these parameters are constrained by climate and soil moisture. (Adapted from Pastor, 1986).

Urban effects can interface with the model presented in Figure 1 in several places. Changes in urban temperature and precipitation constrain both NPP and decomposition. Enhanced N deposition will increase N availability. Air pollution, especially ozone, can decrease litter quality, inhibiting decomposition (Findlay and Jones, 1990). Accumulations of heavy metals in forest soils can decrease the abundance and activity of soil decomposing organisms (Tyler et al., 1989; Bengtsson and Tranvik, 1989).

Detailed soil process models can be used to depict C dynamics within the decomposition compartment of the conceptual model depicted in Figure 1. These models are driven by the nature and amount of litter input and are centered around C and N pools with turnover times ranging from days to centuries (Jenkinson and Rayner, 1977; Parton et al., 1988). Microorganisms and soil fauna move litter-derived C and N into and between different pools and carry out biogeochemical functions such as nutrient mineralization and trace gas fluxes. The different C and N pools respond variably to changes in the nature and amount of C input and to system disturbances such as cultivation, grazing, and harvest.

Our soil process model (Figure 2) has four pools of soil C that differ in their lability or turnover time, and are readily quantifiable. "Readily mineralizable" C is the C respired in a 10 day aerobic incubation at 25° C and represents a pool of C with a turnover time of days to weeks. Labile, or microbial biomass C, is the C respired in a 10 day aerobic incubation of a fumigated and re-inoculated sample (the chloroform fumigation-incubation method, Jenkinson and Powlson, 1976) and has a turnover time of weeks to months. Potentially mineralizable C is the C respired in a 128 day aerobic incubation at 25° C (Nadelhoffer, 1990) and has a turnover time of months to years. Passive C is total soil C measured by loss on ignition at 450° C and has a turnover time of years to decades to centuries.

Urbanization can affect several of the pools and processes in Figure 2. Changes in litter quality caused by air pollution or invasion by exotic species can change the partitioning of C into different pools as litter decomposes. Temperature increases and enhanced N deposition should increase decomposition rates, while air pollution damage and heavy metal decomposition should have the opposite effect.

Our conceptual models (Figures 1 and 2) allow us to pose a series of hypotheses about the effects of urbanization on biogeochemical processes in ecosystems in urban areas. These hypotheses include:

1. Increased temperature and N deposition in urban areas will increase decomposition rates leading to decreases in all soil C pools.

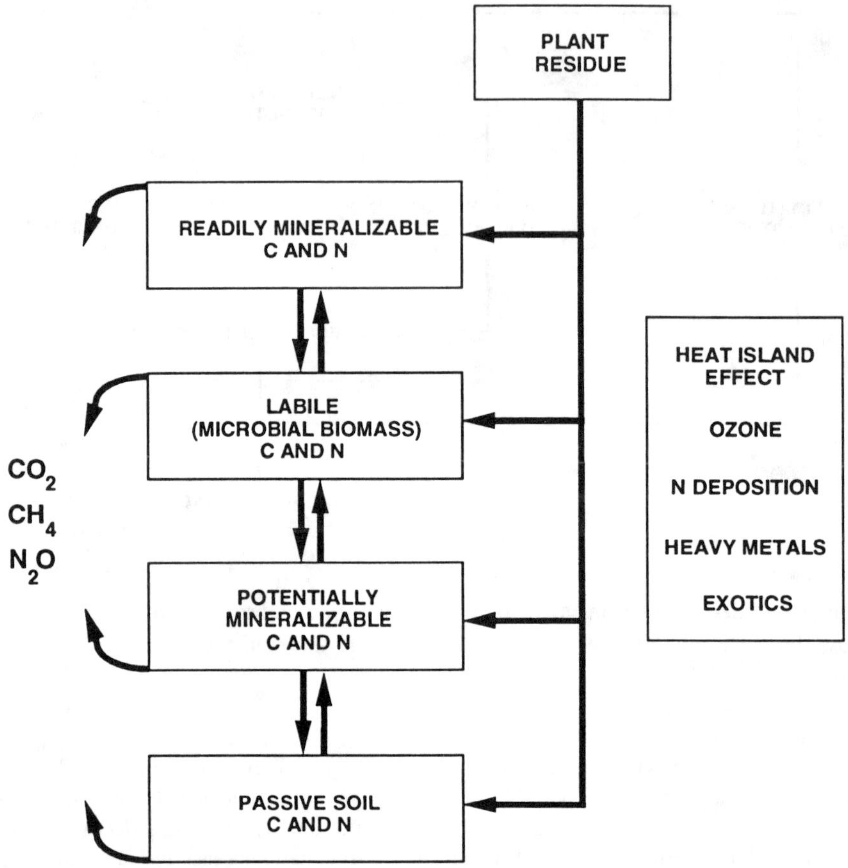

Figure 2. Conceptual model of soil C and N pools and transformations. (Adapted from Parton et al., 1988; and Jenkinson and Rayner, 1977.)

2. Air pollution, especially ozone, will damage tree leaves and reduce litter quality. This will lead to increases in passive C and decreases in the more labile pools of C in urban relative to rural soils.

To test these hypotheses we have begun to monitor soil C pools and to design and carry out field manipulation experiments. Preliminary work and results from several studies are described below.

IV. The New York Urban-Rural Gradient Experiment (URGE)

We have been investigating ecosystem structure and function in forest ecosystems along an urban to rural gradient in the New York City Metropolitan Area for the last 7 years (White and McDonnell, 1988; McDonnell and Pickett, 1990; Pouyat and McDonnell, 1991; McDonnell et al., 1993). The body of past and current URGE research includes detailed characterization of the gradient including quantification of traffic volume, road density, population density, percent land cover, and heavy metal levels in soils, and ecological analysis of plant community structure, soil N cycling, soil fauna, trace gas fluxes, and plant growth.

The URGE study sites are located on a 20 km wide by 140 km long transect that extends from highly urbanized Bronx County, NY (a borough of New York City), through suburban Westchester County, NY to rural Litchfield County, CT (Figure 3). The transect encompasses readily measurable differences in population density, land use, and automobile usage and corresponds closely to the location of gneiss and schist bedrock (McDonnell and Pickett, 1990; Pouyat and McDonnell, 1991). Other radii from the urban sites include sedimentary bedrock or do not support suitable rural sites for comparison. Differences in

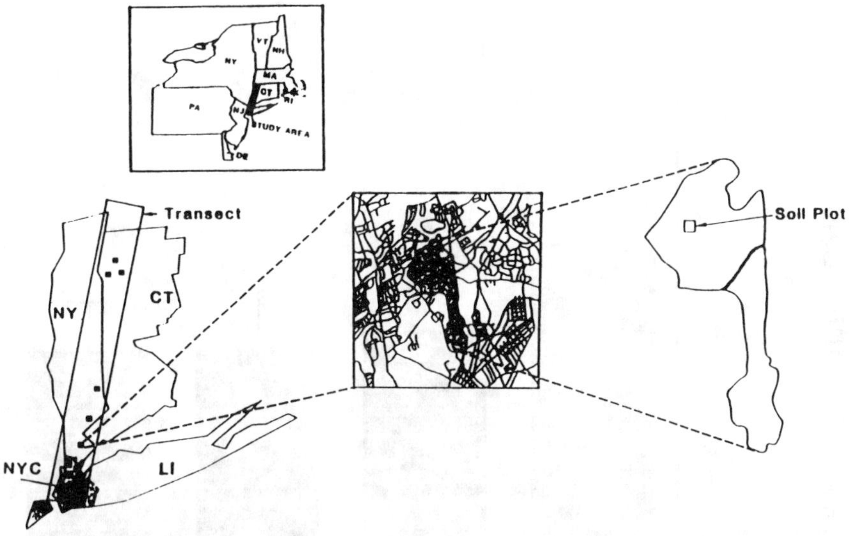

Figure 3. The New York Metropolitan Area showing the urban-rural land use transect with nine research stands. Twenty-seven sites were established among the stands. (Adapted from Pouyat, 1992.)

"urbanization factors" are far more dramatic than variation in soils, temperature or precipitation along the transect, therefore trends with distance along the transect are attributed to urbanization.

Soil types along the transect are classified as Typic or Lithic Dystrochrepts, coarse-loamy, mixed mesic subgroups, and are well-drained, moderate to shallow sandy loams situated on gently sloping terrain (Tornes, 1974; Hill et al., 1980). Forest stands (27 total) were chosen along the transect using the following criteria: (1) location on upland sites with either Charlton or Hollis soil series (Gonick et al., 1970; Tornes, 1974), (2) oak dominated with *Quercus rubra* or *Q. velutina* as major components of the overstory, (3) minimum stand age of 70 yr, and (4) no evidence of recent disturbance. By design, our transect did not include any stands with a significant proportion of non-native tree species.

The soil C pools depicted in Figure 2 were quantified as described above. Trace gas fluxes were measured in laboratory incubations and at nine sites along the transect using field chambers. The chambers (three replicates per site) were constructed from 16.5 cm wide by 20 cm long pieces of PVC pipe fitted with a septum and an air-tight well cap (total volume of 2 L). Samples were taken by syringe every 15 min. for 1 h on August 7 and August 8, 1992 and added to evacuated vials. Concentrations of methane, nitrous oxide and carbon dioxide were measured by gas chromatography. Rates of flux were determined by linear regression analysis of the linear section of the depletion or production curves, which was from 0 to 30 minutes.

A. Preliminary Results and Discussion

As expected, New York City produces a marked "heat island" effect (Bornstein, 1968). Annual mean temperatures over the period 1985-1991 were more than 3° C warmer in the urban core that at the rural end of the transect (Figure 4). Average annual precipitation did not differ systematically between urban and rural sites from 1985-1991 but variation in annual precipitation was higher in the urban areas than in the rural areas (Figure 5). For example, in 1989, there was greater than 50 cm more precipitation in the urban core than at the rural end of the transect, primarily due to differences in summer thunderstorm activity. In the context of the conceptual models presented in Figures 1 and 2, increases in temperature should increase both NPP and decomposition rates in urban sites and should result in reductions in all soil C pools.

Heavy metal levels in soils were higher in urban forest sites than in rural forest sites (Figure 6), reflecting differences in atmospheric deposition along the transect. High levels of heavy metals can decrease

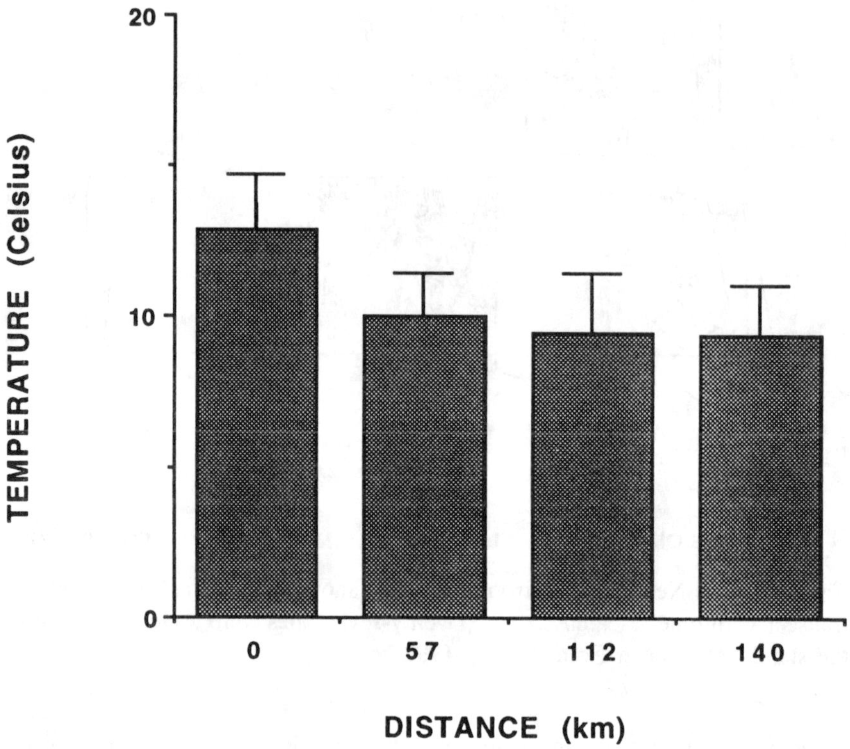

Figure 4. Mean annual temperature versus distance from the urban core (Central Park, New York City) along the New York City urban-rural transect, 1985-1992. Values are mean (range). Because temperature data are temporally autocorrelated, statistical comparisons were not made.

soil decomposition activity, but it is unknown if the concentrations observed in our sites are ecologically significant (Giller and McGrath, 1989; Hopkin, 1993).

A somewhat surprising result was the observation that non-native species of earthworms were abundant in urban sites and were not present at all in rural sites (Figure 7). Earthworms should increase decomposition activity, but they may also increase storage of C in passive pools by fostering physical protection of C in soil aggregates.

Preliminary analysis of the soil C pools depicted in Figure 2 suggest that pools of labile C are lower and pools of passive C are higher in urban relative to rural forest sites (Table 1). While total (passive) soil C was higher in the urban sites, pools of readily mineralizable, labile (microbial), and potentially mineralizable C were higher in rural sites.

The partitioning of soil C among the four pools within the urban and rural forest ecosystems was consistent with the hypotheses presented above. Increased temperature and possibly N deposition tend to increase decomposition rates and depletion of labile C pools in urban forest sites while ozone damage to leaves leads to preservation of passive C. The presence of earthworms in urban forest sites reinforces this pattern. Earthworms stimulate decomposition of fresh litter (Scheu, 1987; Staff, 1987; Lavelle, 1988; Parmelee and Crossley, 1988) speeding the decomposition of readily labile C pools. Earthworms may increase pools of passive organic matter by facilitating physical protection of organic matter in soil aggregates.

Our conceptual models are useful vehicles for evaluating the integrated effect of the different factors associated with urbanization. They provide a basis for examining how changes in ecosystem properties such as litter quality are reflected in soil C pools and microbial processes. The data suggest that in the long-term, urban forest sites should sequester and store more C than rural forest sites. It is important to note that the factors that increase soil C storage in urban forest sites (ozone damage, exotic species) are not normally

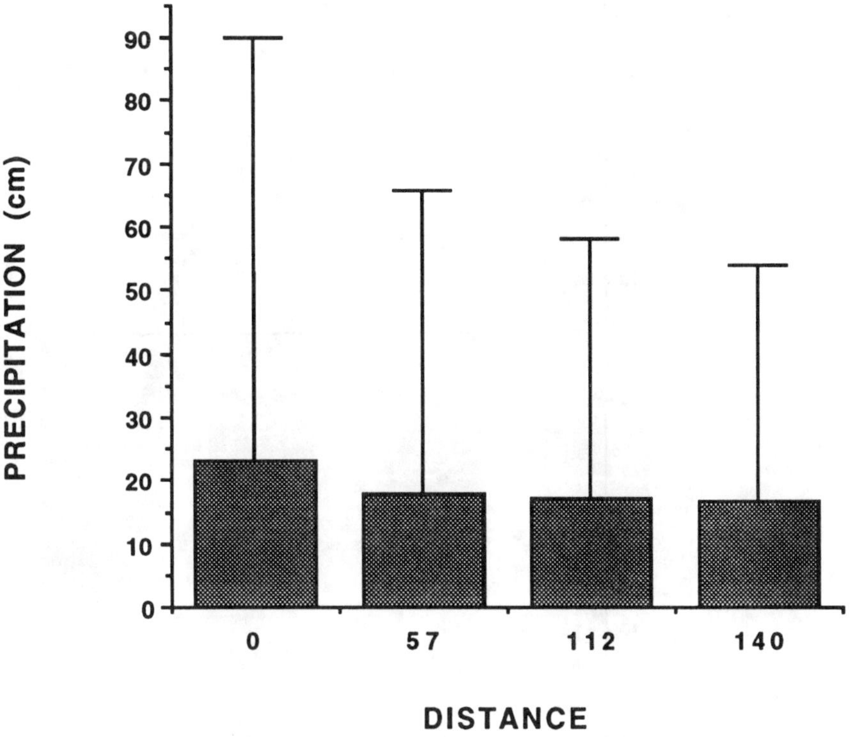

Figure 5. Mean annual precipitation versus distance from the urban core (Central Park, New York City) along the New York City urban-rural transect, 1985-1992. Values are mean (range). Because precipitation data are temporally autocorrelated, statistical comparisons were not made.

associated with global climate change. These results do not suggest that forest ecosystems will store more C in a warmer climate.

Although our data suggest that urban forest sites should store more C than rural forest sites, decreases in labile C pools should result in lower levels of microbial activity in urban forest soils, including trace gas fluxes. We observed significantly lower rates of methane uptake in urban forest soils than in rural forest soils in both laboratory (Figure 8) and field (Table 2) studies. Field fluxes of carbon dioxide did not differ among sites and nitrous oxide fluxes were negligible, but these data are from only one sample date (Table 2). Monthly monitoring of trace gas fluxes along the urban to rural transect are ongoing.

Other ongoing and planned URGE work includes detailed studies of soil N mineralization and nitrification and earthworm dynamics, measurements of variation in N deposition along the transect and analysis of ozone effects on plant growth and litter quality. We are particularly interested in annual and seasonal variation in urban biogeochemistry. Seasonal and annual variation in precipitation and air pollution can cause marked year to year variation in leaf damage, litter quality, and earthworm activity in urban sites. While our analysis of soil C pools suggests that in the long-term the suite of environmental changes associated with urbanization has led to increases in passive and decreases in labile C, specific microbial activities may be much more responsive to seasonal and annual variation in environmental and ecological factors.

V. Summary and Conclusions

Urbanization affects large areas of the earth's surface. Ecosystems within urbanized landscapes are affected by a variety of physical, chemical, and biological factors that have complex and potentially contradictory effects on biogeochemical processes. Study of ecosystems within urban landscapes may be useful for analysis of the effects of global climate change on biosphere processes. We used conceptual models of

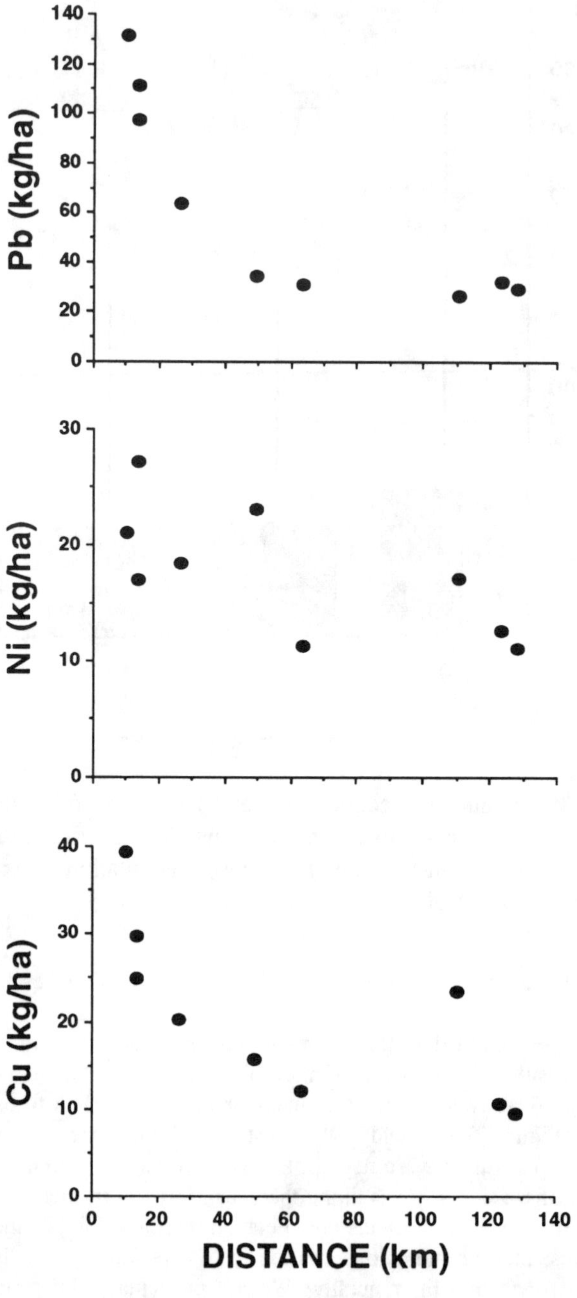

Figure 6. Soil (mineral soil plus forest floor) lead, copper, and nickel concentrations versus distance from the urban core (Central Park, New York City) along the New York City urban-rural transect, 1989, n=3 for each site. (Adapted from Pouyat and McDonnell, 1991.)

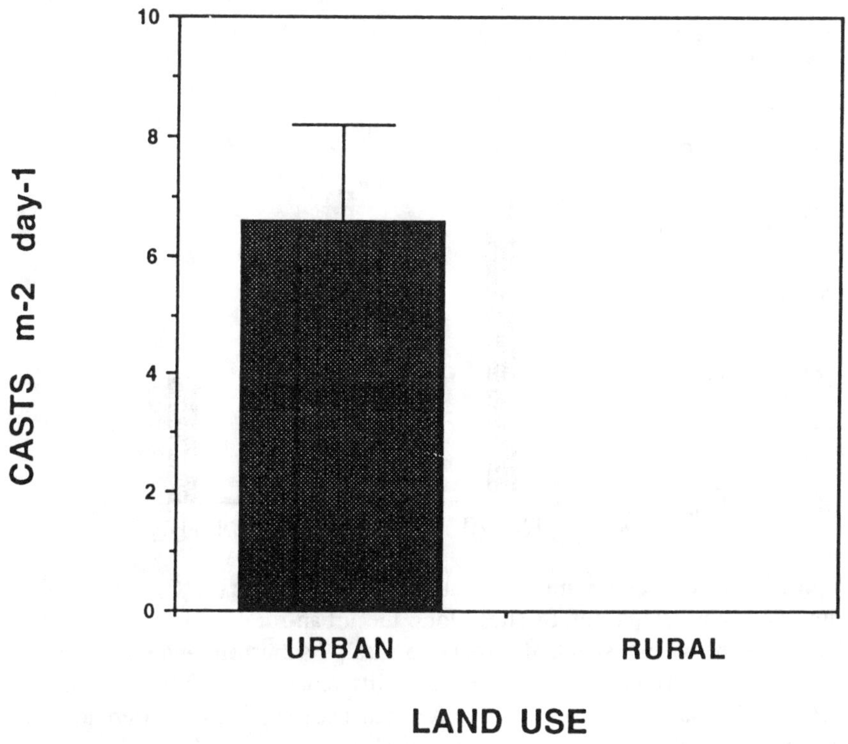

Figure 7. Earthworm casts in urban versus rural forest soils along the New York City urban-rural transect, 1989. Values are means (standard error) of 10 rural and 10 urban sites. Different superscripts indicate a significant difference at $p < 0.05$ in a one-way analysis of variance. (Adapted from Pouyat and Parmelee, submitted.)

Table 1. Soil C pools (0-15 cm) along the New York City urban to rural transect, summer 1993. Values for readily mineralizable, labile, and potentially mineralizable C are the means (standard error) of 5 rural and 6 urban sites sampled in 1993. Values for passive C are the means (standard error) of 9 rural and 9 urban sites sampled in 1989

Pool	Urban		Rural
		-------------mg g^{-1}-------------	
Readily mineralizable	2.1 (.48)	*	7.4 (3.3)
Labile (microbial)	6.9 (1.3)	**	20.9 (2.6)
Potentially mineralizable	4.7 (0.5)	**	6.8 (0.5)
Passive (total)	97 (3.3)	**	73 (4.3)

*Indicates significant difference at $p < 0.05$ in a Wilcoxon rank sum test (non-parametric).

**Indicates significant difference at $p < 0.05$ in a one-way analysis of variance.

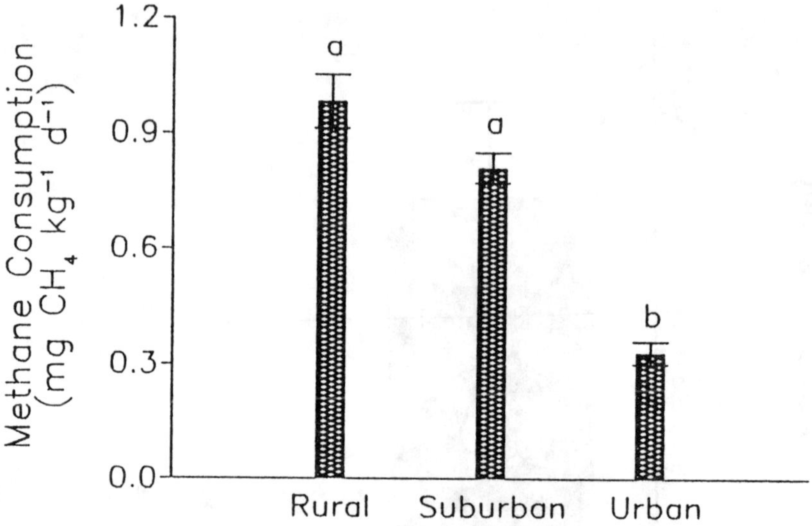

Figure 8. CH_4 consumption rates in laboratory incubations (25°C) of soil spiked with 100 ppmv CH_4 from 18 sites along the urban-rural transect, July 1992. Values are the means (standard error) of 5 rural, 7 suburban, and 6 urban sites. Different superscripts indicate a significant difference at $p < 0.05$ in a one-way analysis of variance with a Fisher's protected least significant difference test. (Adapted from Goldman et al., submitted.)

Table 2. Trace gas fluxes measured in field chambers, August 1992; values are the means (standard error) of three rural and three urban sites

Gas	Urban		Rural
CH_4 ($mg\ m^{-2}\ d^{-1}$)	-2.4 (0.19)	*	-3.1 (0.26)
CO_2 ($g\ C\ m^{-2}\ d^{-1}$)	2.2 (0.24)		2.1 (0.09)
N_2O	0		0

*Indicates significant difference at $p < 0.10$ in a one way analysis of variance. (Methane flux data are from Goldman et al., submitted.)

ecosystem and soil processes to develop hypotheses and design experiments to evaluate the integrated effect of urbanization on soil C pools and microbial activity, including trace gas fluxes. Experiments were carried out in urban and rural forest stands in the New York City metropolitan area. Preliminary results indicate that pools of labile C are lower and pools of passive C are higher in urban relative to rural forest sites. The data suggest that in the long-term, urban forest sites should sequester and store more C than rural forest sites. Decreases in labile C pools should result in decreases in microbial activity in urban relative to rural forest soils. Observed changes in soil C pools were the result of the complex interaction of urbanization factors and should not be used as indicators of changes that will result from global climate change.

VI. Acknowledgements

The authors thank Jennifer Haight, Josh Ulick, and Meadow Goldman for help with field sampling and laboratory analyses. This research was supported by grants from the Andrew W. Mellon Foundation, the U.S. National Science Foundation, the U.S.D.A. Forest Service's Northern Global Change Program, and the Lila Wallace-Reader's Digest Fund.

References

Airola, J.M. and K. Bucholz. 1984. Species structure and soil characteristics of five urban forest sites along the New Jersey Palisades. *Urban Ecology* 8:149-164.

Alig, R.J. and R.G. Healy. 1987. Urban and built-up land area changes in the United States: An empirical investigation of determinants. *Land Economics* 63:215-226.

Bagnall, R.G. 1979. A study of human impact on an urban forest remnant: Redwood Bush, Tawa, near Wellington, New Zealand. *New Zealand Journal of Botany* 17:117-126.

Baker, H.G. 1986. Patterns of plant invasion in North America. p. 44-57. In: H.A.Mooney and J.A. Drake, (eds.), *Ecology of biological invasions of North America and Hawaii.* Springer-Verlag, NY.

Bengtsson, G. and L. Tranvik. 1989. Critical metal concentrations for forest soil invertebrates. *Water, Air, and Soil Pollution* 47:381-417.

Bornkamm, R.D., J.A. Lee, and M.R.D.Seaward. 1982. *Urban Ecology*. Blackwell Scientific Publications, Oxford.

Bornstein, R.D. 1968. Observations of the urban heat island effect in New York City. *J. Applied Meterology* 7:575-582.

Candelino, C.A. and W.L. Chameides. 1990. Natural hydrocarbons, urbanization, and urban ozone. *J. Geophysical Research.* 95:13,971-13,979.

Chang, T.Y., D.P. Chock, R.H. Hammerle, S.M. Jappar, and I.T. Salmeen. 1992. Urban and regional ozone air quality: Issues relevant to the automobile industry. *Critical Reviews in Environmental Control* 22:27-66.

Changnon, S.A., Jr. 1980. Evidence of urban and lake influences on precipitation in the Chicago area. *J. of Applied Meteorology* 19:1137-1159.

Dickinson, R.E. 1966. The process of urbanization. p. 463-478. In: F.F. Darling and J.P. Milton, (eds.), *Future environments of North America.* Natural History Press, Garden City, NY.

Findlay, S.E.G.and C.J. Jones. 1990. Exposure of cottonwood plants to ozone alters subsequent leaf decomposition. *Oecologia* 82:248-250.

French, H.F. 1990. *Clearing the Air: A Global Agenda.* World Watch Paper 94. World Watch Institute, Washington, D.C.

Frey, H.T. 1984. *Expansion of urban area in the United States: 1960-1980.* U.S.D.A. Economic Research Service Staff Report No. AGES830615. Washington, D.C.

Giller, K. and S. McGrath. 1989. Muck, metals and microbes. *New Sci.* 124(1689):31-32.

Graedel, T.E. and P.J. Crutzen. 1989. The changing atmoshpere. *Scientific American* 261:58-68.

Gonick, W.N., A.E. Shearin, and D.E. Hill. 1970. *Soil survey of Litchfield County, Connecticut.* USDA Soil Conservation Service. U.S. Government Printing Office. Washington D.C.

Goldman, M.B., P.M. Groffman, R.V. Pouyat, M.J. McDonnell, and S.T.A. Pickett. Methane uptake and nitrogen availability in forest soils along an urban to rural gradient. Submitted to *Soil Biology and Biochemistry*.

Haub, C. and M.M. Kent. 1989. 1989 World Population Data Sheet. Population Reference Bureau, Inc. Washington, D.C.

Hill, D.E., E.H. Sautter, and W.N. Gonick WN. 1980. *Soils of Connecticut.* Connecticut Agricultural Experiment Station Bulletin No. 787.

Hobbs, E. 1988. Using ordination to analyze the compositionn and structure of urban forest islands. *Forest Ecology and Management* 23:139-158.

Hopkin, S.P. 1993. Ecological implications of "95% protection levels" for metals in soil. *Oikos* 66:137-141.

Jenkinson, D.S. and D.S. Powlson. 1976. The effects of biocidal treatments on metabolism in soil V. A method for measuring soil biomass. *Soil Biology and Biochemistry* 8:209-213.

Jenkinson, D.S. and J.H. Rayner. 1977. The turnover of soil organic matter in the Rothamsted classical experimetns. *Soil Science* 123:298-305.

Kellogg, W.W. and R. Schware. 1981. *Climate Change and Society.* Westview Press, CO.

Landsberg, H.E. 1981. *The urban climate.* Academic Press, NY.

Lavelle, P.I. 1988. Earthworm activities and the soil system. *Biology and Fertility of Soils* 6:237-251.

McDonnell, M.J. and S.T.A. Pickett. 1990. The study of ecosystem structure and function along urban-rural gradients: an unexploited opportunity for ecology. *Ecology* 71:1231-1237.

McDonnell, M.J., S.T.A. Pickett, and R.V. Pouyat. 1993. The application of the ecological gradient paradigm to the study of urban effects. p. 175-189. In: McDonnell, M.J. and S.T.A. Pickett (eds.). *Humans as Components of Ecosystems: Subtle Human Effects and the Ecology of Populated Areas*. Springer-Verlag, New York.

Nadelhoffer, K.J. 1990. Microlysimeter for measuring nitrogen mineralization and microbial respiration in aerobic soil incubations. *Soil Science Society of America Journal* 54:411-415.

Oke, T.R. 1990. The micrometeorology of the urban forest. *Quaterly J. Royal Meteorological Society* 324:335-349.

Parmelee, R.W. and D.A. Crossley, Jr. 1988. Earthworm production and role in the nitrogen cycle of a no-tillage agroecosystem on the Georgia Piedmont. *Pedobiologia* 32:353-361.

Parton, W.J., J.W.B. Stewart, and C.V. Cole. 1988. Dynamics of carbon, nitrogen, phosphorus and sulfur in cultivated soils: A model. *Biogeochemistry* 5:109-131.

Pastor, J. 1986. Reciprocally linked carbon-nitrogen cycles in forests: biological feedbacks within geological constraints. p. 131-140. In: G.I. Agren (ed.), *Predicting Consequences of Intensive Forest Harvesting on Long-Term Productivity*. Swed. Univ. Agric. Sci., Dept. Ecology and Environmental Research Report #26.

Pouyat, R.V. 1992. *Soil characteristics and litter dynamics in mixed deciduous forests along an urban-rural gradient*. Dissertation thesis. Rutgers University.

Pouyat, R.V. and M.J. McDonnell. 1991. Heavy metal accumulation in forest soils along an urban to rural gradient in southern NY, USA. *Water, Air, and Soil Pollution* 57-58:797-807.

Pouyat, R.V. and W.C. Zipperer. 1992. The uses and management of woodlands in urban and suburban environments. p. 26-29. In: P.D. Rodhell (ed.), *Proceedings of the fifth national urban forest conference*. American Forests, Washington, D.C.

Pouyat, R.V. and R.W. Parmelee. Environmental effects of soil invertebrate densities in oak stands along an urban-rural land use gradient. Submitted to *Pedobiologia*.

Rudnicky, J.L. and M.J. McDonnell. 1989. Forty-eight years of canopy change in a hardwood-hemlock forest in New York City. *Bulletin of the Torrey Botanical Club* 116:52-64.

Scheu, S. 1987. Microbial activity and nutrient dynamics in earthworms casts (Lumbricidae). *Biol. Fert. Soils* 5:230-234.

Scheu, S. and V. Wolters. 1991. Buffering of the effect of acid rain on decomposition of C-14 labelled beech leaf litter by saprophagous invertebrates. *Biol. Fert. Soils* 11:285-289.

Schwartz, S.E. 1989. Acid deposition: Unraveling a regional phenomenon. *Science* 243:753-763.

Seinfeld, J.H. 1989. Urban air pollution: State of the science. *Science* 243:745-752.

Sillman, S., J.A. Logan, and S.C. Wofsy. 1990. A regional scale model for ozone in the United States with subgrid representation of urban and power plant plumes. *Journal of Geophysical Research* 95:5731-5748.

Staaf, H. 1987. Foliage litter turnover and earthworm population in three beech forests of contrasting soil and vegetation types. *Oecologia* (Berlin) 72:58-64.

Stearns, F. and T. Montag (eds.). 1974. *The urban ecosystem: a holistic approach*. Dowden, Hutchinson and Ross, Inc. Stroudsburg.

Tornes, L.A. 1974. Soil Survey of the Hemlock Forest of the New York Botanical Gardens, Bronx New York. Unpublished report. USDA Soil Conservation Service in cooperation with the Cornell University Agricultural Experiment Station.

Tyler, G., A.M. Balsberg Pahlsson, G. Bengtsson, E. Baath, and L. Tranvik. 1989. Heavy-metal ecology of terrestrial plants,microorganisms and invertebrates. *Water, Air, and Soil Pollution* 47:189-215.

White, C.S. and M.J. McDonnell. 1988. Nitrogen cycling processes and soil characteristics in an urban versus rural forest. *Biogeochemistry* 5:243-262

CHAPTER **14**

Dynamics of Forest Floor and Soil Organic Matter Accumulation in Boreal, Temperate, and Tropical Forests

Kristiina A. Vogt, Daniel J. Vogt, Sandra Brown, Joel P. Tilley, Robert L. Edmonds, Whendee L. Silver, and Thomas G. Siccama

I. Introduction

Analysis and models of the global carbon budget suggest that soils may play an important role in sequestering carbon in ecosystems. On a global scale, soils contain about twice as much carbon as vegetation (Post et al., 1990) and this carbon is characterized by having the longest residence time compared to the other carbon pools (Anderson, 1991). By discerning which factors may control amounts and rates of soil organic carbon accumulation, management strategies may be developed to enhance the sequestration of carbon in ecosystems. To achieve this goal requires determining which abiotic and biotic factors correlate with the accumulation of carbon in the soil within as well as across climatic zones. Since soil organic matter accumulation is a balance between production and decomposition, an understanding of factors that influence both processes must be obtained if we are going to be able to predict carbon fluxes.

Our approach was to synthesize 93 data sets from 15 different forest climatic types to search for general patterns of detrital carbon accumulation (e.g., soil organic matter [SOM] and forest floor [FF]). These patterns of organic matter accumulation were then correlated with site factors to develop relationships that could be used in models and also for management. The contribution of SOM and FF to total ecosystem carbon budgets were examined for forest ecosystems by climatic zones, climatic forest types, and by soil orders. These large scale analyses were complemented by more specific comparisons to determine factor(s) that predict amounts of SOM and FF accumulation, such as temperature, precipitation, temperature/precipitation ratio, net primary production, litter transfers and N inputs in litter, soil N contents, litter and soil C/N ratios, foliage life-span, and by the soil order and texture index (e.g., sandy = sands, loamy sands and sandy loams; loamy = loam, silt loam, silts, sandy clay loam, silt clay loam; loam-clay = clay loam, sandy clay, silty clay; and clayey = clay).

There are several parameters that can affect SOM accumulation (e.g., topography, drainage) that will, however, not be discussed in this paper since the availability of that type of information is sparse and therefore difficult for statistical comparisons using even nearly-balanced designs. This is not meant to be an exhaustive list of all the studies that have been conducted on SOM or FF accumulations (see Post et al. 1982, Schlesinger 1984). The data sets included here are those where site-specific information existed for the above-mentioned climatic and ecosystem parameters, and total ecosystem organic matter data were available. Statistical analyses on these data were conducted using SAS System for Personal Computers (SAS Institute Inc.; Cary, NC, USA) for all data computation, linear regressions, one way analysis of variance, and multiple comparisons of means using the Student-Newman-Keuls and Scheffe's tests at P = 0.10. If the data were not normal or the variance homogenous, they were "rank" transformed and then analyzed using parametric statistics (Conover and Iman, 1981).

ISBN 1-56670-117-1/95/$0.00+.50

II. Accumulations of Soil Organic Matter and Forest Floor Mass

B. By Climatic Zones, Soil Texture Index, Foliage Life-Span

Forest ecosystems are frequently ranked along natural climatic-zone groupings (i.e., boreal, temperate, subtropical, tropical) for comparisons of structure and function. When comparing forests in tropical, temperate and boreal zones using only climatic life-zone groupings, clear predictive patterns of SOM accumulation do not typically exist (Coleman et al. 1989). This is due, in part, to the high coefficients of variation that exist around mean values obtained using groups based on very general criteria. In such cases, differences will only be significant when a specific factor(s) (e.g., low temperatures, high allophane contents in soil, etc.) have overriding controls on decomposition.

With the data set used in this paper, some significant differences in SOM and FF accumulations did occur using broad climatic index groupings (Table 1). For example, in the boreal region where low temperatures strongly limit decomposition (Van Cleve et al., 1990), SOM accumulations were significantly lower when compared to the temperate climatic zone (Table 1). However, SOM accumulations in the tropics were not significantly different from those in other climatic groupings examined (Table 1). Similar to the boreal zone, decomposition can be limited in the tropics and subtropics, especially in regions dominated by oxisols where the chemical complexation of organics reduce the access to decomposer activity (Coleman et al., 1989).

Unlike the patterns in the soil, clear patterns of carbon accumulation occurred in surface organic layers (FF) that reflected broad-scale temperature gradients: highest accumulations in the boreal region, intermediate values in the temperate region and the lowest values in the tropics (Table 1). Soil texture had no apparent relationship with SOM (Table 1). However, forest floor accumulation on loamy type soils was significantly higher than all other texture classes.

Broad-scale groupings by foliage life-span were quite successful in isolating general patterns of SOM and FF accumulation (Table 1). Forests dominated by evergreen tree species had significantly lower SOM accumulations than sites dominated by a mixture of deciduous and evergreen species. Sites dominated by deciduous tree species had trends of higher mean SOM accumulations than the evergreen forests, but these differences were not significant at this grouping level (Table 1). Similarly, FF mass was significantly higher in forests dominated by a mixture of deciduous and evergreen species than in forests dominated by a single foliage life-span (Table 1).

B. By Forest Climatic Types

Further subdivision of data into groups combining the climatic index and foliage life-spans (e.g., forest climatic type) did not increase the ability to identify clear patterns of SOM accumulation, but did separate the forest floor accumulations into more significant groupings (Table 2). SOM accumulations were not significantly different among the boreal, temperate, and tropical climatic zones; however, the means were different enough to suggest trends. Unlike SOM, forest floor accumulations are more sensitive to climatic and foliage life-span indices and therefore separation of the data into significant groups is possible. Similar to the earlier comparisons using a climatic index (Table 1), combining the climatic index with foliage life-spans resulted in even a greater number of significant groupings for the forest floor. This reflects the greater sensitivity of the decomposition of surface organic layers to climatic factors in contrast to organic matter in the soil.

Even though a greater number of significantly distinct groups were identified for the FF using the forest climatic type groupings, several within-climatic-zone differences could not be discerned using this approach (Table 2). For example, in the boreal and cold temperate zone forests, identifiably significant groupings could not be detected based on the deciduous and evergreen behavior because of the inherent variability in each forest-climatic-type grouping (Table 2). Since FF accumulations in general were much lower in the warm temperate and tropical zones, it was impossible to show significant differences by forest climatic type (Table 2).

Forest floor and profile soil carbon contents are commonly presented using climatic life-zones because it has been useful to scale information up to larger spatial scales (Post et al., 1982; Vogt et al., 1986). This type of informational grouping does have limitations if one is interested in scaling down to specific sites since the averages (especially in the tropics) given are based on only a few data sets and for limited forest

Table 1. Soil organic matter (SOM) and forest floor (FF) accumulations by climatic index, soil texture index, and by foliage life-span; mean (standard deviation); (significant differences in columns are shown by different lower case letters [Student-Newman-Keuls at P = 0.10])

	SOM	FF
	---- $Mg\ ha^{-1}$ ----	
CLIMATIC INDEX		
Boreal	107 (105) a (n = 5)	83 (50) a (n = 5)
Temperate	198 (142) b (n = 56)	42 (39) a (n = 54)
Tropical	155 (186) ab (n = 29)	12 (22) b (n = 25)
SOIL TEXTURE INDEX		
Sandy	167 (142) a (n = 56)	26 (30) ab (n = 50)
Loamy	195 (113) a (n = 23)	71 (46) a (n = 23)
Loam-Clay	90 a (n = 1)	13 ab (n = 1)
Clayey	218 (290) a (n = 10)	10 (12) b (n = 10)
FOLIAGE LIFE-SPAN		
Deciduous + semi-deciduous	197 (71) ab (n = 20)	35 (39) b (n = 21)
Evergreen	166 (173) b (n = 66)	32 (37) b (n = 61)
Mixed (deciduous + evergreen)	299 (159) a (n = 4)	105 (21) a (n = 4)

types within a climatic zone (Brown and Lugo, 1982), and many other factors are important in controlling organic matter accumulation (Coleman et al., 1989) that cannot be incorporated at this scale.

This approach, of grouping data by climatic and vegetative-climatic types, assumes that information on climate and vegetation provide sufficient baseline information to do a rough sorting of the data and it is useful in this context. Even though significant differences are difficult to show, the rough classification of data by vegetative-climatic type may be very useful for identifying factors that should be pursued at site-specific scales. For the data set used in this paper, the larger scale data analyses by groups did suggest some of the patterns that are explicitly shown at the stand level. For example, grouping of SOM data by foliage life-span suggested patterns that were verified at the site specific scale.

Even though mean organic carbon pools in different components are not significantly different at the climatic zone level, the proportions of SOM to total ecosystem organic matter are higher in deciduous than evergreen forests of the higher latitudes (Table 2). For example in the boreal climatic zone, deciduous forests contain about 72% of total ecosystem carbon in the soil in contrast to the evergreen forests with only 17%. Similarly, in the cold temperate zone, deciduous forests had 47% of total ecosystem organic matter in the soil compared to 39% in evergreen forests (Table 2). Again this pattern in higher latitude forests was site specific; sites higher in nutrients (e.g., deciduous-dominated or high-site-quality conifer stands) contained significantly more organic matter in the soils in the cold temperate zone

This pattern observed for higher latitude forests was not found in lower latitude forests (Table 2). For example, in the subtropical zone, the proportion of total ecosystem organic matter in SOM was relatively

Table 2. Total living organic biomass, forest floor biomass, and soil organic matter accumulations, and total organic matter (minus detrital wood) by forest types; mean/(standard deviation; sample size); different lower case letters by columns show significant differences by column (Student-Newman-Keuls at P = 0.10)

Forest climatic type[a]	Total living biomass	Forest floor biomass	Detrital wood mass	Soil organic matter	Total organic matter
	----------------	-Mg ha^{-1}-	----------------		
BOBLDE	78 a	37 abcd	0.11 b	207 ab	322 a
	(91; 2)	(44; 2)	(-;1)	(104; 2)	(239; 2)
BONLEV	93 a	113 a		41 b	247 a
	(70; 3)	(24; 3)		(6; 3)	(80; 3)
CTBLDE	245 a	50 abcd	25 a	223 a	458 a
	(68; 7)	(42; 12)	(6; 2)	(63; 12)	(113; 7)
CTBLEV		10 bcd		165 ab	
		(2; 3)		(20; 3)	
CTNLEV+		105 a		299 a	
BLDE		(21; 4)		(159; 4)	
CTNLEV	367 a	42 abc	64 ab	193 ab	585 a
	(337; 21)	(38; 27)	(86; 11)	(185; 27)	(447; 21)
WTBLDE	170 a	9 bcd		168 ab	357 a
	(0.7; 4)	(4; 3)		(46; 3)	(56; 2)
WTBLEV		18 abcd		196 a	
		(12;2)		(0.07;2)	
WTNLEV	103 a	23 abcd		122 ab	242 a
	(73; 4)	(10; 5)		(45; 5)	(39; 4)
STNLEV	204 a	9 bcd	2 ab	133 ab	259 a
	(168; 3)	(3; 3)	(2; 2)	(63; 2)	(172; 2)
TRBLSDE	359 a	2 d		76 ab	437 a
	(-; 1)	(-; 1)		(; 1)	(; 1)
TRBLEV	332 a	7 bcd	11 ab	188 ab	666 a
	(151; 8)	(4; 6)	(-; 1)	(258; 13)	(462; 5)
TRNLEV		72 ab		2 b	
		(69; 2)		(1; 2)	

[a]BOBLDE = Boreal broadleaf deciduous; BONLEV = boreal needle-leaved evergreen; CTBLDE = cold temperate broadleaf deciduous; CTBLEV = cold temperate broadleaf evergreen; CTNLEV+BLDE = cold temperate needle-leaved evergreen+broadleaf deciduous; CTNLEV = cold temperate needle-leaved evergreen; WTBLDE = warm temperate broadleaf deciduous; WTBLEV = warm temperate broadleaf evergreen; WTNLEV = warm temperate needle-leaved evergreen; STBLEV = subtropical broadleaf evergreen; STNLEV = substropical needle-leaved evergreen; TRBLSDE = tropical broadleaf semi-deciduous; TRBLEV=tropical broadleaf evergreen; TRNLEV = tropical needle-leaved evergreen.

similar in the broadleaf-deciduous forests (50%) and the broadleaf/needle-leaf evergreen forests (55%). Insufficient data do not allow for this comparison to be made for the tropics. These comparisons also cannot be made in the warm temperate zone because the evergreen forests were quite young and would not realistically represent mature forests of that climatic zone. The percent of total organic matter in SOM in the warm temperate broadleaf deciduous forests (50%) is very similar to the values reported in the subtropical forests.

None of the comparisons within forest climatic types were significant because of high within group variability (Table 2), although forests dominated by evergreen trees in the cold temperate, subtropical, and tropical zones tended to have 20-30% higher total ecosystem organic matter than those dominated by deciduous trees. Forests in the boreal zone had the opposite pattern with the deciduous species accumulating about 25% higher total ecosystem organic matter than the evergreen forests. Most of these differences in the boreal zone are caused by the higher accumulation of organic matter in the soil horizons of the deciduous dominated forests (Table 2). The results in the boreal zone reflect the greater ability of the deciduous species to fix carbon under low temperatures and short growing seasons (Viereck et al., 1983).

C. By Soil Order

Another approach that has been used to examine SOM dynamics is to group the data using soil order as a common denominator (see Post et al., 1982; Buringh, 1984). Soil order information should help clarify SOM and FF accumulation patterns if they exist, especially when done in the context of the climatic forest type groupings. Groupings based on soil orders should help reduce the high coefficients of variation within the climatic zones and forest climatic types because this approach indirectly incorporates site water and nutrient availability information. Although soil order approach helped reduce variability so that significant differences could be identified at the extreme ends of the data sets, there was not sufficient reduction in variability to identify distinct groups across soil orders (Tables 3,4). Perhaps the soil suborder level should be used but not everyone includes soil information or uses a common soil taxonomy classification system.

In mature forest stands, Buringh (1984) reported the greatest differences in mean SOM contents between histosols (647 Mg ha^{-1}) and spodosols (224 Mg ha^{-1}) with the other soil orders being relatively similar (397-466 Mg ha^{-1}). In this study, only the ultisols in the cold temperate zone had significantly higher SOM accumulations compared to the other soil groupings (Table 3) and this relationship is strongly influenced by two sites (Grier, 1976). No significant differences could be shown in SOM values between the remaining soil orders (Table 3). The highest SOM accumulations did occur for ultisols in the cold temperate and tropical climatic zones.

When similar SOM values occur within a soil order but across different climatic regions (e.g., ultisols between the tropical and temperate sites), the role of soil chemical factors in stabilizing organic matter in the soil becomes important (Coleman et al., 1989). The generally higher accumulations of SOM in some ecosystems can be related to soil components that strongly react and complex with organics (e.g., allophane) making them unavailable to microbes. In some soil orders (e.g., Andisols), organic matter accumulates in the soil because amorphous Al protects soil carbon from being decayed due to the formation of organo-aluminum complexes (Boudot et al., 1986). Also nutrient deficiencies (e.g., P) may limit the ability of microbes to degrade litter (Sanchez, 1976). Thus the capacity of the soil to sequester organic matter is related to the factors that influence carbon stabilization in that soil.

D. By Soil Texture Index

Other studies have shown soil texture to be an important grouping criterion to use when comparing sites within specific climatic zones. For example, Greenland and Nye (1959) found wide variability in SOM in several tropical forests that were related to soil type: 95 Mg ha^{-1} in soils of sandy texture, 183 in medium textured soils and 679 Mg ha^{-1} in soils high in allophane. These differences are suggested to reflect the influence of soil texture on decomposition rates with fine, clayey soils having slower decomposition rates than coarse, sandy soils (Van Veen and Kuikman, 1990). Sandy soils should have lower SOM levels since (1) the abiotic factors (e.g., temperature and moisture) are important in controlling organic matter accumulation while the soil chemical factors play a more minor role in stabilizing the organic matter and (2) litter inputs into the decomposition system will be lower because of lower primary production.

Table 3. Total living biomass, forest floor biomass, soil organic matter accumulations, and total organic matter (minus detrital wood) grouped by forest climatic zone by soil order; mean/(standard deviation; sample size); different lower case letters show significant differences by column (Student-Newman-Keuls at P = 0.10; Scheffe's test for SOM at P = 0.10))

SOIL ORDER Forest climatic type	Total living biomass	Forest floor biomass	Soil organic matter	Total organic matter
	- - - - - - - - -	Mg ha^{-1}	- - - - - - - - -	
ALFISOL				
Cold temperate	463 a (336; 4)	30 ab (19; 4)	112 b (29; 4)	605 ab (327; 4)
SPODOSOL				
Cold temperate	235 a (176; 8)	76 ab (43; 21)	214 ab (98; 21)	437 ab (253; 8)
Warm temperate	44 a (27; 2)	21 ab (18; 2)	147 b (51; 2)	212 ab (7; 2)
INCEPTISOL				
Boreal	105 a (62; 4)	102 a (30; 4)	101 b (120; 4)	308 ab (138; 4)
Cold temperate	306 a (297; 13)	26 ab (22; 15)	167 b (102; 15)	476 ab (306; 13)
Warm temperate	- -	14 ab (-; 1)	146 b (-; 1)	
ULTISOLS				
Cold temperate	667 a (616; 2)	28 ab (8; 2)	773 a (4; 2)	1,468 a (629; 2)
Subtropical	144 a (95; 8)	8 b (3; 10)	167 b (100; 10)	324 ab (192; 8)
Tropical	365 a (21; 2)	10 b (3; 2)	300 ab (410; 5)	937 a (642; 2)
ENTISOL				
Warm temperate	170 a (-; 1)	6 b (-; 1)	222 ab (-; 1)	397 ab (-; 1)
Subtropical		10 ab (-; 1)	52 b (-; 1)	
Tropical	79 a (-; 1)	49 ab (63; 3)	15 b (23; 3)	124 b (-; 1)
OXISOL				
Subtropical	116 a (21; 2)	6 b (0.27; 2)	126 b (50; 2)	247 ab (29; 2)
Tropical	368 a (132; 6)	5 b (4; 4)	128 b (51; 7)	590 ab (136; 3)

Table 4. Above- (ANPP), below-ground (BNPP) and total (TNPP) net primary production grouped by forest climatic zones by soil order; mean/(standard deviation; sample size); different lower case letters show significant differences by column (Student-Newman-Keuls at P = 0.10)

SOIL ORDER Forest climatic zone	ANPP	BNPP	TNPP
	- - - - - - kg ha^{-1} yr^{-1} - - - - -		
ALFISOL			
Cold temperate	12,050 abc (2,743; 3)	3,910 a (2,085; 3)	15,960 ab (3,925; 3)
SPODOSOL			
Cold temperate	7,490 abc (3,508; 6)	7,095 a (5,708; 4)	14,193 ab (1,599; 4)
Warm temperate	8,895 abc (6,244; 2)	5,530 a (891; 2)	14,425 ab (7,135;2)
INCEPTISOL			
Boreal	2,274 c (1,946;4)	2,375 a (-; 1)	7,539 b (-; 1)
Cold temperate	10,165 abc (4,134; 14)	3,042 a (1,427; 12)	14,126 ab (4,100; 11)
Warm temperate	4,400 bc (-; 1)		
ULTISOLS			
Cold temperate	26,055 c (8,690; 2)	4,100 a (1,980; 2)	30,180 b (10,706; 2)
Subtropical	10,310 ab (3,320; 10)		
ENTISOL			
Warm temperate	6,105 bc (-; 1)	8,720 a (-; 1)	14,825 ab (-; 1)
Subtropical	7,600 abc (-; 1)		
OXISOL			
Subtropical	15,700 ab (4,808; 2)		
Tropical	14,485 ab (2,807; 2)	2,000 a (-; 1)	14,500 ab (-; 1)

Table 5. Soil organic matter accumulation by soil texture indeces by forest climatic zones. mean (standard deviation; sample size); no significant differences (Sheffe's test at P=0.10) by soil texture indeces across all forest climatic zones.

FOREST CLIMATIC ZONE	Soil texture index	Soil organic matter
		- Mg ha^{-1} -
BOREAL	Loamy	121 (138; n = 3)
	Sandy	87 (66; n = 2)
TEMPERATE	Clayey	74 (-; n = 1)
	Loamy	206 (109; n = 20)
	Sandy	204 (162; n = 33)
TROPICAL	Clayey	234 (302; n = 9)
	Loam-clay	90 (-; n = 1)
	Loamy	121 (93; n = 19)

Similarities in soil texture have been used to explain why mean organic matter contents do not vary between the tropics and the temperate zone when comparisons were made using similar soil orders (Sanchez, 1976). However, broad grouping of SOM and FF accumulations using a soil texture index was unable to show significant differences between different soils in our analyses (Table 1). Separating soil texture indices by climatic zone did show possible trends even though none of these comparisons were significant (Table 5). Mean SOM values were the lowest in the sandy soils of the boreal climatic zone but the tropical and the temperate climatic zones did not follow this pattern (Table 5). Furthermore, even though the clayey soils did have the highest SOM accumulations (218 Mg ha^{-1}) as suggested by Greenland and Nye (1959), the loamy soils had very similar values (195 Mg ha^{-1}) (Table 1). The Greenland and Nye (1959) SOM value for clayey soils was considerably higher than that recorded for tropical sites in this study (Table 5), therefore, differences of the magnitude seen by them could not be detected with this data set. The inability to show significant patterns using the soil texture indices again reflects the high variability that exists within data groupings.

III. Primary Production and Litter Transfer Relationships

A. Climatic Zones and Soil Orders

Comparisons of aboveground (ANPP), belowground (BNPP) and total net primary production (TNPP) data were examined for significant variations by forest climatic types (Table 6). Except for the two boreal forest types, which had significantly lower ANPP (Table 6), no other significant differences were observed. No patterns for BNPP data existed among the forest climatic types or by forest climatic types within each soil order (Tables 5,6). Grouping data at the level of forest climatic type and even by forest climatic types within soil orders were ineffective in separating the differences in ANPP and BNPP beyond that suggested by their mean values (Table 6).

Table 6. Aboveground (ANPP), belowground (BNPP), and total (TNPP) net primary production, and above- plus belowground detrital litter transfers by forest type; mean/(standard deviation; sample size); different lower case letters show significant differences by column (Student-Newman-Keuls at P = 0.10); abbreviations given in Table 2

Forest climatic type	ANPP	BNPP	TNPP	Above detrital transfer	Below detrital transfer
			kg ha^{-1} yr^{-1}		
BOBLDE	3,476ab	2,375a	7,539a	2,646ab	
	(2387; 2)	(-; 1)	(-; 1)	(197; 3)	
BONLEV	1,311b			322b	
	(334; 3)			(-; 1)	
CTBLDE	12,094ab	2,568a	15,350ab	9,571ab	1,000a
	(2760; 6)	(752; 5)	(2950; 5)	(12270; 6)	(-; 1)
CTNLEV/	7,800ab			5,038ab	
BLDE	(-; 1)			(-; 1)	
CTNLEV	10,841ab	4,498a	16,243ab	3,163ab	6,152a
	(7107; 18)	(3323; 16)	(7257; 15)	(2629; 20)	(4395; 8)
WTBLDE	6,420ab	8,720a	14,825ab	3,619ab	6,750a
	(2194; 3)	(-; 1)	(-; 1)	(310; 3)	(-; 1)
WTNLEV	8,895ab	5,530a	14,425ab	5,910ab	3,313a
	(6244; 2)	(891; 2)	(7135; 2)	(2197; 5)	(979; 4)
STBLEV	10,509ab		21,400b	7,479a	2,500a
	(4040; 11)		(-; 1)	(2896; 12)	(-; 1)
STNLEV	13,250a		24,500b	6,993ab	
	(2616; 2)		(-; 1)	(5805; 3)	
TRBLEV	14,485a	2,000a	14,500ab	9,521a	
	(2807; 2)	(-; 1)	(-; 1)	(1857; 7)	

This lack of significant trends in ANPP and BNPP is not surprising since this grouping does not incorporate site differences in water and nutrient availability which can significantly control carbon allocation and plant production. The highest TNPP values were reported in the subtropics where TNPP varied from 21.4-24.5 Mg ha^{-1} yr^{-1} whereas the lowest value of 7.5 Mg ha^{-1} yr^{-1} was reported in a boreal forest. A comparison of data within the cold and warm temperate forested zones showed that deciduous-dominated forests had similar TNPP values to that of the evergreen forests (Table 6). This again reflects how a greater proportion of total production can occur belowground in evergreen forests compared to deciduous forests (Vogt et al., 1986).

Similar to the production comparisons above, the variability in litter transfers from above- and belowground plant tissues could not be separated using the forest climatic type groupings. Again the key factors controlling these transfer processes are not directly related to the climatic type grouping (Table 6). It seems that "a priority" patterns in the data do exist but the high variability masks any significant separations. In general, aboveground litter transfers were higher in the deciduous compared to the evergreen dominated sites in the cold temperate and subtropical climatic zones (Table 6). There are too few data points existing for belowground detrital transfers to even begin addressing how they might vary with the evergreen or deciduous behavior of the forest (Table 6). Total litter transfers were very similar within the same climatic zone. These patterns have also been reported in an earlier synthesis of data across a diversity of forest climatic types (Vogt et al., 1986).

B. SOM and FF

Since net primary production is strongly related to site nutrient status (Vogt et al., 1986), one would expect a good relationship between net primary production and SOM. However in this dataset, neither ANPP nor BNPP data were effective in explaining SOM accumulation patterns across all climatic zones (Table 7). When total net primary production data were used, 44% of the variation in SOM was significantly explained across all climatic zones (Table 7).

When comparisons were made within climatic zones, it appeared that the relationship between net primary production and SOM was stronger in some climatic zones than others, and ANPP seemed to explain more of the SOM variation than BNPP (Table 8). For example, in the cold temperate zone, 57% and 42% of the variation in SOM was explained by TNPP and ANPP, respectively. In the boreal zone, 91% of the variation in SOM was explained by ANPP (Table 8). No significant relationships between BNPP and SOM were obtained (Table 8), but as previously mentioned these analyses were data limited. However, BNPP did explain 57% of the variation in the FF accumulation within the cold temperate zone (Table 8).

Above- and belowground litter transfers have been identified to be important in controlling SOM dynamics. Despite this, neither above- nor belowground litter transfers in this dataset correlated well with SOM accumulations when examined within or across all climatic zones (Table 7,8). This is the opposite of what would have been expected based on site specific studies. In a long-term study in Sweden, where different quality and quantity of litter material were added to agricultural plots, the rate and type of organic matter input into the system were the most important factors in predicting SOM accumulation levels (Paustian et al., 1992). Therefore when fertilizing a low site quality stand with nitrogen to increase its site quality, the higher aboveground litterfall that occurs with fertilization is needed if higher SOM accumulations are to be achieved.

Several studies have identified belowground litter inputs into the soil as a factor necessary to know in order to understand SOM accumulation. In an annual herb field in Michigan, Richter et al. (1990) suggested that soil organic C dynamics were controlled by two different factors (1) a rapidly cycling component of the carbon cycle that was dominated by plant inputs (especially roots) and (2) domination by slow cycling pools resistant to decay that were associated with clay-mineral fractions. That study identified root inputs as being important since the elimination of these inputs caused a 75% reduction in total belowground carbon (Richter et al., 1990). Hendrickson and Robinson (1984) also reported that a large fraction of the carbon turning over in the soil is due to root mass, root exudates, and rhizosphere microbial biomass, suggesting that plant roots are in important in maintaining the labile pools of soil carbon. The importance of the contribution of root mortality to detrital inputs varies with forest type; evergreen species may contribute twice as much organic matter annually than deciduous species in some ecosystems (Vogt et al., 1986). In tropical forests, the distinction needs to be made between broadleaf evergreen and needle-leaved evergreen forests; with broadleaf evergreen forests having greater detrital inputs from root turnover than coniferous forests (Lugo 1992).

As previously stated, our analysis showed that belowground litter transfers by themselves (not including exudates etc.) were poorly correlated to SOM accumulations (Tables 7,8). However, there is some suggestion that knowing the amount of N added into the soil horizon from root turnover may be useful in understanding SOM accumulation. An analysis of the data from the cold temperate zone showed 91% of the SOM accumulation could be explained by belowground litter transfers of N (Table 8). How much of this may be due to autocorrelation or to a real relationship is not known. Furthermore, this relationship may not be universally transferable to all ecosystems. For example, studies in a tropical forest suggest that most of the N in root tissues may not be available to microorganisms because of complexation by plant polyphenolics (Bloomfield et al., 1993). The contribution of root-tissue-N to microbial decay of root tissues may be very low and therefore may be a poor indicator of decomposition rates. This is an area that needs more research.

Table 7. Summary statistics for regressions of soil organic matter (SOM) and forest floor (FF) with abiotic and biotic variables across all climatic zones (r^2, (sample size),* = P < 0.01; ** = P < 0.10)

	- - SOM - -	- - FF - -
Annual air temperature, °C		
Minimum	0.04; (66)	0.62; (59)*
Maximum	0.02; (64)	0.004; (57)
Mean	0.05; (78)**	0.54; (74)*
Ratio of mean annual air temperature (°C)/ precipitation (cm)	0.005; (78)	0.22; (74)*
Net Primary Production		
Aboveground	0.24; (50)*	0.17; (50)*
Belowground	0.16; (24)	0.34; (26)*
Total	0.44; (25)*	0.02; (28)
Litter transfers		
Aboveground	0.01; (62)	0.09; (62)**
Total	0.10; (13)	0.07; (13)
Nitrogen input in aboveground litter	0.001; (29)	0.23; (32)*
Total Soil N	0.78; (56)*	0.04; (55)
Soil C/N ratio	0.04; (53)	0.16; (49)*

IV. Time Dimensions

Most of the studies that have examined soil organic matter turnover rates using carbon dating have estimated that the process is very slow: ages of SOM reported in the literature range from around 100 to over 5,700 years (see Skjemstad et al., 1990). Decomposition of organic matter in the soil is slower than surface litter layers because the edaphic environment of the soil causes organic matter to become stabilized with mineral colloids (e.g., clays) that protect them from microbial attack (Anderson, 1991).

Changes in SOM accumulation over time can also be estimated using a series of different aged stands dominated by the same overstory species. This approach suggests that SOM accumulates at low rates under natural conditions and may be quite stable over long-time scales suggesting an inherent site capacity to store carbon (Vogt, 1987). Organic matter contents of soils may not fluctuate signficantly until a natural disturbance impacts the site, the site is harvested or there is a shift in plant species dominance. These SOM fluctuations can be positive, no change if similar species are allowed to reoccupy a site immediately and the harvesting did not result in large inputs of organic matter (Vogt, 1987; Johnson et al. 1991) or negative when entirely different plants are introduced such as under intensive agricultural management (Brown and Lugo, 1990). Depending upon the type of disturbance, significant short-term (5-10 yrs) increases in SOM can occur that are not sustained into later stages of stand development (Vogt, 1987).

There are several reasons why changes in SOM pools are not detected under undisturbed conditions with time. Most forests have low annual accumulations of SOM and our instruments and/or methodologies are not sophisticated enough to detect these small changes. The high within site variability in SOM also requires large sample sizes and spatially explicit sampling which can be financially unrealistic. However, primary

Table 8. Summary statistics for regressins of soil organic matter (SOM) and forest floor (FF) with abiotic and biotic variables within forest climatic zones. (r^2, (sample size), * = $P < 0.01$; ** = $P < 0.10$)

	--- SOM ---	--- FF ---
MINIMUM ANNUAL AIR TEMPERATURE, °C		
Boreal	0.00; (4)*	0.00; (4)*
Cold temperate	0.04; (35)	0.69; (35)*
Warm temperate	0.59; (3)	0.33; (3)
Subtropical	0.13; (13)	0.37; (13)**
Tropical	0.76; (11)*	0.11; (4)
MAXIMUM ANNUAL AIR TEMPERATURE, °C		
Boreal	0.00; (4)*	0.00; (4)*
Cold temperate	0.01; (35)	0.12; (35)**
Subtropical	<0.01; (13)	0.21; (13)
Tropical	0.38; (11)**	0.24; (4)
MEAN ANNUAL AIR TEMPERATURE, °C		
Boreal	0.00; (4)*	0.00; (4)*
Cold temperate	0.08; (42)**	0.53; (42)*
Warm temperate	<0.01; (5)	0.11; (5)
Subtropical	0.10; (13)	0.23; (15)**
Tropical	0.90; (14)*	0.01; (8)
RATIO MEAN AIR TEMPERATURE, °C/ PRECIPITATION, cm		
Boreal	0.18; (4)	0.47; (4)
Cold temperate	<0.01; (42)	0.22; (42)*
Warm temperate	<0.00; (5)	<0.01; (5)
Subtropical	0.07; (13)	0.10; (15)
Tropical	0.08; (14)	0.08; (8)
ABOVEGROUND NET PRIMARY PRODUCTION		
Boreal	0.91; (5)*	0.05; (5)
Cold temperate	0.42; .(25)*	0.13; (25)**
Warm temperate	0.43; (5)	0.65; (5)**
Subtropical	0.03; (13)	<0.01; (13)
BELOWGROUND NET PRIMARY PRODUCTION		
Cold temperate	0.02; (21)	0.57; (21)*
Warm temperate	0.27; (3)	0.08; (3)
TOTAL NET PRIMARY PRODUCTION		
Cold temperate	0.57; (20)*	0.001; (20)
Warm temperate	0.37; (3)	0.63; (3)
ABOVEGROUND LITTER TRANSFERS		
Boreal	0.99; (4)*	0.47; (4)
Cold temperate	0.02; (30)	0.02; (30)
Warm temperate	0.05; (8)	0.25, (8)
Subtropical	0.01; (13)	0.02; (15)
Tropical	0.33; (7)	0.12; (5)
BELOWGROUND LITTER TRANSFERS		
Cold temperate	0.27; (9)	0.26; (9)
Warm temperate	0.11; (3)	0.01; (3)

Table 8.-- continued

TOTAL LITTER TRANSFERS		
Cold temperate	0.10; (8)	0.38; (8)**
Warm temperate	0.01; (3)	0.13; (3)
ABOVEGROUND LITTER TRANSFERS NITROGEN		
Boreal	0.99; (4)*	0.49; (4)
Cold temperate	0.02; (12)	0.04; (12)
Warm temperate	0.07; (4)	0.91; (4)**
Subtropical	0.13; (9)	0.01; (11)
BELOWGROUND LITTER TRANSFER NITROGEN		
Cold temperate	0.91; (3)	0.97; (3)
Subtropical	- -	0.20; (3)
TOTAL LITTER TRANSFER NITROGEN		
Cold temperate	0.89; (3)	0.98; (3)**
TOTAL SOIL N		
Boreal	0.38; (5)	0.02; (5)
Cold temperate	0.84; (27)*	0.04; (27)
Warm temperate	0.72; (4)	0.37; (4)
Subtropical	0.21; (13)	0.01; (15)
Tropical	0.99; (7)*	0.15; (4)
SOIL C/N RATIO		
Boreal	0.76; (5)*	0.34; (5)
Cold temperate	0.00; (27)	0.20; (27)**
Warm temperate	0.65; (4)	0.30; (4)
Subtropical	0.84; (13)*	0.17; (13)
Tropical	0.44; (4)	- -

and secondary successional studies can be used to estimate annual SOM accumulation rates. If the time-scales of SOM accumulation are examined for primary succession, values of organic matter accumulation per year are low and vary from 0.012-0.655 Mg ha^{-1} yr^{-1} (Schlesinger, 1990). Huntington et al. (1988) estimated that changes in soil carbon pools should be detectable when changes in SOM approach over 4 Mg ha^{-1}. Accordingly the annual values estimated in a primary successional sequence would be impossible to detect because the annual changes would be masked by the error terms.

Despite these difficulties, examining a chronosequence of stand ages gives a rough estimation of organic matter accumulation over time. This type of comparison of two different chronosequences of stands in Washington suggests that SOM will accumulate in higher elevation sites over a 160 year-time scale while no SOM accumulation may be detectable over a 400 year-time period in lower elevation sites. For example, in subalpine *Abies amabilis* ecosystems varying in age from 23 to 180 years, organic matter accumulated at a rate of 0.34 Mg ha^{-1} yr^{-1} in the soil, 0.65 Mg ha^{-1} yr^{-1} in the forest floor and at a rate of 3,230 Mg ha^{-1} yr^{-1} for total living biomass (Vogt et al., 1982). In contrast to *A. amabilis*, there was no measurable change in SOM accumulations over a 140-yr time span in low elevation *P. menziesii* stands when comparing similar age sequences of stands within a specific site quality (Vogt, 1987). However, SOM accumulations did vary significantly between low (63 Mg ha^{-1}) and high (110 Mg ha^{-1}) site quality stands with the high site quality stands having significantly higher SOM accumulations. For *P. menziesii* stands, no change in SOM accumulation over a 400-yr time span is suggested since SOM accumulation in the 150-yr-old high site (110 Mg ha^{-1}) was the same as that reported for a high site 450-year-old *P. menziesii* stand (113 Mg ha^{-1}) in Oregon (Grier and Logan, 1977).

The proportion of total ecosystem organic matter in SOM + FF differed between the low and high elevation sites in Washington. The cooler *A. amabilis* sites had 18.9% of their ecosystem organic matter accumulating in the FF + SOM while the low elevation *P. menziesii* stands accumulated less than 1.6% of total ecosystem organic matter in both these compartments. The initial low accumulation of ecosystem organic matter in detritus by *P. menziesii* during the first 150 years of its life-span finally increased to levels comparable to *A. amabilis* by the time it reached 450-years of age - mainly due to increases in forest floor accumulation. The 450 year-old *P. menziesii* stand in Oregon had 30% of total ecosystem organic matter in the detrital pool but 17% of this was in the coarse wood component (Grier and Logan, 1977).

There are few data of carbon accumulation by ecosystem component and stand developmental sequence for boreal and tropical ecosystems. In the tropics, forest ages are not easy to determine and most are classified as being mature.

V. Influence of Abiotic and Biotic Factors

To identify which abiotic and biotic factors are important in regulating net carbon accumulation in the soil, it is useful to examine either the soil formation equation (Brady, 1990) or the driving variables used in models that predict organic matter accumulation (Parton et al., 1987). In the Century model (Parton et al. 1987), inputs include the amounts of above- and belowground plant material being transferred to detrital pools and their chemical quality (e.g., lignin/N ratios), soil texture, and climate. The soil formation equation lists five major factors: climate (temperature and precipitation); organisms (vegetation, soil animals and microorganisms); nature of the parent material; topography; and the time the soil has been subjected to these factors (Brady, 1990).

Few studies determine the abiotic and biotic factors identified above and present them in the published literature. Basic site information, even such as climatic data, is frequently not reported in studies involving SOM accumulations since they are mostly focussed on ecosystem carbon budgets. Furthermore, ecosystems components are inadequately sampled (e.g., belowground biomass and turnover) because of the difficulty and time-consuming nature of collecting that data. The Century model also separates soil organic matter into several pools (e.g., active, slow and passive pools) which have different turnover times (Parton et al., 1987). These pools are rarely considered in field studies except for a few isolated cases (see Spycher et al., 1983). However, there is no doubt that these different pools need to be incorporated into studies examining the dynamics of carbon accumulation in soil. These pools will probably have different magnitudes of SOM accumulating and also different turnover rates.

A. Climatic Variables

There have been many attempts in the past to relate SOM accumulation to climate using mean annual temperature as a surrogate for climate (Jenny et al., 1949). Site specific examples seemed to substantiate a relationship between climatic factors and SOM. For example, a strong relationship between SOM accumulation and mean annual temperature was suggested to explain the variations reported in SOM along an altitudinal gradient in Thailand by Kira and Shidei (1967). In that study, the SOM values ranged from 60 Mg ha^{-1} in a tropical forest at an altitude of 400 m with gradually increasing values to 130 Mg ha^{-1} in a mossy forest at 1700 m.

Initially, these comparisons suggested that climate would be sufficient in itself to predict SOM accumulations, however, these relationships weakened when more data were added (Anderson, 1991). Despite this, climatic variables have been used to group detrital organic carbon accumulations using very large spatial scales. For example, Post et al. (1982) related a very large database on soil carbon pools (2,700 soil profiles) to climate using the Holdridge life-zone classification system. They found that soil carbon accumulations increased with increasing precipitation and decreasing temperature within a given precipitation range. Furthermore, Brown and Lugo (1982) found 83% of the variation in SOM was explained by a temperature/precipitation ratio for six life-zone groups (Brown and Lugo, 1982). However, use of this temperature/precipitation ratio with our data did not give good correlations with SOM across all climatic zones or even within subtropical or tropical climatic zones (Table 7).

Using our database across all climatic zones, very little of the variation in SOM could be explained by climatic factors or ratios thereof (Table 7,8). Within each climatic zone, only the tropical zone had any significant relationship between SOM and climate variables. Data analyzed in this paper suggest similar significant relationships and correlations between between mean annual air temperature (r^2 = 0.90) and minimum air temperature (r^2 = 0.76) and SOM in the tropical zone. However, these data are clustered and more intermediate data points need to be collected to substantiate these patterns. In contrast to SOM, climatic variables and FF accumulation were strongly related to one another across all climatic zones. Here, minimum annual air temperature and mean annual air temperature explained 62% and 54% of the variation in forest floor accumulation, respectively (Table 7). However within climatic zones, minimum air temperature did not explain any of the variation in forest floor accumulation in the tropical or the boreal zones (Table 8).

B. Deciduous, Evergreen Behavior

Unlike the tropical example of Kira and Shidei (1967), altitudinal patterns of SOM accumulation in the cold temperate zone may be related to foliage life-span. For example, along an altitudinal gradient in Vermont, SOM averaged 262, 204, and 376 Mg ha^{-1} in sites dominated by deciduous hardwoods, coniferous evergreens, and mixed evergreen/deciduous tree species, respectively (Siccama, 1968). Deciduous sites had about 20% higher SOM accumulations than the more nutrient poor coniferous evergreen sites while the mixed sites had significantly higher accumulations (30-45% higher) than the other two forest types.

Several other data sets show these differences in SOM accumulation between deciduous and evergreen forests. The hardwood deciduous site (*Alnus rubra*, a nitrogen-fixing species) had 31% higher SOM than the immediately adjacent coniferous evergreen (*P. menziesii*) stand on the same soil in Washington (Turner, 1975). In New Zealand, conifers always accumulated less (by 43-52%) organic matter in the soil than in the adjacent hardwood forests growing on the same soil type (Frederick et al., 1985). These studies suggest that patterns of change in SOM do not generally follow a simple altitudinal gradient (and the associated decreases in temperature) and that these changes may be related to whether the dominant tree species are evergreen or deciduous. Whether similar patterns of higher SOM accumulations in deciduous compared to evergreen forests exist in the wet tropics is not clear because of the small database.

C. Site Nutrient Status

There appears to be a strong relationship between organic matter accumulation and site nutrient status - especially soil N availability. In this paper, 78% of the variation in SOM contents could be directly explained by soil N across all climatic zones (Table 7). However, the importance of soil N in explaining SOM accumulation patterns varied within specific climatic zones - with the strongest relationships being obtained in the cold temperate and tropical climatic zones (Table 8).

This pattern of higher SOM contents in nutrient-rich sites also occurs in evergreen forests where tree species remain constant but the site nutrient status is naturally high or low (Vogt, 1987). For *P. menziesii*-dominated stands, nutrient-poor sites had 61% of the SOM contents of the nutrient-rich sites (Vogt 1987). This pattern of higher SOM levels with higher soil N can be mimicked by fertilizing N-poor sites - thereby maintaining tree species constant but increasing total litterfall and N availability. For example, Van Cleve and Moore (1978) obtained significant increases (12-18%) in surface soil organic matter after N and P fertilization of aspen stands over a 7 yr period. The authors suggested this increase was caused by increased production of the plants growing on the fertilized site.

Several reasons can be given to explain the higher SOM accumulation in deciduous versus evergreen dominated sites or nutrient-rich versus -poor sites. Anderson (1991) suggested that in acid soils, typical of coniferous forests, "SOM undergoes less stabilization". A lower degree of SOM stabilization would explain the lower SOM values reported in coniferous forests but does not account for the fact that SOM accumulation can vary within the coniferous forests based on their site nutrient status. This idea should be generalized to all sites with the degree of SOM stabilization more dependent on site nutrient status. Higher organic matter accumulation in soils in more productive sites is partially explained by the hypothesis put forth by Sollins et al. (1984); nutrient-rich sites have narrower soil C/N ratios and more microaggregates that protect C and N from being attacked by microbes, and therefore organic matter accumulates in the soil.

How readily the C/N ratio can be used to predict the turnover time of SOM is not clear and may vary depending on the climatic region and how the soil C measurements are collected and fractionated (e.g. light and heavy fractions). In this paper, the C/N ratio of the soil was unable to explain any of the variation in SOM accumulation across all climatic zones even though a highly significant correlation existed between soil N and SOM (Table 7). This is contrasted by the individual climatic zone analyses where significant relationships between soil C/N and SOM were obtained in the boreal and subtropical zones (Table 8).

D. Decomposition

Factors controlling decomposition in the soil environment are similar to those occurring in the forest floor. However, the soil has chemical factors that may inhibit microbes from following the predictable patterns of substrate decomposition under particular climatic and tissue chemistry constraints. This results in surface organic layers decomposing differently from organic matter in the soil. For example, surface leaf litter decomposition rates are generally slower on nutrient-poor than in nutrient-rich sites due to poor substrate quality (e.g., higher lignin and secondary chemical compositions) (Horner, 1987) and low nutrient availability that reduces microbial activity (Swift et al., 1979). For soils, the opposite pattern is suggested by the data in this paper. In such cases, low nutrient availability will reduce microbial ability to degrade plant tissues and slower rates of plant tissue decomposition are due to high lignin and secondary chemical contents of tissues in boreal, temperate and tropical ecosystems (Flanagan and Van Cleve, 1983; Anderson, 1991; Bloomfield et al., 1993). Climatic factors are important in accelerating the decomposition rate of tissues in the warm temperate, subtropical, and tropical climatic zones even though the chemical composition of tissues is the dominant controlling factor for decay (Swift et al., 1979).

The chemical quality and subsequent decomposability of litter can also be modified by several abiotic factors (e.g., site nutrient status, temperature, and water availability), creating a complex mosaic of factors affecting decomposition rates of surface litter (Brown et al., 1984; Horner, 1987). For example, when N availability limits plant growth, low N levels typically result in higher tannin and other secondary chemicals concentrations in plant tissues making these tissues more resistant to decay (Horner, 1987). In addition, an inverse relationship exists between the lignin concentration of different canopy tissues and soil N availabilities in several ecosystems (Horner, 1987; Wessmann et al., 1988). Therefore, nutrient-poor sites tend to have greater surface organic matter accumulations than nutrient-rich sites (Vogt et al., 1986; Vogt, 1987), but the opposite pattern is observed in the soil.

When examining the magnitude of litter inputs from above- and belowground tissues, it is important that these tissues are not considered as having similar decomposition rates. In some ecosystems, above- and belowground tissues have been shown to have similar decomposition rates, but this is probably the exception rather than the rule (Vogt et al., 1991). In a comparative decay study of leaves and roots in a subtropical forest, roots always decayed significantly more slowly than leaves because of the higher proportion of complex polyphenolics in root tissues (Bloomfield et al., 1993). The much slower decomposition of roots probably contribute to greater sequestering of carbon in the soil than what might occur due to leaf decay. An insufficient number of root decay studies have been conducted in different forest climatic types to be able to generalize at this point (Vogt et al., 1991).

VI. Conclusions

In conclusion, the data used in this paper show that when examining global carbon budgets, the cold temperate and tropical climatic zones had the greatest potential to sequester carbon compared to the other climatic zones. The cold temperate zone forests had the highest organic accumulations in total detritus (FF, SOM). The tropical and the boreal zones had similar total accumulations of carbon in the detrital components with the major difference being that boreal zone had almost half of its total detrital carbon accumulating in the forest floor and the tropical zone had 90% of the total in the soil. The highest carbon accumulations occurred in highly weathered soils that had higher clay contents (e.g., ultisols, oxisols).

In the boreal and cold temperate climatic zones, foliage life-span and site nutrient status were important criteria in separating sites as to their carbon sequestering potential in the forest floor and soil. For example, the highest total carbon accumulations in detritus occurred when stands were composed of a mixture of evergreen and deciduous tree species. The lowest accumulations of carbon in the forest floor and soil occurred in the evergreen dominated forests with the deciduous forests having intermediate values.

Even though the evergreen forests had less carbon accumulating in the soil, total ecosystem organic matter values were always higher in the evergreen forests compared to the deciduous forests suggesting that evergreen forests sequestered higher total amounts of carbon. Unless one is attempting to specifically increase the amount of carbon sequestered in the soil, a management strategy to increase total carbon sequestration in an ecosystem should consider planting evergreen trees. In the tropics, broadleaf evergreen forests have higher carbon sequestering capabilities than planted needle-leaved evergreen forests. If the management strategy is to increase the amount of carbon in the soil, one should plant a mixture of deciduous and evergreen trees. If a poor site quality evergreen forest is being managed and there is no option for replacing the existing trees, the amount of carbon sequestered in the ecosystem can be increased by fertilizing with nitrogen and increasing its site quality class.

VII. Summary

1. Forest floor accumulations were strongly related to climatic gradients while SOM was not. Forest floor mass followed a temperature gradient with accumulations in the boreal zone > temperate zone > subtropical > tropical climatic zone. Significantly lower SOM accumulations were measured in the boreal climatic zone than in the temperate zone. The tropical climatic zone was characterized by having intermediate SOM values to the boreal and temperate zones but these differences were not significant.
2. SOM and FF accumulations varied significantly with foliage life-span. Deciduous dominated forests had higher SOM accumulations than evergreen while the mixed deciduous/evergreen forests had the highest detrital accumulations. SOM contents were higher (by 48-57%) in hardwood forests compared to coniferous forests located on the same soil type.
3. SOM and FF accumulations were not effectively separated using soil texture indices. Clayey soils had the highest SOM accumulations. Loamy soils had significantly higher FF accumulations compared to the other soil texture indices.
4. In general, a significantly greater proportion of total ecosystem organic matter was located in the soil in deciduous than in evergreen dominated forests. The effect of the deciduous species was greater in the higher latitude forests and decreased in the subtropical and tropical forests.
5. Using the forest-climatic-type grouping, only boreal forests had significantly lower ANPP compared to the other sites. The highest TNPP occurred in the subtropics and the lowest in the boreal zone.
6. Ultisols had significantly higher SOM accumulations than the other soil orders.
7. In cold temperate, needle-leaved forests, SOM accumulations were significantly higher (approximately 40%) in nutrient rich sites than in nutrient poor sites.
8. In low elevation forests of the cold temperate zone, stand developmental sequences suggested that the SOM was relatively stable and may not change over a 400 year-time scale. In the subalpine forests of the cold temperate zone, stand developmental sequences had organic matter annually accumulating in soil (<0.4 Mg ha^{-1} yr^{-1}) over a 160 year-time span.
9. None of the climatic variables or ratios of climatic variables explained the variation in SOM accumulation across or within the climatic zones. However, from 54-62% of the variation in FF accumulation was explained by minimum and mean annual air temperatures. Within the cold temperate zone, 69% of the variation in FF accumulation was explained by minimum air temperature.
10. A strong relationship was found between soil N and SOM accumulation ($r^2 = 0.78$) but not with the soil C/N ratio. TNPP explained only 44% of the variation in SOM.

Acknowledgments

Ideas presented in this paper were generated while conducting research under the auspices of the Northern Global Change Program, U.S.D.A. Forest Service, the Insect and Disease Laboratory, Northeastern Forest Experiment Station in Hamden, Connecticut, and the Ecosystem Program of the National Science Foundation. Additional assistance was obtained as part of the National Science Foundation grant to the Center for Terrestrial Ecology (University of Puerto Rico) and the Institute of Tropical Forestry as part of the Long-Term Ecological Research in the Luquillo Experimental Forest.

References

Anderson, J.M. 1991. The effects of climate change on decomposition processes in grassland and coniferous forests. *Ecol. Appl.* 1:326-347.

Bloomfield, J., K.A. Vogt and D.J. Vogt. 1993. Decay rate and substrate quality of fine roots and foliage of two tropical tree species in the Luquillo Experimental Forest, Puert Rico. *Plant Soil* (in press)

Boudot, J.P., B.A. Bel Hadj, and T. Chone. 1986. Carbon mineralization in andosols and aluminium-rich highland soils. *Soil Biol. Biochem.* 18:457-461.

Brady, N.C. 1990. *The nature and properties of soils.* 10th Ed. MacMillan Publishing Co. NY.

Brown, S. and A.E. Lugo. 1982. The storage and production of organic matter in tropical forests and their role in the global carbon cycle. *Biotropica* 14:161-187.

Brown, P.H., R.D. Graham, and D.J.D. Nicholas. 1984. The effects of manganese and nitrate supply on the levels of phenolics and lignin in young wheat plants. *Plant Soil* 81:437-440.

Brown, S. and A.E. Lugo. 1990. Effects of forest clearing and succession on the carbon and nitrogen content of soils in Puerto Rico and US Virgin Islands. *Plant Soil* 124:53-64.

Buringh, P. 1984. Organic carbon in soils of the worlds. p. 91-109. In: G.M. Woodwell (ed.) *The role of terrestrial vegetation in the global carbon cycle: measurement by remote sensing. SCOPE.* John Wiley & Sons, NY.

Coleman, D.C., J.M. Oades, and G. Uehara (eds.). 1989. *Dynamics of soil organic matter in tropical ecosystems.* NifTAL Project. Department of Agronomy and Soil Science, College of Trop. Agric. & Human Resources. University of Hawaii.

Conover, W.J. and R.L. Iman. 1981. Rank transformations as a bridge between parametric and nonparametric statistics. *Am. Stat.* 35:124-129.

Flanagan, P.W. and K. Van Cleve. 1983. Nutrient cycling in relation to decomposition and organic-matter quality in taiga ecosystems. *Can. J. For. Res.* 13:795-817.

Frederick, D.J., H.A.I. Madgwick, M.F. Jurgensen, and G.R. Oliver. 1985. Dry matter, energy, and nutrient contents of 8-year-old stands of *Eucalyptus regnans*, *Acacia dealbata*, and *Pinus radiata* in New Zealand. *New Zealand J. For. Sci.* 15:142-157.

Greenland, D.J. and P.H. Nye. 1959. Increases in carbon and nitrogen contents of tropical soils under natural fallows. *J. Soil Sci.* 9:284-299.

Grier, C.C. 1976. Biomass, productivity, and nitrogen-phosphorus cycles in hemlock-spruce stands of the central Oregon coast. p. 71-81. In: W.A. Atkinson and R.J. Zasoski (eds.) *Western Hemlock Management.* Institute of Forest Products Cont. #34. University of Washington., Seattle.

Grier, C.C. and R.S. Logan. 1977. Old-growth P*seudotsuga menziesii* communities of western Oregon watershed: biomass distribution and production budgets. *Ecol. Monogr.* 47:373-400.

Hendrickson, O.Q., and J.B. Robinson. 1984. Effects of roots and litter on mineralization processes in forest soil. *Plant Soil* 80:391-405.

Horner, J.D. 1987. *The effects of manipulation of nitrogen and water availability on the polyphenol content of Douglas-fir foliage: implications for ecosystem theory.* Dissertation. N.M.S.U., Albuquerque, New Mexico.

Huntington, T.G., D.F. Ryan, and S.P. Hamburg. 1988. Estimating soil nitrogen and carbon pools in a northern hardwood forest ecosystem. *Soil Sci. Soc. Am. J.* 52:1162-1167.

Jenny, H., S.P. Gessel, and F.T. Bingham. 1949. Comparative study of decomposition rates of organic matter in temperate and tropical regions. *Soil Sci.* 68:419-432.

Johnson, C.E., A.H. Johnson, T.G. Huntington, and T.G. Siccama. 1991. Whole-tree clear-cutting effects on soil horizons and organic-matter pools. *Soil Sci. Soc. Am. J.* 55:497-502.

Kira, T. and T. Shidei. 1967. Primary production and turnover of organic matter in different forest ecosystems of the western pacific. *Jap. J. Ecol.* 17:70-87.

Lugo, A.E. 1992. Comparison of tropical tree plantations with secondary forests of similar age. *Ecol. Monogr.* 62:1-41.

O'Brien, B.J. and J.D. Stout. 1978. Movement and turnover of soil organic matter as indicated by carbon isotope measurements. *Soil Biol. Biochem.* 10:309-317.

Parton, W.J., D.S. Schimel, C.V. Cole, and D.S. Ojima. 1987. Analysis of factors controlling soil organic matter levels in Great Plains grasslands. *Soil Sci. Soc. Am. J.* 51:1173-1179.

Paustian, K., W.J. Parton, and J. Persson. 1992. Modeling soil organic matter in organic-amended and nitrogen-fertilized long-term plots. *Soil Sci. Soc. Am. J.* 56:476-488.

Post, W.M., W.R. Emanuel, P.J. Zinke, and A.G. Stangenberger. 1982. Soil carbon pools and world life zones. *Nature* 298:156-159.

Post, W.M., T.H. Peng, W.R. Emanuel, A.W. King, V.H. Dale, and D.L. DeAngelis. 1990. The global carbon cycle. *Am. Sci.* 78:310-326.

Richter, D.D., L.I. Babbar, M.A. Huston, and M. Jaeger. 1990. Effects of annual tillage on organic carbon in a fine-textured udalf: the importance of root dynamics to soil carbon storage. *Soil Sci.* 149:78-83.

Sanchez, P.A. 1976. *Properties and management of soils in the tropics.* John Wiley & Sons. NY.

Schlesinger, W.H. 1984. Chapter 4. Soil organic matter: a source of atmospheric CO_2. p. 111-127. In: G.M. Woodwell (ed.), *The role of terrestrial vegetation in the global carbon cycle: measurement by remote sensing.* SCOPE. John Wiley & Sons, NY.

Schlesinger, W.H. 1990. Evidence from chronosequence studies for a low carbon-storage potential of soils. *Nature* 348:232-234.

Siccama, T.G. 1968. *Altitudinal distribution of forest vegetation in relation to soil and climate on the slopes of the green mountains.* Dissertation. U. of VT., Burlington, Vermont.

Skjemstad, J.O., R.P. Le Feuvre, and R.E. Prebble. 1990. Turnover of soil organic matter under pasture as determined by ^{13}C natural abundance. *Aust. J. Soil Res.* 28:267-276.

Sollins, P., G. Spycher, and C.A. Glassman. 1984. Net nitrogen mineralization from light- and heavy-fraction forest soil organic matter. *Soil Biol. Biochem.* 16:31-37.

Spycher, G., P. Sollins, and S. Rose. 1983. Carbon and nitrogen in the light fraction of a forest soil: vertical distribution and seasonal patterns. *Soil Sci.* 135:79-87.

Swift, M.J., O.W. Heal, and J.M. Anderson. 1979. *Decomposition in terrestrial ecosystems.* Blackwell Publications, Oxford, England.

Turner, J. 1975. *Nutrient cycling in a Douglas-fir ecosystem with respect to age and nutrient status.* Dissertation. University of Washington, Seattle.

Van Cleve, K. and T.A. Moore. 1978. Cumulative effects of nitrogen, phosphorus, and potassium fertilizer additions on soil respiration, pH, and organic matter content. *Soil Sci. Soc. Am. J.* 42:121-124.

Van Cleve, K., W.C. Oechel, and J.L. Hom. 1990. Response of black spruce (*Picea mariana*) ecosystems to soil temperature modification in interior Alaska. *Can. J. For. Res.* 20:1530-1535.

Van Veen, J.A. and P.J. Kuikman. 1990. Soil structural aspects of decomposition of organic matter by micro-organisms. *Biogeochem.* 11:213-233.

Viereck, L.A., C.T. Dyrness, K. Van Cleve, and M.J. Foote. 1983. Vegetation, soils, and forest productivity in selected forest types in interior Alaska. *Can. J. For. Res.* 13:703-720.

Vogt, D.J. 1987. *Douglas-fir ecosystems in western Washington: biomass and production as related to site quality and stand age.* Dissertation. University of Washington, Seattle.

Vogt, K.A., C.C. Grier, C.E. Meier, and R.L. Edmonds. 1982. Mycorrhizal role in net primary production and nutrient cycling in *Abies amabilis* ecosystems in western Washington. *Ecology* 63:370-380.

Vogt, K.A., C.C. Grier, and D.J. Vogt. 1986. Production, turnover, and nutrient dynamics of above- and belowground detritus of world forests. *Adv. Ecol. Res.* 15:303-377.

Vogt, K.A., D.J. Vogt, and J. Bloomfield. 1991. Input of organic matter to the soil by tree roots. p. 171-190. In: B.L. McMichael and H. Persson (eds.) *Plant roots and their environment.* Elsevier Science Publishers B.V.

Wessmann, C.A., J.D. Aber, D.L. Peterson, and J.M. Melillo. 1988. Remote sensing of canopy chemistry and nitrogen cycling in temperate forest ecosystems. *Nature* 335:154-156.

Data References

Bormann and Likens, 1979 *Pattern and process in a forested ecosystem.* Springer-Verlag.; Brown et al., 1983 *USDA For. Serv. Exp. Sta. Gen Techn.* Rep. SO-44; Brown and Lugo, 1982 *Biotropica* 14:161-187; Cannell, 1982 *World forest biomass and primary production data.* Academic Press; Cole and Rapp 1981 p. 341-409. *Dynamic properties of forest ecosystems.* Cambridge University Press; Covington, 1976 *Forest floor organic matter and nutrient content and leaf fall during secondary succession in northern hardwoods.* Dissertation. Yale University; Covington, 1981 *Ecology* 62:41-48; Cox et al. 1978 *Pedobiolobia, Bd.* 18:264-271; Cromack et al., 1979 p. 449-476. *Forest Soils and Land Use.* 5th North Amer. For. Soils Conf.; DeAngelis et al., 1981 p. 567-672. *Dynamic properties of forest ecosystems.* Cambridge University Press; Edwards, 1982 *J. Ecol.* 70:807-827; Edwards, 1977 *J. Ecol.* 65:971-992; Edwards and Grubb, 1977

J. Ecol. 65:943-969; Edwards and Grubb, 1982 *J. Ecol.* 70:649-666; Ewel et al., 1987 *Can. J. For. Res.* 17:325-329; Ewel et al., 1987 *Can. J. For. Res.* 17:330-333; Federer et al., 1990 *USDA For. Serv. Northeastern For. Exp. Sta. Gen. Tech. Rep.* NE-141; Fogel and Hunt, 1979 *Can. J. For. Res.* 9:245-256; Fogel and Hunt, 1983 *Can. J. For. Res.* 13:219-232; Folster et al., 1976 *Oecol. Plant.* 11:297-320; Frederick et al., 1985 *New Zealand J. For. Sci.* 15:142-157; Friedland et al., 1984 *Water, Air, and Soil Pollution* 21:161-170; Gholz et al., 1986 *Can. J. For. Res.* 16:529-538; Greenland and Kowal, 1960 *Plant Soil* 12:154-173; Grier, 1976 p. 71-81 *Western Hemlock Management.* Institute of Forest Products Cont. #34. University of Washington; Grier and Logan, 1977 *Ecol. Monogr.* 47:373-400; Grier and Ballard, 1981 *Can. J. Bot.* 59:2635-2649; Grier et al., 1981 *Can. J. For. Res.* 11:155-167; Grubb and Tanner, 1976 *J. Arnold Arboretum* 57:513-568; Grubb and Edwards, 1982 *J. Ecol.* 70:623-648; Harris et al., 1975 p. 116-122 *Productivity of world forests.* National Academy of Science, Washington D.C.; Horner, 1987 *The effects of manipulation of nitrogen and water availability on the polyphenol content of Douglas-fir foliage: implications for ecosystem theory.* Dissertation. New Mexico State University; Horner et al., 1987 *Oecologia (Berl.)* 72:515-519; Huntington et al., 1989 *Soil Sci.* 148:380-386; Huttel, 1975 p. 123-130. *Tropical ecological systems.* Ecological Studies 11. Springer-Verlag; Huttel and Bernhard-Reversat, 1975 *Terre Vie* 29:203-228; Johnson et al., 1982 p. 186-232. *Analysis of coniferous forest ecosystems in the western United States.* US/IBP Synthesis Series 14. Hutchinson Ross Publishing Co.; Jordan, 1971 *J. Ecol.* 59:127-142; Kinerson et al., 1977 *Oecologia* 29:1-10; Klinge and Rodriguez, 1968 *Amazonian* 1:287-302; Klinge et al., 1975 p. 115-122. *Tropical ecological systems.* Ecological Studies 11. Springer-Verlag; Lang et al., 1981 *Can. J. For. Res.* 11:388-399; Lodge et al., 1991 *Biotropica* 23:336-342; Lugo, 1992 *Ecol. Monogr.* 62:1-41; Mailly and Margolis, 1991 *For. Ecol. Manage.* 55:259-278; Mabberley, 1992 *Tropical rain forest ecology.* 2nd Ed. Chapman and Hall; Meier, 1981 *The role of fine roots in N and P budgets in young and mature Abies amabilis ecosystems.* Dissertation. University of Washington; Nakane, 1975 *Jap. J. Ecol.* 25:206-216; Nakane et al., 1984 *Bot. Mag. Tokyo* 97:39-60; Odum, 1970 *A tropical rain forest.* U.S. Atomic Energy Commission, Div. of Tech. Inf., Nat. Tech. Inf. Serv., Springfield, Virginia; Ogawa et al., 1965 *Nature Life Southeast Asia* 4:49-81; Ovington and Olson, 1970 Ch. H-2. *A tropical rain forest.* U.S. Atomic Energy Commission, Div. of Tech. Inf., Nat. Tech. Inf. Serv., Springfield, Virginia; Proctor et al. 1983 *J. Ecol.* 71:237-260; Santantonio and Hermann, 1985 *Ann. Sci. for.* 42:113-142; Siccama, 1968 *Altitudinal distribution of forest vegetation in relation to soil and climate on the slopes of the green mountains.* Dissertation. University of Vermont; Singh and Singh, 1987 *The Bot. Rev.* 53:80-192; Sprugel 1984 *Ecol. Monogr.* 54:165-186; Spycher et al., 1983 *Soil Sci.* 135:79-87; Tanner, 1977 *J. Ecol.* 65:883-918; Tanner, 1980 *J. Ecol.* 68:573-588; Tanner, 1980 *J. Ecol.* 68:833-848; Tanner, 1985 *J. Ecol.* 73:553-568; Turner, 1975 *Nutrient cycling in a Douglas-fir ecosystem with respect to age and nutrient status.* Dissertation. University of Washington; Viereck et al., 1983 *Can. J. For. Res.* 13:703-720; Vogt, 1987 *Douglas-fir ecosystems in western Washington: biomass and production as related to site quality and stand age.* Dissertation. University of Washington; Vogt, 1991 Tree Physiol. 9:69-86; Vogt et al., 1982 *Ecology* 63:370-380; Welch and Klemmedson, 1975 p. 159-178. *Forest soils and forest land management.* Les Presses de l'Universite Laval. Quebec; White et al., 1988 *Soil Biol. Biochem.* 20:101-105; Yoda and Kira, 1969 *Nature Life Southeast Asia* 6:88-110; Zielinski, 1984 p. 150-165. *Forest ecosystems in industrial regions.* Springer-Verlag.

CHAPTER 15

CH_4 and N_2O Flux in Subarctic Agricultural Soils

V.L. Cochran, S.F. Schlentner, and A.R. Mosier

I. Introduction

Collectively, methane (CH_4) and nitrous oxide (N_2O) are responsible for about 20% of the anticipated global warming (Mosier and Schimel, 1991). These two gases are increasing in the atmosphere at annual rates of 1.1 and 0.3 %, respectively. Both are produced by microbial activity in the soil and are influenced by man's activities. Nitrous oxide is produced during the oxidation of ammonium (nitrification) (Bremner and Blackmer, 1978; Smith and Chalk, 1980; Cochran et al., 1981).and the reduction of nitrate (denitrification) (Bremner and Blackmer, 1978; Ryden and Lund, 1980). McElroy and Woofsy (1985) estimate that between 0.5 and 1.5% of mineral N fertilizers are emitted as N_2O, and Eichner (1990) estimated that in 1984 the worldwide contribution of mineral N fertilizers to atmospheric N_2O was between 0.2 and 2.1 Tg of N. This compares to an estimated 0.023 to 0.315 Tg N_2O-N from N added to the soil by leguminous crops.

About 70% of global CH_4 emissions are thought to be of biological origin (Arah 1991). Methane is produced by anaerobic bacteria in the presence of organic substrate and the absence of oxygen. Rice paddies, wetlands, and the intestines of ruminants and termites are major sources. Several researchers have found that upland soils consume CH_4 (Steudler et al., 1989; Mosier et al., 1991; Nesbit and Breitenbeck, 1992). However, they found that ammoniacal fertilizers reduced the rate of CH_4 oxidation in these same soils. Mosier et al. (1991) found that cultivated fields adjacent to native land consumed less CH_4 than the field in native grasses. The processes involved in the inhibition have not been elucidated. Microorganisms capable of oxidizing CH_4 are very diverse and include obligate and facultative CH_4 oxidizers, free living N_2-fixing organisms, and organisms that oxidize NH_4^+ (Higgins et al., 1981: Bedard and Knowles, 1989). NH_4^+ can also serve as a competitive inhibitor (Whittenbury et al., 1970). Nesbit and Breitenbeck (1992) suggest that NH_4^+ may act as an irreversible inhibitor of CH_4 oxidation. With such a diverse group of organisms involved in oxidation of CH_4, it is difficult to predict how management of our agriculture lands will affect CH_4 oxidation. The climates of agricultural lands range from tropical to subarctic and from humid to xeric. Relatively few field studies have been conducted on quantifying the consumption of CH_4 in agricultural systems and even fewer have looked at N fertilizer management effects on CH_4 consumption.

The objectives of this study were to: 1) quantify the flux of N_2O and CH_4 gases in subarctic agricultural soils, 2) evaluate the effects of N fertilizer on the long term and short term consumption of CH_4 by soil in permanent grass.

II. Methods and Materials

Two sites were selected in areas which had been in grass for 10 or more years to monitor the effects of urea application on CH_4 and N_2O flux. One site was on Fairbanks silt loam (*Alfic Cryochrepts*) with a south facing slope of about 15% (hillside). This site had been in mixed Kentucky bluegrass (*Poa pratensis* L.) and

ISBN 1-56670-117-1/95/$0.00+.50

smooth brome grass (*Bromus inermis* Leyss.) for over 10 years and had not been fertilized. The other site was a Tanana silt loam (*Pergelic Cryaquept*) with less than 1% slope (flat) and had been in smooth brome grass hay production for 10 or more years and received about 90 kg N ha^{-1} yr^{-1}. These sites were picked because of their different N fertilization history but similar plant history. Twelve plots (3m by 3m) were laid out in each site and the grass mowed and removed , but no fertilizer was applied in 1991. Base rings for vented gas flux chambers (20 cm dia.) were driven into the ground in August 1991 for use the following spring and summer. Gas flux measurements began as soon as the soil thawed and the plots were accessible in the spring (April 28 for the hillside plots and May 22 for the plot on the flat). Gas flux was measured by placing a vented chamber onto the preinstalled base rings and measuring the concentration of CH_4 and N_2O in the chamber at 0, 15, and 30 minutes after the cover was installed. The vented chambers were modifications of the system described by Hutchinson and Mosier (1981). Inside dimensions were 20 cm dia. by 6 cm high. The chamber also had a vent tube that was 0.5 cm dia. by 15 cm long which allowed the inside pressure to equalize with that outside the chamber. Diffusion of gas through the vent was calculated to take longer than the time the chambers were in place. Gas samples were collected by removing 50 ml of gas from the chamber with a 60 ml polypropylene syringe equipped with a gas tight valve to seal the syringe for transporting and storage until analysis. Gas analysis was performed within 24 h. Analysis of N_2O was by the procedures described by Mosier and Mack (1980). Gas samples were introduced into a 5.0 cm^3 sample loop and brought to ambient pressure. A 4 port valve and a 10 port valve were used in conjunction with a 1 m precolumn and 3 m (2 mm dia.) column of Porapak Q (80/100 mesh) to flush water vapor from the system prior to the gas entering a ^{63}N ECD. The flow rate for methane-argon (5% methane) carrier gas was 25 cm^3 min^{-1}, and the oven and ECD temperatures were 50 and 250 °C, respectively. Analysis of CH_4 was by introducing the gas sample into a 0.5 cm^3 sample loop attached directly to a 10 port valve. The sample was then transported to a FID via a 2 m stainless steel column (2 mm dia.) packed with Porapak N (80/100 mesh) using a flow rate of 40 cm^3 with oven temperature of 50°C and a detector temperature of 200 °C. For both analyses the pressure in the sample loop was adjusted to ambient level by use of an electronic manometer. Gas flux for each gas was computed by the procedures described by Hutchinson and Mosier (1981).

On May 26 1992, the chamber bases at the hillside site were covered and urea was broadcast at a rate of 90 kg N ha^{-1} to 6 plots while the other 6 plots received no N fertilizer. The plots were arranged in a randomized complete block design. The covers were removed from the chamber bases and urea solution was uniformly sprayed over the soil within each base ring of the fertilized plots to apply the equivalent of 90 kg N ha^{-1}. This was followed with 0.5 l of distilled water to wash the urea solution from the foliage and into the soil. The next day, gas samples were obtained for flux measurements at mid-day. A similar procedure was followed on the flats on May 29. Samples were obtained twice a week from each site until the soil froze in September. Samples were taken at mid-day at all times to assure uniform sampling conditions. Mid-day sampling also avoided the night low and afternoon high temperatures. All plots were mowed weekly to simulate grazing and to avoid excessive plant material in the chambers.

Air and soil temperatures were taken at the time of gas sampling with an electronic thermometer equipped with thermocouple detectors. Soil temperatures were taken at a depth of 5 cm and air temperature was taken at the height of the vented chambers. Soil samples (0-15 cm) were taken outside of the chamber bases of each plot once a week. The soil samples were immediately sealed in a polyethylene bag and frozen for later analysis. Prior to analysis the samples were thawed for 0.5 h, and 5 g subsamples were extracted with 0.05 l of 2 M KCl after shaking for 1 h. The suspension was then filtered through a Whatman #610 filter and analyzed colorimetrically for NO_3-N and NH_4-N in a segmented flow autoanalyzer (Am. Public Health Assoc., 1975). A subsample was used to determine gravimetric water content of each sample which was then used to express the N values based on dry weight. A carbon analyzer was used to determine organic C values by dry combustion on subsamples of samples taken from each plot at the start of the experiment. The hillside soil contained 1.85 %C and the flats had 2.96% C. Soil pH values of 6.27 and 6.46 for the hillside and flat, respectively, were determined on 2:1 water to soil suspension of subsamples from the first sampling.

Statistical analysis for each sample date was done by comparing the standard error of the mean for each day. Those that overlapped were then analyzed for significance (0.05) by analysis of variance.

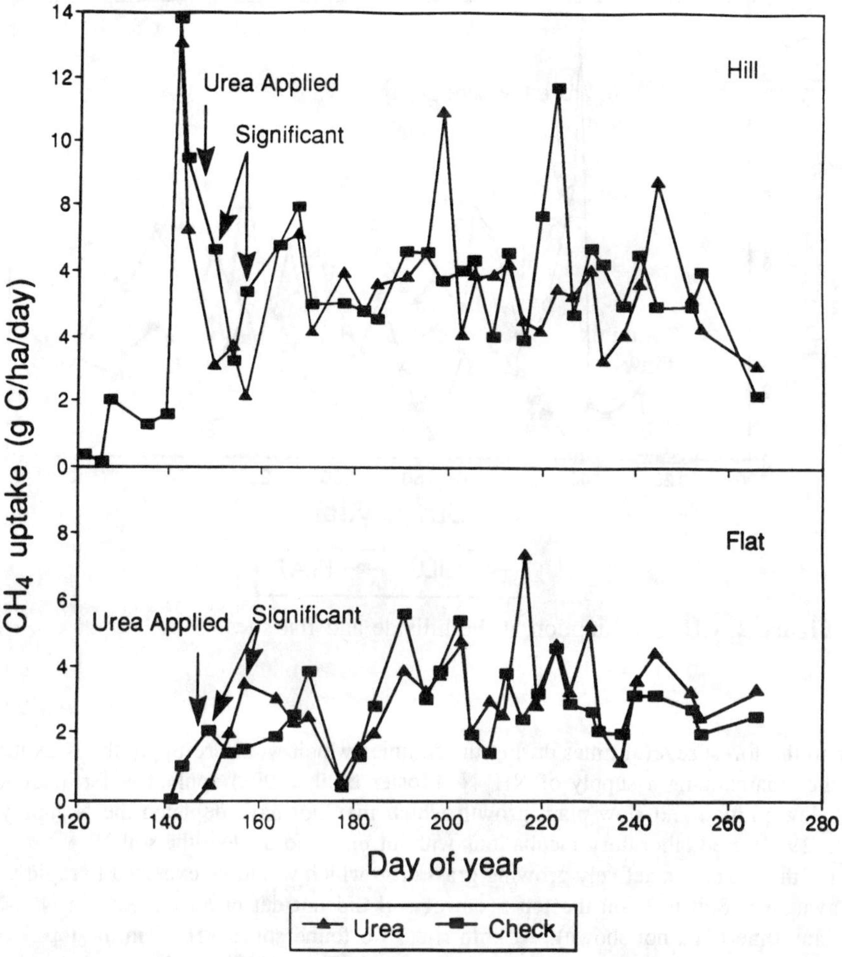

Figure 1. Influence of urea application on CH_4 consumption at the hillside and flat sites.

III. Results and Discussion

A. Methane Consumption

The hillside site began to thaw in mid-April and went through a series of freeze-thaw events before it completely thawed. During this period some plots produced CH_4 while most consumed methane. This was followed by a short period of high consumption and then a sharp decline (Figure 1). After the urea was applied, consumption immediately dropped in those plots that received urea. The difference in consumption between those plots which received urea and those that did not was significant at only two sampling dates indicating a very short period of inhibition. The pattern was somewhat different at the flat site. There was a short period of time when production of CH_4 occurred, but there was no sharp increase in consumption as occurred at the hillside site. Urea caused a decrease in consumption at only one sampling. This was followed by no differences due to urea, and then a one-time increase in consumption from the plots fertilized with urea (Figure 1).

Ammonium is known to interfere with the enzymes involved with CH_4 consumption by methanotrophic bacteria (Whittenbury et al. 1970) which would account for the immediate decrease in consumption after the urea was applied. Similar but longer term inhibition has been observed after the application of ammoniacal fertilizer to native forest (Steudler et al., 1989) and to native grasses (Mosier et al., 1991). The reasons for the differences between this and the other studies is not obvious. Steudler et al. (1989) applied

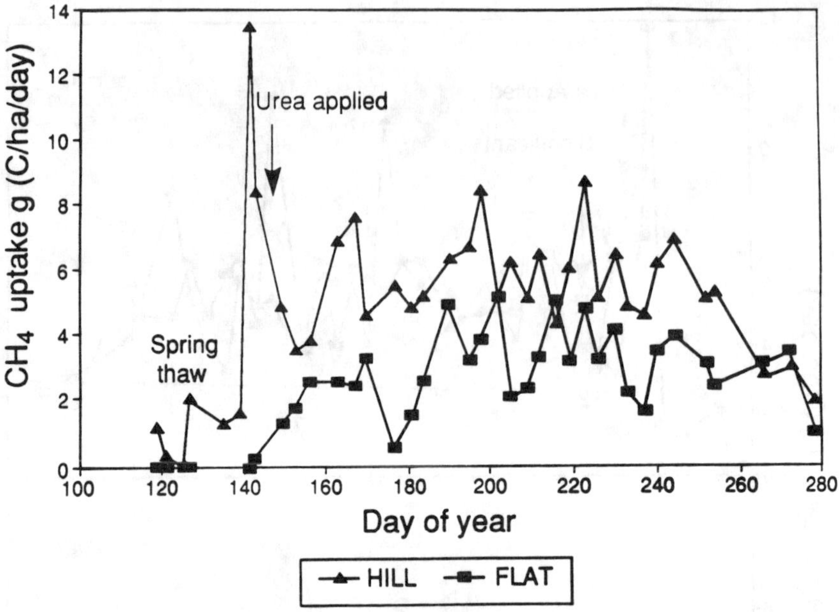

Figure 2. CH_4 consumption at the hillside and flat sites.

N fertilizer to the forest several times during the summer which would resupply the N as it was taken up by plants: thus maintaining a supply of NH_4-N. Mosier et al. (1991) applied N fertilizer to short grass prairie with low rainfall and slow plant growth which may not have depleted the N supply. Nesbit and Breitenbeck (1992) used laboratory incubations without plants to deplete the soil N.

We applied the urea to an actively growing grass crop which would be expected to rapidly remove N as it became available. Soil tests on the top 5 cm of soil did not detect an increase in NO_3-N due to the fertilizer at any time (data not shown). At both sites, we found some NH_4-N in the top 5 cm of soil for several weeks after the urea was applied (data not shown). Rainfall during the summer in interior Alaska is insufficient to meet plant requirements, resulting in a moisture deficit and little opportunity for leaching of soil N. Therefore, some of the recovered NH_4-N was likely due to slow nitrification because it was in the surface soil which tended to dry between rains. Consumption of CH_4 occurred in soil cores taken from our plots at all depths to 50 cm with the greatest consumption occurring between 20 and 30 cm (Sparrow and Cochran, 1993). Thus, inhibition at the soil surface would have little effect on the overall system. Whalen et al. (1992), using $^{14}CH_4$, determined that about 40% of the CH_4 consumed went into microbial biomass. This indicates a dynamic population of CH_4 consuming organisms that could adjust to changes in substrate. Therefore, inhibition at the surface could be compensated for by increased numbers at soil depths below that influenced by the application of urea. Ammonium N is held near the application site on the soil colloids. Thus, it is unlikely that it leached to lower depths. The fact that we had consumption to 50 cm indicates that a resident population was present to respond to additional CH_4. It is also possible that we had lower mineral N levels at depths at which consumption occurred than was in the soils of those studies which found long term inhibition of CH_4 consumption.

The differences in CH_4 consumption between the hill and the flat are shown in Figure 2. In this case, the values for both fertilizer treatments are combined to reduce variability. The hillside site consumed greater amounts of CH_4 during much of the summer. The hillside had no known history of N fertilization and the flat site had received annual applications of about 90 kg N yr^{-1} for the last 10 or more years. This difference in long term fertilizer history may account for the differences in consumption between the two sites, similar to that reported by Mosier et al. (1991) in the short grass prairie; however it was more likely due to differences in the physical conditions at each site. The hillside was about 5 °C warmer during the time the differences in CH_4 consumption were the greatest (data not shown). However, CH_4 consumption did not increase with warming of the soil above about 15 °C (Figure 3). Laboratory incubation studies showed that, in these soils, there was a large increase in CH_4 consumption between 5 and 15 °C, but much less increase

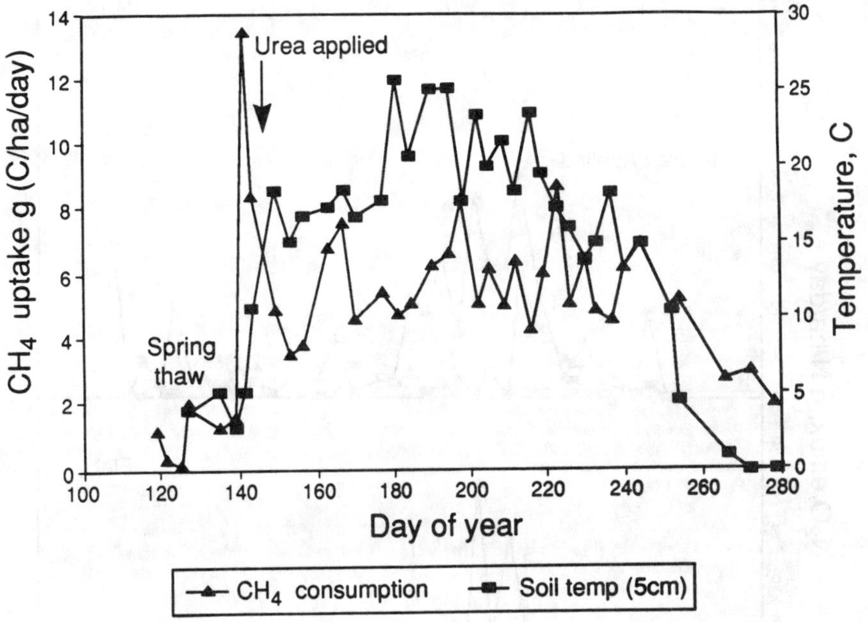

Figure 3. Soil temperature at 5 cm and CH_4 consumption at the hillside site.

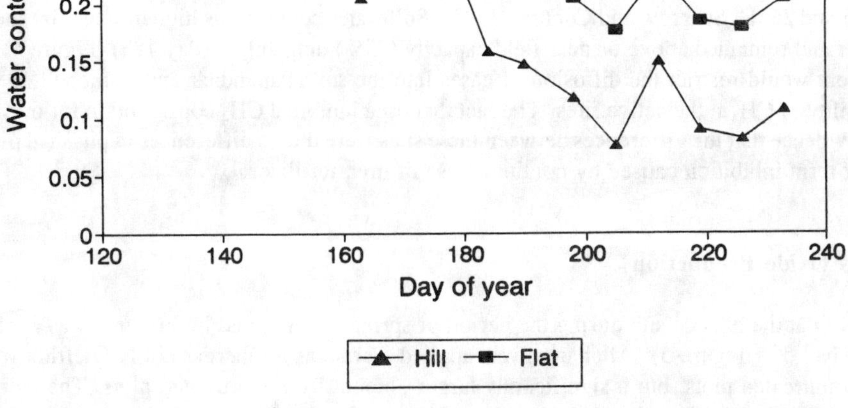

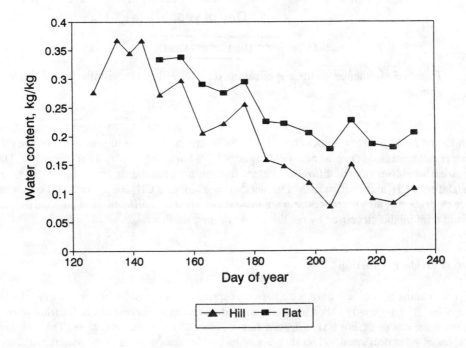

Figure 4. Soil water content at the hillside and flat sites during the summer.

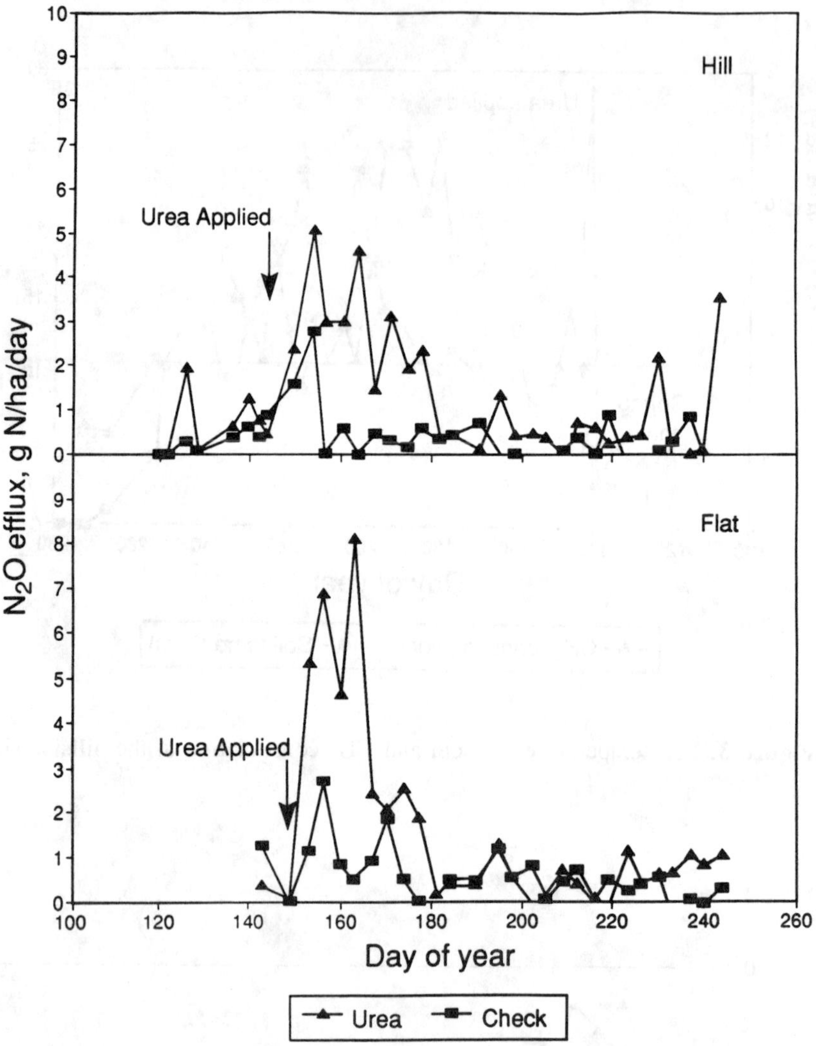

Figure 5. Influence of urea application on N_2O efflux at the hillside and flat sites.

between 15 and 25 °C (Sparrow and Cochran 1993). Soil water content was higher in the flat site throughout the summer and remained above or near field capacity (25%) until July 3 (day 183) (Figure 4). The higher water content would restrict the diffusion of gases into the soil (Papendick and Campbell 1981) reducing the availability of CH_4 at the active sites. The fact that urea inhibited CH_4 consumption for only a brief time is further evidence that the differences between these sites were due to differences in physical properties and not to long term inhibition caused by continued use of urea fertilizer.

B. Nitrous Oxide Production

Efflux of N_2O at the hillside site during the period of spring thawing and before urea was applied was less than 1 g N ha^{-1} d^{-1} (Figure 5). After urea was applied there was an increase in N_2O efflux from both the treated and untreated plots, but a significantly larger amount from the treated plots. The increase in N_2O from the check plots decreased to less than 1 g N ha^{-1} d^{-1} for the duration of the summer. Efflux of N_2O from the fertilized plots continued at higher rates for about 30 d with a maximum of 5 g N ha^{-1} d^{-1} after the urea was applied, and then dropped to that of the check plots. We were not able to measure N_2O flux from the flat until just before the urea was applied. The pattern of N_2O efflux of the flat is the same as that of the hill except that there was a greater efflux of N_2O immediately after the urea was applied and more rapid drop back to the level of the check while the hillside sustained a lower level of efflux somewhat longer (Figure 5). Both sites exhibited a short burst of N_2O efflux just after the soil thawed which is similar in time of event to that observed by others (Cates and Keeney, 1987; Christensen and Tiedje, 1989). However, their rates were about tenfold higher than those we measured. The increase in N_2O efflux due to the application of urea coincides with the detection of small amounts of NH_4-N near the soil surface. Thus, it is reasonable to assume that it is the result of the nitrification as reported by others (Cochran et al, 1981; Bremner and Blackmer, 1978). Concentrations of NO_3-N were <2 g kg^{-1} soil throughout the season (data not shown), the grasses were growing rapidly depleting the soil water and mineral N. Therefore, it is unlikely that denitrification accounted for much if any of the N_2O efflux.

The amount of N lost to the atmosphere as N_2O was estimated to be less than 100 g ha^{-1} during the growing season. This is about 0.1% of the N applied which is in the middle of the range reported by Eichner (1990) for urea applications to agricultural soils in warmer climates. Soil NO_3-N concentrations were low (<2 g kg soil) for both fertilizer treatments at both sites throughout the season with no differences due to the application of urea. If the low NO_3-N concentrations at the end of this season are indicative of the amount of NO_3-N available during spring thaw, we do not expect denitrification to contribute significantly to the efflux of N_2O during the thawing period next spring when the soil may become anaerobic due to saturated conditions. Neither of these sites received N fertilizer the previous year; therefore, we expect that the available NO_3-N for denitrification was low during spring thaw which would account for our measured N_2O emission levels being much lower than those reported by Cates and Keeney (1987) and Christensen and Teidje (1989).

IV. Summary and Conclusions

The application of urea to actively growing grasses reduced CH_4 consumption for a very short time at sites with or without previous applications of N fertilizer. Where urea had been routinely applied for many years the reduction in consumption was followed by a short period of increased consumption. The cause of the increased consumption was not ascertained, but may have been due to the stimulation of nitrifiers which have been shown to consume CH_4 (Ward 1978). The short duration of inhibition of consumption was attributed to both the rapid uptake of N by actively growing grasses and to the fact that consumption of CH_4 occurs at lower depths than where ammoniacal fertilizer had moved in the profile. The hillside site which had no known history of N fertilization consumed more CH_4 than the site with a history of annual applications of urea. However, there were physical differences between these two sites that likely caused the differences in CH_4 consumption. These were cooler soil temperatures and higher soil water content in the site with a history of N fertilization. Both of these factors will decrease CH_4 consumption. Additional studies need to be conducted on these sites to evaluate the effects of continued use of urea fertilizer.

Emission of N_2O during the spring thaw and after application of urea were similar to results of studies at warmer climates with other crops. The amount of urea N lost as N_2O-N was also similar to that reported for warmer climates. Very little NO_3-N was found at any time so we do not expect denitrification to contribute much additional N_2O during spring thaw.

References

American Public Health Association. 1975. Automated laboratory analysis. p. 620-624. In: Rand M.C. (ed.), *Standard Methods of the Examination of Water and Waste Water.* Am. Public Health Assoc. Washington D.C.

Arah, J. 1991. Biological sinks of methane and nitrous oxide. *Geography*. 76-79.

Bedard, C., and R. Knowles. 1989. Physiology, biochemistry and specific inhibitors of CH_4, NH_4 and CO oxidation by methanotrophs and nitrifiers. *Microbiol. Rev*. 53:68-84.

Bremner. J.M., and A.M. Blackmer. 1978. Nitrous oxide: Emission from soils during nitrification of fertilizer nitrogen. *Science*. 199:295-296.

Cates, R.L.Jr. and D.R. Keeney. 1987. Nitrous oxide production throughout the year from fertilized and manured maize fields. *J. Environ. Qual.* 16:443-447.

Christensen, S. and J.M. Tiedje. 1990. Brief and Vigorous N_2O production by soil at spring thaw. *J. Soil Sci.* 41:1-4.

Cochran, V.L., L.F. Elliott, and R.I. Papendick. 1981. Nitrous oxide emissions from a fallow field fertilized with anhydrous ammonia. *Soil Sci. Soc. Am. J.* 45:307-310.

Eichner, M.J. 1990. Nitrous oxide emissions from fertilized soils: Summary of available data. *J. Environ. Qual.* 19:272-280.

Higgins, I.J., D.J. Best, R,C. Hammond, and D. Scott. 1981. Methane-oxidizing microorganisms. *Microbiol. Rev.* 45:556-590.

Hutchinson, G.L. and A.R. Mosier. 1981. Improved soil cover methods for field measurements of nitrous oxide fluxes. *Soil Sci. Soc. Am. J.* 45:311-316.

McElroy, M.B. and S.C.Woofsy. 1989. Nitrous oxide sources and sinks. p. 81. In: *World Meteorological Organization, Atmospheric Ozone 1985*. Vol. 1. NASA, Washington, D.C.

Mosier,A, D. Schimel, D. Valentine, K. Bronson, and W. Parton. 1991. Methane and nitrous oxide fluxes in native, fertilized and cultivated grasslands. *Nature* 350:330332.

Mosier, A.R., and L. Mack. 1980. Gas chromatographic system for precise, rapid analysis of nitrous oxide . *Soil Sci Soc. Am. J.* 44:1121-1123.

Mosier. A.R. and D.S. Schimel. 1991. Influence of agricultural nitrogen on atmospheric methane and nitrous oxide. *Chemistry & Industry* 23:874-877.

Nesbit S.P. and G.A. Breitenbeck. 1992. A laboratory study of factors influencing methane uptake by soils. *Agric. Ecosystems and Environment* 41:39-54.

Papendick, R.I. and G.S. Campbell. 1981. Theory and measurement of water potential. p. 1-22. In: J.F. Parr, W.R. Gardner, and L.F. Elliott (eds.), *Water Potential Relations in Soil Microbiology*. Soil Sci. Soc. Am. Spec. Publ. No. 9. Madison WI.

Ryden, J.C., and L.J. Lund. 1980. Nature and extent of directley measured denitrification losses from some irrigated vegetable crop production units. *Soil Sci. Soc. Am. J.* 44:505-511.

Smith C.J., and P.M. Chalk. 1980. Gaseous nitrogen evolution during nitrification of ammonia fertilizer and nitrate transformations in soils. *Soil Sci. Soc. Am. J.* 44:277-282.

Sparrow, E.B. and V.L. Cochran. 1993. Effect of soil depth and temperature on CH_4 oxidation in subarctic soils. (This proceedings.).

Steudler, P.A., R.D. Bowden, J.M. Melillo, and J.D. Aber. 1989. Influence of nitrogen fertilization on methane uptake in temperate forest soils. *Nature* 341:314-316.

Whalen, S.C., W.S. Reeburgh, and V.A. Barber. 1992. Oxidation of methane in boreal forest soils: A comparison of seven measures. *Biogeochemistry* 16:181-211.

Ward, B.B. 1987. Kinetic studies on ammonia and methane oxidation by *Nitrosococcus oceanus. Arch. Microbiol.* 147:126-225.

Whittenbury, R., K.C. Phillips, and J.F. Wilkinson. 1970. Enrichment, isolation and some properties of methane-utilizing bacteria. *J. Gen. Microbiol.* 61:205-218.

CHAPTER **16**

Dynamics of Soil C and N in a Typic Cryoboroll and a Typic Cryoboralf located in the Cryoboreal Regions of Alberta

N.G. Juma

I. Introduction

In order to predict the rates and amounts of gaseous emission from soils, it is necessary to quantify dynamics of C and N in soils. Over the past 100 years, four main approaches have been used to study the nature of soil organic matter: (i) acid hydrolysis; (ii) classical humic fractionation techniques; (iii) particle size fractionation; and (iv) the kinetic approach, which categorizes heterogeneous organic materials by their decomposition rates (Juma and McGill, 1986). The kinetic approach, as used by Paul and Juma (1981), advanced the pioneering work of Jansson (1958) and Jenkinson (1966) by dividing soil organic N into four moieties: (i) microbial biomass (mainly bacteria and fungi); (ii) non-microbial active N; (iii) stabilized fraction, and (iv) old organic matter. This approach has been successful at predicting mineralization-immobilization turnover of N in soils (Juma and McGill, 1986). The advantage of this method of organic matter study is that the chloroform fumigation method for determination of microbial C and N (Jenkinson and Powlson, 1976) can be used to link the decomposer food web directly with elemental cycling. Use of ^{14}C and ^{15}N tracers allows the dynamics and partitioning of amended C and N to be followed among various components in the plant-soil system (Paul and Juma, 1981). We have continued to use these methods in our studies. The advantages of these methods and representative examples have been reviewed by Juma and McGill (1986) and Juma (1993a).

Most investigations aimed at understanding the dynamics of C and N in agroecosystems in the Cryoboreal regions of Alberta have been conducted in the field at a scale of 1 to 100 m^2 and have included annual crops and long-term rotations established in 1930. The paper reports on a synthesis of information about soil C and N cycling obtained at contrasting spatial and temporal scales. Specifically, detailed short-term data from laboratory and field experiments have been used to provide insight into mechanisms of C and N cycling in soil while the data from long-term experiments have been used to assess the impact of management on C and N reserves in soils.

II. Description and Management of the Experimental Sites

A. Description of the Experimental Sites

Field experiments were conducted on two pedogenically different soils, a Typic Cryoboroll (a Black Chernozem) at the Ellerslie research station (53° 25' N, 113° 33' W) and a Typic Cryoboralf (a Gray Luvisol) at the University of Alberta Breton plots (53° 07' N, 114° 28' W). Typic Cryoborolls are naturally endowed with a thick mollic epipedon which has granular structure, high nutrient and base status, and

ISBN 1-56670-117-1/95/$0.00+.50

Table 1. Properties of soil at Ellerslie and Breton

Depth (cm)	Total C ($g\ kg^{-1}$)	Total N ($g\ kg^{-1}$)	pH[a]	Bulk Density ($Mg\ m^{-3}$)	Texture[b]
		Ellerslie (Typic Cryoboroll)			
0-10	64.6	5.34	6.1	0.86	SiCL
10-20	63.2	4.91	6.0	1.06	SiCL
20-30	52.3	4.19	6.0	1.17	SiC
		Breton (Typic Cryoboralf)			
0-10	21.7	1.85	6.2	1.11	SiL
10-20	19.2	1.64	6.3	1.30	CL
20-30	11.3	1.04	6.1	1.55	CL

[a]1:2 soil water (mass:volume); [b]particle size was determined by the hydrometer method.

neutral pH. In contrast, Typic Cryoboralfs are acid, leached, degraded soils with an eluviated horizon. These soils are important for cereal production in Alberta and represent two extremes of natural fertility, biological, physical and chemical properties (Table 1), and crop productivity. Ellerslie receives 452 mm of precipitation annually and the average maximum and minimum temperatures in July are 22.4 and 9.6 °C. Breton is wetter and cooler than Ellerslie, with 547 mm of precipitation annually, and average maximum and minimum temperatures in July of 21.2 and 8.8 °C (Izaurralde et al., 1993).

B. Management of Plots used for Short-Term Studies

Field experiments were established in 1986 at Ellerslie and Breton to study above-ground productivity and selected soil properties using three crops (barley (*Hordeum vulgare* L.), barley-field pea (*Pisum sativum* L.) intercrop and fababean (*Vicia faba* L.)). The crops were grown continuously for three years and the treatments were replicated four times at each site. Plot sizes were 24.7 by 9.3 m at Ellerslie and 12 by 6.8 m at Breton. The crops were seeded between the 15 and 20 of May at Ellerslie and between the 20 and 25 of May at Breton each year. Seeding rates used for barley, field pea, and fababean were those recommended for the cultivar and soil zone. In 1987 and 1988, seeding rates for intercrops were reduced by approximately 30% with respect to their sole crop rates.

Barley and barley-field pea plots at both locations were fertilized with N at 75, 70, and 70 kg ha^{-1}, as urea or ammonium nitrate, in 1986, 1987, and 1988, respectively. Fababean plots were fertilized with N at 75 kg ha^{-1} only in 1986. Fertilizer P as triple super phosphate was placed near the seed of all crops each year at a rate of 17 kg ha^{-1}. Legume seeds were inoculated prior to seeding with *Rhizobium leguminosarum* (culture C or Q for field pea and fababean, respectively).

The plots were hand-weeded only during the 1986 growing season, but herbicides were used during the second and third year of the study. Glyphosate [N-(phosphonomethyl) glycine] at a product rate of 1.0 L ha^{-1} was applied pre-plant in the fall or spring to annual crops for perennial weed control. Barley and barley-field pea plots received post-emergent applications of MCPA+MCPB, a 15:1 mixture of the sodium salts of MCPB [4-(2-methyl-4 chlorophenoxy) butyric acid] and MCPA [2-methyl-4 chlorophenoxyacetic acid], at a product rate of 1.5 L ha^{-1}. Fababean plots received post-emergent applications of bentazon [3-(1-methyl-ethyl)- (1H)-2,1,3-benzothiadiazin-4(3H)-one 2,2-dioxide] at a product rate of 1.5 L ha^{-1}.

During the last week of August and the first of September, six one-m^2 areas from annual-crop plots were harvested by hand for determination of dry matter and seed yield (Mg ha^{-1}). Plants were separated and counted (spikes m^{-2} or no. plants m^{-2}), dried at 60 °C for 72 h, and threshed. Mechanical harvest with swathers, combining, and straw spreading followed. Detailed information on seeding dates, fertilization rates, weed control and sampling procedures have been described by Izaurralde et al. (1993).

Studies on microbial, faunal, and plant interactions, and on C and N cycling in barley plots at Ellerslie and Breton were conducted by Rutherford and Juma (1989a and b) and Dinwoodie and Juma (1988a and b). Pertinent results from these studies are reported in this paper.

C. Management of Plots used for Long-Term Studies

The Breton Plots were established in 1930 to find "a system of farming suitable for the wooded soil belt" (Robertson, 1979). The original plots (31.5 m x 8.5 m each) were designed to compare two cropping systems and to test several soil amendments and fertilizers. The original cropping systems were continuous wheat (*Triticum aestivum* L.) and a four-crop rotation with three years of cereal grains and one year of legumes. In 1938 the continuous wheat system was converted to a wheat-fallow rotation to help control weeds. In 1939, the four-crop rotation was changed to a five-crop rotation of wheat, oat (*Avena sativa* L.), barley, forage, forage. The forage crop has varied over the years but has always included a legume. For the 1939-1954 period, the forage component consisted of "mixed legumes". For the 1955-1966, the forage component was a five-crop mixture (alfalfa (*Medicago sativa* L.), red clover (*Trifolium pratense* L.), bromegrass (*Bromus inermis* Leyss.), creeping red fescue (*Festuca rubra* L.), timothy (*Phleum pratense* L.)). In 1967, the forage mixture was changed to alfalfa and bromegrass to reflect commonly recommended forage mixture for the area. All phases of the two rotations and fertilizer and soil amendment are present every year, however, the individual treatments are not replicated.

The soil amendments included several combinations of the nutrients (nitrogen, phosphorus, potassium, and sulfur) as well as lime and farmyard manure (Robertson and McGill, 1983). In this paper, the results of three treatments (Check, NPKS, and Manure) are discussed. Fertilizer application methods varied over the years. Initially, all fertilizers were annually broadcast. From 1946 to 1966, fertilizers were added every second year. In 1964, annual applications were resumed and phosphate was drilled with the seed. Average fertilizer application rates for N, P, K, and S were 10, 6, 16, and 10 kg ha^{-1} yr^{-1}, respectively. Manure was applied once every five years at the rate of 44 Mg ha^{-1} in both rotations and was added after the second forage crop was harvested in the 5-yr rotation. The annual equivalent N, P, K, and S rates based on estimates of manure applied from 1976-1986 were 76, 42, 91 and 20, kg ha^{-1} yr^{-1}, respectively (Cannon et al., 1984). In 1972, lime was added to the east half of all plots of the 5-year rotation and to the entire area of the 2-year rotation. The check or control plots have not received fertilizers since 1930.

Major revisions were introduced in 1980 (Cannon et al., 1984; Robertson and McGill, 1983). In this paper, the data from the Breton plots for the period of 1930-1979 have been used. Crop yield trends for the 2-yr and 5-yr rotation up to 1989 have been presented by Juma (1993b). The latter data show the impact of added nutrients through fertilizers on crop productivity.

III. Net Mineralization of Carbon and Nitrogen from Soil Samples

One of the simplest methods to study the kinetics of mineralization of C and N from soil is to incubate soil samples under aerobic or anaerobic conditions for extended periods. This approach, was originally proposed by Stanford and Smith (1972) has been widely used to assess the effect of cropping practices (Campbell et al., 1991), soil types (Juma et al., 1984) and different environments (Gharous et al., 1990). Stanford and Smith (1972) proposed that there existed a universal rate constant which could be used for different soils, therefore, in order to predict mineralization of N from soils it was only necessary to measure the pool size of mineralizable N. Talpaz et al. (1981) reanalyzed the net N mineralization data from the 39 soils used by Stanford and Smith (1972) and showed each soil had a specific net mineralization rate constant and a pool size.

An alternate way to examine the mineralization data is to express it on an absolute and a relative basis. Results from laboratory incubation experiments showed that the amount of N mineralized over 40 weeks under laboratory conditions from surface soil samples from Ellerslie (120 $\mu g\ g^{-1}$ soil) was almost three times that of Breton (48 $\mu g\ g^{-1}$ soil), however the proportion of total N mineralized from Breton (3.2%) was higher than that at Ellerslie (2.4%). Expression of data as a proportion of total N showed that the N of Typic Cyroboralf was being mineralized at a higher rate than that in the richer Typic Cyroboroll (Juma, 1993a). The ratio of CO_2 mineralized/net N mineralized was 10/1 in these unamended soils. This approach yields a rough estimate of turnover of C and N in different soils. In order to gain a further insight into the mechanisms responsible for these observations, detailed studies, such as those described in the next section, were conducted.

IV. Carbon and Nitrogen Dynamics in Soil Samples Obtained from Barley Plots

To determine the effects of soil type on the dynamics of soil organic matter, detailed studies on microbial, faunal, and plant interactions, and on C and N cycling were conducted by Rutherford and Juma (1989a and b) and Dinwoodie and Juma (1988a and b). Data relevant to this paper are given below.

Differences in specific activity of the various below ground pools revealed that an average of 17% of the microbial C was active at Ellerslie while 43% was active at Breton (Table 2). Active microbial C (g m^{-2}) was the not significantly different between sites because total microbial C was lower at Breton than at Ellerslie (Dinwoodie and Juma, 1988b). Carbon flow rates (respiration) were higher at Breton compared to Ellerslie (Table 2). These results suggest that the soil with lower amount of organic matter had a higher turnover rate, therefore it would be more difficult to use these soils to sequester C from the atmosphere.

In order to gain a further insight into the mineralization of C and N in these soils, Rutherford and Juma (1989a and b) studied food web dynamics, and shoot and root growth over the growing season. Although microbial C (g m^{-2}) was greater at Ellerslie, microbial C made up a larger proportion of soil C (mg g^{-1} soil C) at Breton (Table 3). Soil C at Breton supported proportionally greater microbial C and protozoan populations and equal nematode and microarthropod populations compared to Ellerslie (Rutherford and Juma, 1989b). Ten-day incubations of non-fumigated soil showed that CO_2-C evolved was greater at Ellerslie (120 μg g^{-1} soil) than Breton (97 μg g^{-1} soil), however there was no significant difference between soils in CO_2-C evolution when expressed on an area basis (g m^{-2}) (Table 3). CO_2-C evolution was greater from Breton soil samples when expressed as a proportion of soil C (mg g^{-1} soil C) or as a proportion of microbial C (mg g^{-1} microbial C) (Table 3).

The dynamics of N were similar to those of C. Microbial N made up a larger proportion of soil N at Breton compared to Ellerslie (Table 3). Net N mineralization was not different between soils when expressed as a proportion of microbial N but it was greater in soil from Breton when expressed as a proportion of total soil N. Net ^{15}N mineralization did not significantly differ between soils when expressed as a proportion of soil ^{15}N (Table 3).

Rutherford and Juma (1989b) proposed that the food web was more active at Breton compared to Ellerslie in 1986. Differences in detrital trophic web structure affect C and N turnover in the soil and the size and activity of the active N pools. Expressed proportionally, high C activity and a large decomposer food web kept a larger proportion of soil N in the soil organisms at Breton compared to Ellerslie (Table 3). Rapid C turnover and greater microbial and faunal activity resulted in a greater proportional net N mineralization rate at Breton.

Table 2. Microbial C, active microbial C, and carbon flow rate (respiration) in soil samples obtained over a depth of 0-30 cm from Ellerslie and Breton

	--------------Site--------------	
Variable	Ellerslie	Breton
Total microbial C (g/m^2)	116.0a[a]	64.0b
Active microbial C (% of total)	17.2a	42.8b
Active microbial C (g/m^2)	20.0a	27.1a
Carbon flow rate (g m^{-2} d^{-1})	1.7a	2.5b

[a] Means followed by different letters are significantly different at $P < 0.05$. (Adapted from Dinwoodie and Juma, 1988b.)

Table 3. Absolute and relative comparisons of selected variables in 0-10 cm depth at the two sites

Variable	Site: Ellerslie	Site: Breton
Absolute (per m^2 basis)		
Soil C (g m^{-2})	5530a[a]	2380b
Microbial C (g m^{-2})	45.9a	26.4b
CO_2-C evolved (g m^{-2} 10 $days^{-1}$)	10.3a	10.7a
Soil N (g m^{-2})	456a	206b
Microbial N (g m^{-2})	5.5a	4.1a
Net N mineralized (μg g^{-1} soil 10 $days^{-1}$)	10.4a	7.2b
Net ^{15}N mineralized (μg 100 g^{-1} soil 10 $days^{-1}$)	4.2a	3.0b
Relative		
Microbial C (mg g^{-1} soil C)	8.2a	11.1b
CO_2-C evolved (mg g^{-1} soil C 10 $days^{-1}$)	1.9a	4.5b
CO_2-C evolved (g g^{-1} microbial C 10 $days^{-1}$)	0.23a	0.41b
Microbial N (% of soil N)	1.2a	2.0b
Net N mineralized (mg g^{-1} microbial N 10 $days^{-1}$)	17a	20a
Net N mineralized (mg g^{-1} soil N 10 $days^{-1}$)	0.20a	0.39b
Net ^{15}N mineralized (% of soil ^{15}N 10 $days^{-1}$)	4.8a	3.2a

[a] Means followed by different letters are significantly different at $P < 0.05$.
(Adapted from Rutherford and Juma, 1989 a and b.)

V. Net Above-Ground C and N Production at Ellerslie and Breton

The above-ground productivity of barley, barley intercropped with field pea, and faba bean grown continuously from 1986 to 1988 at Ellerslie and Breton showed that the 3-year average C yield over all crops was higher at Ellerslie (316 g m^{-2}) than at Breton (282 g m^{-2}) under similar levels of management. The average barley C yield from fertilized plots at Breton was only 10% lower than that at Ellerslie (Izaurralde et al., 1993). In contrast, barley yield differences of up to 50% were observed between sites on unfertilized plots at Breton (M. Nyborg, personal communication), emphasizing the importance of N inputs if comparable barley yields are to be attained. In the case of the intercrop, the C yield differences between sites was 14% (Izaurralde et al., 1993). In contrast, the 3-year average C yield of faba bean (382 g m^{-2}) was higher than at Ellerslie (328 g m^{-2}). These results also show that the net C fixation potential under adequate fertilization for the two sites ranged from 280 to 382 g m^{-2}, but the amount of N ranged from 7.9 in barley to 20.2 g m^{-2} in fababean. Without soil-water stress, the faba bean/*Rhizobium* symbiosis is capable of fixing high amounts of dinitrogen at Breton. Using enriched-^{15}N isotope techniques, Gu (1988) reported up to 20 g m^{-1} of N fixed by faba bean at Breton.

The amount of N uptake by barley represents N uptake from fertilizer N and from N mineralized from soil organic matter. Assuming half of N uptake in the barley crop is from fertilizer N, the amount of N which was mineralized from soil organic matter and taken by barley (3.9 to 5.7 g m^{-2}) represents microbial and faunal activity. The amount of total N in the 0-10 cm depth at Ellerslie and Breton was 456 and 206 g m^{-2} (Rutherford and Juma, 1989b). Assuming that most of N was mineralized in the 0-10 cm depth, then the proportion of total N taken up by barley was 1.24% at Ellerslie and 1.89% at Breton. These data are consistent with those of Dinwoodie and Juma (1988a and b) and Rutherford and Juma (1989a and b) obtained with incubation of soil samples in the laboratory. Under field conditions, the primary productivity is affected by a number of variables. Except for faba bean, annual fertilization with N and P as well as prevailing climatic conditions, masked to some extent the influence of inherent soil properties on crop yield and demonstrated the importance of hierarchical constraints (e.g., irradiance, temperature > water > nutrients) on crop productivity under Cryoboreal conditions. The biological effects in agroecosystems are generally masked by management practices.

VI. Average Crop Yields and Long-Term Trends of Soil Organic Matter at Breton

A. Average Crop Yields

The 5-year running averages for crop yields in the 2-yr and 5-yr crop rotations under three treatments (check, manure and NPKS) have been reported by Juma (1993b). The impact of increasing the fertilizer rates on crop yields on these plots from 1980 onward have also been presented by Juma (1993b). Soil sampling was last conducted in 1979, therefore crop yield data and soil N in different rotations and management practices up to 1979 will be presented here. The C values for grain and dry matter yields were calculated from crop yields and first cut of forages assuming that the C content of above-ground material was 40%. The average values of grain and dry matter yields and their standard deviations are presented in Tables 4 and 5.

Table 4. Long-term dry matter C yield averages (g m^{-2}) and their standard deviations at the Breton plots (1930-1979)

		Treatments		
Rotation	Crop	Check	NPKS	Manure
2-yr	Wheat	49 (21)	65 (31)	77 (33)
5-yr	Wheat	48 (21)	90 (35)	88 (32)
	Oat	50 (21)	77 (28)	88 (32)
	Barley	33 (18)	54 (21)	60 (27)

Table 5. Long-term dry matter C yield averages (g m^{-2}) and their standard deviations at the Breton plots (1930-1979)

		Treatments		
Rotation	Crop	Check	NPKS	Manure
2-yr	Wheat	135 (57)	180 (78)	220 (74)
5-yr	Wheat	122 (47)	257 (94)	230 (81)
	Oat	129 (38)	206 (68)	223 (84)
	Barley	77 (32)	124 (44)	136 (46)
	Hay (1)	54 (33)	160 (94)	118 (50)
	Hay (2)	51 (40)	169 (79)	120 (61)

The grain C yields in the NPKS and manure treated plots were similar and up to 2-fold higher than the check plots (Table 4). Grain yields of cereals in the 5-yr rotation showed that the yields of wheat > oat > barley in all treatments. The benefit of the legume in the forage crop was minimal for the third cereal crop in the rotation and the barley yields were considerably lower than the wheat yields (Table 4). The wheat grain C yield from NPKS and manured treatments of the 5-yr rotation were consistently higher than those of the 2-yr rotation. This suggests a synergistic effect of legumes and manure on wheat yields in the 5-yr crop rotation than in the 2-yr rotation.

In these experiments the forages have been removed for hay in early to mid-July and the second year forages have been "plowed down" in late July without much regrowth. The trends in the total dry matter production were similar to those of the grain yields. The forage yields in the check and manure treatments were lower than those of barley (Table 5). In contrast, forage yields were higher than barley yields in the NPKS treatment. This may be due to the addition of S in these deficient soils. The yields of forages in the NPKS and manure-treated plots were almost 2 to 3-fold higher than in the check plots. The amounts of nitrogen, phosphorus, potassium, and sulfur added through manure were several times larger than those added in the fertilizers.

It is important to note that fertilizer N in the NPKS treatment was applied at a rate of 10 kg ha^{-1} yr. The total dry matter yield for cereal crops on these plots ranged from 124 for third-year barley to 257 g m^{-2} for first year wheat (Table 5). Above-ground yields for fertilized barley at Breton using current technologies was 290 g m^{-2} which shows that addition of nutrients through fertilizers and crop rotations in the long-term plots were only 12% lower than wheat yields. Analysis of data presented in Tables 4 and 5 also show that the straw C/grain C ratio for wheat and oat ranged from 1.54 to 1.85, and that for barley ranged from 1.26 to 1.33. With changes in crop breeding and fertilization practices, this ratio was 1.14 for barley grown at Breton using current technologies (Izaurralde et al., 1993). Although the above-ground crop residues were not returned to the soil, this ratio may have a bearing on the amount of photosynthates transferred below ground.

B. Organic Matter Trends

Linear regression analyses showed that the slope of total N in the three treatments of the 2-yr rotation were not significantly different from zero, therefore, there were no significant changes of total soil N over 42 years (Table 6, Figure 1). In contrast, soil total N increased in all treatments of the 5-year rotation. The increase was the highest for the manure treatment followed by NPKS and check treatments (Table 6, Figure 1).

After 42 years of cropping at the Breton Plots, the soil organic N content was about 37, 42, and 42 per cent higher in the soil of Check, NPKS and Manure treatments, respectively, of the 5-yr rotation plots than in those of the 2-yr (wheat-fallow) rotation plots (Figure 1). The main factor affecting C and N sequestration in these soil was the input of C during the growing season through roots and root products. In the wheat-fallow rotation, the plots were fallowed every alternate year. In contrast, crops were grown more continuously in the 5-yr rotation and therefore more plant material (stubble and roots) was added than in the 2-yr rotation plots. Cereal production on these crops was at a frequency of 60% (3 out of 5 crops) compared to 50% in the wheat-fallow rotation. Further, the 5-yr rotation plots were bare for a smaller portion of the time so that decomposition processes were probably slower and erosion losses were smaller. A greater amount of nitrogen was probably left by forage crop roots than cereal crop roots and this would be conducive to higher soil organic matter. After 42 years of cropping at the Breton Plots, the soil organic N content in the NPKS treatments of the 2-yr and 5-yr rotations were 6 and 10 percent higher than in their respective Check plots (Figure 1). The increase can be explained by higher crop yields (forages and grain) on fertilized plots. Thus the amount of stubble and root material added to the soil was greater on these plots although all forages were removed as hay and the straw of cereals was also removed. The soil organic N content in the Manure treatments of the 2-yr and 5-yr rotations were 36 and 41 percent higher than in their respective Check plots (Figure 1). This result arises from the fact that manure is organic matter and the average annual addition was approximately 9 t ha^{-1}. In addition, manure served as a fertilizer and improved crop growth and hence organic matter additions via roots and stubble.

VII. Summary and Conclusions

Although the soil at Ellerslie has more organic C and N, microbial biomass and faunal populations than at Breton, there was a greater proportional mineralization from soil samples at Breton. The higher activity of C and N at Breton than at Ellerslie can be explained by a higher activity of the web. Isotopic dilution studies also showed that a large proportion of the organic matter at Ellerslie is stabilized and unavailable for rapid microbial decomposition. The inherent low fertility of the soil at Breton can be almost overcome with proper fertilizer management.

The data from the Breton Plots clearly show that nutrient inputs through commercial fertilizer and manure, especially N, have increased the productivity of Typic Cryoborolls. The addition C through roots and input of N through biologically fixed N or fertilizers has improved this soil and C from the atmosphere has been sequestered. The formation of organic matter is a continual process mediated by microorganisms and soil fauna and is directly dependent on the quantity, quality, and placement of above- and below ground residues. Data from the Breton plots suggest that the impact of root C inputs on organic matter content be re-evaluated because all above-ground crop residues were removed and not returned to the plots.

Table 6. Trends of total soil N (g kg^{-1} soil) in check, manure and NPKS treatments of the 2-yr and 5-yr rotation at Breton

Rotation	Treatment	Equation	r^2
2-yr	Check	Total N = 1.04 (±0.0676) - 0.00182 (±0.0234) x Year	0.16 ns
	NPKS	Total N = 1.09 (±0.0732) - 0.000664 (±0.00254) x Year	0.02 ns
	Manure	Total N = 1.14 (±0.121) + 0.00687 (±0.00421) x Year	0.47 ns
5-yr	Check	Total N = 0.940 (±0.0539) + 0.00942 (±0.00187) x Year	0.89**
	NPKS	Total N = 0.881 (±0.112) + 0.0163 (±0.00391) x Year	0.85**
	Manure	Total N = 0.926 (±0.117) + 0.0248 (±0.00407) x Year	0.92**

(Adapted from Juma, 1993b.)

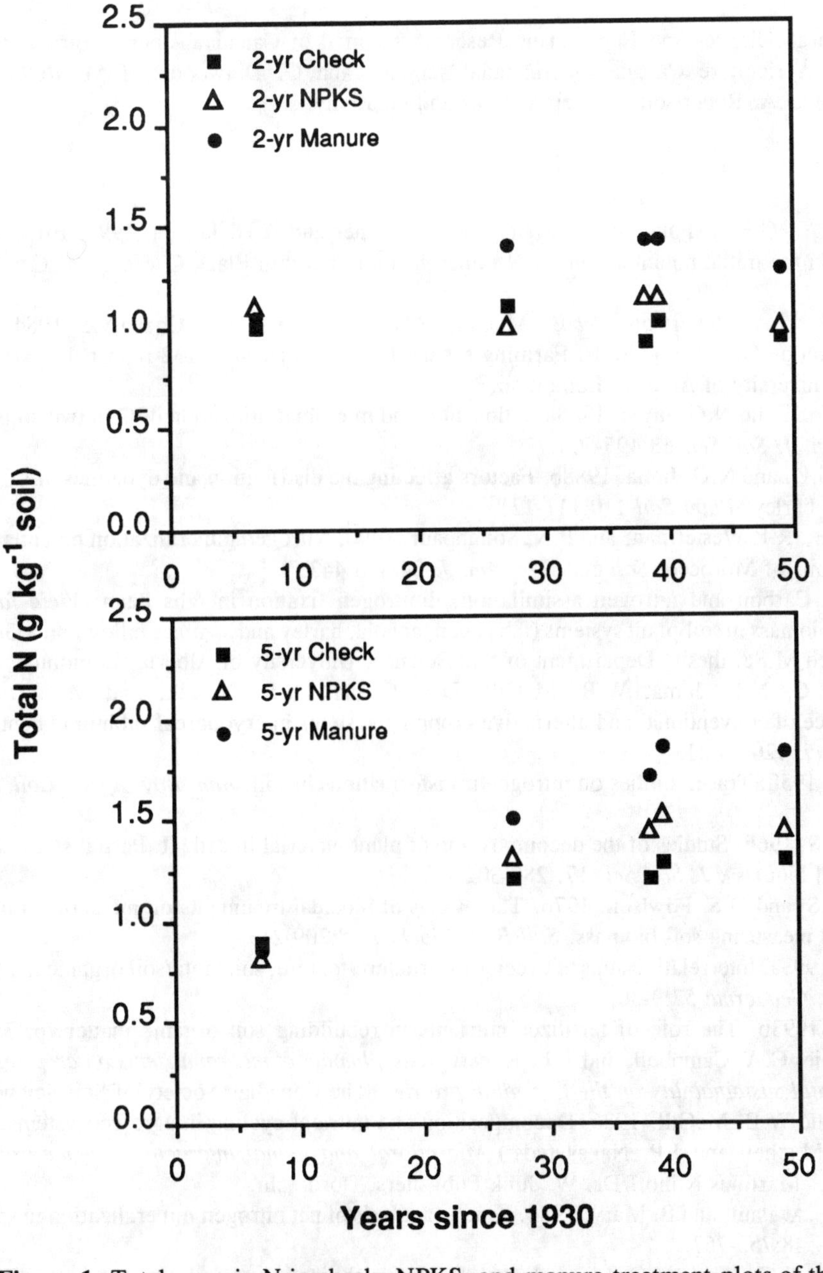

Figure 1. Total organic N in check, NPKS, and manure treatment plots of the 2-yr and 5-yr rotation plots between 1937 and 1979. (Adapted from Juma, 1993b.)

Acknowledgements

I thank: Natural Sciences and Engineering Research Council of Canada, Alberta Agriculture Research Institute and Agriculture Canada for financial support, and G. Dinwoodie, P.M. Rutherford, R.C. Izaurralde, and J.A. Robertson for their support and creativity.

References

Campbell, C.A., G.P. LaFond, A.J. Leyshon, R.P. Zentner and H.H. Janzen. 1991. Effect of cropping practices on the initial potential rate of N mineralization in a thin Black Chernozem. *Can. J. Soil Sci.* 71:43-54.

Cannon, K.R., J.A. Robertson,, W.B. McGill, F.D. Cook, and D.S. Chanasyk. 1984. Production Optimization on Gray Wood Soils. Farming for the Future Project 79-0132 report, Department of Soil Science, University of Alberta, Edmonton.

Dinwoodie, G.D. and N.G. Juma. 1988a. Allocation and microbial utilization of C in two soils cropped to barley. *Can. J. Soil Sci.* 68:495-505.

Dinwoodie, G.D. and N.G. Juma. 1988b. Factors affecting the distribution and dynamics of ^{14}C in two soils cropped to barley. *Plant Soil* 110:111-121.

Gharous, M.E., R.L.Westerman, and P.N. Soltanpaur. 1990. Nitrogen mineralization potential of arid and semiarid soils of Morocco. *Soil Sci. Soc. Am. J.* 54:438-443.

Gu, J. 1988. Carbon and nitrogen assimilation, dinitrogen fixation in faba bean (*Vicia faba* L.) and microbial biomass in soil-plant systems (faba bean, canola, barley and summer fallow) on a Gray Luvisol. Unpublished M.Sc. thesis. Department of Soil Science, University of Alberta, Edmonton, Canada.

Izaurralde, R.C, N.G. Juma, W.B. McGill, D.S. Chanasyk, S. Pawluk, and M.J. Dudas. 1993. Performance of conventional and alternative cropping systems in cryoboreal subhumid central Alberta. *J. Agric. Sci.* 120:33-41.

Jansson, S.L. 1958. Tracer studies on nitrogen transformations in soil. *Ann. Roy. Agric. Coll. Sweden* 24: 101-361.

Jenkinson, D.S. 1966. Studies of the decomposition of plant material in soil. II. Partial sterilization of soil and the soil biomass. *J. Soil Sci.* 17: 280-302.

Jenkinson, D.S. and D.S. Powlson. 1976. The effects of biocidal treatments on metabolism in soil. V. A method for measuring soil biomass. *Soil Biol. Biochem.* 8:209-213.

Juma, N.G., 1993a. Interrelationships between soil structure/texture, soil biota/soil organic matter and crop production. *Geoderma* 57:3-30.

Juma, N.G., 1993b. The role of fertilizer nutrients in rebuilding soil organic matter. p. 363-387. In: D.A. Rennie, C.A. Campbell, and T.L. Roberts (eds.) *Impact of macronutrients on crop responses and environmental sustainability on the Canadian prairies.* The Canadian Society of Soil Science, Ottawa.

Juma, N.G. and W.B. McGill. 1986. Decomposition and nutrient cycling in agro-ecosystems. p. 74-136. In: M.J. Mitchell and J.P. Nakas (eds.) *Microfloral and faunal interactions in natural and agro-ecosystems.* Martinus Nijhoff/Dr. W. Junk Publishers, Dordrecht.

Juma, N.G., E.A. Paul, and B. Mary. 1984. Kinetic analysis of net nitrogen mineralization in soil. *Soil Sci. Soc. Am. J.* 48:753-757.

Paul, E.A. and N.G. Juma, N.G. Mineralization and immobilization of soil nitrogen by soil microorganisms. *Ecol. Bull. (Stockholm)* 33:179-195.

Robertson, J.A. 1979. Lessons from the Breton Plots. *Agriculture and Forestry Bulletin (University of Alberta, Edmonton)* 2(2):8-13.

Robertson, J.A. and W.B.McGill. 1983. New Directions for the Breton Plots. *Agriculture Forestry Bulletin (University of Alberta, Edmonton)* 6(1):41-45.

Rutherford, P.M. and N.G. Juma. 1989a. Dynamics of microbial biomass and soil fauna in two contrasting soils cropped to barley (*Hordeum vulgare* L.). *Biol. Fertil. Soils* 8:144-153.

Rutherford, P.M. and N.G. Juma. 1989b. Shoot, root, soil and microbial nitrogen dynamics in two contrasting soils cropped to barley (*Hordeum vulgare* L.). *Biol. Fertil. Soils* 8:134-143.

Stanford, G. and S.J. Smith. 1972. Nitrogen mineralization potentials of soils. *Soil Sci. Soc. Am. J.* 36:465-472.

Talpaz, H., P. Fine, and B. Bar-Yosef. 1981. On the estimation of N-mineralization parameters from incubation experiments. *Soil Sci. Soc. Am. J.* 45:993-996.

CHAPTER 17

Effect of Soil Depth and Temperature on CH_4 Consumption in Subarctic Agricultural Soils

E.B. Sparrow and V.L. Cochran

I. Introduction

Approximately 25% of the earth's soil C is sequestered in boreal regions (Hobbie and Melillo 1984). Most global circulation models predict increasing temperatures and some predict drier summer conditions for high latitude areas (Houghton et al., 1990). Additionally, some biological models (Lashof, 1989; Khalil and Rasmussen, 1989; Guthrie, 1986; Hameed and Cess, 1983) project that a warmer climate will increase rates of CH_4 emission through increased rates of CH_4 production by methanogens. However, these models do not consider CH_4 oxidation which can be a modulating factor on CH_4 emissions (Nesbit and Breitenbeck, 1992; Whalen and Reeburgh, 1990). Global soil methane oxidation rates have been estimated to range from 5 to 50 Tg per year (Born et al., 1990; Crutzen, 1991) representing 1 to 10% of current estimates for net CH_4 flux to the atmosphere (Crutzen, 1991). Should significant global warming occur, large areas of land in subarctic regions may become more suitable for arable agriculture. This could affect rates of CH_4 consumption, a positive control on atmospheric CH_4 increases, and the contribution of this process to CH_4 budgets in the future. Oxidation of CH_4 in temperate, tropical, wetland, tundra, and boreal soils has been documented (Crill, 1991; Keller et al., 1990; King and Adamsen, 1992; Whalen et al., 1992; Whalen and Reeburgh, 1990) but not in subarctic agricultural soils. Impact of large scale land use changes and soil management practices on rates of CH_4 uptake, could be important. The objectives of this study were to : 1) quantify CH_4 consumption rates in subarctic agricultural soils, and 2) determine CH_4 consumption rates in soils from fertilized and unfertilized grass fields, at different soil depths and under different temperature regimes.

II. Materials and Methods

A. Description of Study Sites

Two sites near Fairbanks, Alaska (64° 49'N, 147° 52'W) were selected based on their different N fertilization but similar plant history. One site (hillside) was on Fairbanks silt loam (*Alfic cryochrepts*) with a south facing slope of about 15%, pH of 6.27 and 18.5 g C kg^{-1} soil. This site had been in mixed Kentucky bluegrass (*Poa pratensis* L.) and smooth bromegrass (*Bromus inermis* Leyss) for over 10 years and had not been fertilized. The other site (flat) was on Tanana silt loam (*Pergelic cryaquept*) with less than 1% slope, 29.6 g C kg^{-1}soil and pH of 6.46. This site had been in smooth bromegrass hay production for 10 or more years and received about 90 kg N ha^{-1} yr^{-1}. Twelve plots (3 m by 3 m) were established in each site. The grass was mowed and clippings removed, with no fertilizer applied in 1991. Urea was broadcast at a rate of 90 kg N ha^{-1} to 6 plots at the hillside on May 26, 1992, while the other 6 received no N fertilizer. The

ISBN 1-56670-117-1/95/$0.00+.50

plots were arranged in a randomized complete block design. A similar procedure was followed on the flats on May 29, 1992.

B. Laboratory and Field Experiments

For the laboratory studies, soil cores, 50 cm in length were collected in plastic tubes (2.3 cm diameter, 1 mm wall) lining the soil corer. The soil cores enclosed in the plastic tubes were cut into 10-cm segments, capped at the bottom, and the tubed cores placed in 450 ml jars with lids fitted with a septum for syringe sampling of headspace gas. After each sampling of the headspace (daily for a week), the same volume, as was removed in the sampling, was replaced by injecting hydrocarbon-free air into the covered jar. Three 10-cm cores from the same depth were placed in each jar and three replicate jars used per depth. Experiments involved time-course measurement of headspace CH_4 after exposing soil cores to atmospheric CH_4 concentrations (1.8 $\mu L\ L^{-1}$). Rates of CH_4 oxidation were calculated from the soil mass (oven dry soil cm^{-3}), jar headspace volume, specific volume of CH_4, and decrease in CH_4 concentration. Soil core samples collected on July 7 were used on the same day for jar experiments and incubated at 10 °C. Those collected on September 29, 1992 were capped and immediately put in a -20 °C freezer. Later, the frozen cores were cut into 10-cm sections, the bottom-capped tubed cores put into jars. The jars with cores were equilibrated for 8 h at the temperature at which they were going to be incubated, with the lids laying loosely on the jars. A set of samples from the hillside were incubated at 5 °C, 15 °C and 25 °C while another set of cores from both hillside and flat, were incubated at 15 °C.

For *in situ* field soil depth CH_4 measurements, a 55-cm long PVC pipe (1.9 cm diameter) with the bottom capped and sealed with high resin epoxy, was used. Holes were drilled in the PVC tube at 5, 10, 20, 30, 40 and 50 cm from the top. A 0.32 cm diameter copper tubing was placed at the hole at each depth. The tubing was then run from each depth through the center of the PVC pipe up to the end of the pipe above the soil surface, which was also sealed with the high resin epoxy. The upper end of the copper tubing had a swagelok fitting and septum attached through which gas sampling was done with a 60 ml polypropylene syringe equipped with a gastight valve to seal the syringe for transport and storage until analysis.

Methane analysis was performed within 24 hr after sampling, following the procedure described by Cochran et al. (1993). Soil pH were determined on 2:1 water to soil suspension of subsamples. A carbon analyzer was used to determine organic C content by dry combustion of soil subsamples.

Statistical analysis of data consisted of conventional analysis of variance procedures (Gomez and Gomez, 1984). The Waller-Duncan BLSD (Bayes Least Significant Difference) test (Peterson, 1985) was used for mean separation when significant main effects or interactions occurred.

III. Results and Discussion

A. Laboratory Studies

Methane uptake rates in the first week of July, were highest in soils from the top 0-30 cm depth in both hill and flat sites (Figure 1). There was a significant ($P \leq 0.05$) site and N treatment interaction effect on methane consumption rates. Rates were highest in soil cores from unfertilized hill plots (9.95 g CH_4 ha^{-1}) followed by those in fertilized hill plots (6.40 g ha^{-1}), and lowest in soil cores from the flat site, with no differences between fertilizer treatments (4.35 g ha^{-1}). CH_4 consumption rates for each 10 cm core section were summed to give area-based rates for each plot or site.

Similar to results obtained in early July, highest CH_4 consumption rates occurred in the top 30 cm depth in soil cores collected from both sites at the end of September (Figure 2). In contrast there was no significant fertilizer effect on CH_4 uptake rates; however, there was a significant site by depth interaction effect ($P \leq 0.05$). CH_4 consumption rates were highest in the 10-20 cm soil depth of the flat site which exceeded the highest rates found in the surface 0-30 cm depth in the hillside. Negative values for CH_4 uptake in the 30-50 cm soil depths of the flat site suggests CH_4 production may have occurred.

The effect of incubation temperature on CH_4 consumption rates in hill soil cores are shown in Figure 3. There was a significant temperature by depth interaction ($P \leq 0.01$). Methane uptake rates in the different soil depths were similar at 5 °C. Increasing temperature from 5 to 25 °C generally increased rates in the surface to 30 cm depths. A different optimum temperature requirement in the 30-40 and 40-50 cm depths

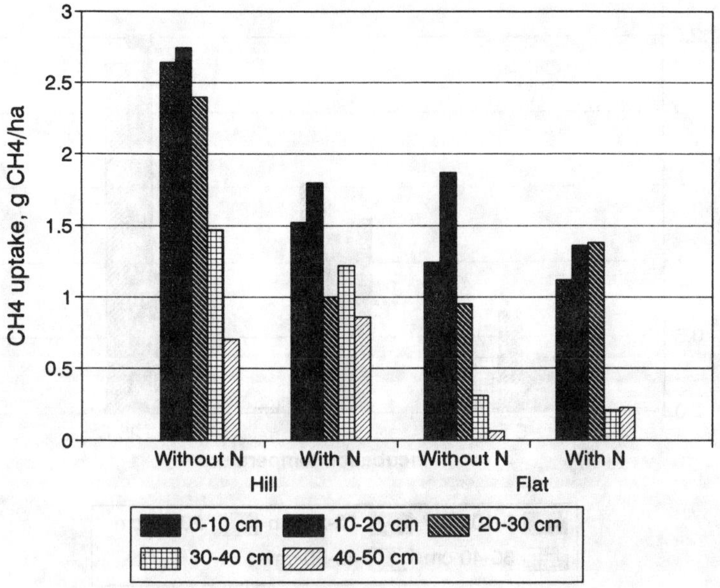

Figure 1. Cumulative CH_4 consumption (72 hours) in soil cores collected in early July in unfertilized and fertilized plots at the hill and flat sites. (Site x Treatment interaction, $P \leq 0.001$, BLSD = 0.35; Depth, $P \leq 0.01$, BLSD = 0.37.)

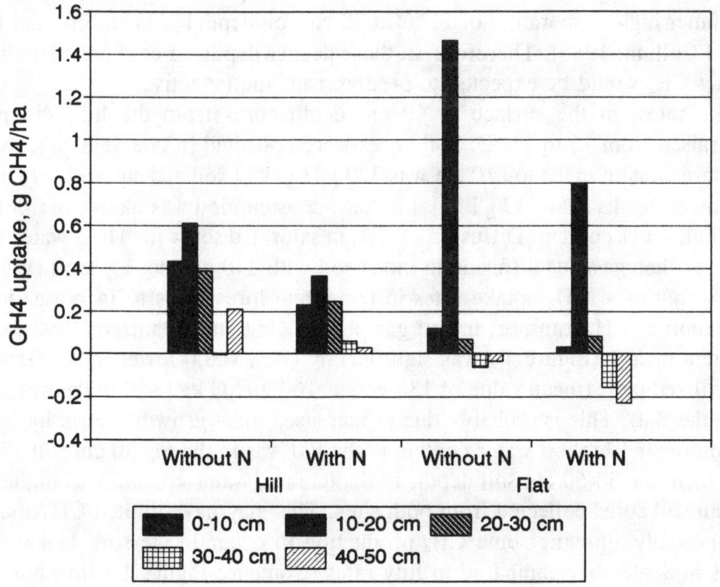

Figure 2. Cumulative CH_4 consumption (72 hours) in soil cores collected late September in unfertilized and fertilized plots at the hill and flat sites. (Site x Depth interaction, $P \leq 0.05$, BLSD = 0.35.)

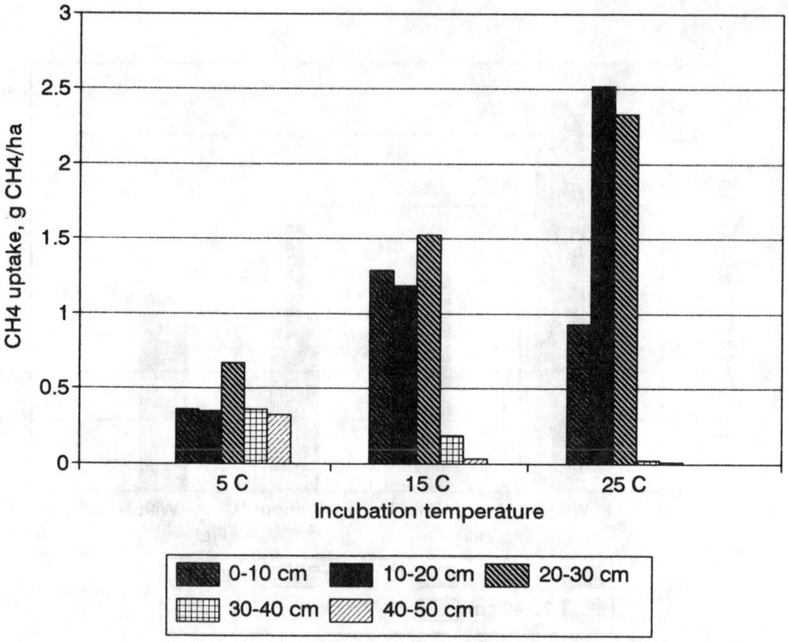

Figure 3. Cumulative CH_4 consumption (72 hours) in soil cores from unfertilized hill plots and incubated at different temperatures. (Temperature x Depth interaction, $P \leq 0.01$, BLSD = 1.07.)

for CH4 consumption is suggested by the non-significant but apparent decrease in rates when temperature was raised. A summer high temperature of 11 °C at 30 cm soil depth has been reported for soils in interior Alaska (Lewis and Cullum, 1986). Therefore, at these deeper depths, a cold adapted microbial population capable of utilizing CH_4 would be expected to predominate and be active.

Methane uptake rates, in the surface to 10 cm depth cores from the hill, did not increase when temperature was raised from 15 to 25 °C. Soil cores were collected in late September when snow was on the ground. Moisture content in the top 10 cm was 320 g H_2O kg^{-1} soil and significantly higher ($P \leq 0.001$) than those of the lower depths (111-117 g kg^{-1}). Methane consumption was likely not limited by temperature but restricted by high water content. Diffusion of CH_4 in saturated soil with 410 g water kg^{-1} soil was found to be 10^4-fold slower than gaseous diffusion in moist soil with 110 g water kg^{-1} soil (Whalen et al., 1990). A nonsignificant response of CH_4 uptake rates in temperate forest soils to increases in temperature was attributed to limitation of CH_4 transport in soil gas phase (King and Adamsen, 1992).

Soil water content in July (Figure 4), was significantly ($P \leq 0.05$) lower in the fertilized plots of both sites than in unfertilized plots (mean value of 130 versus 160 g H_2O kg^{-1} soil in the hill and 170 versus 230 g H_2O kg^{-1} soil in the flat). This is probably due to increased plant growth, hence increased water uptake from the soil. Additionally, highest water content in the flat was in the top 30 cm soil while in the hillside, it was in samples from the 30-60 cm soil depth. In September, water content was highest (390 g H_2O kg^{-1} soil) in the 0-10 cm soil cores collected from both sites. This may have limited CH_4 and oxygen exchange with atmosphere possibly allowing some CH_4 production to occur in the soil, hence decreased net CH_4 consumption rates in September compared to July rates (Compare Figure 1 with Figure 2).

B. Field Soil Depth CH_4 Distributions

In situ CH_4 concentrations (average of 15 weekly measurements in soil gaseous phase, from June to September 17) decreased with depth at both sites (Figure 5). Surface soil CH_4 concentrations were about 1.80 $\mu L\ L^{-1}$ (ppm) reflecting ambient atmospheric concentration. Soil CH_4 concentrations ranged from 1.43 $\mu L\ L^{-1}$ at 5 cm depth to 0.62 $\mu L\ L^{-1}$ at 50 cm.

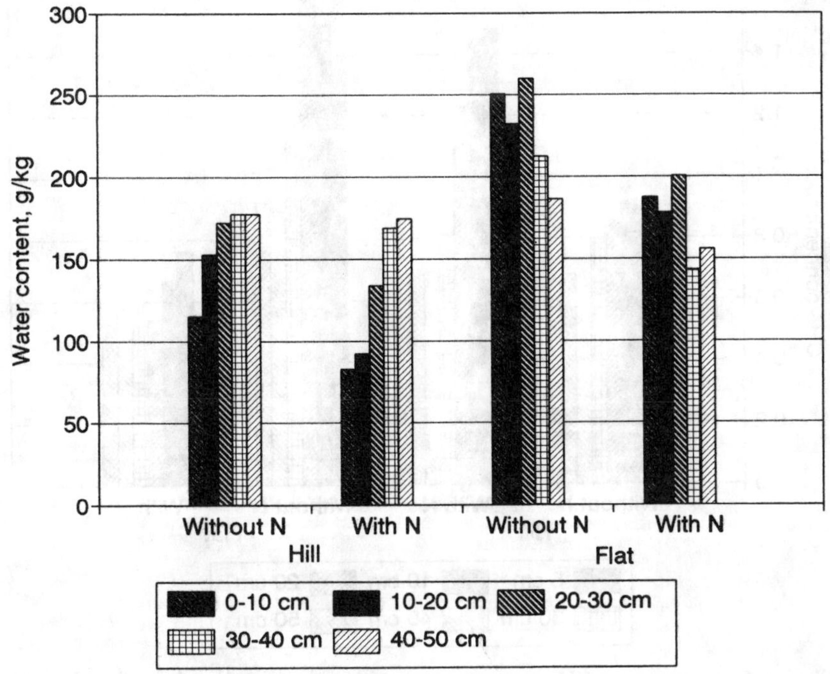

Figure 4. Soil water content of soil cores collected in early July in unfertilized and fertilized plots at the hill and flat sites. (Site x Depth interaction, $P \leq 0.0001$, BLSD = 2.81.)

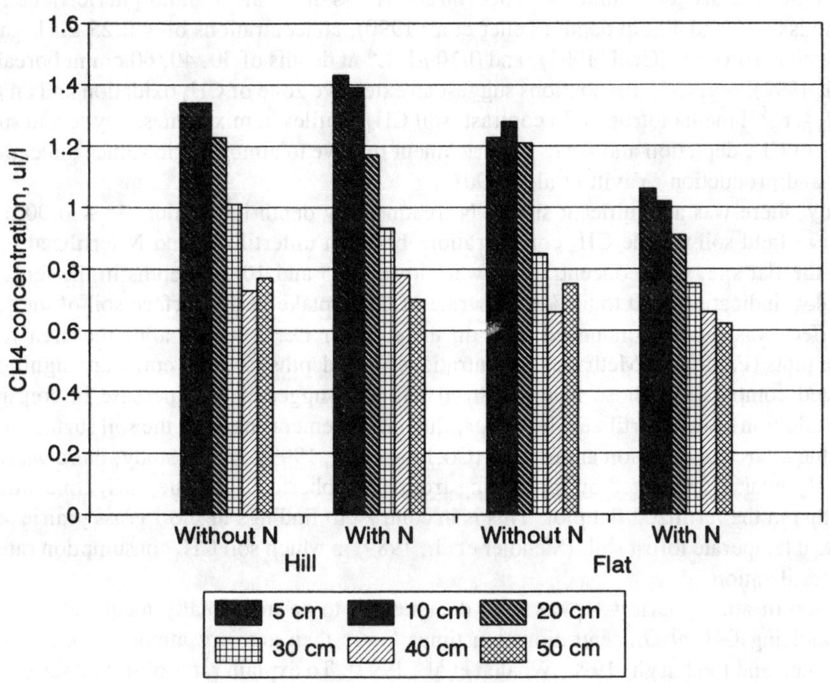

Figure 5. *In situ* summer soil CH_4 concentration in unfertilized and fertilized plots. (Site x Treatment x Depth interaction, $P \leq 0.0005$, BLSD = 0.13.)

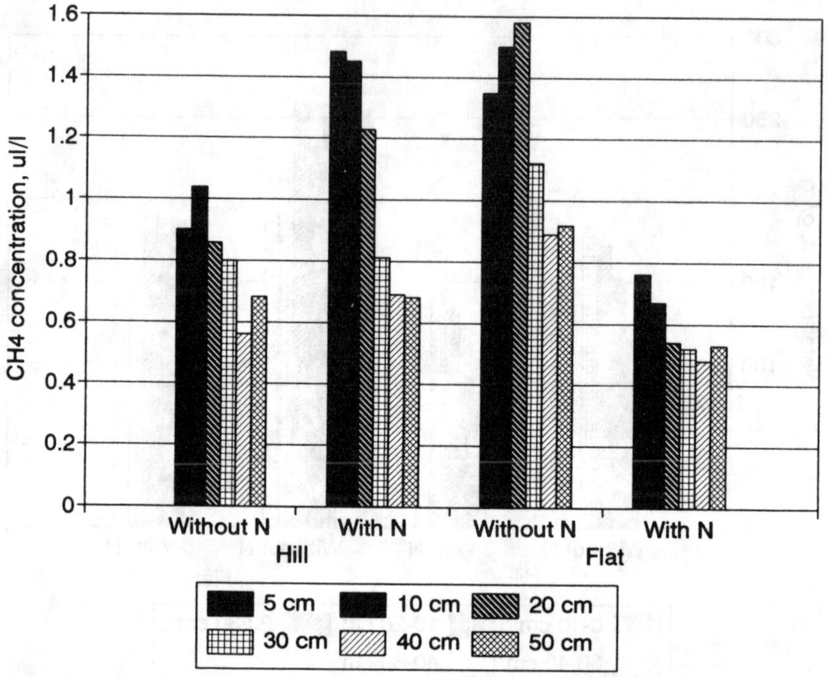

Figure 6. *In sutu* soil CH_4 concentrations in unfertilized and fertilized plots two weeks after fertilizer application. (Site x Treatment x Depth interaction, $P \leq 0.005$, BLSD = 0.13.)

Similarly, other researchers found CH_4 concentrations less than half of atmospheric value in forest and agricultural soils at 20 and 40 cm depth (Keller et al. 1990), concentrations of $<0.25\ \mu L\ L^{-1}$ at 15 cm soil depth in a mixed hardwood, (Crill, 1991), and 0.10 $\mu L\ L^{-1}$ at depths of 30, 40, 60 cm in boreal forest soils (Whalen et al. 1992). Soil CH_4 distributions suggest an extensive zone of CH_4 oxidation and an atmospheric source of CH_4 for soil methanotrophs. In contrast, soil CH_4 profiles in mixed mesophytic and spruce forests showed zones of CH_4 depletion and zones of enrichment relative to atmospheric values indicating both CH_4 consumption and production (Yavitt et al., 1990).

In our study, there was a significant site by N treatment by depth interaction ($P \leq 0.005$). There was no difference in field soil profile CH_4 concentrations between unfertilized and N fertilized hillside plots. However, in the flat site, CH_4 concentrations were lower at 5 and 10 cm depths in the fertilized than in unfertilized plots indicating a 30 to 50% higher rate of CH_4 uptake in the surface soil of the fertilized flat plots. This effect was more pronounced early in the summer i.e. 2 weeks after the urea fertilizer was applied to the plots (Figure 6). Methane concentrations at all depths (5 to 50 cm) were significantly lower in the fertilized compared to those in unfertilized flat plot suggesting an increase in consumption rate following the addition of urea fertilizer. However, flux measurements made at the soil surface did not detect a difference due to urea application at this time (Cochran et al., 1993). In this study, there was a 2 to 5-fold increase in CH_4 uptake (calculated by subtracting from atmospheric CH_4 value, CH_4 concentrations at the different depths) in the fertilized flat plot. This is in contrast to findings in short grass prairie soils (Mosier et al., 1991) and temperate forest soils (Steudler et al., 1989) in which soil CH_4 consumption rates decreased following N fertilization.

Soil oxidation of atmospheric CH_4 has been documented to be microbially mediated and a population capable of oxidizing CH_4 at concentrations ten times lower than ambient atmospheric concentrations is indicated (Whalen and Reeburgh, 1990; Whalen et al., 1992). To explain rates observed for the flux of CH_4 from the atmosphere into soil and/or the high consumption rates in the soil profile, it has been proposed (Conrad, 1984) that microorganisms must not only have high affinities for CH_4 (low K_m) but should also be present in relatively high numbers in the soil. Studies have been done mainly on methanotrophic bacteria with low-affinity for CH_4 activity characterized by high K_m (1290-680 ppm in air), high V_{max} and threshold values (Megraw and Knowles, 1987a; Megraw and Knowles, 1987b; King, 1990; Whalen et al., 1990;

Bender and Conrad, 1992). Current methods used for enumerating and isolating methanotrophic organisms favor these microorganisms with low affinity activity (exposed to high CH_4 concentrations) rather than those with high affinity activity (exposed to ambient air CH_4 concentrations) (Bender and Conrad, 1992), which probably reflect the difficulty in isolating and counting the latter microorganisms.

IV. Summary

In this study, differences in response of CH_4 consumption rates in soils with contrasting history of N fertilization, (the hillside with no known fertilization while the flat site was fertilized yearly for the past 10 or more years) may reflect a difference in microbial populations capable of oxidizing CH_4. These populations may be affected differently by ammoniacal N fertilizer application. In the flat site it is reasonable to expect a dense and active nitrifying microbial population to have been enriched by the long history of yearly applications of ammoniacal fertilizers. Nitrifiers have been reported to oxidize methane (Ward, 1987) and ammonia-oxidizing bacteria have been suggested to be significant methane oxidizers (Bedard and Knowles, 1989). Addition of N fertilizer to the flat site in the spring may have increased microbial biomass of nitrifiers which in turn could have consumed more CH_4. Alternatively, N fertilizer addition may have increased a population of methanotrophs (not necessarily nitrifiers) adapted to high soil N concentrations.

On the other hand, in the hillside with no known N fertilization history, a different population of methanotrophs probably existed whose CH_4-oxidizing activity was affected adversely by ammonia, an inhibitor of methane monooxygenase (Bedard and Knowles, 1989), soon after the ammoniacal fertilizer was applied. The inhibitory effect as measured by surface flux was of a very short duration (Cochran et al., 1993).

V. Conclusions

The results of this study indicate that there was no longterm inhibition of CH_4 oxidation by a one-time application of urea fertilizer to subarctic soils planted to grass. Potential CH_4 consumption rates were highest in the surface to 30 cm soil depth and appeared to be influenced by soil moisture content. The effect of increasing soil temperature on CH_4 uptake rates varied with soil depth, strongly suggesting the presence of microbial populations with different temperature optimum for CH_4 oxidizing activity.

More studies are required to better understand the basic microbiology and controls of CH_4 oxidation in subarctic terrestrial and other ecosystems to predict future trends in atmospheric oxidation by soils.

Acknowledgement

We thank S.K. Pike and W.M. Karidis for technical assistance with this project.

V. References

Bedard, C. and R. Knowles. 1989. Physiology, biochemistry and specific inhibitors of CH_4, NH_4 and CO oxidation by methanotrophs and nitrifiers. *Microbiol. Rev.* 53:68-84.

Bender, M. and R. Conrad. 1992. Kinetics of CH_4 oxidation in oxic soils exposed to ambient air or high CH_4 mixing ratios. *FEMS Microbiol. Ecol.* 101:261-270.

Born, M., H. Dorr, and L. Ingeborg. 1990 Methane consumption in aerated soils of the temperate zone. *Tellus* 42(B):68-84.

Cochran, V.L., S.F. Schlentner, and A.R. Mosier. 1995. CH_4 and N_2O flux in subarctic agricultural soils. p. 179-186. In: R. Lal, J. Kimble, E. Levine, and B.A. Stewart. *Soil Management and Greenhouse Effect*, Advances in Soil Science, CRC Press, Inc., Boca Raton, FL.

Conrad, R. 1984. Capacity of aerobic microorganisms to utilize and grow on atmospheric trace gases (H2, CO, CH_4). p. 461-467. In: M.J. Klug and C.A. Reddy (eds.), *Current perspectives in microbial ecology*. American Society for Microbiol. Washington, D.C.

Crill, P.M. 1991. Seasonal patterns of methane uptake and carbon dioxide release by a temperate woodland soil. *Global Biogeochem. Cycles* 5:319-334.

Crutzen, P.J. 1991. Methane's sinks and sources. *Nature* 350:380-381.

Cullum, R.F. and C.E. Lewis. 1986. Conservation tillage and residue management systems for interior Alaska: An update. p. 346-367. In: H. Steppuhn, and W. Nicholaichuk (eds.), *Proceedings of the Symposium on Snow Management for Agriculture*. Great Plains Agricultural Council Publication No. 120. University of Nebraska, NE.

Gomez, K.A. and A.A. Gomez. 1984. *Statistical procedures for agricultural research*. 2nd Ed. J. Wiley & Sons, NY.

Guthrie, P.D. 1986. Biological methanogenesis and the CO_2 greenhouse effect. *Journal of Geophys. Research* 91 (D10):10847-10851.

Hameed, S. and R.D. Cess. 1983. Impact of a global warming on biospheric sources of methane and its climatic consequences. *Tellus* 35B:1-7.

Hobbie, J.E. and J.M. Melillo. 1984. Role of microbes in global carbon cycling. p. 389-393. In: M.J. Klug and C.A. Reddy (eds.), *Current perspectives in microbial ecology*. American Society for Microbiol. Washington, D.C.

Houghton, J.T., G.J. Jenkins, and J.J. Ephramus, (eds.). 1990. *Climate change, The IPCC scientific assessment*. Cambridge University Press, NY.

Keller, M., M.E. Mitre, and R.F. Stallard. 1990. Consumption of atmospheric methane in soils of central Panama. *Global Biogeochem. Cycles* 4:21-27.

Khalil, M.A.K. and R.A. Rasmussen. 1989. Climate-induced feedbacks for the global cycles of methane and nitrous oxide. *Tellus*. 41B:554-559.

King, G. 1990. Dynamics and controls of methane oxidation in a Danish wetland sediment. *FEMS Microbiol. Ecol.* 74:309-324.

King, G.M. and A.P.S. Adamsen. 1992. Effects of temperature on methane consumption in a forest soil and in pure cultures of the methanotroph *Methylomonas rubra*. *Appl. and Environ. Microbiol.* 58:2758-2763.

Lashof, D.A. 1989. The dynamic greenhouse: Feedback processes that may influence future concentrations of atmospheric trace gases and climatic change. *Climatic Change* 14:213-242.

Megraw, S.R. and R. Knowles. 1987a. Active methanotrophs suppress nitrification in a humisol. *Biol. Fertil. Soils* 4:205-212.

Megraw, S.R. and R. Knowles. 1987b. Methane production and consumption in a cultivated humisol. *Biol. Fertil. Soils* 5:56-60.

Mosier, A., D. Schimel, D. Valentine, K. Bronson, and W. Parton. 1991. Methane and nitrous oxide fluxes in native, fertilized and cultivated grasslands. *Nature* 350:330-332.

Nesbit S.P. and G.A. Breitenbeck. 1992. A laboratory study of factors influencing methane uptake by soils. *Agric. Ecosystems and Environment* 41: 39-54.

Peterson, R.G. 1985. *Design and analysis of experiments*. Dekker, NY.

Steudler, P.A., R.D. Bowden, J.M. Melillo, and J.D. Aber. 1989. Influence of nitrogen fertilization on methane uptake in temperate forest soils. *Nature* 341:314-316.

Ward, B.B. 1987. Kinetic studies on ammonia and methane oxidation by *Nitrosococcus oceanus*. *Arch. Microbiol.* 147:126-225.

Whalen, S.C. and W.S. Reeburgh. 1990. Consumption of atmospheric methane by tundra soils. *Nature* 346:160-162.

Whalen, S.C., W.S. Reeburgh, and V.A. Barber. 1992. Oxidation of methane in boreal forest soils: A comparison of seven measures. *Biogeochem.* 16:181-211.

Whalen, S.C., W.S. Reeburgh, and K.A. Sandbeck. 1990. Rapid methane oxidation in a landfill cover soil. *Appl. Environ. Microbiol.* 56:3405-3411.

Yavitt, J.B., D.M. Downey, G.E. Lang, and A.J. Sextone. 1990. Methane consumption in two temperate forest soils. *Biogeochem* 9:39-52.

CHAPTER 18

Wetlands and Global Change

William J. Mitsch and Xinyuan Wu

I. Introduction

The role of wetlands in the landscape, particularly in providing functions that are valuable to humans, has been extensively reported (see, e.g., Mitsch and Gosselink, 1993). These values include the role of wetlands at the ecological population level where wetlands have been long recognized as preservers of biodiversity and havens for commercially valuable furbearers, waterfowl, and timber. At the ecosystem level, wetlands have become recognized for their role in preventing flooding and droughts and improving water quality. More recently, researchers have begun to look at wetlands on the global scale as perhaps providing a role in the cycling of chemicals such as carbon, sulfur, and nitrogen throughout the biosphere. It is clear that wetlands exchange all of these chemicals with the atmosphere in varying amounts; what is not clear is how much that exchange is on a global scale or how important wetlands might be in both the generation and retention of the so-called "greenhouse gases."

Thus, wetlands may play a key and pivotal role in the stabilization of climate. The enormous volume of peat deposits in the world's wetlands has the potential to contribute significantly to atmospheric carbon dioxide, depending on the balance between draining and oxidation of the peat deposits and their formation in active wetlands. Wetlands are a recognized source of methane, considered a greenhouse gas and a possibly important gas in the stratosphere. The high productivity of many wetland systems suggests that wetlands may also be a major sink for carbon, turning the gases into long-stored peat. Indeed, the carbon that we now are reintroducing to the atmosphere by burning fossil fuels was stored for millennia in the peats of swamps, marshes, and marine sediments of the Carboniferous Period.

Humans are not passive players in the recent process. There are indications that the global carbon balance of wetlands has shifted due to agricultural drainage of wetlands throughout the world and combustion of peat mined from peatlands in Russia and other European countries. Wetlands, as with tropical rain forests, may have shifted from being a net sink to a source of carbon to the atmosphere during the past 100 years.

Wetlands are also affected by the potential changes that would result from climatic change. Sea level rise and temperature increases will change the hydrologic characteristics of both coastal and inland wetlands. These effects also include some possible feedbacks, both positive and negative. Global change causes wetlands to change, which in turn affects their role in the global carbon budget.

Finally, while we usually consider management options inappropriate in the face of changing climate, experiments in rice paddies and some approaches to coastal wetland management may offer some alternatives to minimize both the negative role of some wetlands in global change and the effects that global change has on our coastal wetlands. This is the essence of ecological engineering, allowing us to bend but not break in the face of major environmental change while utilizing nature's energy sources.

This paper will review the role of wetlands in the global carbon cycle and global change, outlining the global extent of wetlands, the significant carbon processes in wetlands, particularly methanogenesis, and the estimates on a global scale. The paper will then review the effects that global climate change may have on

ISBN 1-56670-117-1/95/$0.00+.50

wetlands and discuss the possible feedbacks. Finally, we will review some management options that may minimize emissions of methane and protect coastal wetlands in the advance of sea level rise.

II. Present Role of Wetlands

A. Global Extent and Distribution of Wetlands

Wetlands include the swamps, bogs, marshes, mires, fens, and other wet ecosystems found throughout the world under many names. They are ubiquitous, found on every continent except Antarctica and in every clime from the tropics to the tundra (Figure 1). Wetlands are a very diverse group, with each wetland type having its own structure and function. It is much easier to generalize about oceans, lakes, or even forests and their role in global carbon budgets, despite regional differences, than it is to generalize about wetlands. Wetland types are decidedly different depending of their peat accumulation, nutrient cycling, and, most important, hydrology.

While any estimate of wetland extent in the world is difficult and depends on an accurate definition, one of the earliest approximations of wetland coverage in the world was made by the Russian scientists Bazilivich et al. (1971) and interpreted by Maltby and Turner (1983) that over 6% of the land surface of the world, or 8.56 million km^2, is wetland (Figure 2a). Wetlands are found in arid regions as inland salt flats; in humid, cool regions as bogs, fens and tundra; along rivers and streams as riparian wetland forests and back swamps; and along temperate, subtropical, and tropical coastlines as salt marshes and mangrove swamps. Almost 56% of the total wetland area in the world by this estimate is found in tropical and subtropical regions. Boreal regions account for 30 percent of the wetlands and temperate wetlands only 12 percent.

Using global digital data bases at 1° resolution, Matthews and Fung (1987) of the NASA Goddard Space Flight Center, estimated 5.26 million km^2 of wetlands in 5 classes: (1) forested bogs, (2) non-forested bogs, (3) forested swamps, (4) non-forested swamps, and (5) alluvial formations. They found that about half of the wetland area was in the boreal region between 50-70°N, and 90 percent of wetlands in this region were defined by them as bogs (Matthews, 1990a). Tropical and subtropical wetlands made up a total of about 30 percent of the world's wetlands in the Matthews and Fung study.

Aselmann and Crutzen (1989, 1990) from Max Planck Institute in Germany, estimated a global wetland area of 5.57 million km^2 from regional wetland surveys and monographs (Figure 2c) rather from maps as used by Matthews and Fung. Their wetland categories included (1) bogs, (2) fens, (3) swamps, (4), marshes, and (5) floodplains. They found that what they called bogs and fens accounted for about 60 percent of the world's wetlands (3.35 km^2), very close to the estimated 3.46 million km^2 of northern boreal and subarctic peatlands estimated by Gorham (1991). Yet Aselmann and Crutzen (1989) describe bogs and fens as also occurring in temperate latitudes (40-50° N) and in tropical latitudes. Both Matthews and Fung (1987) and Aselmann and Crutzen (1989) show a much lower percentage of wetlands in tropical and subtropical regions than did the earlier Maltby and Turner (1983) estimate. Their definitions of zones also differ. Each research group estimated coverage by rice paddies but did not included it in their total. Matthews et al. (1991) estimated 1.5 million km^2 while Aselmann and Crutzen (1989) estimated 1.3 million km^2 of rice paddies.

B. Carbon Transformations in Wetlands

While biodegradation of organic matter by aerobic respiration is limited by the reduced conditions in wetland soils, several anaerobic processes can degrade organic carbon. The major processes of carbon transformation under aerobic and anaerobic conditions are shown in Figure 3.

1. Fermentation

Fermentation of organic matter, which occurs when organic matter itself is the terminal electron acceptor in anaerobic respiration by microorganisms, forms various low molecular weight acids and alcohols, and CO_2 such as lactic acid

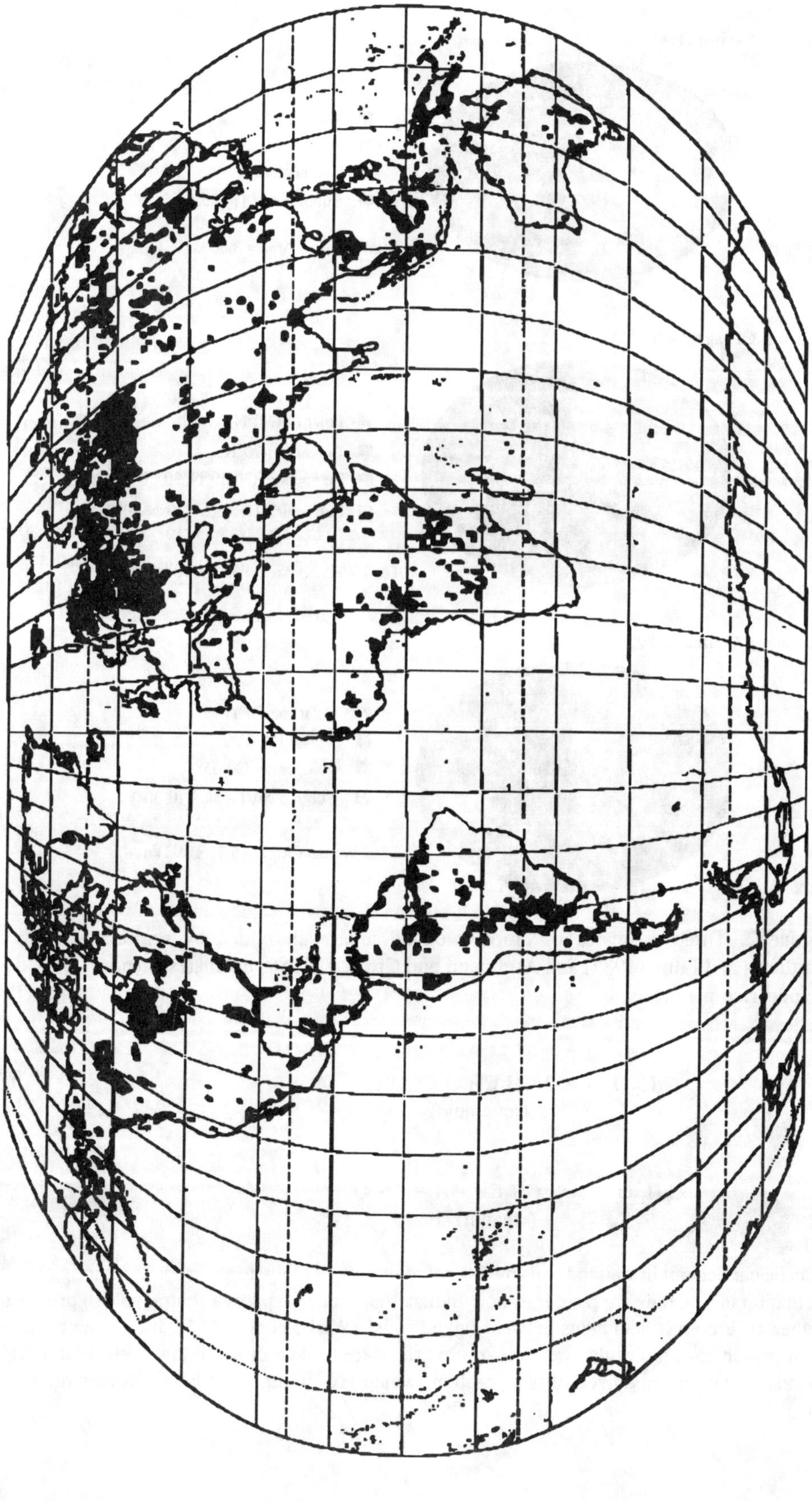

Figure 1. Approximate global extent of wetlands, based on a map of forested wetlands prepared by Matthews (1990b) and published in Lugo et al. (1990). Copyright© 1990 Elsevier Science Publishers, BV, Amsterdam, (Reprinted with permission.)

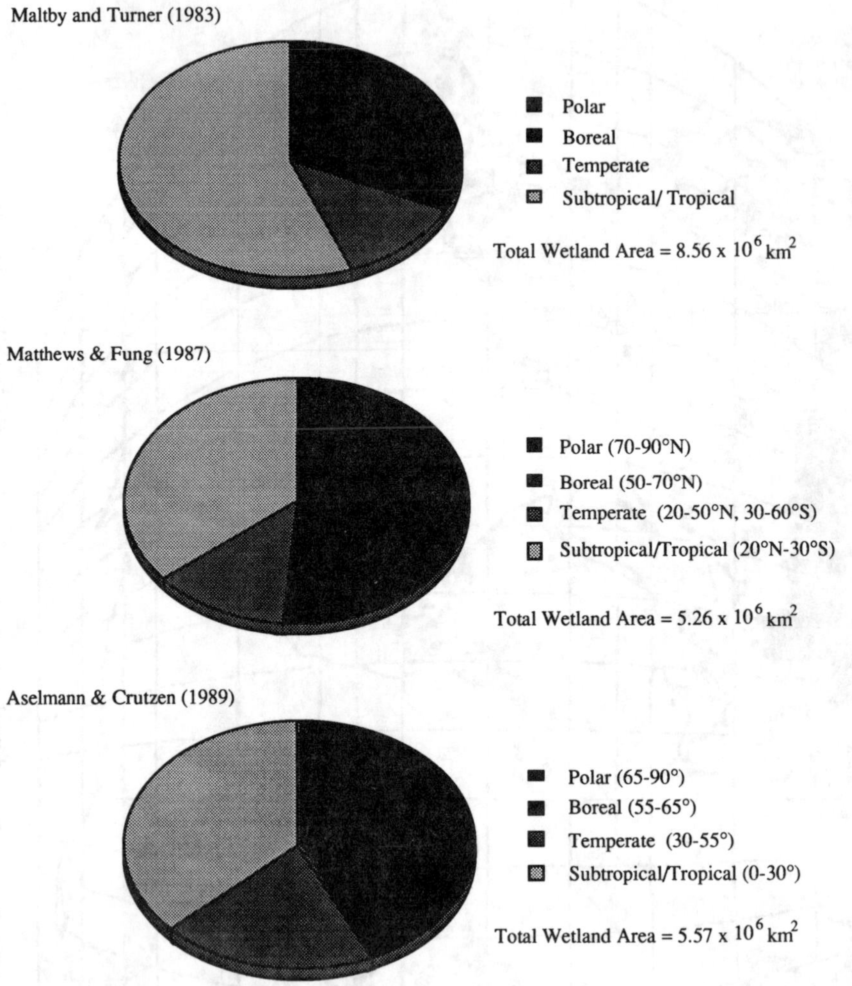

Figure 2. Three estimates of global wetland area, subdivided by region. Matthews and Fung (1987) and Aselmann and Crutzen (1989) estimates do not include rice paddies.

$$C_6H_{12}O_6 \dashrightarrow 2\ CH_3CH_2O\ COOH \quad \text{(lactic acid)} \tag{1}$$

or ethanol

$$C_6H_{12}O_6 \dashrightarrow 2\ CH_3CH_2OH + 2\ CO_2. \quad \text{(ethanol)} \tag{2}$$

Fermentation can be carried out in wetland soils by either facultative or obligate anaerobes. Although *in situ* studies of fermentation in wetlands are rare, it is thought that "fermentation plays a central role in providing substrates for other anaerobes in sediments in waterlogged soils" (Wiebe et al., 1981). It represents one of the major ways in which high molecular weight carbohydrates are broken down to low molecular weight organic compounds, usually as dissolved organic carbon, which are, in turn, available to other microbes (Valiela, 1984).

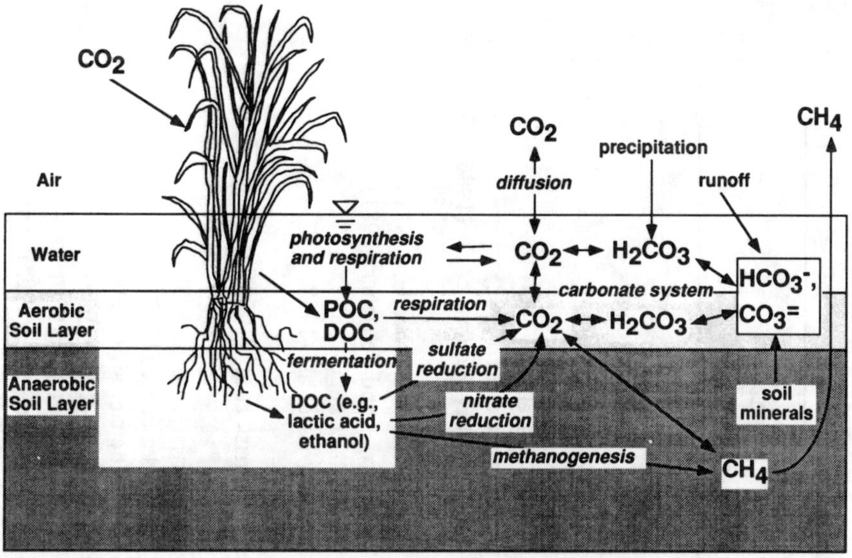

Figure 3. Major carbon transformations in wetlands. (From Mitsch and Gosselink, 1993; reprinted with permission.)

2. Methanogenesis

Methanogenesis occurs when certain bacteria (methanogens) uses CO_2 or the organic matter itself as the terminal electron acceptor at about -250 mv, producing low molecular weight organic compounds and methane gas, as for example:

$$CO_2 + 8\,H^+ \dashrightarrow CH_4 + 2\,H_2O \tag{3}$$

or alternatively, a low weight organic compound such as one from a methyl group:

$$CH_3COO^- + 4\,H_2 \dashrightarrow 2\,CH_4 + 2\,H_2O. \tag{4}$$

Methane, which can be released to the atmosphere when sediments are disturbed, is often referred to as "swamp gas" or "marsh gas." Methane production requires extremely reduced conditions, with a redox potential between -250 and -350 mv, after other terminal electron acceptors (O_2, NO_3^-, and $SO_4^=$) have been reduced.

The rates of methanogenesis from both saltwater wetlands and freshwater wetlands, as well as from domestic wetlands such as rice paddies, have a considerable range (Table 1). The number in Table 1 represent independent measurements of methane generations from 139 sites (or treatments) in boreal, tropical, and temperate regions of the world as reported in 52 published papers. Comparison of rates of methane production from different studies is difficult, because different methods are used and because the rates depend on both temperature and hydroperiod. Methanogenesis is seasonal in temperate zone wetlands. Harriss et al. (1982) noted maximum methane production in a Virginia freshwater swamp in April-May, while Wiebe et al. (1981) found methane production to generally peak in late summer in a Georgia salt marsh. Furthermore, Harriss et al. (1982) measured a net uptake of methane by the wetland during a drought when the wetland soil was exposed to the atmosphere. Researchers in boreal wetlands have found beaver ponds to have much higher methane flux rates than other wetland types (Naiman et al., 1991; Roulet et al., 1992a).

We summarized the data in Table 1 by major wetland type and climate in Table 2, illustrating the range of the means for the various studies. Our review suggests that most of the studies to date have been in

Table 1. Methane emission rates for various freshwater and saltwater wetlands and rice paddies

Type of Wetland	Methane emission, mg C m^{-2} day^{-1}		Sampling season	Reference
	Mean (±SE)	Range		
Freshwater wetlands				
Amazon floodplain, Brazil				
flooded forest	83 ± 13	1-570	Jul-Aug (85)	Devol et al. (1988)
macrophyte	443 ± 20	1-3900	Jul-Aug (85)	Devol et al. (1988)
Central Amazonian floodplain, Brazil				
flooded forest	144 ± 20	0-918	Jul-Sep (85)	Bartlett et al. (1988)
grass mats	173 ± 54	0-2248	Jul-Sep (85)	Bartlett et al. (1988)
Reed wetland, Vercelli, Italy		18	May-Sep (84)	Holzapfel-Pschorn et al. (1986)
Freshwater wetlands, Florida				
cypress dome, fertilized; Gainsville	726		Summer	Harriss and Sebacher (1981)
Corkscrew Swamp	50 ± 39		Summer	Harriss and Sebacher (1981)
Corkscrew Swamp	96 ± 59			Crill et al. (1988)
wet prairies and sawgrass marsh, S. Florida	46 ± 13		~Year-round (80-85)	Harriss et al. (1988)
swamp forest, S. Florida	44 ± 13		Jan, Feb, Jun, Dec (80, 84,85)	Harriss et al. (1988)
impoundments and disturbed	56 ± 8		Jan-Feb (84-85)	Harriss et al. (1988)
wetlands, S.Florida Everglades		1-1950	Mar,Jun,Aug,Nov (85-87)	Burke et al. (1988)
sawgrass marsh, >1m, Everglades	29 ± 2		Winter (84-85)	Bartlett et al. (1989)
sawgrass marsh, >1m, Everglades	54 ± 17		Winter (84-85)	Bartlett et al. (1989)
swamp forest, Everglades	52 ± 17		Winter (84-85)	Bartlett et al. (1989)
Tidal freshwater marsh, Louisiana		440	Year round (80-81)	DeLaune et al. (1983), Smith et al. (1982)

Table 1. -- continued

Type of Wetland	Methane emission, mg C m^{-2} day^{-1} Mean (±SE)	Range	Sampling season	Reference
Four Holes Swamp, South Carolina	7.4 ± 2.6		Summer	Harriss and Sebacher (1981)
Creeping Swamp, North Carolina	54	0-180	Year-round	Mulholland (1981)
Freshwater wetlands, Georgia				
Okeefenokee Swamp	69 ± 50		Summer	Harriss and Sebacher (1981)
Okeefenokee Swamp	106 ± 31			Crill et al. (1988)
swales and sloughs, Ogeechee floodplain	64	-14-256	Year-round (87-89)	Pulliam (1993)
backswamp, Ogeechee floodplain	19	-14-248	Year-round (87-89)	Pulliam (1993)
Swamps, Virginia				
waterlogged condition, Dismal Swamp		1-15	Year-round (80-81)	Harriss et al. (1982)
drought condition, Dismal Swamp		-1-4	Year-round (80-81)	Harriss et al. (1982)
constantly flooded, Newport News Swamp	98	0-754	Year-round(85-86)	Wilson et al. (1989)
bank site, Newport News Swamp	88	0-356	Year-round (85-86)	Wilson et al. (1989)
Mountain bogs, West Virginia	188 ± 59			Crill et al. (1988)
Subalpine/alpine marshes, Colorado				
Lake Dillon	0.1 ± 0.1		Jun-Oct (89-90)	Smith and Lewis (1992)
Long Lake	10 ± 2.3		Jun-Oct (89-90)	Smith and Lewis (1992)
Pass Lake	1.2 ± 0.5		Jul-Oct (89-90)	Smith and Lewis (1992)
Red Rock Lake	66 ± 6.6		Jun-Oct (89-90)	Smith and Lewis (1992)
Rainbow Lake	29 ± 4.2		Jun-Oct (89-90)	Smith and Lewis (1992)
Swamp, Michigan	255	90-458	Jul-Sep (75)	Baker-Blocker et al. (1977)

Table 1. -- continued

Type of Wetland	Methane emission, mg C m^{-2} day^{-1} Mean (±SE)	Range	Sampling season	Reference
Northern peatlands, Minnesota				
bog	86	14-351	Aug (83)	Harriss et al. (1985)
fen	3	2-4	Aug (83)	Harriss et al. (1985)
fen and bog	314	45-1457	Aug (83)	Harriss et al. (1985)
sedge meadow	498		Aug (83)	Harriss et al. (1985)
wild rice bed	370	95-662	Aug (83)	Harriss et al. (1985)
shoreline fen	124	119-128	Aug (83)	Harriss et al. (1985)
forest bog	58 ± 16	8-521	Jun, Aug (86)	Crill et al. (1988)
forest fen	107 ± 14	51-197	Jun, Aug (86)	Crill et al. (1988)
open bog	221 ± 23	14-650	Jun, Aug (86)	Crill et al. (1988)
circumneutral fen	244 ± 23	114-533	Jun, Aug (86)	Crill et al. (1988)
acid fen	77 ± 10		Jun, Aug (86)	Crill et al. (1988)
permanently wetted zone		22-30	~Year-round	Naiman et al. (1991)
occasionally inundated zone		0.6-1.1	~Year-round	Naiman et al. (1991)
forest bog	20	-1-362	Nov-Mar (88-90)	Dise (1992)
open bog	89	0-792	Nov-Mar (88-90)	Dise (1992)
junction bog	135	8-575	Nov-Mar (88-90)	Dise (1992)
forest bog	67		Jun-Sep (88-90)	Kelly et al. (1992)
junction fen	280		Jun-Sep (88-90)	Kelly et al. (1992)
Boreal wetland, Alaska				
wet-meadow tundra	54 ± 4.6		Jul (91)	Torn and Chapin (1993)
wet-meadow tundra, darkened	75 ± 14		Jul (91)	Torn and Chapin (1993)
wet tussock tundra	1 ± 0.8		Jul (91)	Torn and Chapin (1993)

Table 1. -- continued

Type of Wetland	Methane emission, mg C m^{-2} day^{-1}		Sampling season	Reference
	Mean (±SE)	Range		
wet tussock tundra, graminoids removed	0.4 ± 0.2		Jul (91)	Torn and Chapin (1993)
wet tundra at Prudhoe Bay, flooded	700		May-Aug (90-91)	Vourlitis et al. (1993)
wet tundra, Prud. Bay, well drained	100		May-Aug (90-91)	Vourlitis et al. (1993)
wet tundra at APL 133-3, flooded	1500		Jun-Aug (90-91)	Vourlitis et al. (1993)
wet tundra at APL 133-3, moist	1100		Jun-Aug (90-91)	Vourlitis et al. (1993)
wet tundra at APL 133-3, dry	200		Jun-Aug (90-91)	Vourlitis et al. (1993)
alpine fen	217 ± 11		Aug (84)	Sebacher et al. (1986)
boreal marsh	80 ± 3.5		Aug (84)	Sebacher et al. (1986)
meadow tundra	30 ± 15		Aug (84)	Sebacher et al. (1986)
moist tundra	3.7 ± 2		Aug (84)	Sebacher et al. (1986)
wet coastal tundra	89 ± 21		Aug (84)	Sebacher et al. (1986)
fen and marsh	147 ± 21			Crill et al. (1988)
Boreal wetlands, Canada				
a transitional fen	30		Jun-Aug (84)	Moore and Knowles (1987)
a very rich fen	14		Jun-Aug (84)	Moore and Knowles (1987)
a transitional fen	14		Jun-aug (84)	Moore and Knowles (1987)
a rich fen	35		Jun-Aug (84)	Moore and Knowles (1987)
patterened fen	26		Jun-Aug (89)	Moore et al. (1990)
open fen	45		Jun-Sep (89)	Moore et al. (1990)
open fen-vegetated area	59		Jul-Aug (90)	Whiting and Chanton (1992)
conifer swamps		<5	May-Oct (90)	Roulet et al. (1992a)
beaver ponds		23-68	May-Oct (90)	Roulet et al. (1992a)

Table 1. -- continued

Type of Wetland	Methane emission, mg C m^{-2} day^{-1} Mean (±SE)	Range	Sampling season	Reference
thicket swamps		0-66	May-Oct (90)	Roulet et al. (1992a)
bogs		5-16	May-Oct (90)	Roulet et al. (1992b)
Subarctic mire, Sweden				
ombrotrophic-hummocks	0.6	0.3-1.8	Jun, Jul, Sep (74)	Svenson and Rosswall (1984)
ombrotrophic-between hummocks	2.0	1.5-3	Jun, Jul, Sep (74)	Svenson and Rosswall (1984)
ombrotrophic-shallow depressions	14	2.2-22	Jun, Jul, Sep (74)	Svenson and Rosswall (1984)
ombrotrophic-deep depressions	17	6.2-42	Jun, Jul, Sep (74)	Svenson and Rosswall (1984)
intermediate ombro/minerotrophic	54	6.5-70	Jun, Jul, Sep (74)	Svenson and Rosswall (1984)
minerotrophic	325	41-713	Jun, Jul, Sep (74)	Svenson and Rosswall (1984)
Saltwater wetlands				
Salt marsh and mangrove, Florida				
salt marsh, Panacea	2.5 ± 0.6	188 ± 59	Sep, Nov, Dec (81)	Bartlett et al. (1985)
mangroves, South Florida	3 ± 0.3	-2.4-4.8	Dec (85)	Harriss et al. (1988)
red mangrove, Everglades	3.2 ± 0.3		winter (84-85)	Bartlett et al. (1989)
dwarf red mangrove, Everglades	61 ± 14		Winter (84-85)	Bartlett et al. (1989)
Salt marsh, Louisiana				
salt	12		~Year-round (80-81)	DeLaune et al. (1983), Smith et
brackish	0.8		~Year-round (80-81)	al. (1982) [same for both]
Salt marsh, Georgia				
salt marsh	2.1			Atkinson and Hall (1976),
tall *Spartina*	0.8		Year-round (75-90)	Wiebe et al., (1981), King and
intermediate *Spartina*	29		Year round (75-76)	Wiebe (1978) [same for
short *Spartina*	109		Year-round (75-76)	all three]
salt marsh, Sapelo Island	10.1 ± 3.4	1.7-14.4	Nov (81)	Bartlett et al. (1985)

Table 1. -- continued

Type of Wetland	Methane emission, mg C m^{-2} day^{-1} Mean (±SE)	Range	Sampling season	Reference
Salt marsh, Georgetown, S. Carolina	0.9 ± 0.4	-2.9-7.2	Apr, Aug, Nov (81-82)	Bartlett et al. (1985)
Salt marsh, Virginia				
creek bank, Bay Tree Creek	3.9 ± 0.6	-0.8 12.2	Year-round (81-83)	Bartlett et al. (1985)
short *Spartina*, Bay Tree Creek	2.3 ± 0.5	0-10.4	Year-round (81-83)	Bartlett et al. (1985)
high marsh meadow, B. Tree Creek	1.5 ± 0.5	-2-16	Year-round (81-83)	Bartlett et al. (1985)
salt marsh, Wallops Island	-0.2 ± 0.4	-1.5-1.3	Jun (82)	Bartlett et al. (1985)
high salinity, Queen's Creek	12	0-35	Year-round (83-84)	Bartlett et al. (1987)
intermediate salinity, Queen's Creek	46		Year-round (83-84)	Bartlett et al. (1987)
low salinity, Queen's Creek	37	0-194	Year-round (83-84)	Bartlett et al. (1987)
Salt marsh, Lewes, Delaware	0.4 ± 0.5	-1-2.9	Jun (82)	Bartlett et al. (1985)
Salt marsh, Southern California		0-3.4	~Year-round (79-80)	Cicerone and Shetter (1981)
Rice paddies				
Davis, California				
unfertilized	24	10-51	Late summer (80)	Cicerone and Shetter (1981)
fertilized	113	40-225	Late summer (80)	Cicerone and Shetter (1981)
fertilized	188	1-3705	Jun-Sep (82)	Cicerone et al. (1983)
fertilized	74		Jul-Oct (83)	Cicerone et al. (1992)
no added organic matter	10		May-Oct (85)	Cicerone et al. (1992)
250 g/m^2 added rice straw	90		May-Oct (85)	Cicerone et al. (1992)
500 g/m^2 added rice straw	218		May-Oct (85)	Cicerone et al. (1992)
Beaumont, Texas				
Lake Charles field	157 ± 12		Jun-Sep (89)	Sass et al. (1990)

Table 1. -- continued

Type of Wetland	Methane emission, mg C m^{-2} day^{-1} Mean (±SE)	Range	Sampling season	Reference
Beaumont field	45 ± 13		Jun-Sep (89)	Sass et al. (1990)
Lake Charles field	273 ± 30		Jun-Sep (90)	Sass et al. (1991a)
Beaumont field	101 ± 12		Jun-Sep (90)	Sass et al. (1991a)
no straw incorporated	217-329		Apr-Oct (90)	Sass et al. (1991b)
straw incorporated	228-423		Apr-Sep (91)	Sass et al. (1991b)
normal flood	80 ± 21		Apr-Sep (91)	Sass et al. (1992)
normal flood-midseason drain	42 ± 11		Apr-Sep (91)	Sass et al. (1992)
normal flood-multiple aeration	10 ± 2.6		Apr-Sep (91)	Sass et al. (1992)
late flood	113 ± 20		Apr-Sep (91)	Sass et al. (1992)
Vercelli, Italy				
fertilized and unfertilized	288	2-918	Apr-Sep (83)	Holzapfel-Pschorn and Seiler (1986)
rice	369		Jul, Aug (84)	Holzapfel-Pschorn et al. (1985)
rice	212		May-Sep (84)	Holzapfel-Pschorn et al. (1986)
weeds	149		May-Sep (84)	Holzapfel-Pschorn et al. (1986)
unfertilized	210	120-285	May-Sep (84-86)	Schutz et al. (1989)
Andalusia, Spain	72	36-252	Jul-Oct (82)	Seiler et al. (1984)
Ibaraki, Japan				
alluvial soil	51		May-Sep (88)	Yagi and Minami (1990)
alluvial soil w/rice straw	173		May-Sep (88)	Yagi and Minami (1990)
peaty soil w/rice straw	293		May-Sep (88)	Yagi and Minami (1990)
volcanic ash soil	23		May-Sep (88)	Yagi and Minami (1990)
volcanic ash soil w/rice straw	66		May-Sep (88)	Yagi and Minami (1990)

Table 1. -- continued

Type of Wetland	Methane emission, mg C m^{-2} day^{-1} Mean (±SE)	Range	Sampling season	Reference
Hangzhou, China				
early rice	135	1.8-900		Schutz et al. (1991), Wang et al. (1990) [same for both]
late rice	453	1.8-3600		
Nanjing, China	47-257		Jul-Sep (90)	Chen et al. (1993)
Sichuan, China	486	36-4320		Schutz et al. (1991), Khalil et al. (1990)
Beijing, China	263-880	188±59	Jul-Sep (90)	Chen et al. (1993)
Nadia, India				
with cowdung manure		52-155	Aug (86)	Saha et al. (1989)
Jute-retting tanks, India	2664	14-14022	Aug-Oct (88-91)	Banik et al. (1993)

Table 2. Ranges of mean CH_4 emission rates (number of sites/treatments) for major wetland types

	CH_4 emission rates (mg-C m^{-2} day^{-1})			
	Polar	Boreal	Temperate	Sub-/Tropical
Tundra	-	3.7-1500 (12)	-	-
Bog	-	0.7-17 (5)	20-221 (7)	-
Fen	-	14-325 (11)	3-314 (8)	-
Freshwater marsh	-	23-80 (2)	0.1-498 (17)	29-443 (7)
Forest swamp	-	5-66 (2)	7.4-106 (6)	44-144 (7)
Salt marsh	-	-	0-109 (17)	2.5 (1)
Mangrove	-	-	-	3-61 (3)
Rice paddy	-	-	10-880 (34)	47-486 (9)

northern peatlands (bogs and fens) and freshwater marshes. Many more studies have been undertaken in the temperate zone as compared to studies in boreal or tropical climates. A comparison of methanogenesis between freshwater wetlands (marsh and swamp) and marine wetlands (salt marsh and mangrove) shows that the rate of methane production is higher in the former (up to 500 mg C m^{-2} d^{-1}) compared to the latter (up to 100 mg C m^{-2} d^{-1}), apparently because of the lower amounts of sulfate in freshwater systems. In general, methane production is low in reduced soils if sulfate concentrations are high (Gambrell and Patrick, 1978; DeLaune et al., 1983; Valiela, 1984). Possible reasons for this phenomenon include: (1) competition for substrates that occurs between sulfur and methane bacteria; (2) the inhibitory effects of sulfate or sulfide on methane bacteria; (3) a possible dependence of methane bacteria on products of sulfur-reducing bacteria; or (4) a stable redox potential that does not drop low enough to reduce CO_2 because of an ample supply of sulfate. Evidence suggests that methane may actually be oxidized to CO_2 by sulfate reducers (Valiela, 1984).

The data in Tables 1 and 2 can be compared with generalizations that have been assumed to estimate global methane production from wetlands. Gorham (1991) assumed an average of 77.5 mg C m^{-2} d^{-1} in estimating the global contributions of northern peatlands. Table 2 suggests that the means of several studies ranges from 0.7 to 17 mg C m^{-2} d^{-1} for boreal bogs and a much higher 14 to 325 mg C m^{-2} d^{-1} for the more mineral-rich fens. Thus depending on the mix of bogs and fens in northern peatlands, Gorham's assumption could be low or high. Matthews and Fung (1987) used cross-latitude assumptions to determine the following average emission rates of methane: 150 mg-C m^{-2} d^{-1} for peatlands (called bogs); 53 mg-C m^{-2} d^{-1} for forested swamps; 90 mg-C m^{-2} d^{-1} for marshes (called non-forested swamps); and 23 mg-C m^{-2} d^{-1} for riparian wetlands (called alluvial).

Aselmann and Crutzen (1989), in their review of the literature available at the time, suggested a geometric mean emission rates of 15 to 310 mg CH_4 m^{-2} d^{-1} (11-230 mg C m^{-2} d^{-1}) in increasing order from bogs, fens, swamps, marshes, and rice paddies. They estimated overall emission rates as follows: bogs 11 mg C m^{-2} d^{-1}; fens 60 mg C m^{-2} d^{-1}; swamps 63 mg C m^{-2} d^{-1}; floodplains 75 mg C m^{-2} d^{-1}; marshes 190 mg C m^{-2} d^{-1}; and rice paddies 230 mg C m^{-2} d^{-1}. They fail to make the distinction between freshwater and saltwater marshes and swamps. The data show fens to have much higher emissions in both boreal and temperate regions than forested swamps and marshes (forest swamp and freshwater marsh in Table 2). Methane emissions from freshwater systems are much higher than saltwater systems (salt marsh and mangrove).

Recent estimates suggest that some of the above measurements may have led to estimates of methane emission that are too high. Moore and Roulet (1994) suggest that most annual flux measurements in Canada are less than 10 g CH_4 m^{-2} yr^{-1} with the primary controlling mechanisms being soil temperature, water table position, or a combination of both. Roulet et al. (1992a), in a detailed field study of 12 minerotrophic wetlands and 3 beaver ponds in the low boreal region of central Canada, estimated a habitat-weighted average emission of 1.633 g CH_4 m^{-2} yr^{-1}, which would translate to a daily rate of only 3.35 mg C m^{-2} d^{-1}. If their study is accurate, it suggests that emissions of methane from northern peatlands may be much less than originally estimated.

3. Sulfur Reduction as a Source of Carbon

The sulfur cycle is important in some wetlands for the oxidation of organic carbon. This is particularly true in most coastal wetlands where sulfur is abundant. Sulfur reduction often "competes" with the generation of methane. Sulfur-reducing bacteria require an organic substrate, generally of low molecular weight, as a source of energy in converting sulfate to sulfide. Conveniently, the process of fermentation described above can supply these necessary low molecular weight organic compounds, such as lactate or ethanol (see Equations 1 and 2). Equations of sulfur reduction, also showing the oxidation of organic matter, are as follows:

$$2\ CH_3CHOHCOO^- + SO_4^= + 3\ H^+ \dashrightarrow 2\ CH_3COO^- + 2\ CO_2 + 2\ H_2O + HS^- \qquad (5)$$

and

$$\underset{\text{(acetate)}}{CH_3COO^-} + SO_4^= \dashrightarrow 2\ CO_2 + 2\ H_2O + HS^-. \qquad (6)$$

The importance of this fermentation-sulfur reduction pathway in oxidation of organic carbon to CO_2 in saltwater wetlands was demonstrated for a New England salt marsh by Howarth and Teal (1979, 1980) and Howes et al. (1984, 1985). These studies showed that more than half of the carbon dioxide evolution from salt marshes was due to the fermentation-sulfur reduction pathway, with aerobic respiration accounting for much of the rest. Only a small percent of carbon release was due to methanogenesis in these saltwater systems. A similar study using the same methods in a salt marsh in Georgia yielded sulfate reduction rates that were one-third those measured in the New England salt marsh, explained by the authors as due to less underground organic productivity in the Georgia marsh (Howarth and Giblin, 1983).

There have been few if any studies that have estimated the importance of sulfate reduction on the oxidation of organic carbon in freshwater wetlands. However several studies of freshwater lakes suggest that methanogenesis results in a greater oxidation of organic carbon than does sulfate reduction. Smith and Klug (1981) found that 2.5 times more organic carbon was mineralized by methanogenesis than by sulfate reduction in a lake in Michigan. Ingvorsen and Brock (1982) estimated that four times more carbon was mineralized via methanogenesis than by sulfate reduction in anaerobic lake sediments in Lake Mendota, Wisconsin. Estimates of the rates of these processes in freshwater systems, when compared with those for the salt marsh, support the generalizations that oxidation of organic carbon by methane production is dominant in freshwater wetlands while oxidation of organic carbon by sulfate reduction is dominant in saltwater wetlands (Capone and Kiene, 1988).

4. Peat Accumulation

If wetlands are to serve as sinks of carbon, the uptake of carbon dioxide by macrophytes, mosses, and microbiota, followed by death and partial decay to permanent peat deposits, is the mechanism by which that carbon retention will occur. Bazilivich and Tishkov (1982) presented a detailed breakdown of energy flow and peat accumulation in a mesotrophic bog in the Russia (Figure 4). The bog is a sphagnum-pine community with an estimated gross primary productivity of 490 g C m^{-2} yr^{-1}, with about 60% of that consumed by plant respiration. Mitsch and Gosselink (1993) estimated that about 43 g C m^{-2} yr^{-1} accumulated permanently in the peat after considerable microbial enrichment and some surface oxidation and outflow (Figure 4). Assuming an average density of peat of 0.1 g/cm^3 for peat (Ovenden, 1990) and 50% carbon, this accumulation is equivalent to the 0.86 mm/yr, at the high end of the range of peat accumulation reported for European bogs by Moore and Bellamy (1974) of 0.2 - 0.8 mm/yr but lower than the accumulation of 1 to 2 mm/yr reported by Cameron (1970) for North American bogs. In a detailed analysis of peat accumulation in Canada, Ovenden (1990) estimates that peat accumulation ranges from 0.1 -1.4 mm/yr and a typical range of carbon accumulation of 10 to 35 g C m^{-2} yr^{-1}. Gorham (1991) compiled more recent unpublished data from ^{14}C measurements and suggested an overall accumulation of 0.5 mm/yr as reasonable for northern peatlands; this translates to an accumulation of 29 g C m^{-2} yr^{-1}.

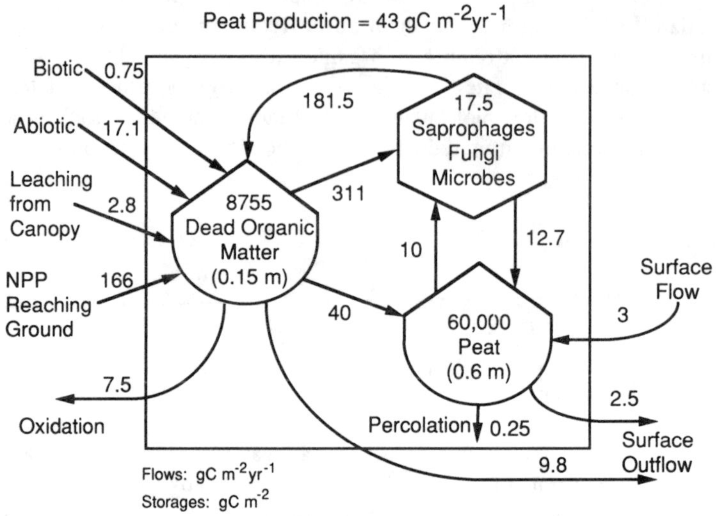

Figure 4. Fluxes of carbon in mesotrophic bog in Russia, based on work of Bazilivich and Tishkov (1982). (From Mitsch and Gosselink, 1993; reprinted with permission.)

C. Wetlands Role in the Global Carbon Cycle

The enormous volume of peat deposits in the world's wetlands has the potential to contribute significantly to worldwide atmospheric carbon dioxide levels, depending on the balance between draining and oxidation of the peat deposits and their formation in active wetlands. An overall global carbon budget, with an emphasis on wetland fluxes, is shown in Figure 5. Armentano and Menges (1986) estimated that before recent human disturbance of wetlands, net global retention of carbon in wetland peats was 0.057-0.083 x 10^{15} g-C/yr (Pg-C/yr), most in boreal peatlands. Gorham (1991) estimated an accumulation of 0.076 Pg-C/yr of carbon in northern peatlands. This annual retention is small compared to the estimated 1,400 Pg of organic carbon stored in the world's soils (Post et al., 1982); about one-third of that, or 455 Pg is in boreal and subarctic peat (Gorham, 1991).

About one-fourth of the carbon sequestered by wetlands is re-released by methane emissions (Figure 5). For all wetlands in the world, Cicerone and Oremland (1988) estimated a release of 0.09 Pg-C/yr of methane from wetlands and 0.08 Pg-C/yr from rice paddies, out of a total release of about 0.4 Pg-C/yr of methane from all sources. Aselmann and Crutzen (1989), in some of the most widely quoted values of methane emissions from wetlands, suggest a range of 0.03-0.12 Pg-C/yr from wetlands and 0.04-0.10 Pg-C/yr from rice paddies (shown in Figure 5). Gorham estimated the emissions of CH_4 from northern peatlands to be 0.046 Pg-C/yr.

There are some indications that the global carbon balance of wetlands has shifted, primarily because of agricultural conversion of peatlands. By 1900 there had been no significant change in North America but in European countries anywhere from 20 to 100 percent of the stored carbon had been lost to the atmosphere. By 1980 the total carbon shift due to agricultural drainage was estimated by Armentano and Menges (1986) to be 0.063-0.085 Pg-C/yr, with a further 0.032-0.039 Pg-C/yr released from peat combustion. Gorham (1991) estimated the following carbon losses for northern peatlands: oxidation due to drainage of 0.0085-0.042 Pg-C/yr (low number is for long-term; high number for short-term); and combustion of 0.026 Pg-C/yr of peat. In Figure 5 we estimate the long-term contributions of burning and draining from Gorham's estimates (0.035 Pg-C/yr).

Thus wetlands, as with tropical rain forests, may be close to having shifted from being a net sink to a net source of carbon to the atmosphere during the past 100 years, if we include the effects of draining, peat combustion, and methane emissions from natural and managed wetlands. Currently wetlands are estimated

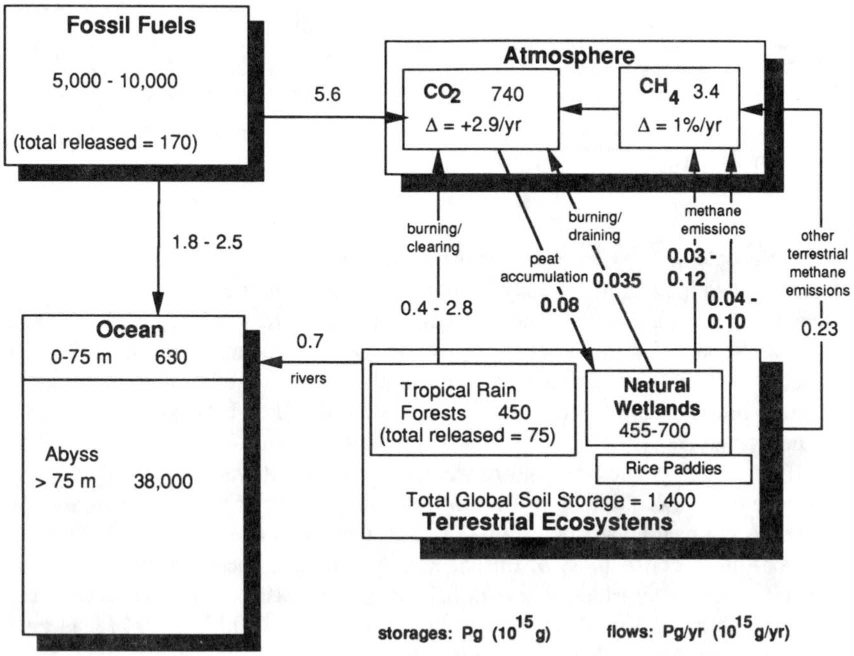

Figure 5. Estimated global carbon budget, including wetlands. All fluxes are in 10^{15} g C/year (Pg/year) while storages are in 10^{15} g C (Pg). Peat accumulation and burning/draining based on estimates of Gorham (1991). Wetlands and rice paddy methane emissions from estimates of Aselmann and Crutzen (1989); other terrestrial methane emissions from Cicerone and Oremland (1988). Fossil fuel production, ocean uptake, and release from tropical rain forest based on estimates of Detwiler and Hall (1988) and Houghton (1990).

to be a sink of about 0.08 Pg-C/yr and a source of 0.055 Pg-C/yr. To put these numbers in perspective (see Figure 5), the burning of fossil fuel contributes an estimated 5.6 Pg-C/yr and deforestation of tropical rain forests an additional 0.4-2.8 Pg-C/yr of carbon to the atmosphere (Detwiler and Hall, 1988; Houghton, 1990). These estimates do not take into account the greater relative importance of methane as a greenhouse gas, although we agree with Lashof and Ahuja (1990) and Gorham (1991) that the multiplying factor of 20 for methane's effect is probably not justified because of short residence time of CH_4 in the atmosphere (14 years) relative to CO_2 (230 years).

III. Possible Effects of Global Warming on Wetlands

A. Effects on Extent and Distribution of Wetlands

1. Sea Level Rise

One of the major impacts of possible global changes on wetlands is the effect that sea level rise will have on coastal wetlands. Estimates of sea level rise over the next century range from 50 to 200 cm (Titus, 1986; Titus et al., 1991). These estimates are based, to a certain degree, on sea-level rise of 10 to 25 cm over the past century, some of which is attributed to global warming of the last century. If the rise in sea level is not accompanied with accretion of sediments, then there would be a gradual disintegration of coastal marshes due to excessive inundation, erosion, and saltwater intrusion. Since much of the coastline of the world is developed, efforts to protect dry upland from inundation by the construction of bulkheads or dikes

Table 3. Percent coastal wetland loss in the United States with sea level rise

	Sea level rise		
	0.5m	1m	2m
If no shores are protected	17-43%	26-66%	29-76%
If densely developed dryland is protected	20-45%	29-69%	33-80%
If all dryland is protected	38-61%	50-82%	66-90%

(From Titus, 1991.)

would exacerbate the problem. In essence, the wetlands would be trapped between the rising sea and dry land being protected, a situation that has already happened over the centuries in The Netherlands and China (Titus, 1991). In most of our regions where coastal wetlands exist, "the slope above the wetland is steeper than that of the wetlands; so a rise in sea level causes a net loss of wetland acreage" (Titus, 1991).

Estimates on the loss of coastal wetlands vary, with much of the variability dependent on the assumed sea-level rise. Table 3 presents some estimated scenarios from the U.S. EPA for coastal wetlands of the United States. If there is no shoreline protection, seal level rise of 1-m could reduce coastal wetlands by 26 to 66 percent. If densely-populated dryland is protected, the loss of wetlands is only slightly higher at 29 to 69 percent. If the policy would be to protect all dryland, the loss of wetlands increases dramatically to 50 to 82 percent. It is not well understood if these figures could be extrapolated to the rest of the world. In long developed coastlines such as those of Europe and the Far East, the losses would probably be less.

A number of specific studies have looked at regional changes in coastal wetlands that would result in the event of dramatic sea level rise. A spatial cell-based simulation model named SLAMM (Sea Level Affecting Marshes Model) has been used by Park and colleagues (Park et al., 1991; Lee et al., 1992) to illustrate the effects of sea level rise on coastal region. For example, Lee et al. (1992) predicted a 32 to 40 percent loss of wetlands in northeastern Florida with a rise of 1 to 1.25 m. Most of that loss is the low intertidal salt marsh.

Day and Templet (1989) and McKee and Mendelssohn (1989) discuss some of the implications of the apparent sea level rise that is already occurring in the Mississippi River delta in Louisiana at the rate of 1.2 m/100 years (Mitsch and Gosselink, 1993), primarily because of sediment subsidence rather than actual sea rise. It is estimated that coastal loss rates as high as 130 km^2 per year are being experienced now in Louisiana. In Louisiana, as could be the case in global sea-level rise, the loss of wetlands is exacerbated by the reduced sediment inputs to the wetlands and flood prevention strategies such as river dikes and levees (Day and Templet, 1989). By contrast, Patrick and DeLaune (1990) measured the accretion and subsidence of sediments in the salt marshes of San Francisco Bay, California, and found that because of an adequate supply of sediments, the salt marshes of south San Francisco Bay would probably be able to have a net accretion rate despite sea level rise.

McKee and Mendelssohn (1989) showed in lab and field experiments in Louisiana that freshwater tidal plant species are quickly replaced with increased water levels and resulting salinity. The effect of water level increase in experimental field plots was a more reduced soil environment, which caused a reduction in the above-ground biomass and stem density of the macrophytes. They suggest that salt-water intrusion is usually cited as the primary factor for the present deterioration of tidal freshwater marshes in Louisiana. Actually the cause of wetland loss is a combination of water level rise and increased salinity.

Lefeuvre (1990) presented a summary of the ecological effects of sea level rise in coastal France near Mont-Saint-Michel Bay. The salt marshes in this region are currently accreting at a rate of about 30 ha per year due to a positive sediment influx of 2cm/yr. He hypothesized that fragmentation would occur to the salt marshes and polders in the event of a maximum rise in sea level (1 to 2 m per century) but that this fragmentation would enhance marine ecology by increasing the marsh-tidal flat interfaces.

2. Warming Effects on Inland Wetlands

In addition to the effects of climate change on coastal wetlands through sea-level rise, the change in climate will undoubtedly affect the distribution of inland wetlands. Very little has been published on this effect. In the tundra, any melting of the permafrost would result in the loss in wetlands. In boreal and temperate areas, climate change would result in changing rainfall patterns, thus affecting runoff and groundwater

Table 4. Estimated changes in CO_2 and CH_4 emissions (10^{15} g/yr = Pg/yr) from northern wetlands with change in climatic conditions (assumes 5 °C temperature rise)

	Warm/wet		Warm/dry	
	CO_2	CH_4	CO_2	CH_4
Tundra	1.3	0.1	1.6	0.1
Boreal Peatland	-	0.12	0.83	0.12

(From Post, 1990.)

inflows to wetlands. In general, a decrease in precipitation or increase in evapotranspiration would result in less frequent flooding of existing wetlands, although formal designation of wetlands may not change. Greater precipitation patterns, of course, would increase the length and depth of flooding of inland wetlands. Most susceptible to these effects are depressional wetlands that have very small watersheds and that are in regions near the edge of a dry versus wet climate now. The prairie pothole region of mid-continent North America would be among the most susceptible.

B. Feedbacks—Effects on Wetland Ecosystem Structure and Function

The effects of climate change on wetlands can lead to both positive and negative feedbacks in climate change. Almost all of the research on this topic has been done in boreal and arctic peatlands, with very little if any research on the topic in temperate and tropical wetlands.

One example of a possible positive feedback is increased temperatures expected from global change leading to increased emissions of greenhouse gases as a result of increased metabolism. Estimates of the increased net flux rates of CO_2 and CH_4 from tundra and boreal peatlands from one study are illustrated in Table 4. Comparing these figures with the present fluxes shown in Figure 5 suggest that this feedback—climate change increasing the release of greenhouse gases from wetlands—may be significant. In these estimates, an increase of 5 °C average temperature was assumed and was projected for two moisture change—wetter and drier. An increase in CO_2 emission of 0.83 Pg/yr was estimated for boreal peatlands in warm/dry conditions due to increased aerobic respiration that results from an increase in the thickness of the aerobic surface layer. In the tundra system, CO_2 emissions were projected to increase by 1.3 to 1.6 Pg/yr due to both the increased aerobic respiration rate at higher temperatures, but also due to the increased length of the growing season. Christensen (1991) predicts that, as a result of the warming, the tundra would change from being a net sink of CO_2 to becoming a net source of up to 1.25 Pg/yr with an increase of 5 °C due to a combination of thermokarst erosion, deepening of the active layer in permafrost areas, lowering of the water table, and higher temperatures.

Increases in methane emissions are expected to be most significant in warmer and wetter conditions although there is much uncertainty about the mechanisms that might be affected (Christensen, 1991). Whalen and Reeburgh (1990) and others suggested that, in warmer, drier conditions, microbial oxidation of methane may actual decrease methane emissions, providing a negative feedback to the global emission of methane. Prior to those papers, methane emissions from boreal peatlands were estimated by Post (1990) to increase by 0.12 Pg/yr due to higher metabolism at higher temperatures for both wet and dry conditions (Table 4). This increase assumes a methane production rate of 200 mg C m^{-2} d^{-1}, a high rate for assumptions based on our analysis above. For example, Gorham (1991) estimated a carbon accumulation of only 77 mg C m^{-2} d^{-1} for northern peatlands. Comparing the Post (1990) and Gorham (1991) studies, we would conclude that methane emissions from northern peatlands would increase by 260 percent from boreal wetlands as a result of an increase in temperature of 5 °C. Burke et al. (1990) suggest that by the year 2080, annual emissions from wetlands could be 0.28 Pg/yr, approximately 150 percent more than the current rate of 0.11 Pg/yr. This estimate assumes a 3 °C warming, primarily as a result of the doubling of the CO_2 concentrations.

In general, both the increase in temperature and the changes in water levels are the important variables in the production of methane and carbon dioxide from wetlands but their relative importance for methane generation has only recently been investigated (Moore and Knowles, 1989; Whalen and Reeburgh, 1990; Roulet et al., 1992b; Moore and Roulet, 1994). Using a model with inputs of 3 °C rise in temperature and a decrease in the water table of between 14 and 22 cm for a subarctic fen, Roulet et al. (1992b) estimated that increased temperature raised the methane flux between 5 and 40 percent, but the lowered water table decreased the methane flux by 74 to 81 percent. This decrease in methane flux in drier conditions was due to a decrease in the zone of active methanogenesis and to the an increase in methane oxidation in the aerobic layer. They suggest that the methane emissions are much more influenced by the moisture regime than by the increase in temperature for climatic conditions resulting from a doubling of CO_2 in the atmosphere. This decrease in methane production with drier conditions is in keeping with previous laboratory column work reported by Moore and Knowles (1989) where methane production decreased as water table was lowered.

IV. Wetland Management with Global Warming

There may be few opportunities for managing wetlands in the face of global change. Certainly we cannot say if wetlands are significant global sinks or sources of carbon with much certainty. Therefore managing wetlands on some global scale without knowing how significant that management is would not be appropriate. Nevertheless, some limited experimentation, especially in rice paddies, suggest some management alternatives that might be appropriate for those systems, especially in the reduction of methane emissions. For coastal wetlands, there are some other management strategies that take advantage of natural energies and might therefore be appropriate.

A. Rice Paddy Management

1. Water Level Management

Water level alteration is one of the easiest and most effective ways to control methane emissions from managed wetlands. Sass et al. (1992) measured the effects on methane emissions of four different water management methods in some rice fields in Texas (Figure 6a). They found that temporary drainage (midseason drainage and multiple aeration) decreased methane emission due to both increased CH_4 consumption in the aerobic layer and decreased CH_4 production. This may only be practical in flat systems with sufficient control of water supply (Bouwman, 1991). Sass et al. (1992) found that the multiple aeration treatment required 2.7 times more water than did the normal flood treatment. Broadcasting rice seeds directly in the field instead of transplanting reduces the period of inundation and CH_4 formation. But it may reduce the yield by 15-25% and has other disadvantages.

2. Fertilization

There may be a relationship between the C/N ratio of the organic matter in wetlands and the CH_4 emission (Tsutsuki and Ponnamperuma, 1987; Yagi & Minami, 1990). Yagi and Minami (1990) found the compost with a low C/N ratio cause lower emissions of methane than uncomposted rice straw with high C/N ratio in rice paddies in Japan (Figure 6b). Schutz et al. (1989), on the other hand, found high emission in field applied with compost.

Because of its competition with methanogenesis, enhancement of sulfate reduction is often suggested as a management alternative to reduce methane emissions. Sulfate in the mineral fertilizer, when incorporated in the soil, may suppress CH_4 emissions. This has long been known as one of the primary reasons that methanogenesis is lower in saltwater wetlands than in freshwater wetland. Techniques to reduce emissions in rice paddies will be acceptable only if they do not adversely affect yields (Bouwman, 1991).

B. Coastal Wetland Management

The management possibilities for coastal wetlands in view of sea level rise offer some possibilities of management to save coastal wetlands. For example, in a revision of the original drawings of Titus (1991), Figure 7 shows two future conditions. In Future 1 the house is protected with a bulkhead in the face of rising sea level and the salt marsh is lost. In Future 2, the house is moved upland to accommodate the

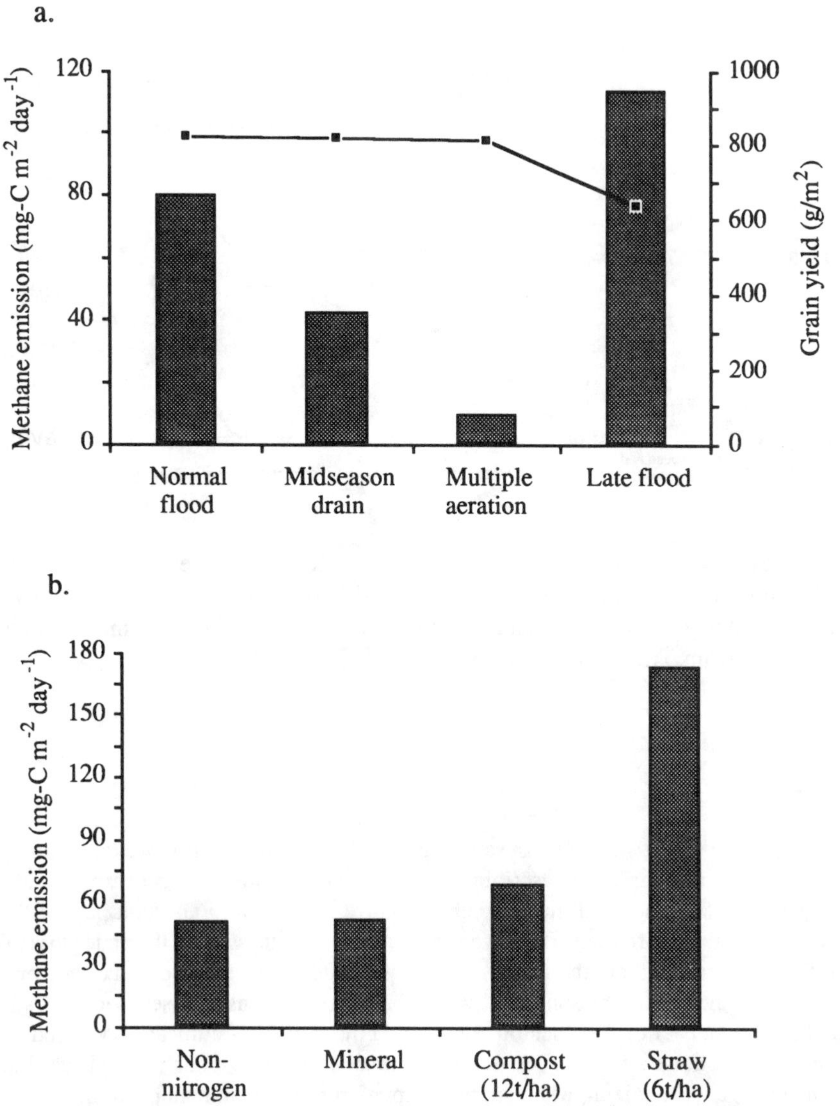

Figure 6. Results of experiments with rice paddy management for methane generation as a function of a) hydrologic regime (Sass et al., 1992), and b) nitrogen fertilization (Yagi and Minami, 1990).

wetland which would begin to form if a gentle slope and adequate sediment sources are available. This would model the wetlands of the Great Lakes which, for centuries, have been "wetlands on skateboards" moving inland and lakeward with the frequent (over periods of decades) water level changes in the Great Lakes (Mitsch, 1992). It has only been with the stabilization of the coastline that diking of the remaining wetlands along the Great Lakes was necessary for their survival. A better model is to give the wetlands room to expand.

Day and Templet (1989) conclude, after extensive investigation of the apparent sea level rise in coastal Louisiana, that we can manage coastal wetlands in periods of rising sea level through comprehensive, long-range planning and through the application of the principles of ecological engineering by using nature's energies such as vegetation productivity, winds, currents, and tides as much as possible. They cite the example of using vegetation to enhance sediment accretion such as with brush fences developed by the Dutch.

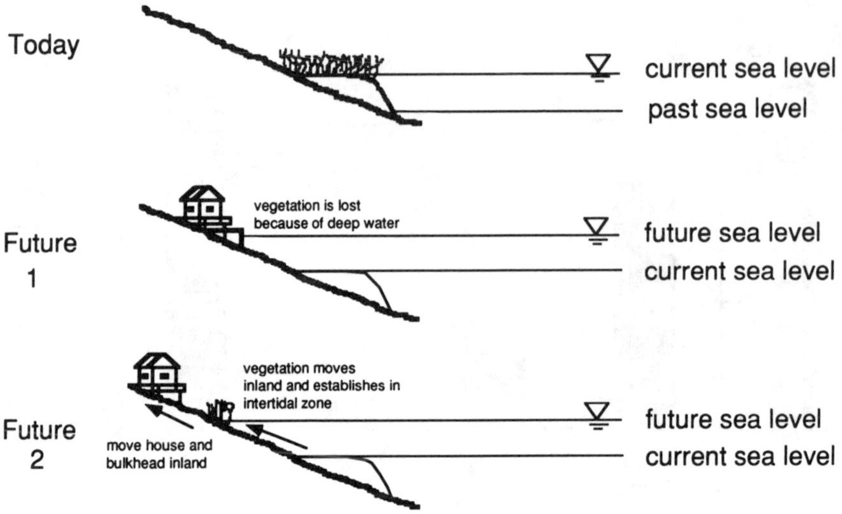

Figure 7. Illustration of coastal wetland management in the face of sea level rise. Future 1 is without moving human habitation inland. Future 2 involves moving human activity inland to allow room for wetland to move inland (adapted from Titus, 1986, 1991).

V. Conclusions

Wetlands are among our most valued ecosystems and are known for much more than their role in global change. We must take particular care in ascribing both negative (e.g. methane generation) and positive (CO_2 sequestering) values to wetlands and must rather recognize them as an important part of the biosphere's respiratory system. Wetlands are easily destroyed and altered, as witnessed by the major loss of the resource in the United States and throughout the world over the past 100 years. It is clear that they provide a major role in the carbon balance of the biosphere. It would be prudent for us to assess feedbacks by continuing to investigate both the impact of wetlands on climate and the impact of climate on wetlands together—the system does indeed constitute a two-edged sword. We must also manage our wetlands carefully in face of possible climate change, recognizing when simple adaptation rather than manipulation is most appropriate and keeping in mind that capturing natural energies to enhance wetland function is much more effective than are expensive, highly-structured solutions.

References

Armentano, T.V. and E.S. Menges. 1986. Patterns of change in the carbon balance of organic soil wetlands of the temperate zone. *Journal of Ecology* 74:755-774.

Aselmann, I. and P.J. Crutzen. 1989. Global distribution of natural freshwater wetlands and rice paddies, their net primary productivity, seasonality and possible methane emissions. *Journal of Atmospheric Chemistry* 8:307-358.

Aselmann, I. and P.J. Crutzen. 1990. A global inventory of wetland distribution and seasonality, net primary productivity, and estimated methane emissions. p. 441-449. In: A.F. Bouwman (ed.), *Soils and the Greenhouse Effect*. John Wiley and Sons, Chichester.

Atkinson, L.P. and J.R. Hall. 1976. Methane distribution and production in the Georgia salt marsh. *Estuarine and Coastal Marine Science* 4:677-686.

Baker-Blocker, A., T.M. Donahue, and K.H. Mancy. 1977. Methane flux from wetlands areas. *Tellus* 29:245-250.

Banik A., M. Sen, and S.P. Sen. 1993. Methane enission from jute-retting tanks. *Ecological Engineering* 2:73-79.

Bartlett, K.B., R.C. Harriss, and D.I. Sebacher. 1985. Methane flux from coastal salt marshes. *Journal of Geophysical Research* 90:5710-5720.

Bartlett, K.B., D.S. Bartlett, R.C. Harriss, and D.I. Sebacher. 1987. Methane emissions along a salt marshe salinity gradient. *Biogeochemistry* 4:183-202.

Bartlett, K.B., P.M. Grill, D.I. Sebacher, R.C. Harriss, J.O. Wilson, and J.M. Melack. 1988. Methane flux from the central Amazonian floodplain. *Journal of Geophysical Research* 93:1571-1582.

Bartlett, D.S., K.B. Bartlett, J.M. Hartman, R.C. Harriss, D.I. Sebacher, R. Pelletier-Travis, D.D. Dow and D.P. Brannon. 1989. Methane emissions from the Florida Everglades: patterns of variability in a regional wetland ecosystem. *Global Biogeochemical Cycles* 3:363-374.

Bazilevich, N.L. and A.A. Tishkov. 1982. Conceptual balance model of chemical element cycles in a mesotrophic bog ecosystem. p. 236-272. In: *Ecosystem Dynamics in Freshwater Wetlands and Shallow Water Bodies*, Vol. II. SCOPE and UNEP, Center of International Projects, Moscow, USSR.

Bazilevich, N.L., L.Y. Rodin, and N.N. Rozov. 1971. Geophysical aspects of biological productivity. *Soviet Geography* 12:293-317.

Bouwman, A.F. 1991. Agronomic aspects of wetland rice cultivation and associated methane emissions. *Biogeochemistry* 15:65-88.

Burke, M.K., R.A. Houghton, and G.M. Wodwell. 1990. Progress toward predicting the potential for increasing emissions of CH_4 from wetlands as a consequences of global warming. p. 451-455. In: A.F. Bouwman (ed.), *Soils and the Greenhouse Effect*. John Wiley and Sons, Chichester.

Burke, R.A., Jr.,.T.R. Barber, and W.M. Sackett. 1988. Methane flux and stable hydrogen and carbon isotope composition of sedimentary methane from the Florida Everglades. *Global Biogeochemical Cycles* 2:329-340.

Cameron, C.C. 1970. *Peat deposits of northeastern Pennsylvania*. U.S. Geological Survey Bulletin 1317-A. 90 pp.

Capone, D.G. and R.P. Kiene. 1988. Comparison of microbial dynamics in marine and freshwater sediments: contrasts in anaerobic carbon metabolism. *Limnology and Oceanography* 33:725-749.

Chen, Z., D. Li, K.S. and B. Wang. 1993. Features of CH_4 emission from rice paddy fields in Beijing and Nanjing. *Chemosphere* 26:239-245.

Christensen, T. 1991. Arctic and sub-Arctic soil emissions: Possible implications for global climate change. *The Polar Record* 27:205-210.

Cicerone, R.J. and J.D. Shetter. 1981. Source of atmospheric methane: measurements in rice paddies and a discussion. *Journal of Geophysical Research* 86:7203-7209.

Cicerone, R.J., J.D. Shetter, and C.C. Delwiche. 1983. Seasonal variation of methane flux from a California rice paddy. *Journal of Geophysical Research* 88:11022-11024.

Cicerone, R.J. and R.S. Oremland. 1988. Biogeochemical aspects of atmospheric methane. *Global Biogeochemical Cycles* 2:299-327.

Cicerone, R.J., C.C. Delwiche, S.C. Tyler and P.R. Zimmerman. 1992. Methane emissions from California rice paddies with varied treatments. *Global Biogeochemical Cycles* 6:233-248.

Crill, P.M., K.B. Bartlett, R.C. Harriss, E. Gorham, E.S. Verry, D.I. Sebacher, L. Madzar and W. Sanner. 1988. Methane flux from Minnesota peatlands. *Global Biogeochemical Cycles* 2:371-384.

Day, J.W., Jr. and P.H. Templet. 1989. Consequences of sea level rise: implications from the Mississippi Delta. *Coastal Management* 17:241-257.

Delaune, R.D., C.J. Smith, and W.H. Patrick. 1983. Methane releases from Gulf coast wetlands. *Tellus* 35B:8-15.

Detwiler, R.P., and C.A.S. Hall. 1988, Tropical forests and the global carbon cycle, *Science* 239:42-47.

Devol, A.H., J.E. Richey, W.A. Clark, S.L. King, and L.A. Martinelli. 1988, Methane emissions to the atmosphere from the Amazon floodplain. *Journal of Geophysical Research* 93:1583-1592.

Dise, N.B. 1992. Winter fluxes of methane from Minnesota peatlands. *Biogeochemistry* 17:71-83.

Gambrell, R.P. and W.H. Patrick, Jr. 1978. Chemical and microbiological properties of anaerobic soils and sediments. p. 375-423. In: D.D. Hook and R.M.M. Crawford (eds.), *Plant Life in Anaerobic Environments*. Ann Arbor Sci. Pub. Inc., Ann Arbor, Mich.

Gorham, E. 1991. Northern peatlands: Role in the carbon cycle and probable responses to climatic warming. *Ecological Applications* 1:182-195.

Harriss, R.C. and D.I. Sebacher. 1981. Methane flux in forested freshwater swamps of the southeastern United States. *Geographical Research Letters* 8:1002-1004

Harriss, R.C., D.I. Sebacher, and F.P. Day, Jr. 1982. Methane flux in the Great Dismal Swamp. *Nature* 297:673-674.

Harriss, R.C., E. Gorham, D.I. Sebacher, K.B. Bartlett, and P.A. Flebbe. 1985. Methane flux from northern peatlands. *Nature* 315:652-654.

Harriss, R.C., D.I. Sebacher, K.B. Bartlett, D.S. Bartlett, and P.M. Crill. 1988. Source of atmospheric methane in the south Florida environment. *Global Biogeochemical Cycles* 2:231-243.

Holzapfel-Pschorn, A., R. Conrad, and W. Seiler. 1985. Production, oxidation, and emission of methane in rice paddies. *FEMS Microbiology Ecology* 31:343-351.

Holzapfel-Pschorn and W. Seiler. 1986. Methane emission during a cultivation period from an Italian rice paddy. *Journal of Geophysical Research* 91:11803-11814.

Holzapfel-Pschorn, A., R. Conrad and W. Seiler. 1986. Effects of vegetation on the emission of methane from submerged rice paddy soil. *Plant and Soil* 92:223-233.

Houghton, R. A. 1990. The global effects of tropical deforestation. *Environmental Science and Technology* 24:414-424.

Howarth, R.W. and A. Giblin. 1983. Sulfate reduction in the saltmarshes at Sapelo Island, Georgia. *Limnology and Oceanography* 28:70-82.

Howarth, R.W. and J.M. Teal. 1979. Sulfate reduction in a New England salt marsh. *Limnology and Oceanography* 24:999-1013.

Howarth, R.W. and J.M. Teal. 1980. Energy flow in a salt marsh ecosystem: the role of reduced inorganic sulfur compounds. *The American Naturalist* 116:862-872.

Howes, B.L., J.W. Dacey, and G.M. King. 1984. Carbon flow through oxygen and sulfate reduction pathways in salt marsh sediments. *Limnology and Oceanography* 29:1037-1051.

Howes, B.L., J.W. Dacey, and J.M. Teal. 1985. Annual carbon mineralization and below ground production of *Spartina alterniflora* in a New England salt marsh. *Ecology* 66:595-605.

Ingvorsen, K. and T.D. Brock. 1982. Electron flow via sulfate reduction and methanogenesis in the anaerobic hypolimnion of Lake Mendota. *Limnology and Oceanography* 27:559-564.

Kelly, C.A., N.B. Dise, and C.S. Martens. 1992. Temporal variations in the stable carbon isotopic composition of methane emitted from Minnesota peatlands. *Global Biogeochemical Cycles* 6:263-269.

Khalil, M.A.K. and R.A. Rasmussen. 1990. Atmospheric methane: recent global trends. *Environmental Science and Technology* 24:549-553.

Khalil, M.A.K., R.A. Rasmussen, M.X. Wang, and L.Ren. 1990. Emissions of trace gases from Chinese rice fields and biogas generators: CH_4, NO_2, CO, CO_2, Chlorocarbons and hydrocarbons. *Chemosphere* 20:207-226.

King, G.M. and W.J. Wiebe 1978. Methane release from soils of a Georgia salt marsh. *Geochimica et Cosmochimica Acta* 42:343-348.

Lashof, D.A. and D.R. Ahuja. 1990. Relative contributions of greenhouse gas emissions to global warming. *Nature* 344:529-531.

Lee, J.K., R.A. Park, and P.W. Mausel. 1992. Application of geoprocessing and simulation modeling to estimate impacts of sea level rise on the northeast coast of Florida. *Photogrammetric Engineering and Remote Sensing* 58:1579-1586.

Lefeuvre, J.C. 1990. Ecological impact of sea level rise on coastal ecosystems of Mont-Saint-Michel Bay. p. 139-153. In: J.J. Beukema et al. (eds.) *Expected Effects of Climatic Change on Marine Coastal Ecosystems*. Kluwer Academic Publishers.

Maltby and Turner. 1983. Wetlands of the world. *Geographical Magazine* 55:12-17.

Matthews, E. 1990a. Global data bases for evaluating trace gas sources and sinks. p. 311-325. In: A.F. Bouwman (ed.), *Soils and the Greenhouse Effect*. John Wiley and Sons, Chichester.

Matthews, E. 1990b. Global distribution of forested wetlands. Addendum to: A.E. Lugo, M. Brinson, and S. Brown (eds.) *Forested Wetlands*. Elsevier, Amsterdam.

Matthews, E. and I. Fung. 1987. Methane emission from natural wetlands: global distribution, area, and environmental characteristics of sources. *Global Biogeochemical Cycles* 1:61-86.

Matthews, E., I. Fung, and J. Lerner. 1991. Methane emission from rice cultivation: geographic and seasonal distribution of cultivated areas and emissions. *Global Biogeochemical Cycles* 5:3-24.

McKee, K.L. and I.A. Mendelssohn. 1989. Response of a freshwater marsh plant community to increased salinity and increased water level. *Aquatic Botany* 34:301-316.

Mitsch, W.J. 1992. Combining ecosystem and landscape approaches to Great Lakes wetlands. *Journal of Great Lakes Research* 18:552-570.

Mitsch, W.J. and J.G. Gosselink. 1993. *Wetlands, 2nd Ed.* Van Nostrand Reinhold, New York. 722 pp.

Moore, P.D. and D.J. Bellamy. 1974. *Peatlands*. Springer-Verlag, New York. 221 pp.

Moore, T.R and R. Knowles. 1987. Methane and carbon dioxide evolution from subarctic fens. *Canadian Journal Soil Science* 67:77-81.

Moore, T.R and R. Knowles. 1989. The influence of water table levels on methane and carbon dioxide emissions from peatland soils. *Canadian Journal Soil Science* 69:33-38.

Moore, T.R, N.T. Roulet, and R. Knowles. 1990. Spatial and temporal veriations of methane flux from subarctic/northern boreal fens. *Global Biogeochemical Cycles* 4:29-46.

Moore, T.R. and N.T. Roulet. 1994. Methane Emissions from Canadian Peatlands. (This volume)

Mulholland, P.J. 1981. Organic carbon flow in a swamp-stream ecosystem. *Ecological Monographs* 52:307-322.

Naiman, R.J., T. Manning, and C.A. Johnston. 1991. Beaver population fluctuations and tropospheric methane emissions in boreal wetlands. *Biogeochemistry* 12:1-15.

Ovenden, L. 1990. Peat accumulation in northern wetlands. *Quaternary Research* 33:377-386.

Park, R.A., J.K. Lee, P.W. Mausel, and R.C. Howe. 1991. Using remote sensing for modeling the impact of sea level rise. *World Resource Review* 3:184-205.

Patrick, W.H., Jr. and R.D. DeLaune. 1990. Subsidence, accretion, and sea level rise in south San Francisco Bay marshes. *Limnology and Oceanography* 35:1389-1395.

Post, W.M. 1990. Report of a workshop on climate feedbacks and the role of peatlands, tundra, and boreal ecosystems in the global carbon cycle. ORNL, Environmental Science Division, Publication 3289.

Post, W.M., W.R. Emanuel, P.J. Zinke, and A.G. Stangenberger. 1982. Soil carbon pools and world life zones. *Nature* 298:156-159.

Pulliam, W.M. 1993. Carbon dioxide and methane exports from a southeastern floodplain swamp. *Ecological Monographs* 63:29-53.

Roulet, N.T., R. Ash, and T.R. Moore. 1992a. Low boreal wetlands as a source of atmospheric methane. *Journal of Geophysical Research* 97:3739-3749.

Roulet, N.T., T.R. Moore, J. Bubier, and P. Lafleur. 1992b. Northern fens: CH_4 flux and climate change. *Tellus* 44B:100-105.

Saha, A.K., J. Rai, V. Raman, R.C. Sharma, D.C. Parashar, S.P. Sen, and B. Sarkar. 1989. Methane emission from inundated field in a monsoon region. *Indian Journal of Radio and Space Physics* 18:215-217.

Sass, R.L., F.M. Fisher, and P.A. Harcombe. 1990. Methane production and emission in a Texas rice field. *Global Biogeochemical Cycles* 4:47-68.

Sass, R.L., F.M. Fisher, P.A. Harcombe, and F.T. Turner. 1991a. Mitigation of methane emissions from rice fields: possible adverse effects of incorporated rice straw. *Global Biogeochemical Cycles* 5:275-287.

Sass, R.L., F.M. Fisher, F.T. Turner, and M.F. Jund. 1991b. Methane emission from rice fields as influenced by solar radiation, temperature, and straw incorporation. *Global Biogeochemical Cycles* 5:335-350.

Sass, R.L., F.M. Fisher, Y.B. Wang, F.T. Turner, and M.F. Jund. 1992. Methane emission from rice fields: the effect of floodwater management. *Global Biogeochemical Cycles* 6:249-262.

Schutz, H., A. Holzapfel-Pschorn, R. Conrad, H. Rennenberg, and W. Seiler. 1989. A three years continuous record on the influence of daytime, season and fertilizer treatment on methane emission rates from an Italian rice paddy field. *Journal of Geophysical Research* 94:16405-16416.

Schutz, H., P. Schroder, and H. Rennenberg. 1991. Role of plants in regulating the methane flux to the atmosphere. p. 9-63. In: T.D. Sharkey, E.A. Holland, and H.A. Mooney (eds.), *Trace Gas Emissions by Plants*. Academic Press, Inc.

Sebacher, D.I., R.C. Harriss, K.B. Bartlett, S.M. Sebacher, and S.S. Grice. 1986. Atmospheric methane sources: Alaskan tundra bogs, an alpine fen, and a subarctic boreal marsh. *Tellus* 38B:1-10.

Seiler, W., A. Holzapfel-Pschorn, R. Conrad, and D. Scharffe. 1984. Methane emission from rice paddies. *Journal of Atmospheric Chemistry* 1:214-268.

Smith, C.J., R.D. Delaune and W.H. Patrick, Jr. 1982. Carbon and nitrogen cycling in a *Spartina alterniflora* salt marsh. p. 97-104. In: J.R. Freney and I.E. Galcally (eds.), *The Cycling of Carbon, Nitrogen, Sulfur, and Phosphorus in Terrestrial and Aquatic Ecosystems*. Springer-Verlag, NY.

Smith, L.K. and W.M. Lewis, Jr. 1992. Seasonality of methane emissions from five lakes and associated wetlands of the Colorado Rockies. *Global Biogeochemical Cycles* 6:323-338.

Smith, R.L. and M.J. Klug. 1981. Electron donors utilized by sulfate-reducing bacteria in eutrophic lake sediments. *Applied and Environmental Microbiology* 42:116-121.

Svensson, B.H. and T. Rosswall 1984. In situ methane production from acid peat in plant communities with different moisture regimes in a subarctic mire. *Oikos* 43:341-350.

Titus, J.G. 1986. Greenhouse effect, sea level rise, and coastal zone management. *Costal Zone Management Journal* 14:147-171.

Titus, J.G. 1991. Greenhouse effect and coastal wetland policy: how Americans could abandon an area the size of Massachusetts at minimum cost. *Environmental Management* 15:39-58.

Titus, J.G., R.A. Park, S.P. Leatherman, J.R. Weggel, M.S. Greene, P.W. Mausel, S. Brown, G. Gaunt, M. Trehan, and G. Yohe. 1991. Greenhouse effect and sea level rise: the cost of holding back the sea. *Coastal Management* 19:171-204.

Torn, M.S. and F.S. Chapin, III. 1993. Environmental and biotic controls over methane flux from arctic tundra. *Chemosphere* 26:357-368.

Tsutsuki, K. and F.N. Ponnamperuma. 1987. Behaviour of anaerobic decomposition products in submerged soils. *Soil Science and Plant Nutrition* 33:13-33.

Valiela, I. 1984. *Marine Ecological Processes*. Springer-Verlag, New York. 546 pp.

Vourlitis, G.L., W.C. Oechel, S.J. Hastings, and M.A. Jenkins. 1993. The effect of soil moisture and thaw depth on CH_4 flux from wet coastal tundra ecosystems on the North Slope of Alaska. *Chemosphere* 26:329-337.

Wang, M., A. Dai, R. Shen, H. Wu, H. Schutz, H. Rennenberg, and W. Seiler. 1990. CH_4 emission from a Chinese rice paddy field. *Acta Meteorologica Sinica* 4:265-275.

Wasson, J.G., J.L. Guyot and H. Sanejouand. 1991. First estimation of organic carbon transport in the Rio Desaguadero (Altiplano, Bolivia). *Rev. Sci. Eau.* 4:363-379.

Whalen, M.N. and W.S. Reeburgh. 1990. Consumption of atmospheric methane by tundra soils. *Nature* 346:160-162.

Whalen, M.N. and W.S. Reeburgh. 1992. Interannual veriations in tundra methane emission: A 4-year time series at fixed sites. *Global Biogeochemical Cycles* 6:139-159.

Whiting, G.J. and J.P. Chanton. 1992. Plant-dependent CH_4 emission in a subarctic Canadian fen. *Global Biogeochemical Cycles* 6:225-231.

Whiting, G.J., J.P. Chanton, D.S. Bartlett, and J.D. Happell. 1991. Relationship between CH_4 emission, biomass, and CO_2 exchange in a subtropical grassland. *Journal of Geophysical Research* 96:13,067-13,071.

Wiebe, W.J., R.R. Christian, J.A. Hansen, G. King, B. Sherr and G. Skyring. 1981. Anaerobic respiration and fermentation. p. 137-159. In: L.R. Pomeroy and R.G. Wiegert (eds.), *The Ecology of Salt Marsh*. Springer-Verlag, NY.

Wilson, J.O., P.M. Crill, K.B. Bartlett, D.I. Sebacher, R.C. Harriss and R.L. Sass. 1989. Seasonal variation of methane emission from a temperate swamp. *Biogeochemistry* 8:55-71.

Yagi, K. and K. Minami. 1990. Effect of organic matter application on methane emission from some Japanese paddy fields. *Soil Science and Plant Nutrition* 36:599-610.

CHAPTER 19

Emission, Production, and Oxidation of Methane in a Japanese Rice Paddy Field

Kazuyuki Yagi, Katsumi Kumagai, Haruo Tsuruta, and Katsuyuki Minami

I. Introduction

The concentration of atmospheric methane (CH_4) has been increasing at a rate of about 1% per year (Steel et al., 1987; Blake and Rowland, 1988). Methane is an important greenhouse gas in the climate system because of its infrared absorption spectrum. Methane also has a strong influence on chemistry in the troposphere and stratosphere. The rapid increase, therefore, could be of significant environmental consequence (Crutzen, 1987; Cicerone and Oremland, 1988).

Most of the atmospheric CH_4 is produced by the bacterial activities in extremely anaerobic ecosystems such as natural and cultivated wetlands, sediments, sewage, landfills, and the rumen of herbivorous animals. Of the wide variety of the sources, rice paddy fields are considered as one of the most important sources, taking into account the recent increase in the harvest area in the world. For this reason, field measurements of CH_4 emission from paddy fields have been performed in various part of the world to evaluate the amount of CH_4 emitted from paddy fields to the atmosphere (see Watson et al., 1992). These field studies showed that there were strong diurnal and seasonal variations of CH_4 emission from paddy fields during the cultivation periods, and that different processes were involved in the variations of the CH_4 emission. The most important process for CH_4 emission is production of CH_4 by methanogenic bacteria in the anaerobic layer of paddy soils. A positive correlation was observed between the seasonal variation of CH_4 production and that of CH_4 emission in paddy fields (Yagi and Minami, 1991). Another important process is oxidation of the produced CH_4 by methanotrophic bacteria taking place in the aerobic layers of soil and the rhizosphere of rice plants. The amount of CH_4 emitted from paddy fields to the atmosphere is mainly controlled by the balance of the production and oxidization of CH_4 in paddy soils.

The importance of the CH_4 oxidation has been demonstrated in laboratory experiments (de Bont et al., 1978; Holzapfel-Pschorn et al., 1985; Conrad and Rothfus, 1991). Field measurements showed that the proportion of CH_4 emission to CH_4 production ranged between 3-56% (Schütz et al., 1989a) and 10-81% (Sass et al., 1990). The remainder of the produced CH_4 is considered to be oxidized in the plowed layer of the paddy soil or leached to the lower layer. It is not clear whether this large range results from seasonal variability or from differences in agricultural practices, and which processes are controlling this variation. A laboratory incubation experiment demonstrated that the oxidation activity was much higher than the production activity in paddy soils when a sufficient amount of oxygen was present (Inubushi et al., 1990). However, there have been no field studies measuring the magnitude and distribution of CH_4 oxidation in paddy fields.

In this paper, we report results of the simultaneous measurements of emission, production, and oxidation of CH_4 in a Japanese rice paddy field during an entire cultivation period. We focused on seasonal changes of CH_4 production and oxidation in different depths of paddy soil, and their relationship to CH_4 emission, in order to understand factors controlling this process in paddy fields.

ISBN 1-56670-117-1/95/$0.00+.50

II. Experimental Design and Methods

A. Study Site and Soil

A field study was carried out in an experimental paddy field of the Ryugasaki Branch of the Ibaraki Agricultural Experiment Station (at latitude 35° 61' N, longitude 140° 13' E) during the 1991 growing season. The Station is located in part of the intensive rice cultivation area in the alluvial lowland of the Tone River. The soil of the paddy field is classified as Gley soil (Fluvaquent).

Two plots with an individual size of 100 m^2 were used for this study: 1) no rice straw (control), and 2) rice straw applied (straw). At these plots, a long-term field experiment on fertilization including mineral fertilizer with/without rice straw has been performed since 1978. Percolation loss from the ponded water amounted to 7.5 and 11.0 mm/day at the control and straw plots, respectively. The thickness of the plowed layer in the paddy soil was about 12 cm at both the plots. Some physical and chemical properties of the soils are listed in Table 1.

Rice (*Oryza sativa*, type *japonica*, cultivar Koshihikari) was cultivated according to conventional methods of the area. To the rice straw plot, rice straw from the previous growing season was cut and plowed back at a rate of 600 g/m^2. Prior to flooding, the compound mineral fertilizer (ammonium-potassium--chloride-phosphate) was incorporated into both of the control and straw plots at rates of 6.0 g N/m^2, 2.6 g P/m^2, and 5.0 g K/m^2. The fields were flooded on May 7 (day 0). Rice seedlings at the 3-4 leaf stage were transplanted on May 10 (day 3) being spaced at 15 x 30 cm. Paddy water level was maintained between 0 and 10 cm after flooding until the middle of June. From the end of June through the beginning of July, the paddy was drained for 3-5 days. This practice known as midsummer drainage is commonly performed over Japan for supplying molecular oxygen to the roots of the rice plants. After that, the irrigation water was supplied every 3-5 days until the maturing stage of rice plants. The mineral fertilizer at rates of 3.0 g N/m^2 and 2.5 g K/m^2 was topdressed to both of the plots on July 16 (day 70). The plots were drained on August 20 (day 105) and the rice was harvested on September 15 (day 131), yielding about 550 g/m^2 of brown rice. There were no significant differences in the growth of rice plants and yields between the mineral and the rice straw plot.

B. Methane Emission

The closed chamber method was used for the measurements of the CH_4 emissions from paddy fields into the atmosphere. The details of the chambers and the sampling system are described elsewhere (Yagi and Minami, 1993). The chamber was constructed from polycarbonate plates and has a cross-sectional area of 0.36 m^2 and a height of 0.5 or 1.0 m, which covers eight rice plants. Methane emissions were determined by measuring the temporal increase of the CH_4 concentration of the air inside the chamber during 30 min. The measurements were usually performed once a week throughout the flooding periods. The air inside the chamber was drawn through sampling tube by an air pump (Shibata, MP-2N) at a flow rate of 2 L/min, and 1 L of the air was collected in a 1 L Tedlar Bag. Air sampling was usually performed at three times (0, 15, and 30 min). All the measurements were performed in duplicate at each plot between 10 a.m. and 1 p.m.

The CH_4 concentration of the samples collected in the Tedlar Bag was determined by using a gas chromatograph (Shimadzu GC-12A) equipped with a gas sampler (Shimadzu MGS-5) and a flame ionization detector. Five milliliters of air sample was injected and separated on a molecular sieve 5A column (2.6 mm i. d. x 2.0 m) with 40 ml/min N_2 of carrier flow at 70 °C. Calibration was performed by using the CH_4 standard gases at the concentrations of 10 and 100 ppmv (Takachiho Kogyou Co., Ltd.).

C. Methane Production and Oxidation

Methane production rates and oxidation potential were measured six times during the cultivation period (five times during the flooding period and once after the drainage) by the *in situ* incubation method. Paddy soil was collected by using a PVC core sampler, having an inner diameter of 7 cm and a length of 14 cm. Secondary sampling holes (12 mm diameter) were drilled in both the sides of the sampler, which were covered by tape for sampling. Soil was sampled by inserting the sampler into paddy soil to the depth of 12 cm. After soil was sampled with paddy water at the top of the sampler, the bottom end of the sampler was

Table 1. Properties of the soil collected before flooding

	Control plot	Straw plot
Texture	SCL	SCL
pH (H_2O) (1:2.5)	5.9	6.1
Total C (%)	1.36	1.60
Total N (%)	0.13	0.15
C/N	10	11
Available N (μg/g)*	98	112
Free Fe (mg/g)¶	8.67	8.54

* Amount of nitrogen mineralized during 28 days under anaerobic incubation at 30°C: ¶ amount of iron extracted with hot water in the presence of sodium hydrosulfite and EDTA. (From Asami and Kumada, 1959.)

sharpened and covered with plastic film. Then the sampler was transported immediately to the laboratory adjacent to the field to take subcores for the incubation.

In the laboratory, the tape was removed and subcores at depths of 0-1, 2-3, 4-5, and 9-10 cm were taken by using a plastic syringe (5 ml) from which the needle end had been cut off. The subcore was transferred to a test tube (16 mm i. d. x 18 cm). The tubes were immediately flushed with N_2 in the case of measuring CH_4 production. The tubes were then evacuated and refilled with either N_2 (for CH_4 production) or a mixture of CH_4 and air (1:4, for CH_4 oxidation). After being closed with a butyl rubber stopper under the stream of the filling atmosphere, the tubes were transported to the field and buried back in paddy soil at the same depth collected. The tubes were incubated in the field for 7 days in the case of CH_4 production and for 2 or 3 days in the case of CH_4 oxidation. All of the incubations were made in duplicate at each depth. Preliminary study showed that both the rates of production and consumption of CH_4 were almost linear as time elapsed during those periods, and that the CH_4 consumption rate increased linearly with the increase in the initial CH_4 concentration in the range of 0-20% (Kumagai et al., 1993).

After those periods had elapsed, the tubes were collected from the field and stored on ice during transport to the laboratory for CH_4 analysis. Head space air samples were taken from each tube and CH_4 concentration was analyzed by using the same GC/FID system as for the emission measurements. The concentrations of CO_2 and O_2 in the head space air were also measured by using a gas chromatograph (Shimadzu GC-4A) equipped with a thermal conductivity detector. The total amount of the produced CO_2 was calculated according to Tsutsuki and Ponnamperuma (1987).

D. Temperature and Soil Eh

Temperature of air and paddy soil at the depths of 0, 2, 5, and 10 cm was continuously recorded during the cultivation periods by using Pt-resistance thermometers and a data logger (KADEC U-6). The redox potential (Eh) of paddy soil was measured by using Pt-tipped electrodes (Hirose Rika Co., Ltd.) and an oxidation-reduction potential meter (Toa RM-1K). For the measurements of soil Eh, the electrodes were inserted into the soil at the depths of 2, 5, and 10 cm, and kept in place throughout the cultivation periods. The potentials measured against AgCl reference electrode were reported in reference to the hydrogen electrode. All the Eh measurements were made at least in triplicate at each depth.

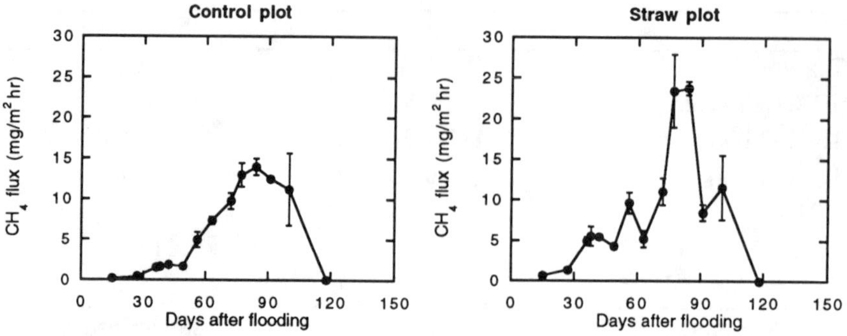

Figure 1. Seasonal variations of CH_4 fluxes from the control (left) and straw (right) plots during 1991 cultivation period. Bars indicate the deviation between duplicate measurements.

III. Results

A. Methane Emission and Soil Eh

Figure 1 displays seasonal variations of CH_4 fluxes from the control and straw plots during 1991 cultivation period. In both of the plots, pronounced variations in the CH_4 fluxes were observed. The fluxes began to increase 3-4 weeks after flooding, and gradually increased till July, although there were fluctuations, especially in the straw plot. At the end of July, corresponding to heading period of rice plants, the fluxes showed a rapid rate of increase and had peak fluxes on July 30 (day 84) of 14.0 and 23.8 mg/m^2 hr in the control and straw plots, respectively. During this period, the daily mean soil temperature at the 2 cm depth had a peak value of 27.4 °C. Following this peak, the fluxes decreased and were almost negligible on September 2 (day 118). The straw plot always showed higher CH_4 fluxes than the control plot except on July 9 (day 63) and August 6 (day 91). The averaged fluxes of CH_4 during the flooding period (from May 15 to August 20) were 5.97 for the control and 8.61 mg/m^2 hr for the straw plots. Total emission rates of CH_4 during the cultivation period were estimated to be 15.3 and 20.5 g/m^2 for the control and straw plots, respectively. The total emission rates during the previous three cultivation periods at the same site ranged between 8.2-40.6 and 19.9-43.1 g/m^2 in the control and straw plots, respectively, and the values in the 1991 cultivation period were in the range of that observed in previous years (Yagi and Minami, 1990, 1993). Mean soil temperatures during the flooding period (from May 15 to August 20) were 23.4, 23.1, 23.0, and 22.6 °C at the 0, 2, 5, and 10 cm depths, respectively.

Figure 2 displays the seasonal variations in soil Eh at the 2, 5, and 10 cm depths in the control and straw plots. The Eh in the plowed layer of the soil gradually decreased during the first five weeks after flooding. After that, significantly high values were observed at the straw plot on June 25 (day 49) and July 9 (day 63). These periods coincided with the periods of midsummer drainage and the following week. At the control plot, the Eh at the 2 and 5 cm depths was relatively stable in the range between -261 and -152 mV during the period that large CH_4 emissions were observed (from July 2 to August 15, or day 56 to 100). The Eh at the 10 cm depth in the control plot was always higher than those at the shallower depths ranging between -192 and +87 mV during the same period. In the straw plot, the Eh at the 10 cm was lower than that in the control plot, ranging between -250 and -129 mV during this period. The Eh at all the depths rapidly increased at the end of August after the irrigation was interrupted and the paddy field was intermittently drained. The period of low redox potential in the paddy soil corresponded closely to the period of CH_4 emission from the paddy field to the atmosphere. Soil pH was about 6.0 prior to flooding and ranged from 6.0 to 6.8 during the flooding periods.

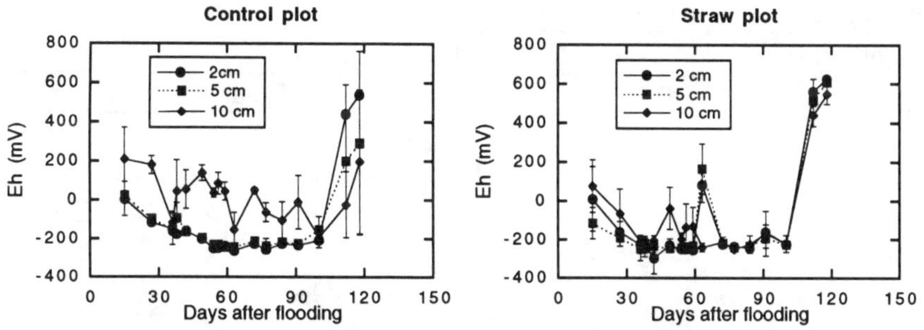

Figure 2. Seasonal variations of soil Eh (redox potential) at depths of 2, 5, and 10 cm in the control (left) and straw (right) plots during 1991 cultivation period. Bars indicate the standard deviation amongst triplicate measurements at each depth.

Table 2. Vertical distribution of CH_4 production rates and their seasonal variation during the cultivation period

		CH_4 production rate (μg/g soil day)				
Plot	Incubation	0-1 cm	2-3 cm	4-5 cm	9-10 cm	Average
Control	June 18	<u>0.26</u>	0.08	0.20	0.07	0.15
	July 2	<u>2.20</u>	0.95	0.67	0.25	1.02
	July 18	1.19	<u>1.48</u>	0.72	0.41	0.95
	July 30	3.29	<u>4.37</u>	2.14	1.34	2.79
	August 13	<u>4.90</u>	1.26	0.99	0.86	2.00
	September 2	0.46	0.01	0.08	<u>0.95</u>	0.38
Straw	June 18	<u>1.50</u>	1.42	0.19	0.24	0.84
	July 2	3.18	<u>3.52</u>	2.85	1.16	2.68
	July 18	0.44	2.67	<u>2.99</u>	2.20	2.08
	July 30	1.95	1.74	<u>3.16</u>	2.64	2.37
	August 13	2.30	2.86	3.15	<u>3.95</u>	3.06
	September 2	0.01	0.00	0.02	<u>0.25</u>	0.07

The underline indicates the maximum value in the vertical profile.

B. Methane Production

The vertical distribution of CH_4 production rates and its seasonal variation in the paddy soils are listed in Table 2. The production rates during the flooding period ranged from 0.07 to 4.90 μg/g soil day and from 0.19 to 3.95 μg/g soil day for the control and straw plots, respectively. The production rates after the drainage (incubated from September 2) were significantly lower than those before drainage except in the case of the deepest layer (9-10 cm). For the control plot, the rates in the upper layer (0-3 cm) were higher than those in the lower layer (4-10 cm) in most of the cases. In contrast, the zone of maximum production shifted to the deeper layer as time elapsed after flooding of the straw plot. The data from the incubation of July 18 at 0-1 cm showed lower values, which possibly was an influence of the topdressing of mineral fertilizer performed on July 16.

Table 3. Daily average soil temperature and its range in the paddy soil during the incubation experiments

Incubation period	Temperature (°C) 0-1 cm	2-3 cm	4-5 cm	9-10 cm
June 18-25	22.0 ± 3.1	22.0 ± 1.8	21.8 ± 1.5	21.8 ± 0.8
July 2-9	23.4 ± 6.0	23.1 ± 2.9	23.1 ± 2.5	22.7 ± 1.3
July 18-25	25.0 ± 5.0	24.4 ± 2.6	24.3 ± 2.1	23.6 ± 1.2
July 30-Aug. 6	25.3 ± 4.4	25.3 ± 2.3	25.2 ± 2.0	25.0 ± 1.3
August 13-20	22.7 ± 5.2	22.8 ± 2.4	22.6 ± 2.0	22.5 ± 1.0
September 2-9	25.3 ± 6.4	25.2 ± 3.2	25.1 ± 2.6	24.7 ± 1.4

Table 4. Vertical distribution of the amount of carbon mineralized (CO_2-C + CH_4-C) during the incubation periods, and percentage of CH_4-C

Plot	Incubation start	CO_2-C + CH_4-C (μg C/g soil day) (precentage of CH_4-C)				
		0-1 cm	2-3 cm	4-5 cm	9-10 cm	Average
Control	June 18	0.96 (21)	0.47 (13)	0.67 (22)	0.69 (7)	0.70 (16)
	July 2	3.10 (53)	1.33 (53)	0.98 (51)	0.84 (23)	1.56 (45)
	July 18	1.63 (55)	1.69 (66)	1.17 (46)	0.92 (34)	1.35 (50)
	July 30	3.47 (71)	3.90 (84)	2.14 (75)	1.54 (66)	2.76 (74)
	August 13	4.30 (86)	1.25 (76)	1.06 (70)	1.02 (64)	1.91 (74)
Straw	June 18	2.09 (54)	1.68 (64)	0.98 (14)	1.00 (18)	1.44 (38)
	July 2	3.11 (77)	3.11 (85)	2.66 (80)	1.53 (57)	2.60 (75)
	July 18	1.07 (31)	2.45 (82)	2.74 (82)	2.36 (70)	2.16 (66)
	July 30	1.95 (75)	1.68 (78)	2.79 (85)	2.46 (80)	2.22 (80)
	August 13	2.37 (73)	2.47 (87)	2.75 (86)	3.37 (88)	2.74 (84)

The underline indicates the maximum value in the vertical profile.

Daily average soil temperature and its range at the 0, 2, 5, and 10 cm depths during the incubation periods are listed in Table 3. The shallower layer always showed higher temperatures and a wider diurnal variation. However, the differences in the average temperature between the depths of 0 and 10 cm were within 1.4 °C. Therefore, soil temperature may not have a significant effect on the vertical distribution and seasonal change in CH_4 production in the soil.

The amounts of CO_2 produced during the incubation periods were also measured. The total amounts of CO_2-C and CH_4-C can be considered as the amount of carbon mineralized during the incubation periods. These total amounts, and the proportion of CH_4-C to the total amounts during the flooding period, are listed in Table 4. The total amounts of mineralized carbon ranged from 0.47 to 4.30 μg C/g soil day and from 0.98 to 3.37 μg C/g soil day for the control and straw plots, respectively. The vertical distribution and seasonal change were similar to those of CH_4 production as shown in Table 2. The upper layer showed higher values for the control plot throughout the flooding period. On the other hand, the vertical distribution of the values was more uniform and the maximum values shifted to the deeper layers as time after flooding elapsed for the straw plot. The proportion of CH_4-C also showed significant difference between the two plots. Especially, CH_4-C occupied more than 70% below 2 cm depth after the middle of July for the straw plot, while it ranged from 34 to 84% for the control plot.

Table 5. Vertical distribution of CH_4 oxidation potential and its seasonal variation during the cultivation period

Plot	Incubation start	CH_4 oxidation potential (μg/g soil day) 0-1 cm	2-3 cm	4-5 cm	9-10 cm	Average
Control	June 18	82.6	44.2	35.2	40.6	50.7
	July 2	116.9	41.7	25.9	16.8	50.3
	July 18	277.9	80.3	15.6	15.0	97.2
	July 30	85.1	17.7	25.2	12.1	35.0
	August 13	71.9	37.5	27.9	26.0	40.8
	Sept. 2	83.4	48.3	30.9	31.4	48.5
Straw	June 18	74.7	41.0	41.8	39.7	49.3
	July 2	123.9	60.8	49.3	44.0	69.5
	July 18	264.9	59.9	14.4	38.5	94.4
	July 30	70.6	30.0	33.3	8.2	35.5
	August 13	136.9	31.5	34.3	45.1	62.0
	Sept. 2	28.6	11.6	14.0	3.0	14.3

The underline indicates the maximum value in the vertical profile.

C. Methane Oxidation

In paddy soils, molecular oxygen is rapidly consumed after flooding. Oxygen is present only in the thin surface layer of paddy soils and probably in the rhizosphere of the rice plants during the flooding period. Therefore we refer to the CH_4 consumption rates measured in the incubation under the mixture of CH_4 and air as CH_4 oxidation potential.

The vertical distribution of CH_4 oxidation potential in the paddy soil and its seasonal changes are listed in Table 5. The highest value of the oxidation potential was always observed in the surface layer (0-1 cm), ranging from 28.6 to 277.9 μg/g soil day. The values in the deeper layers were less than 58% of the value in the surface layer. There was no significant difference in the oxidation potential between the control and straw plots. The maximum values were observed in the incubation right after the topdressing in the surface layer for both the control and straw plots.

D. Comparison among Emission, Production, and Oxidation

Methane production rates and oxidation potential per unit area were calculated by multiplying the data of Table 2 and 5 by the bulk density of the soil and listed in Table 6, along with CH_4 emission rates. Methane production rates per unit area were calculated by integrating the data of Table 2 for 1-12 cm. Only the data of the surface layer (0-1 cm) were used for CH_4 oxidation potential because aerobic conditions are limited to only the surface layer in flooded soils except for rhizosphere of plant roots. This partitioning of the plowed layer into production and oxidation layers is based on the differentiation of the reduced and oxidized layers in paddy soils. However, there are very few studies on the thickness of these two layers in paddy soils and their seasonal changes, especially for the surface oxidized layer. Therefore, it should be noted that the values for the oxidation in Table 6 do not precisely represent the real oxidation rates in the surface layer of the paddy soil, but represent the potentials for CH_4 oxidation.

Methane production rates per unit area ranged from 0.012 to 0.224 g/m^2 day and from 0.017 to 0.404 g/m^2 day for the control and straw plots, respectively. The values for the straw plot were always higher than those for the control plot during the flooding period. For both plots, the production rates increased as time elapsed following flooding and showed maximum values from the end of July to August. The period of the

Table 6. Comparison among production rate, oxidation potential, and emission rate of CH_4 in the paddy field

Plot	Incubation start	Production rate (P) 1-12 cm (g/m² day)	Oxidation potential (O) 0-1 cm (g/m² day)	Emission rate (E) (g/m² day)	O/P	E/P
Control	June 18	0.012	1.33	0.046	110.8	3.83
	July 2	0.053	1.89	0.119	32.6	2.25
	July 18	0.073	4.48	0.235	55.3	3.22
	July 30	0.224	1.37	0.335	5.6	1.50
	August 13	0.107	1.16	0.269	10.8	2.51
	Sept. 2	0.066	1.35	0.001	25.0	0.02
Straw	June 18	0.045	1.20	0.131	22.4	2.91
	July 2	0.220	2.00	0.231	8.5	1.05
	July 18	0.280	4.27	0.265	15.6	0.95
	July 30	0.299	1.14	0.571	4.1	1.91
	August 13	0.404	2.21	0.277	5.8	0.69
	Sept. 2	0.0174	0.46	0.000	32.9	0.00

peaks in the production coincides with the period of the peaks observed in the CH_4 emission. The production rates significantly decreased after the drainage.

Methane oxidation potentials in the surface layer of the paddy soil per unit area ranged from 1.16 to 4.48 g/m² day and from 0.46 to 4.27 g/m² day for the control and straw plots, respectively. The potentials showed maximum values in the middle of July for both plots. There was no obvious seasonal pattern in the potential except for the maximum values. No significant difference was observed in the oxidation potential between the plots. In spite of limiting the CH_4 oxidation potential to 0-1 cm layer, the oxidation potentials were always much higher than the production rates. The ratio of the oxidation potential to the production rates ranged from 6.1 to 110.8 and from 3.8 to 27.1 for the control and straw plots, respectively. The ratio for the straw plot was always lower than that for the control plot, due to the higher production rate in the straw plot.

The relation between CH_4 production rate and CH_4 emission rate was shown in Table 6 by expressing the ratio of the emission rate to the production rate. The ratio ranged from 1.50 to 3.83 and from 0.69 to 2.9 for the control and straw plots, respectively, during the flooding period. In most cases, the straw plot showed a lower value than the control plot. The highest values were observed in June at each plot. The Ch_4 emission rates exceeded the CH_4 production rates in soil between 1 and 12 cm depth in all cases at the control plot and in some cases at the straw plot during the flooding period. After the drainage (data of the incubation from September 2), almost all the CH_4 produced in soil was apparently oxidized in soil layer.

Figure 3 plots CH_4 production rates against the CH_4 emission rates (Figure 3a), and the production rates divided by the oxidation potential against the emission rates (Figure 3b) for the data obtained during the flooding period. The Figures show positive correlations between the production or production/oxidation ratio and the emission. Figure 3b shows a better correlation ($r = 0.844$) than Figure 3a ($r = 0.733$). If the data of the incubation from July 18 and September 2 were eliminated from the correlation because of the possible influence of topdressing, respectively, the correlation coefficients become 0.713 and 0.953 for the production rates against the emission rates and for the production/oxidation ratio against the emission rates, respectively.

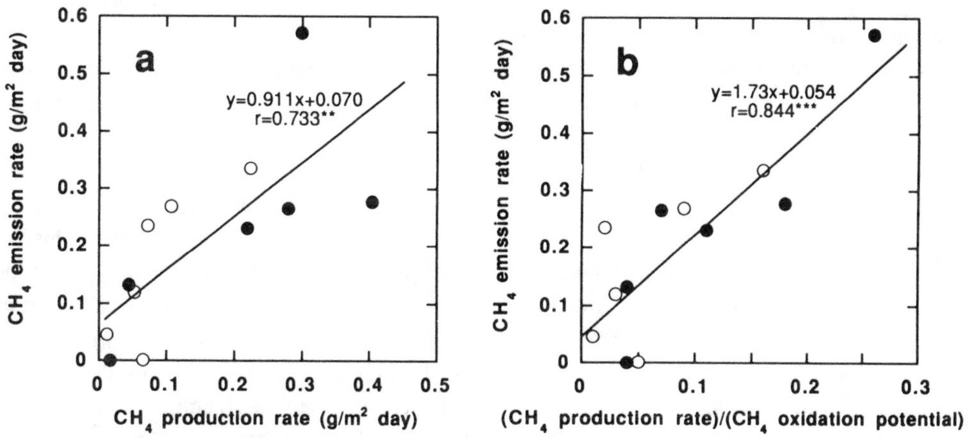

Figure 3. Relationship amongst production, oxidation, and emission of CH_4 in the paddy field: a. plotted CH_4 production rates to the CH_4 emission rates, b. plotted CH_4 production rates divided by CH_4 oxidation potential to CH_4 emission rates for the data obtained during the flooding period. Open and closed circles represent the data for the control and straw plots, respectively.

IV. Discussion

Methane emissions from the rice paddy field in Ryugasaki, Japan showed a pronounced seasonal variation (Figure 1). Although there was an interannual variation on total emission rates, the pattern of the seasonal variation in Ryugasaki in 1991 was almost identical to those of the previous years, showing small emissions in the early season of cultivation, large emissions during mid- to late-summer, and negligible after the fields were drained (Yagi and Minami, 1990, 1993). This pattern was similar to those observed at rice paddy fields in some other locations (e.g. Cicerone et al., 1983; Schütz et al., 1989b; Sass et al., 1990; Yagi et al., 1994). Several factors are known to control temporal variation of CH_4 emission from paddy fields, such as temperature, availability of organic substrates, plant activity, and agricultural practices (Conrad, 1989).

The other important factor controlling emission and production of CH_4 in paddy fields is redox potential (Eh) of the paddy soil, because the development of anaerobic conditions in soil is essential for the activity of methanogenic bacteria. In our measurements, the variation in CH_4 emission was negatively correlated with the variation in soil Eh in the plowed layer as shown in Figures 1 and 2. A significant difference in soil Eh at the 10 cm depth was observed between the control and straw plot. This difference in soil Eh indicated that extremely anaerobic conditions developed throughout the plowed layer in the soil of the straw plot during the latter stage of the cultivation, while the extremely anaerobic conditions were limited to the upper part of the plowed layer in the soil of the control plot. The progress of soil reduction is controlled by the relative abundance of electron donors and electron acceptors in the soil (Takai and Kamura, 1966, Patrick et al., 1985). In the absence of O_2, the main electron-accepting chemical species are Fe^{3+}, Mn^{4+}, NO_3^-, and SO_4^{2-}. The main electron-donor is readily decomposable organic matter. The data listed in Table 4 show that the amount of carbon mineralized during the anaerobic incubation decreased with depth in the plowed layer of the control plot, while it was more uniform within the plowed layer of the straw plot. The vertical and seasonal changes of the amount of mineralized carbon generally coincide with the changes in Eh and the CH_4 production rates in the soil. Therefore, it is suggested that the higher CH_4 production and emission in the straw plot than the control plot resulted from widely developed anaerobic layer in the soil of the straw plot due to higher amount of readily decomposable organic matter, being combined with higher amount of readily decomposable organic matter itself as the substrate for CH_4 production.

Methane production in soils of Texas paddy fields decreased with depth throughout the cultivation period, similar to observations of the control plot at Ryugasaki (Sass et al., 1990). In contrast, the results in an Italian paddy field showed the reverse distribution pattern (Schütz et al., 1989a). They found that CH_4 production below 2 cm depth increased with depth and CH_4 was still produced at a depth of 25 cm in the Italian paddy field. Studies on redox potentials of soils in Japanese paddy fields showed that the Eh of the B horizon of the soil lying above the gley horizon was only slightly affected by flooding and drainage. The oxidized condition persisted throughout the year due to the lack of readily decomposable organic matter (Aomine, 1962). In Ryugasaki site, the gley horizon exists below 30 cm and the subsoil above this horizon seems to be the oxidizing layer. The difference in the vertical distribution of CH_4 production may be due to different hydrological conditions and a different distribution of readily decomposable organic matter among the fields.

The CH_4 emission rates to the atmosphere often exceeded the CH_4 production rates in soil between 1 and 12 cm depthe during the flooding period (Table 6). In contrast to our results, CH_4 emitted from Texas paddy fields averaged 42% of CH_4 produced in soil (Sass et al., 1990). In laboratory and field studies conducted on an Italian paddy, the percentage decreased by up to 3% during the season and was paralleled by a decrease in CH_4 ebullition and an increase in plant-mediated CH_4 emission (Holzapfel-Pschorn et al., 1985; Schütz et al., 1989a). There are several possible reasons for our high emission/production ratio compared with the previous studies: a) a substantial CH_4 production in soil below 12 cm depth, b) effects of long incubation period on CH_4 production, and c) effects of diurnal variation of soil temperature by *in situ* incubation. It is unlikely that the deep soil layer has a large contribution to total CH_4 production in soil, because of low contents of readily decomposable organic mattter and of small extension of rice roots especially in early stage of the cultivation. The second possibility is difficult to interpret. We have confirmed a linear increase in CH_4 betweeen 2 and 9 days by anaerobic incubation of soil samples. Schütz et al. (1989a) observed a linear production of CH_4 within 24 hours. It is possible that the soil sampling procedure, especially cutting rice roots, decrease CH_4 production rate in the incubation tube as time elapsed due to inactivation of rice rhizosphere. It is also possible, on the other hand, that it accelerates the production rates in short period after the sampling by liberation of the substrates.

It is known that microbial-mediated CH_4 oxidation ubiquitously occurs in soil and aquatic environments, where it modulates CH_4 emission. It has been demonstrated that half of the CH_4 production in moss areas of tundra is oxidized in the soil (Whalen et al., in press). In landfill cover soil, 10 to 50% of gross landfill CH_4 production was consumed by oxidation (Whalen et al., 1990). Aerated soils act as a sink for atmospheric CH_4. A global soil sink estimate at 29 Tg/yr results from parameterizing soil CH_4 oxidation with soil texture (Dörr et al., 1993). The gross global microbial oxidation of CH_4 in the environments was estimated to be about 700 Tg/yr, which is about 200 Tg/yr larger than global emission to the atmosphere (Reeburgh et al., 1993).

In this study, we calculated CH_4 oxidation, assuming the thickness of the oxidized layer to be 1 cm. However, the thickness of the surface oxidized layer is controlled by the balance of the diffusion rate of molecular oxygen through the flooding water and its consumption rate in the soil. Soil having a high consumption rate of oxygen due to a high content of readily decomposable organic matter will form a very thin oxidized layer, whereas soil having a low consumption rate of oxygen due to a low content of readily decomposable organic matter will form a thick oxidized layer. The thickness of the oxidized layer also varies with water management during the cultivation period. Patrick et al. (1985) have pointed out that oxygen is usually present in a layer extending from a few millimeters to 1 cm below the soil-water interface in paddy fields. Therefore, the oxidation potentials listed in Table 6, which are calculated with assuming the thickness to be 1 cm, are likely the upper limit of the oxidation rate in the surface oxidized layer. However, even if it is assumed that the oxidation of CH_4 was taking place within the top 1 mm layer in the paddy soil, the ratio of the oxidation rates to the production rates ranged between 0.4-11 during the flooding period. This likelyhood indicates that the oxidation of CH_4 in the surface layer of paddy soil has a very high potential for oxidizing CH_4. It is known that, due to the high oxidation potential of the surface layer, CH_4 transport through the rice plants greatly exceeds CH_4 escape by the diffusion across the surface layer (Cicerone and Shetter, 1981; Schütz et al., 1991).

Another possible site for CH_4 oxidation in paddy soils is the rhizosphere of rice plants. The aquatic plants have an aerenchyma system which allow gas exchange between the roots and the atmosphere. By this system oxygen diffuses into the roots and from there eventually into the soil (Schütz et al., 1991). It was demonstrated that a greater number of methanotrophic bacteria were present in the rhizosphere soil than in other parts of the soil in paddy fields (de Bont et al., 1978). However laboratory culture experiments using acetylene as a inhibitor of methanotrophic activity showed that bacterial oxidation of CH_4 in the rhizosphere was of minor importance (de Bont et al., 1978; Miura et al., 1992).

The importance of rhizosphere in dynamics of CH_4 in rice paddy fields is also obvious from the comparison between the production rates and the emission rates. Further studies such as direct tracer experiments of CH_4 production and oxidation, and the inhibition experiments are needed to determine the rates of production and oxidation of CH_4 oxidation in the rhizosphere and their contribution to the CH_4 dynamics in rice paddy fields.

V. Summary

We have made the simultaneous field measurements of emission, production, and oxidation of CH_4 in a Japanese rice paddy field. The average emission rates of CH_4 during the flooding period were 5.97 and 8.61 mg/m^2 hr for the control plot and the rice straw applied plot, respectively. The calculated annual emission rates were 15.3 g/m^2 (control) and 20.5 g/m^2 (straw). Production rates of CH_4 ranged between 0.07 and 4.90 μg/g soil per day, corresponding to 0.01 and 0.40 g/m^2 per day in the plowed layer (0-12 cm). The highest productions were observed in the 0-3 cm layer throughout the flooding period in the control plot. In contrast, the zone of maximum production shifted to the deeper layer as time elapsed after flooding in the straw plot, due to large amount of readily decomposable organic matter and low soil Eh. The CH_4 emission rates accounted for 7-40% of the CH_4 production rates in the paddy soil. Oxidation potentials for CH_4 ranged between 8.2 and 278 μg/g soil per day. The top 1 cm layer always showed the highest oxidation potential in both of the plots. Positive correlations between the production or production/oxidation ratio with the emission were observed. Assuming that oxidation of CH_4 was taking place in the top 1 cm layer in the paddy soil, the ratios of the oxidation rates to the production rates were 4-111 during the flooding period, indicating that the paddy soil has a high potential for oxidizing CH_4 in the presence of oxygen. These results suggest that the greater part of the produced CH_4 in paddy soil is probably oxidized either in the surface layer of the paddy soil or in the rhizosphere of rice plants.

Acknowledgments

We thank Mr. Mitsuru Kubota and Mr. Jiro Aida of the Ibaraki Agricultural Experiment Station for their help with the management of the paddy field in Ryugasaki.

References

Aomine, S. 1962. A review of research on redox potentials of paddy soils in Japan. *Soil Sci.* 94:6-13.

Asami, T. and K. Kumada. 1959. A new method for determining free iron in paddy soils. *Soil Plant Food* 5:141-146.

Blake, D.R. and F.S. Rowland. 1988. Continuing worldwide increase in tropospheric methane. 1978 to 1987. *Science* 239:1129-1131.

Cicerone, R.J. and J.D. Shetter. 1981. Sources of atmospheric methane: measurements in rice paddies and a discuss. *J. Geophys. Res.* 86:7203-7209.

Cicerone, R.J., J.D. Shetter, and C.C. Delwiche. 1983. Seasonal variation of methane flux from a California rice paddy. *J. Geophys. Res.* 88:11022-11024.

Cicerone, R.J. and R.S. Oremland. 1988. Biogeochemical aspects of atmospheric methane. *Global Biogeochem. Cycles* 2:299-327.

Conrad, R. 1989. Control of methane production in terrestrial ecosystems, p. 39-58. In: M.O. Andreae and D.S. Schimel (eds.), Exchange of trace gases between terrestrial ecosystems and the atmosphere. John Wiley and Sons Ltd., Chichester.

Conrad, R. and F. Rothfuss. 1991. Methane oxidation in the soil surface layer of a flooded rice field and the effect of ammonium. *Biol. Fertil. Soils* 12:28-32.

Crutzen, P.J. 1987. Role of the tropics in atmospheric chemistry. p. 107-130. In: R.E. Dickinson (ed.), The geophysiology of Amasonia: vegetation and climate interactions. John Wiley and Sons, NY.

de Bont, J.A.M., K.K. Lee, and D.F. Bouldin. 1978. Bacterial oxidation of methane in a rice paddy. *Ecol. Bull.* 26:91-96.

Dörr, H., T. Moore, J. Bubier and B. Lafleur. 1993. Soil texture parameterization of methane uptake in aerated soils. *Chemosphere* 26:697-714

Holzapfel-Pschorn, A., R. Conrad, and W. Seiler. 1985. Production, oxidation and emission of Methane emission in rice paddies. *FEMS Microbiol. Ecol.* 31:343-351.

Inubushi, K., M. Umebayashi and H. Wada. 1990. Methane emission from paddy fields. *Trans. 14th Int. Congr. Soil Sci.* 2:249-254.

Kumagai, K, K. Yagi, H. Tsuruta, and K. Minami. 1993. The measurement of methane production and oxidation in paddy soils. *Jpn. J Soil Sci. Plant Nutr.* 64:431-434.

Miura, Y., A. Watanabe, M. Kimura, and S. Kuwatsuka. 1992. Methane emission from paddy field (Part 2) main route of methane transfer through rice plant, and temperature and light effects on diurnal variation of methane emission. *Environ. Sci.* 5:187-193

Patrick, Jr., W.H., D.S. Mikkelson, and B.R. Wells. 1985. Plant nutrient behavior in flooded soil. p. 197-228. In: *Fertilizer technology and use*, 3rd Ed., Soil Sci. Soc. Am., Madison.

Reeburgh, W.S., S.C. Whalen, and M.J. Alperin. 1993. The role of methylotrophy in global methane budget. In: Proc. 7th Inter. Symp. Microbial Growth C_1 Compounds.

Sass, R.L., F.M. Fisher, P.A. Harcombe, and F.T. Turner. 1990. Methane production and emission in a Texas rice field. *Global Biogeochem. Cycles.* 4:47-68.

Schütz, H., W. Seiler, and R. Conrad. 1989a. Processes involved in formation and emission of methane in rice paddies. *Biogeochem.* 7:33-53.

Schütz, H., A. Holzapfel-Pschorn, R. Conrad, H. Rennenberg, and W. Seiler. 1989b. A 3-year continuous record on the influence of daytime, season, and fertilizer treatment on methane emission rates from an Italian rice paddy. *J. Geophys. Res.* 94:16405-16416.

Schütz, H., P. Schröder, and H. Rennenberg. 1991. Role of plants in regulating the methane flux to the atmosphere. p. 29-63. In: T.D. Sharkey, E.A. Holland, and H.A. Mooney (eds.), Trace gas emission by plants. Academic Press, NY.

Steel, L.P., P.J. Fraser, R.A. Rasmussen, M.A.K. Khalil, T.J. Conway, A.J. Crawford, R.H. Gammon, K.A. Masarie, and K.W. Thoning. 1987. The global distribution of methane in the troposphere. *J. Atmos. Chem.* 5:125-171.

Takai, Y. and T. Kamura. 1966. The mechanism of reduction in waterlogged paddy soil. *Folia Microbiol.* 11: 304-313.

Tsutsuki, K. and F.S. Ponnamperuma. 1987. Behavior of anaerobic decomposition products in submerged soils - effects of organic material amendment, soil properties, and temperature. *Soil Sci. Plant Nutr.* 33: 13-33.

Watson, R.T., L.G. Meila Filho, E. Sanhueza, and A. Janetos. 1992. Greenhouse gases: Sources and Sinks, p. 29-46. In: J.T. Houghton, B.A. Callander, and S.K. Varney (eds.), Climate change 1992, the supplementary report to the IPCC scientific assessment. Cambridge University Press, Cambridge.

Whalen, S.C., W.S. Reeburgh, and K.A. Sandbeck. 1990. Rapid methane oxidation in a landfill cover soil. *Appl. Environ. Microbiol.* 56:3405-3411.

Whalen, S.C., W.S. Reeburgh, and C.E. Reimers. In press. Processes controlling methane fluxes from tundra environments. In: J.F. Reynords and J.D. Tenhunen (eds.), Landscape function: implications for ecosystem response to distribution. A case study in arctic tundra. Springer-Verlag, NY.

Yagi, K. and K. Minami. 1990. Effect of organic matter application on methane emission from some Japanese paddy fields. *Soil Sci. Plant Nutr.* 36: 599-610.

Yagi, K. and K. Minami. 1991. Emission and production of methane in the paddy fields of Japan. *Jpn. Agric. Res. Quart.* 25:165-171.

Yagi, K. and K. Minami. 1993. Spatial and temporal variations of methane flux from a rice paddy field, p. 353-368. In: R.S. Oremland (ed.), Biogeochemistry of global change: radiatively active trace gases. Chapman and Hall Inc., NY.

Yagi, K., P. Chairoj, H. Tsuruta, W. Cholitkul and K. Minami. 1994. Methane emission from rice paddy fields in the central plain of Thailand. *Soil Sci. Plant Nutr.* 40:29-37.

CHAPTER 20

Methane Emission in a Flooded Rice Soil with and without Algae

Z.P. Wang, C.R. Crozier, and W.H. Patrick, Jr.

I. Introduction

Flooded rice (*Oryza sativa L.*) cultivation emits significant quantities of CH_4 produced by microbial, anaerobic decay of organic matter, contributing about 20% of the global CH_4 budget (Bouwman and Sombroek, 1990). Methane emission from a particular source is the result of both production and consumption of CH_4 (Frenzel et al., 1992). Methane production in flooded rice fields is a strict anaerobic microbiological process performed by CH_4 generating bacteria. The activity of methanogens is closely related to soil biochemical and physical properties such as carbon sources, soil pH, soil Eh, temperature and fertilization (Holzapfel et al., 1986., Yagi and Minami, 1990; Lindau et al., 1991; Wang et al., 1993a). Methane consumption by soil is associated with microbiochemical oxidation of CH_4 by methanotrophs in soil-water system and with the soil physical entrapment capacity for CH_4 (Cappenberg, 1972; Wang et al., 1993a). Before release into the atmosphere, biogenic CH_4 formed in the anaerobic layer of flooded soil passes through the surface oxidized zone which is formed after submergence of soil (Ponnamperuma, 1972). It is assumed that large methanogenic populations, associated with anaerobic conditions low in the soil profile, generate CH_4 which is utilized by methanotrophic bacteria as it passes through the upper oxidized layer (Wagatsuma et al., 1992).

The presence of algae on the surface of a flooded soil may also affect CH_4 emission because of the release of O_2 during algal photosynthesis. Harrison (1914) observed that the gases which occur in rice soils consist mainly of CH_4 and N_2 together with small amounts of CO_2 and H_2. Aiyer (1920) showed that, with an exception of N_2, the other gases found in rice soil were almost undetectable as long as the surface of the soil is not disturbed. An observation of an "organized film" which covered the soil surface was considered to possess the power of arresting and assimilating these gases as well as lead to an increased output of O_2 at the surface of the soils.

The possible material of the "organized film" could be a thin layer of algae which are commonly observed in most rice fields. The relationship between CH_4 emission and the photosynthetic activity of algae in rice soil has been not demonstrated. Undoubtedly, the O_2 released during algal photosynthesis affects the biochemical characteristics of soils. Consequently, CH_4 emission may be affected via increased CH_4 oxidation.

Rice and other aquatic plants also play a significant role in both CH_4 emission and consumption because of their special physiological characteristics (Kimura et al., 1991). The development of aerenchyma in hygrophytes serves as conduits of both CH_4 to the atmosphere and O_2 to the root zone (Seiler et al., 1984). Theses aerenchyma could lead to either CH_4 emission or increased CH_4 oxidation.

The present study describes greenhouse and laboratory experiments with the initial objectives of evaluating: 1) the effects of rice plant development stage on CH_4 emission; 2) methane entrapment in the soil profile; and 3) effects of living rice plants roots on soil Eh. During the experiments unexpected algal

ISBN 1-56670-117-1/95/$0.00+.50

growth was associated with a reduction in CH_4 emission, so additional manipulations were performed to test this relationship.

II. Materials and Methods

The soil used was a silt loam soil (Typic Albaqualf) sampled from Crowley, Louisiana, USA. The slightly acidic soil (pH 5.7) is commonly used for rice production. The soil contains 1.57% of organic matter and its cation exchange capability is 9.4 meq.100 g^{-1}. The soil was air-dried and ground to pass a 1 mm sieve then amended with rice straw (above-ground parts of rice plant after harvest, C:N ratio 57) which had been ground to pass 0.2 mm sieve.

A. Greenhouse Experiment

1. Soil Microcosm

Transparent PVC containers (21 cm in height and 7.5 cm in diameter) were used as soil microcosms. Four hundred grams of Crowley silt loam soil amended with 0.5% of rice straw which was mixed with the soil. Four hundred ml of deionized water was added, flooding the soil and resulting in a 2 cm layer of standing water. Soil samples were pre-incubated for one month so that the added rice straw would undergo rapid initial transformations prior to the transplanting of rice seedlings. Average greenhouse temperatures were 32 °C (day) and 27 °C (night).

2. Rice Plants

Rice seeds were soaked in water for two days and sowed into a sand bed at room temperature (about 25 °C). After 18 days (July 9-July 27,1992), seedlings were big enough to be transplanted.

The greenhouse experiment consisted of two treatments: (1) with rice and (2) without rice. Two healthy seedlings were transplanted into each of 15 replicate microcosms (with rice), and 15 replicate microcosms left unplanted (without rice). Urea was added 7 days after transplanting to both treatments at a rate of 100 mg N kg^{-1} of soil. Microcosms were temporarily covered with another PVC container fitted with a rubber septum. Methane accumulated in the headspace (810 ml) in 1 hour was measured by Perkin-Elmer 900 gas chromatograph (Perkin-Elmer Corp., Norwalk, CT) equipped with a flame ionization detector. Methane emission was measured at 36, 43, 50, 57, 64, 71, 78, 85, and 92 days after transplanting.

B. Laboratory Experiments

1. Methane Entrapment and Rice Roots Distribution in Soil Profile

Three replicate microcosms from each of the greenhouse experiments were frozen after 57 and 92 days of rice growth. The frozen soil cores (10 cm in length and 7.5 cm in diameter) were sliced into 4 depth layers (0-2, 2-4, 4-6, and >6 cm). Soil samples of each depth layer were weighed and loaded into 250 ml plastic beakers. About 60 ml deionized water was added to adjust the soil suspension volume to 150 ml and headspace volume to 100 ml. The container was sealed with a plastic cap fitted with a robber septum to allow gas sampling. After thawing, soil samples were shaken for 1 hour to release the trapped CH_4, which was measured as described previously. Rice roots in each soil layer were washed, air-dried, and oven-dried for 4 hours at 60 °C prior to weighing.

2. Effects of Living Rice Roots on Soil Redox Potential (Eh)

Ninety two days after the transplanting, rice shoots were cut at the soil surface. Three soil containers of each treatment were capped and shaken for 1 hour. Two platinum electrodes and one calomel electrode were installed in the sealed soil suspension container and soil Eh was monitored at 2, 24, 48, and 72 hours after shaking at room temperature (about 25°C).

III. Results and discussion

A. Greenhouse Experiments

1. Without Rice

Methane emission rate in the microcosms without rice plants were highest initially (10.45 $\mu g\ g^{-1}\ d^{-1}$), and decreased thereafter (Figure 1). By 50 days, a substantial decrease in CH_4 emission had occurred (2.58 $\mu g\ g^{-1}\ d^{-1}$). Meanwhile, there was a notable growth of blue-green algae (*Oscillatoria lynbia*), which had formed an algal layer about 0.5 mm thick on soil surface.

To study the effects of algae on CH_4 emission, these containers were split into two groups: (1) algae undisturbed, and (2) algae removed. After 1 week (57 days total) CH_4 emission increased substantially (7.24 $\mu g\ g^{-1}\ d^{-1}$) in the microcosms where algae had been removed (Figure 1). Methane emission persisted at the low rate throughout the experiment in microcosms where algae were undisturbed. Apparently the algal layer at the soil surface contributed to the reduction of CH_4 emission. This could have been a combined effect of a physical mechanism, such as a diffusion barrier and a biological mechanism, such as releasing O_2 during algal photosynthesis stimulating methanotrophs and/or inhibiting methanogens.

2. With Rice

Figure 2 shows CH_4 emission in soil microcosms with rice. Initial CH_4 emission rates (36 and 43 days) were lower from microcosms with rice than from microcosms without rice (Figure 1 and 2). This is probably due to soil disturbance by transplanting. Methane flux peaked during 50-60 days after transplanting, and subsequently declined with no substantial change in CH_4 emission during the last 20 days of growth (Figure 2). Fifty days after transplanting, algae layer present on the soil surface removed following the third measurement. Algae had no substantial effects on reduction in CH_4 emission in presence of rice. The higher emission rate from soil growing rice indicates transportation of CH_4 formed in the anaerobic zone through rice arenchyma. Subsequent decrease in CH_4 emission may have been due to limited substrate (C) concentration and/or increased O_2 transport throughout the soil microcosm which could inhibit methanogenesis and stimulate CH_4 oxidation.

B. Laboratory Experiments

1. Methane Entrapment and the Distribution of Rice Roots in Soil Profile

During the initial growth stage (57 days after transplanting), rice roots were mainly concentrated in the upper layers (1-4 cm), and a clear gradient of CH_4 entrapment was observed (Table 1). Because of the shallow root system at this stage, atmospheric O_2 transported through the roots system into deeper zones of the soil microcosm would be limited. Therefore, methanogenesis could occur in the deeper anaerobic zones, with CH_4 diffusing to the upper layer and emitted through plant arenchyma. This may be a reason for the peak CH_4 emission observed during 50-60 days after transplanting. However, at a later stage of rice growth (92 days), mature rice root densities had increased in the lower soil depths. By this time, the amount of trapped CH_4 was substantially less and only a slight gradient down the profile was noted. The more extensive root network may have played a role in inhibiting methanogenesis since both CH_4 emission rates (Figure 1) and amounts of CH_4 trapped (Table 1) were lower at 92 days.

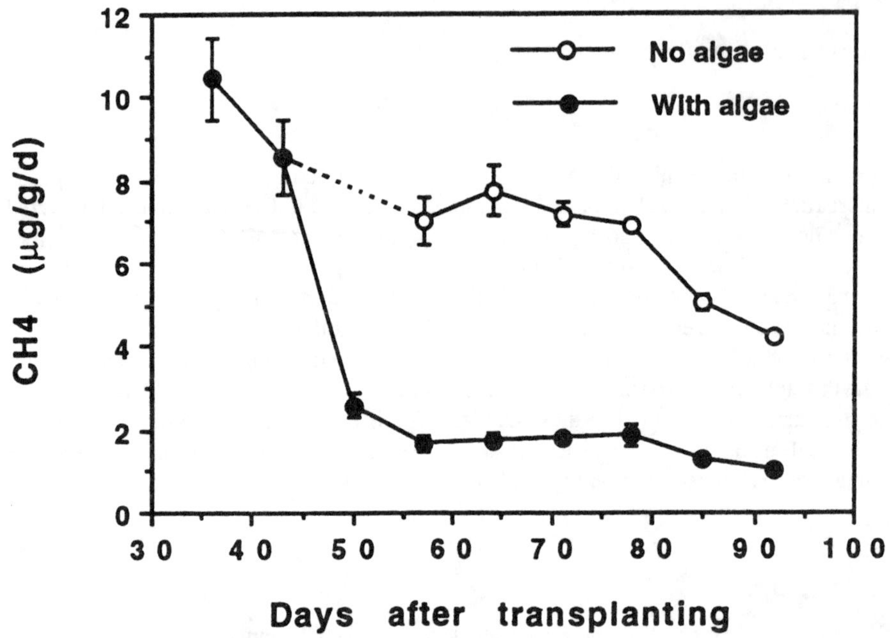

Figure 1. Methane emission in soil microcosms without rice.

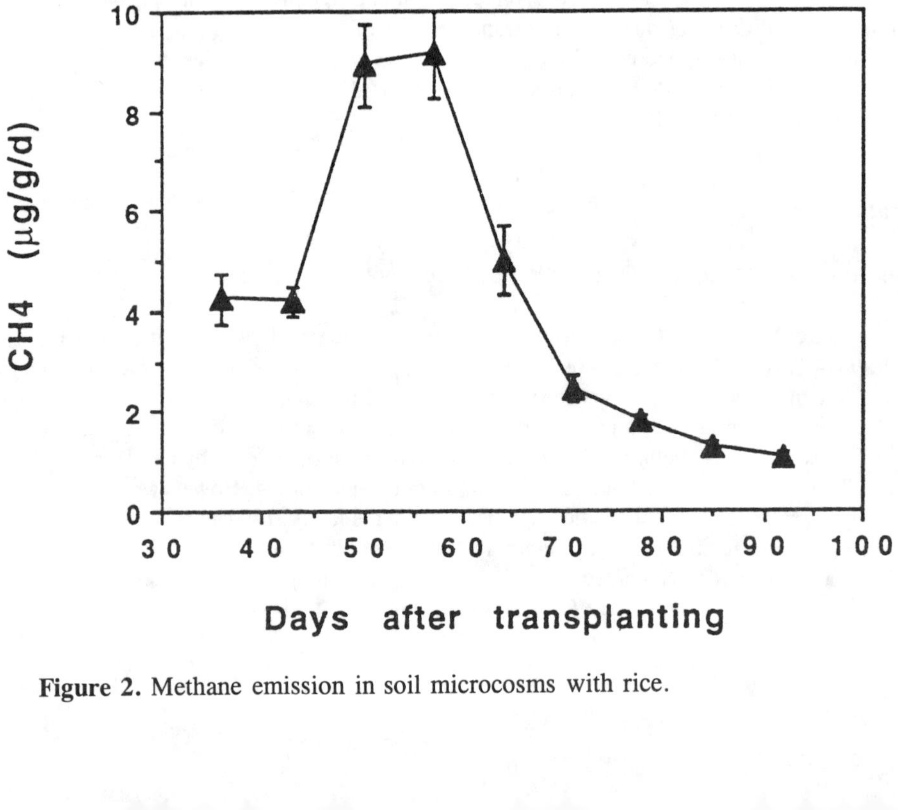

Figure 2. Methane emission in soil microcosms with rice.

Table 1. Methane entrapment and dry weight of rice roots in soil profile at 57 and 90 days after transplanting (data in brackets are standard deviations, n = 3)

Soil depth	------------57 days------------		-------------90 days-------------	
(cm)	CH_4 (ng/g)	Dry weight (g)	CH_4 (ng/g)	Dry weitht (g)
0-2	389 (24.2)	1.74 (0.32)	16 (3.90)	1.89 (0.45)
2-4	498 (46.3)	0.21 (0.05)	25 (3.09)	0.60 (0.21)
4-6	697 (78.9)	0.09 (0.03)	30 (4.12)	0.52 (0.16)
>6	1953 (356.8)	0.05 (0.01)	33 (4.71)	0.54 (0.12)

Table 2. Effects of living rice roots on soil Eh (see Table 1)

	Soil Eh (mV)			
Treatment	2 h.	24 h.	48 h.	72 h.
With rice	360 (42)	40 (5)	-210 (23)	-240 (10)
Without rice	-125 (7)	-170 (5)	-200 (12)	-200 (11)

2. Effects of Living Rice Roots on Soil Eh

Soil Eh was much higher for treatments with rice (360 mV) than without rice (-125 mV) 2 hours after shaking soil in an inert gas (Ar)(Table 2). This was attributed to the atmospheric O_2 which diffused into the soil profile through rice roots. However, soil Eh decreased gradually with incubation duration and reached the lowest level at 72 hours after shaking (-240 mV). These reduced levels may have been due to the reduced supply of atmospheric O_2 transported through living rice roots and to the contribution of greater C sources provided by dead roots. In the control treatment soil Eh was relatively constant throughout the entire incubation period.

In presence of live roots, soil Eh probably exceeded the critical level (about -150 mV) for methanogenesis (Wang et al., 1993b). Consequently the activity of methanogenic bacteria was inhibited.

IV. Summary and Conclusions

Methane emission decreased substantially in the presence of a thin layer of algae in microcosms without rice plants. Decreased CH_4 emissions may have been the result of a physical barrier to diffusion or to O_2 release during photosynthesis and a subsequent stimulation of methanotrophs and/or inhibition of methanogense. In the presence of rice, algae layer did no reduce CH_4 emission substantially. This suggests that CH_4 emission through rice plant arenchyma was greater than emission through soil surface. During the early stage of rice plant development, a distinct gradient of CH_4 entrapped in the soil profile was observed, with substantially higher amount of CH_4 trapped in deeper layers. This indicates a greater activity of methanogens in deeper soils during the early growth period. At the later growth stage, greater profusion on roots in lower soil layers, decreased soil-entrapped CH_4 concentrations and a reduction in CH_4 emission were all noted. Lower CH_4 emission rate with rice plant than without rice plant may have been due to O_2 diffusion through the root network throughout the soil microcosm, resulting in an increased soil redox potential, increased CH_4 oxidation, and reduced CH_4 production.

Acknowledgements

Although the research described in this article has been funded by U.S. Environmental Protection Agency agreement CA817426 to the International Rice Research Institute, it has not been subject to the Agency's review and therefore does not necessarily reflect the view of the Agency, and no official endorsement should be inferred.

References

Aiyer, P.A.S. 1920. The gases of swamp rice soils V: A methane-oxidizing bacterium for rice soils. *Memoirs of the Department of Agriculture in India.* 5:173-180.

Bouwman, A.F. and W.G. Sombroek. 1990. Input to climatic change by soil and agriculture related activities. In: Scharpenseel H.W., Schomaker M and Ayoub A. (eds.), Soils on a warmer earth. *Proceedings of an international Workshop on effects of expected climate change on soil processes in the tropics and sub-tropics, 12-14 February 1990, Nairobi.* p. 15-29.

Cappenberg, Th.E. 1972. Ecological observations on heterotrophic, methane oxidizing and sulfate reducing bacteria in a pond. *Hydrobiologia* 40:471-485.

Frenzel, P., F. Rothfuss, and R. Conrad. 1992. Oxygen profiles and methane turnover in a flooded rice microcosm. *Biol. Fertil. Soils* 14:84-89.

Harrison, W.H. 1914. The gases of swamp rice soils II. their utilization for the aeration of the roots of the crop. *Memoirs of the Department of Agriculture in India.* 4:1-17.

Holzapfel-Pschorn, A. and W. Seiler. 1986. Methane emission during a cultivation period from an Italian rice paddy. *J. Geophy Res.* 90:11803-11814.

Kimura, M., H. Murakami, and H. Wada. 1991. CO_2, H_2, and CH_4 production in rice rhizosphere. *Soil Sci. Plant Nutr.* 37:55-60.

Lindau, C.W., P.K. Bollich, R.D. Delaune, W.H. Patrick, Jr., and V.J. Law. 1991. Effect of urea fertilizer and environmental factors on CH_4 emission from a Louisiana, USA rice field. *Plant and Soil* 136:195-203.

Ponnamperuma, F.N. 1972. The chemistry of submerged soils. *Adv. Agr.* 24:29-96.

Seiler, W., A. Holzapfel-Pschorn, R. Conrad, and D. Scharffe. 1984. Methane emission from rice paddies. *J. Atmospheric Chemistry* 1:241-268.

Wagatsuma, T., K. Jujo., K. Tawaraya., T. Sato, and U. Atssuko. 1992. Decrease of methane concentration and increase of nitrogen gas concentration in the rhizosphere by hygrophytes. *Soil Sci. Plant Nutr.* 38:467-476.

Wang, Z.P., R.D. Delaune., C.W. Lindau, and W.H. Patrick, Jr. 1993a. Methane emission and entrapment as affected by soil properties. *Biol. Fertil. Soils* 16:163-168.

Wang, Z.P., R.D. Delaune., P.H. Masscheleyn, and W.H. Patrick, Jr. 1993b. Soil redox potential and pH effects on methane production in a rice soil. *Soil Sci. Soc. Amer.* 57:383-385.

Yagi, K. and K. Minami. 1990. Effect of organic matter application on methane emission from some Japanese paddy fields. *Soil Sci. Plant Nutr.* 36:599-610.

CHAPTER 21

Soil Management in Semiarid Regions

B.A. Stewart

I. Introduction

U.S. Soil Taxonomy (Soil Survey Staff, 1975) classified all soils into 10 orders. The orders are differentiated by the presence or absence of diagnostic horizons or features that mark differences in the degree and kind of the dominant sets of soil-forming processes that have occurred. Stewart et al. (1990) estimated where six of the soil orders are dominant, the worldwide area occupied by these orders, and their major constraints (Figure 1). The Entisols, Histosols, Inceptisols, and Spodosols not shown occupy 11.1, 1.2, 11.7, and 5.6 million km^2, respectively. More recently (Soil Survey Staff, 1992), Andisols have been added to the classification scheme. The Aridisols make up the largest group of soils in semiarid regions. The information in Figure 1 is useful for gaining a very general overview of where major soil orders occur worldwide with respect to climatic conditions. This information can also be useful for assessing the impacts of management on soil carbon.

Mollisols are noted for their high organic matter levels, and they are located primarily in cool humid regions. Aridisols, the most prevalent of all soil orders, occur mostly in hot, dry regions and are generally low in organic matter. These comparisons are clearly shown in Table 1 that lists recent estimates by Eswaran et al. (1993a,b) of the area of each of the 11 soil orders and their organic carbon contents. While the estimated areas of the soil orders in Table 1 differ somewhat from those shown in Figure 1, they are still useful in gaining insight about the magnitude and characteristics of major soil orders.

Research by the Soil Management CRSP/North Carolina State University (1993) have shown that different soils lose different amounts of soil organic carbon when subjected to controlled oxidizing environments. The clay mineralology appears to be a dominant factor. A 30-month incubation test of 19 surface soils revealed the following range of soil organic carbon loss: 22-33% from soils dominated by kaolinitic clay materials; 16-33% from soils of smectite mineralogy; 13-20% from Oxisols; and 2% from Andisols. This information clarifies the importance of designing soil-type specific strategies for managing crop residues for carbon sequestration. It also allows scientists and policy makers to focus attention on soils that potentially emit the highest levels of carbon dioxide when cultivated.

Barnwell et al. (1992) report that just over 2000 Gt carbon are contained in terrestrial pools, about 550 Gt carbon in living vegetation and 1500 Gt carbon in soil organic matter. In comparison, the size of the atmospheric pool is estimated as 750 Gt carbon. They further reported that for much of the last 100 years, carbon emissions from land use change (shift in biota type and loss of carbon in soils) have exceeded emissions from fossil fuel combustion. Fossil fuel emissions of carbon surpassed those from biota and soils in the 1950s and have increased dramatically since to the point that the annual flux of carbon from fossil fuels is about double that from biota and soils. The atmospheric pool is increasing at a rate of 3 Gt carbon yr^{-1}.

The conversion of forest and grass lands into cropland is generally associated with a sharp decline in the content of soil organic matter. Johnson (1992) reviewed the effects of forest management on soil carbon storage. He reported that there was no general trend toward lower soil carbon with forest harvesting, unless

ISBN 1-56670-117-1/95/$0.00+.50

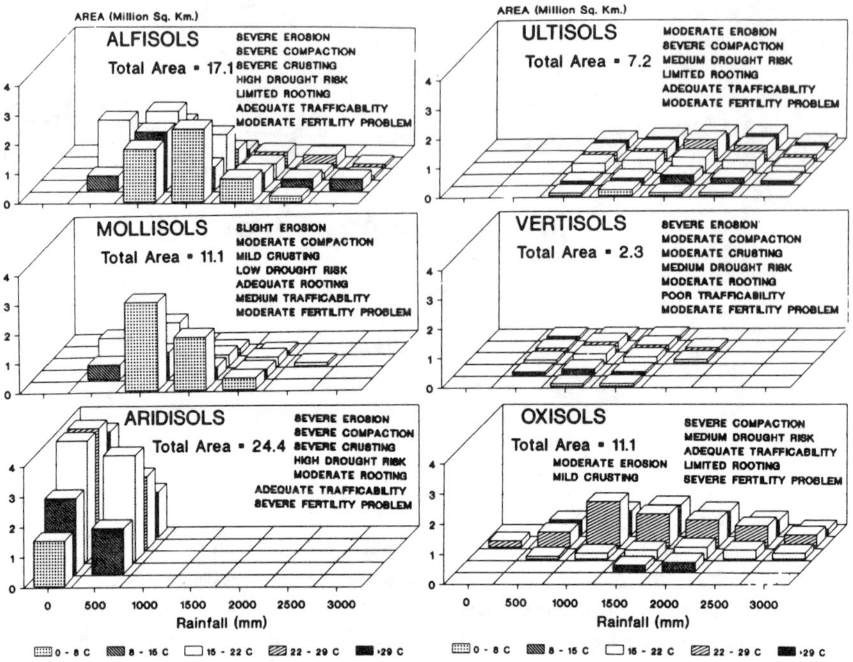

Figure 1. Approximate climatic distribution, area, and chief constraints of selected soil orders. (From Stewart et al., 1990.)

Table 1. Organic C mass in soils of the world.

Order	Area[1]		Organic C		
	Global	Tropical	Global	Tropical	Tropical
	----- 10^3 km^2-------		------- Pg -------		% of global
Histosols	1,745	286	357	100	28
Andisols	2,552	1,683	78	47	60
Spodosols	4,878	40	71	2	3
Oxisols	11,772	11,512	119	119	100
Vertisols	3,287	2,189	19	11	58
Aridisols	31,743	9,117	110	29	26
Ultisols	11,330	9,018	105	85	81
Mollisols	5,480	234	72	2	3
Alfisols	18,283	6,411	127	30	24
Inceptisols	21,580	4,565	352	60	17
Entisols	14,921	3,256	148	19	13
Misc. land	7,644	1,358	18	2	11
Total	135,215	49,669	1,576	506	32

[1] Most recent estimates by USDA Soil Conservation Service (Eswaran et al., 1993).

harvesting was followed by intense burning or cultivation. However, cultivation resulted in up to 50% loss in soil carbon in most cases. Soil organic carbon losses of as much as 50% have also been documented in the U.S. Great Plains as a result of converting grasslands to cropland (Haas et al., 1957). Carbon losses were strongly dependent on management regime and regional location. Burke et al. (1989) concluded that the extent of soil organic matter depletion was dependent on the same variables as those controlling soil organic matter formation: climate, soil texture, and landscape position.

Hobbs and Thompson (1971) studied the effect of cultivation on the nitrogen and organic carbon contents of a Kansas Argiustoll. Results of a longtime study showed that:

1. Nitrogen and organic-carbon contents of surface soils decreased with continued cultivation.
2. Magnitude of these losses from cultivated land was related directly to the quantity of organic matter in the soil originally and to cultivation intensity.
3. Organic-matter losses, as evidenced by losses of both nitrogen and organic carbon, followed a curvilinear trend, being large early in the cultivation period and becoming smaller with continued cultivation.
4. Losses of organic matter were greatest from plots where row crops were produced continuously or alternately with summer fallow; losses were considerably less from plots that produced small grains.
5. Applying straw or manure decreased the rate of organic matter loss.
6. Organic matter losses from the plow layer were demonstrated, but no consistent losses from subsurface layers were noted.

Jenny (1933) also reported a curvilinear loss of nitrogen and organic matter with time for cultivated soils in midwestern U.S. About 25% was lost during the first 20 years of cultivation, and 35% in 60 years. These data are representative of much other data obtained in the same region and extending north into Canada (Allison, 1973).

The above findings suggest that after the first 20 to 30 years of cultivation, most soils reach a new equilibrium, and soil organic matter remains fairly stable. However, if the organic matter level of these soils could be increased to a point nearer that at the time the land was converted to cropland, significant amounts of carbon could be stored. Since most organic matter originates from plant material, raising organic matter levels of the soil would decrease the amount of carbon in the atmosphere because plants obtain their carbon from atmospheric CO_2. Furthermore, since soil organic matter is the key to soil productivity, crop yields would likely increase resulting in an upward spiral of benefits.

II. Management Practices for Increasing Soil Organic Matter

Hornick and Parr (1987) reported that, for most agricultural soils, degradative processes and conservation practices occur simultaneously. In semiarid regions, the dominant degradative processes are loss of soil organic matter and erosion by wind and water. As soil degradative processes proceed and intensify, there is a concomitant decrease in soil productivity. Therefore, the productivity level of an agricultural soil at any time is a result of the interaction of degradative processes and conservation/reclamation practices such as water management, enhancing soil fertility, and crop residue management. In natural ecosystems, productivity and sustainability are achieved through the efficient but delicate balance between all necessary inputs and outputs. Recent efforts, for example alley cropping in the tropics, attempt to duplicate in cropping systems the achievement of the balance that occurs in natural ecosystems.

Using the Hornick and Parr (1987) concept, a system that results in an increase in soil organic matter must utilize practices that will return more organic matter to the soil than is lost by the soil degradative processes. Practices that are generally considered positive can result, in some instances, result in soil degradation. For example, the use of organic wastes increases organic matter, improves soil structure, enhances soil water storage, and reduces erosion. Under some circumstances, however, use of organic wastes could result in toxic accumulations or nutrient depletion caused by increased leaching. The important point illustrated is that degradative processes and soil improvement processes always occur simultaneously, and the net result can be positive or negative.

Allison (1973) stated that it is the moldboard plow and the various systems of cultivation that keep the soil bare for a considerable portion of the year that are primarily responsible for marked decreases in soil organic matter following the breaking of virgin land. The conditions responsible for the steady state of soil organic matter, built up over centuries to the maximum level possible under the prevailing climatic and soil conditions, are suddenly and drastically changed. Destructive oxidation processes now greatly outweigh the constructive ones. The harmful effect of continuous clean cultivation on soil organic matter can be greatly reduced by cropping systems that reduce the number of cultivations and keeps the soil protected by vegetation as much of the time as possible. Allison and Sterling (1949) showed in field experiments at Mandan, North Dakota, a 34% loss of nitrogen (and similar amounts of carbon) for continuous corn or for corn alternated with summer fallow. The loss with continuous small grains was 14%, and for small grains alternated with fallow 26%.

Fertilizers applied as a supplement to proper cropping systems can greatly enhance the build-up of soil organic matter. Because fertilizers increase plant growth, more crop residues including roots will be returned to the soil. It is essential that adequate nitrogen be applied because this element constitutes about 5% of the weight of soil organic matter. Bartholomew (1965) indicates that nearly all organic matter in soils, other than that added during the past two or three years, consists of the residues of microorganisms. Certainly all, or nearly all, of the plant protein and other nitrogenous constituents are utilized rather rapidly by microorganisms and converted into microbial protein. Most of the carbon compounds are oxidized to carbon dioxide and water but a small proportion is built into the cells of microorganisms. Therefore, it is essential that nitrogen be available for converting a portion of the carbon in roots and plant residues into soil organic matter.

Climate is often the most critical factor determining the sustainability of agricultural systems. A generalized view of the effect of varying temperature and moisture regimes on the difficulty of achieving sustainability in an agricultural system was developed by Stewart et al. (1991). As temperatures increase and the amounts of precipitation decrease, the maintenance of soil organic matter becomes more difficult. The reasons for these effects are well understood. As temperatures increase, organic matter decline is greatly accelerated, (particularly in frequently tilled soils). Not only is the rate of organic matter decomposition accelerated under these conditions, but the production of biomass is decreased so there are smaller amounts of crop residues and roots available that are necessary to replenish the soil organic matter reserve.

Allison (1973) stated that increases in soil organic matter in the United States have been largely limited to the humid region, or where irrigation is practiced, because obviously if water is the limiting factor in plant growth, there is an inadequate supply of crop residues for increases in soil organic matter levels. Allison concluded that for most of the well-fertilized arable soils in the United States the organic matter level was either remaining at a constant level or actually increasing at a slow but significant rate. He pointed out that new equilibrium levels had been reached since most soils had been in use for 50-300 years, and the mining process is usually completed after 50 years. Heavy fertilization, together with the general adoption of improved farm management practices, have doubled or tripled the amount of crop residues produced and returned to the land.

The return of additional residues to the soil, however, may not have a significant impact on soil organic matter content if intensive tillage is practiced. Westerman (1992) summarized the data from the Magruder plots in Oklahoma that have been continuously cropped to wheat since 1892. Figure 2 was prepared from data Westerman presented to illustrate how the organic matter content had decreased over the years in spite of the fact that there had been very significant increases in yield of wheat grain. Even where manure has been added, organic matter continued to decrease although it was fairly constant for the last 10 years indicating that a new equilibrium was reached. There were also plots that received nitrogen, phosphorus, and potassium fertilizers, and the yields from those plots were similar to those shown for the manured plots. Although it is somewhat surprising that manure did not increase the organic matter, it is important to state that only enough manure was added to supply adequate nitrogen for wheat production. Beef manure was applied at a rate sufficient to add 134 kg ha^{-1} N and 268 kg ha^{-1} N every fourth year for periods 1930-1967 and 1967-1991, respectively. The important point is that the organic matter content continued to decrease even though increasing amounts of crop residue were being returned to the soil. Intensive tillage, however, was used which strongly indicates that if organic matter levels are going to be maintained, and enhanced, cropping systems that use less intensive tillage practices are going to be necessary.

The effect of tillage on organic matter decline was further illustrated by Lamb et al. (1985). They cultivated a native grassland site in western Nebraska for winter wheat production in a crop-fallow rotation under three tillage systems; no-till, stubble mulch, and plow. After 12 yr of cultivation, losses of soil N from the 0-30 cm depth were 3% for the no-till, 8% for the stubble mulch, and 19% for the plow tillages. Although not reported, it is assumed that comparable losses of organic carbon occurred.

Crop residues can be managed in manners that will lead to increased organic matter levels, thereby sequestering carbon. No-till systems often show increased soil organic matter contents within the first few years of practice. W. D. Kemper (USDA Agricultural Research Service, Beltsville, MD, personal communication) stated that one of the gratifying consequences of no-tillage management is associated increases in organic matter of the soils which has ranged from 40 to over 400 kg ha^{-1} yr^{-1}. The higher rates

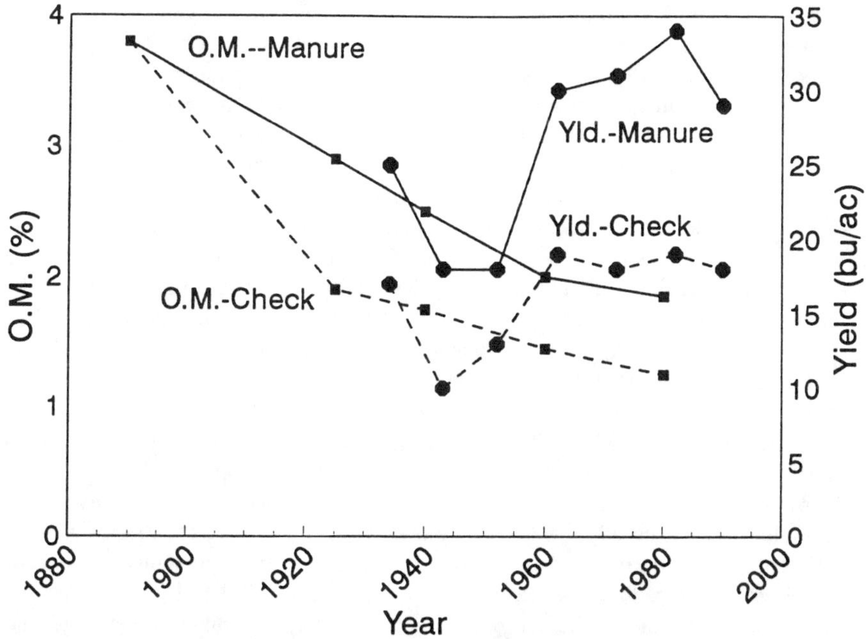

Figure 2. Changes in organic matter and wheat grain yields in a continuous wheat cropping system on manured and control plots in Oklahoma. (Drawn from data of Westerman, 1992.)

are usually associated with leguminous winter cover crops whose residues were left on the soil. This buildup of organic matter is restoring the "nutrient bank" that can help tide plants over periods of deficiency. Also, the sequestration of nitrogen into this accumulating organic matter is probably one of the factors causing the nitrate concentration of water percolating below no-till fields to be less than the nitrate concentration below conventionally tilled soils.

Even in semiarid regions where only limited amounts of crop residues are produced, significant increases in organic matter occur when no-till cropping systems are adopted. Unger (1991) evaluated the distribution with depth of organic matter in a wheat-grain sorghum-fallow rotation. The work was done in the Texas High Plains where the annual precipitation averages 465 mm. The no-till fields had higher organic matter levels than the stubble mulch field, although the differences were relatively small and confined mostly to the top 1- and 2-cm depths.

III. Discussion

Soil organic carbon dynamics are influenced by many factors. As a simplification, Lucas et al. (1977) estimated that 30% of all carbon in plant residues became part of the soil humus. The carbon content of plant residues is approximately 40%.

The rate of microbial decay of soil carbon is related to the proportions of new to old humus, aeration, moisture and temperature (Lucas et al., 1977). Again, Lucas et al. (1977) simplified the dynamics by assuming that the annual depletion rate of the total soil humus found in the top 2 million kg of soil per ha (approximately 15 cm depth) was 50,000 kg yr^{-1}. This was based on the assumption that the ratio of old soil humus to new was 20 to 1 and the rate of decay was 1.6% for old humus and 20% for new. Although these values will be different for various soils and climatic conditions, they are reasonable estimates that are useful as guidelines.

Lucas et al. (1977) estimated soil carbon production for several field crops grown in the Cornbelt of the United States. For a corn crop yielding 6,270 kg ha^{-1}, approximately 806, 175, and 262 kg ha^{-1} of soil carbon would be produced by the stover, roots, and "others", respectively. "Others" include root turnover, exudates, insects, etc. Therefore, a total of about 1,243 kg ha^{-1} carbon was added to the soil humus. Assuming that the soil organic carbon in the top 15 cm was 40,000 kg ha^{-1}, and 2.5% was decomposed, there would be a net gain of 243 kg ha^{-1} of soil carbon. If wheat was grown and yielded 3,025 kg ha^{-1}, the carbon additions would be 605, 134, and 202 for straw, roots, and others, respectively. This would result in a slight loss of soil carbon. Soybeans yielding 2,150 kg ha^{-1} would only add 430, 81, and 121 for straw, roots, and others so there would be a significant loss of soil organic carbon.

Using the above information as a guide, some generalizations can be made about semiarid regions. The inherent levels of soil organic carbon are usually low, generally less than 10,000 kg ha^{-1} in the surface 15 cm and often in the range of 5,000 kg ha^{-1}. The annual additions from cropping are also much lower because the rainfall in these regions is significantly less than in the more humid regions, and the water use efficiency values are also lower. The water use efficiency will always be higher in humid areas than in semiarid regions because it takes less water to produce the same amount of dry matter in a humid region than in an arid or semiarid region. Therefore, there is less carbon being returned to the soil. For example, wheat grain yields in semiarid regions are often less than 1,000 kg ha^{-1}, so the amount of soil carbon added would only be in the range of 300 kg ha^{-1}. This would not be sufficient to maintain the soil organic carbon level of any soil that contained more than about 0.6% organic carbon.

The soil organic carbon situation in many semiarid regions is made even worse by the fact that crop residues are utilized for fuel or fodder. This is particularly true in many developing countries where populations of both people and animals are increasing at rapid rates. Cooper et al. (1987) stated nearly all the crop residues in the dry areas of the Mediterranean region were used as fodder. They further stated that the organic carbon levels are generally low and in some soils, particularly silts and sands, structural stability is poor. The failure to leave crop residues on the soil also eliminates the opportunity to use a surface mulch which is a valuable means of reducing water evaporation from the soil thereby increasing the amount of water available for transpiration by the crop. A mulch can increase soil water storage and water use efficiency that lead to higher yields and to higher levels of crop residues, creating an upward spiral of benefits.

Even though the amount of carbon that can be sequestered in semiarid regions is severely limited by low rainfall and high temperatures, it can still be very significant because of the large amount of lands located in these fragile areas. Cole et al. (1993) concluded that the extent to which soil management can influence gains or losses of soil carbon is highly variable and difficult to predict. They postulated that under optimum soil management it might be feasible to increase the carbon level of existing arable soils in the temperate zones by 1 kg m^{-2}. This would represent about a 10% increase of soil organic carbon in these soils and it would take about 50 years for this change to occur. They stated that similar increases are not likely for all soils, particularly Aridisols and other soils located in arid and semiarid regions. They concluded that the likely potential for soil management to sequester carbon ranges from 0 to about 1 kg m^{-2}, and assumed an average of 0.5 kg m^{-2} as a reasonable goal for agricultural cropland. The potential for semiarid regions would likely range from 0 to less than 0.5 kg m^{-2}. Cole et al. (1988) used the CENTURY simulation model to predict the effect of management practices on soil organic carbon in the semiarid Great Plains of the United States. Their results indicated that soil organic carbon could be increased about 0.1 to 0.3 kg m^{-2} depending on soil texture and location within the Great Plains which have a considerable range in precipitation and temperature. This change would require 40 years and was based on management practices changed to include higher yielding varieties in common use, application of N and P fertilizers, no straw removal, and stubble mulching. Stubble mulching is a form of tillage that uses v-shaped blades to undercut the soil surface about 10 cm deep and leave most of the crop residues on the soil surface. Typically, only about 15% of the crop residue is buried by sweep tillage, so even after 3 or 4 operations, about 50% of the crop residue will remain on the soil surface.

The most important management practice for increasing the organic carbon content of soils in semiarid regions is to use the least amount of tillage feasible. Intensive and frequent tillage buries most of the crop residues and hastens the decomposition of crop residues and soil organic matter. Cultivation increases biological activity in the soil, often as a result of better soil aeration. But cultivation also exposes fresh topsoil to rapid drying and, after each drying, a burst of biological activity occurs for a few days following rewetting. This is because the drying process releases organic compounds, probably from the breakdown of humic materials that bind soil aggregates.

Follett (1993) reviewed the practices that he felt were most important for storing more carbon in cropland soils. These were improving soil fertility, using animal or other wastes containing carbon, minimizing dryland fallow, using conservation tillage, maintaining crop residues, controlling soil erosion, and developing alternate uses for marginal lands. The goal of these practices is to produce as much soil carbon as feasible and them minimize its loss and slow its rate of decomposition.

Croplands offer an important sink for carbon. An increase in soil organic carbon will be highly beneficial for nearly all soils, and it will also help mitigate climate change by stashing CO_2 as carbon in soil. Cole et al. (1993), assuming an average increase of 0.5 kg m^{-2} soil organic carbon as a goal for agricultural cropland, estimated that about 7 Gt of carbon sequestration could be achieved. However, this increase would take at least 50 years and could be achieved only once. Although this is a very small amount relative to the total carbon balance, it is significant in relation to agricultural use.

IV. Summary

Cropland offers a huge potential for sequestering carbon. However, changes must occur in how crop residues are managed. Historically, soils were regenerated by the use of manures or legume cover crops. These practices added both carbon and nitrogen to the soil so the needs of both the plants and soil were addressed. Following World War II, commercial fertilizers became abundant and were in most cases more economical that hauling manure or growing cover crops. Also, farm machinery became bigger, and tillage was intensified in many cases. In recent years, there has been a concerted effort by scientists and policy makers to convince farmers that residues should be managed in such a way that much of the residue remains on the soil surface. This greatly reduces the potential for wind and water erosion, improves infiltration, and in water deficient areas leads to increased amounts of soil water storage. The evidence is clear that such practices also lead to increased soil organic matter levels. In many cases, the increased soil organic matter will lead to increased crop production. Therefore, more CO_2 will be used by the crops and more of the carbon used by the plants will be stored in the soil as organic matter. The amounts, however, will vary greatly from region to region. The potential for sequestering carbon decreases as annual precipitation amounts decrease, and generally as mean temperatures increase. Even in semiarid regions, significant amounts of carbon can be retained in organic matter crop residues are carefully managed, and tillage is used as sparingly as feasible.

References

Allison, F.E. 1973. p. 134. In: *Soil organic matter and its role in crop production*. Elsevier, Amsterdam, The Netherlands.

Allison, F.E. and L.D. Sterling. 1949. Nitrate formation from soil organic matter in relation to total nitrogen and cropping practices. *Soil Sci.* 67:239-252.

Barnwell, T.O., Jr., R.B. Jackson, IV, E.T. Elliott, I.C. Burke, C.V. Cole, K. Paustian, E.A. Paul, A.S. Donigian, A.S. Patwardhan, A. Rowell, and K. Weinrich. 1992. An approach to assessment of management impacts on agricultural soil carbon. p. 423-425. In: J. Wisniewski and A.E. Lugo (eds.), *Natural Sinks of CO_2*. Kluwer Academic Publishers, Dordrecht, The Netherlands.

Bartholomew, W.V. 1965. Mineralization and immobilization of nitrogen in the decomposition of plant and animal residues. p. 287-306. In: W.V. Bartholomew and F.E. Clark (eds.), *Soil Nitrogen*. Agronomy 10, American Society of America, Madison, WI.

Burke, I.C., C.M. Yonker, W.J. Parton, C.V. Cole, K. Flach, and D. S. Schimel. 1989. Texture, climate, and cultivation effects on soil organic matter content in U.S. grassland soils. *Soil Sci. Soc. Am. J.* 53:800-805.

Cole, C.V., I.C. Burke, W.J. Parton, D.S. Schimel, D.S. Ojima, and J.W.B. Stewart. 1988. Analysis of historical changes in soil fertility and organic matter levels of the North American Great Plains. p. 436-438. In: P.W. Unger et al. (eds.), *Challenges in dryland agriculture: A global perspective*. Proc. Int. Cong. on Dryland Farming, Amarillo/Bushland, TX. 15-19 Aug. Texas Agric. Exp. Stn., College Station.

Cole, C.V., K. Flach, J. Lee, D. Sauerbeck, and B. Stewart. 1993. Agricultural sources and sinks of carbon. *Water, Air, and Soil Pollution*. (In press.)

Cooper, P.J.M., P.J. Gregory, D. Tully, and H.C. Harris. 1987. Improving water use efficiency of annual crops in the rainfed farming systems of West Asia and North Africa. *Expl. Agric.* 25:113-158.

Eswaran, H., E. Van Den Berg, and P. Reich. 1993a. Organic carbon in soils of the world. *Soil Sci. Soc. Am. J.* 57:192-194.

Eswaran, H., N. Bliss, D. Lytle, and D. Lammers. 1993b. Major soil regions of the world. USDA-SCS. U.S. Govt. Print. Office, Washington, D.C.

Follett, R. F. 1993. Global climate change, U.S. agriculture, and carbon dioxide. *J. Prod. Agric.* 6:181-190.

Haas, H.J., C.E. Evans, and E.R. Miles. 1957. Nitrogen and carbon changes in soils as influenced by cropping and soil treatments. USDA Tech. Bull 1164. U.S. Gov. Print. Office, Washington, D.C.

Hobbs, J.A. and C.A. Thompson. 1971. Effect of cultivation on the nitrogen and organic carbon contents of a Kansas Argiustoll (Chernozem). *Agron. J.* 63:66-68.

Hornick, S.B. and F.F. Parr. 1987. Restoring the productivity of marginal soils with organic amendments. *American Journal Alternative Agriculture* 2:64-68.

Jenny, Hans. 1933. *Soil fertility losses under Missouri conditions.* Mo. Agric. Exp. Stn. Bull. 324. 10 pp.

Johnson, D.W. 1992. Effects of forest management on soil carbon storage. p. 83-120. In: J. Wisniewski and A.E. Lugo (eds.), *Natural Sinks of CO_2*. Kluwer Academic Publishers, Dordrecht, The Netherlands.

Lamb, J.A., G.A. Peterson, and C.R. Fenster. 1985. Wheat fallow tillage systems' effect on a newly cultivated grassland soils' nitrogen budget. *Soil Sci. Soc. Am. J.* 49:352-356.

Lucas, R.E., J.B. Holtman, and L.G. Connor. 1977. Soil carbon dynamics and cropping practices in agriculture and energy. p. 333-351. In: W. Lockeretz (ed.), *Agriculture and energy*. Academic Press, New York.

Soil Management CRSP Communications. 1993. Daniels Hall, North Carolina State University, Raleigh, NC. June 1993.

Soil Survey Staff. 1992. *Keys to Soil Taxonomy*, 5th Ed. SMSS technical monograph No. 19. Blacksburg, Virginia: Pocahontas Press, Onc. 566 pp.

Stewart, B.A., R. Lal, S.A. El-Swaify, and H. Eswaran. 1990. Sustaining the soil resource base of an expanding world agriculture. *Transactions 14th International Congress Soil Science*, Kyoto, Japan, Aug. 12-18, 1990. Vol. VII:296-301.

Stewart, B.A., R. Lal, and S.A. El-Swaify. 1991. p. 125-144. Sustaining the resource base of an expanding world agriculture. In: R. Lal and F.J. Pierce (eds.), *Soil Management for Sustainability*, Soil and Water Conservation Society, Ankeny, IA.

Soil Survey Staff. 1975. Soil Taxonomy. *A Basic System of Classification for Making and Interpreting Soil Surveys,* Handbook 436. U.S. Department of Agriculture, Soil Conservation Service, Washington, DC.

Unger, P. W. 1991. Organic matter, nutrient, and pH distribution in no- and conventional-tillage semiarid soils. *Agron. J.*:186-189.

Westerman, Robert L. 1992. *Efficient use of fertilizers.* Agronomy 92-1. Oklahoma State University, Stillwater, OK.

CHAPTER 22

Simulation of Soil Organic Matter Dynamics in Dryland Wheat-Fallow Cropping Systems

A.K, Metherell, C.A. Cambardella, W.J. Parton, G.A. Peterson, L.A. Harding, and C.V. Cole

I. Introduction

Management interacts with natural driving variables in an agroecosystem to affect the processes that control the ecosystem properties (Cole, 1987). Soil organic matter is an important component of the ecosystem, influencing both physical and chemical soil properties. Soil organic matter is also a major pool in the global C cycle. Management of agricultural systems has had a major impact on soil organic matter with the cultivation of soils in the Great Plains of North America resulting in large losses of C and N (Hass et al., 1957; Burke et al., 1989). This raises questions as to whether the loss of organic matter will continue and if alternative agricultural management practices can reduce the rate of loss or even reverse the trend.

The CENTURY soil organic matter model (Parton et al., 1987; Parton et al., 1988) integrates the effects of climate and soil driving variables with agricultural management to simulate C, N, and water dynamics in the soil-plant system, with an emphasis on long-term soil organic matter dynamics. The soil organic matter submodel is based on multiple compartments for C and N, and is similar to other models of soil organic matter dynamics (Jenkinson and Rayner, 1977; Jenkinson, 1990; van Veen and Paul, 1981). The pools and flows of C are illustrated in Figure 1. Above- and belowground plant residues are partitioned into structural and metabolic pools as a function of the lignin to N ratio in the residue. With increases in the ratio, more of the residue is partitioned to the structural pools which have much slower decay rates than the metabolic pools. The decomposition of both plant residues and soil organic matter are assumed to be microbially mediated with an associated loss of CO_2 as a result of microbial respiration. Decomposition products flow into a surface microbe pool or one of three soil organic matter pools, each characterized by different maximum turnover rates. The actual decomposition rate is reduced by multiplicative functions of soil moisture and soil temperature and may be increased as an effect of cultivation. The active pool represents soil microbes and microbial products and has a turnover time of months to a few years depending on the environment and sand content. The slow pool includes resistant plant material, derived from the structural pool, and soil stabilized microbial products derived from the active and surface microbe pools. It has a turnover time of 20 to 50 years, while the passive pool which is very resistant to decomposition has a turnover time of 400 to 2000 years. The proportions of the decomposition products from the slow and active pools, which enter the passive pool, increase with increasing soil clay content. A fraction of the flow from the active pool is lost as leached organic matter.

The soil organic matter pools in the CENTURY model are conceptual, kinetically defined pools which do not correspond exactly with chemical or physical soil organic matter fractions. However, soil organic matter fractionation schemes based on combinations of physical methods (Elliott and Cambardella, 1991) are likely to yield fractions related to the aggregate structure and soil organic matter quality, and hence, its rate of turnover. A new agroecosystem version of the CENTURY model (Metherell, 1993) has been designed for the

ISBN 1-56670-117-1/95/$0.00+.50

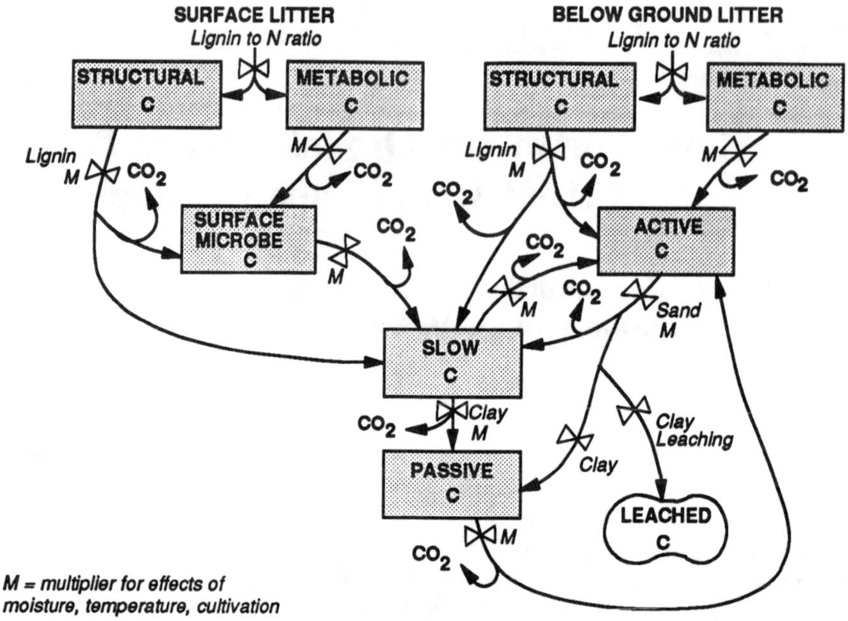

Figure 1. Carbon pools and flows in the CENTURY model.

simulation of complex agricultural management systems including crop rotations, tillage practices, fertilizer, and irrigation. Crops and management options are selected using a preprocessor called EVENT100. A unique feature is that simulations can have a series of repeating sequences reflecting historical management changes or experimental treatments.

The majority of dryland farmers in the semi-arid western half of the North American Great Plains use summer fallow to conserve moisture for the succeeding crop. In the wheat (*Triticum aestivum* L.)-fallow system a crop is only grown every second year. During the fallow, in the alternate years, all weedy plant growth is controlled by cultivation or herbicides. Cultivation methods have evolved from the "dust mulch" with moldboard plowing and frequent harrowing after each significant rainfall, through "conventional bare fallow," which utilizes shallow disking and bare rodweeders, to "stubble-mulch," which uses implements to undercut the soil surface, killing the weeds but leaving crop residues on the soil surface (Greb, 1979). The fallow water storage efficiency is greatest with no-till management where herbicides replace tillage to control weed growth (Smika and Wicks, 1968). Summer fallow, particularly when associated with intensive cultivation, increases the mineralization of soil organic matter (Haas et al., 1957) resulting in degradation of soil structure, reduced water infiltration, and increased susceptibility to erosion. Furthermore, because a crop is only grown every second year, inputs of C to the soil are also considerably lower compared to grassland or continuous cropping, which reduces the formation of new soil organic matter.

The prediction of the effects of different agricultural management strategies on soil organic matter levels is one of the main uses of the CENTURY model (Parton et al., 1987). However, the model had not been formally validated for the effects of tillage methods on crop production and soil organic matter dynamics. The objective of this study was to compare CENTURY model predictions with data from a set of long-term wheat-fallow tillage experiments at Sidney, Nebraska. We were particularly interested in the comparison of physically separated organic matter fractions with the CENTURY model's conceptual, kinetically defined pools.

II. Methods

A. The Sidney Tillage Experiments

In 1969 and 1970 experiments were initiated at two locations on the High Plains Agricultural Laboratory research area near Sidney, Nebraska to compare the effects of fallow tillage methods on crop production, soil moisture and nutrient cycling in the wheat-fallow cropping system (Fenster and Peterson, 1979). The first location (Sites A and B) was on an Alliance silt loam (fine silty, mixed, mesic Aridic Argiustoll). The land had been farmed from 1920 to 1957, when it was reseeded to crested wheatgrass (*Agropyron desertorum* (L.) Gaertn.). The experimental site was broken out of crested wheatgrass sod with a moldboard plow in 1967. The second location (Sites C and D) was on a Duroc loam, (fine silty, mixed, mesic Pachic Haplustoll) derived from mixed loess and alluvium. The land remained in native grasses until moldboard plowed in 1970. The experiments at the second location have a unique feature in that strips of native sod were left intact and incorporated into the experimental design. Each location was divided into two experimental sites, which were fallowed or grew wheat in alternate years.

At both locations bare fallow (plow tillage) was compared with stubble-mulch and no-till treatments (Fenster and Peterson, 1979). At each site (A, B, C and D) there were three replicates of the tillage treatments in a randomized block design. On the Alliance soil (sites A and B), each tillage treatment plot was split with one half receiving ammonium nitrate fertilizer at 45 kg N/ha during wheat growth in April.

B. Physical Separation of Soil Organic Matter

Soil samples collected from site C during the fallow in the summer of 1990 were dispersed in sodium hexametaphosphate and fractionated with a 53 μm sieve. The organic matter fractions retained on the sieve and in the soil slurry which passed through it were termed particulate organic matter (POM) and mineral associated organic matter, respectively (Cambardella and Elliot, 1992). The non-dispersed samples and the soil slurry were analyzed for total organic C (Snyder and Trofymow, 1984) and total Kjeldahl N (Nelson and Sommers, 1980). C and N contents of POM were calculated by difference. The POM fraction was examined using a scanning electron microscope and analyzed for lignin and cellulose contents (Goering and Van Soest, 1970).

Another fractionation scheme characterised size-density classes by separation of the macroaggregates by wet-sieving, sonication to disrupt the aggregates, sieving of particle size classes and density separation with sodium polytungstate (Cambardella and Elliot, 1993). Total organic C, total Kjeldahl N, and N mineralized in a 28 day incubation were determined for the size-density fractions and for intact macroaggregates.

C. Soil Texture

Soil samples to 20 cm depth were collected for soil texture analyses on site C in 1990 and site D in 1991. There was a distinct difference between site C and D (Table 1), even though the plots are adjacent to each other. To parameterize the CENTURY model for sites A and B, data for 0 to 15 cm samples published by Mielke et al. (1986) were used.

Table 1. Soil texture (%, 0-20 cm) on the long-term tillage plots at the High Plains Agricultural Laboratory near Sidney, Nebraska.

	Sites		
Textural class	A&B	C	D
Sand	37.1	37.5	47.8
Silt	40.0	31.8	31.2
Clay	22.9	30.7	21.0

D. Simulation Strategy

Sites C and D, which incorporate the native sod plots, have been the subject of many soil organic matter studies (Broder et al., 1984; Lamb et al., 1985b; Mielke et al., 1986; Follet and Peterson, 1988; Cambardella and Elliott, 1992; J. Doran, pers. comm.). Therefore, these plots were used to validate the CENTURY model's predictions of soil organic matter dynamics, while sites A and B, which incorporate the N fertilizer treatment, were used to calibrate the wheat growth sub-model. No information from any of these sites was used for parameterization of the soil organic matter decomposition sub-model. Estimates for decomposition parameters had been made by simulating a variety of grassland and forest ecosystems (Parton et al., 1993). The various tillage implements have parameters for the transfer of material between the various plant and litter pools and multipliers for the effect of cultivation on decomposition rate which were set on the basis of a subjective assessment of the degree of stirring for each implement. Management operations were scheduled using the CENTURY model's event scheduler, EVENT100, to simulate, as closely as possible, the management of the field experiments.

The mean monthly precipitations and temperatures used as the climate driving variables for the CENTURY model were based on the weather station located at the High Plains Agricultural Laboratory, 2 km from sites A and B, and 6 km from sites C and D. A value of 0.5 g N/m²/yr for total atmospheric N deposition was used (National Atmospheric Deposition Program (NRSP-3)/National Trends Network, 1991; Meyers et al., 1991).

E. Soil Organic Matter Initialization

The CENTURY model was run for 5000 years to reach equilibrium under native grassland conditions for each site, using the stochastic precipitation generator and mean monthly temperatures. For sites C and D, no grazing was simulated after 1940. Because the historical grazing intensity is largely unknown, this factor was used to calibrate the model to closely match the soil organic matter C and N measured on sites C and D in 1990 (Cambardella and Elliott, 1992) and 1991 (Cambardella, unpublished), respectively.

For sites A and B, a sequence of historical crop management events was simulated from 1920 to 1956 followed by 10 years in ungrazed grass. The historical management was chosen to be appropriate to the period, to give grain yields similar to the county average yields, and to result in soil organic matter levels similar to those measured on the experimental plots. The simulated soil organic matter C and N levels prior to wheat planting in 1969 were 2791 g C/m² and 275 g N/m².

III. Results

A. Simulation of Sites A and B

The primary variables used for calibration of the plant production sub-model were wheat yields and grain N content. However, simulation results were also compared to field data for crop residue levels, soil water, and mineral N in the soil profile.

The wheat parameters were calibrated such that the model matched field data for grain yield (12% moisture and 40% C assumed for the conversion of grain yield to carbon) and grain N content for the plus N fertilizer treatment. However, CENTURY underestimated both yields (Table 2) and N content for the zero fertilizer treatment. The field grain yield results showed no response to N during the first ten years of the experiment, while in the second ten years there were 11 and 14% responses in the stubble mulch and no-till treatments respectively (Brown et al., 1991). Similarly, the effect of fertilizer on grain N content only reached statistical significance in the second ten years. However, the CENTURY model predicted a large response to the application of 4.5 g N/m² from the beginning of the experiment. A test simulation using the automatic fertilizer option in CENTURY predicted that fertilizer applications averaging 3.0 g N/m²/crop would be required to achieve maximum wheat yields at minimum plant N concentrations.

The N deficit in the CENTURY simulations was also reflected in the results for soil mineral N (Figure 2). Field results (Lamb et al., 1985a) showed large accumulations of nitrate in the soil profile during fallow,

Table 2. Simulated versus actual mean 1970-1990 grain yields for three tillage treatments by two nitrogen treatments for the Sidney A and B sites

Tillage	Plow		Stubble mulch		No-till	
Nitrogen kg N/ha	0	45	0	45	0	45
	----------g C/m^2----------					
Field	104	102	98	103	99	107
CENTURY	85	106	84	106	82	103

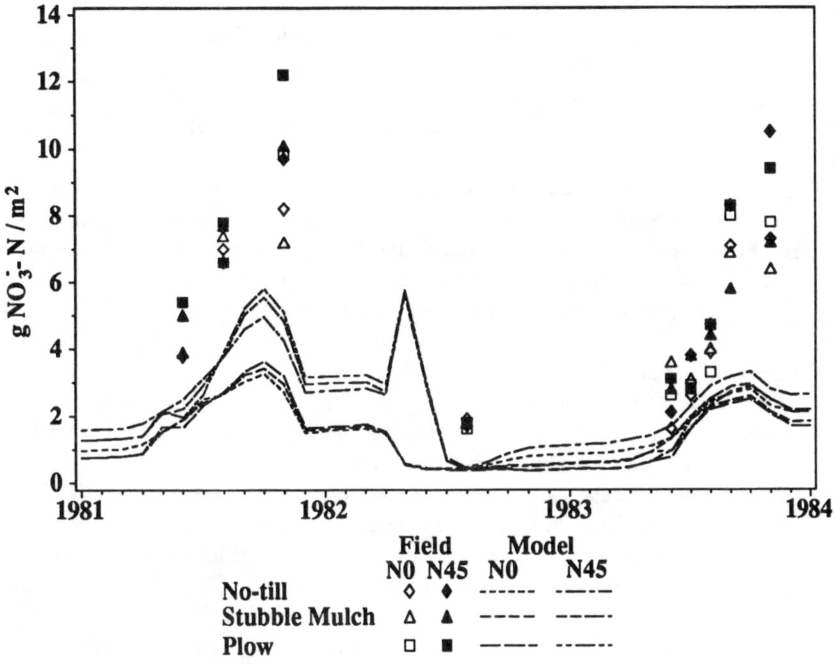

Figure 2. Simulated versus measured mineral nitrogen in the soil profile at Sidney site A. (From Lamb et al., 1985a.)

but the model only predicted a small accumulation of mineral N. Typically, nitrate in the soil profile at the time of wheat planting in the fall, was underestimated by about 2 to 9 g N/m^2.

The simulated crop residue levels (standing dead plus surface litter) showed the seasonal patterns expected from the different tillage treatments and were of a similar magnitude to the field data (Fenster and Peterson, 1979).

B. Simulation of Sites C and D

Sites C and D had a much higher fertility status, having been in native sod until the start of the experiment and it seemed unlikely that the simulation would predict a N deficiency. Thus, despite the problem with N deficits in the simulations of sites A and B, we decided to use the same parameters and schedules of management events with sites C and D to validate the CENTURY model's predictions of soil organic matter dynamics.

Again, the model underpredicted the accumulation of mineral N during fallow, indicating that the problem was not associated with initial soil organic matter levels at the beginning of the experiment. However, simulated wheat yields were not markedly reduced by N deficiency, except in the first crop on site C immediately after breaking the sod. Simulated yields tended to be slightly higher than observed yields and were only weakly correlated ($r^2 = 0.36$).

The CENTURY model predicted a gradual decline in soil organic matter in all three wheat fallow tillage treatments (Figure 3). The relative change was greater for C than for N resulting in a narrowing of the C/N ratio. The decline was greatest in the plow treatment and least in the no-till treatment. Almost all of the field data gave the same ranking of the treatments, but there was considerable variation between sampling dates in absolute C and N levels measured. The data was compiled from a variety of published and unpublished sources and this variation is assumed to be due to four factors: different sampling methods; some differences in depth sampled, although the results presented are all nominally for 20 cm depth; differences in sample preparation, especially the amount of residue and roots removed; and differences in analytical methods. There were two data sets in which we had more confidence in the results. The first set were for samples collected in 1981 and 1982 with results first published by Lamb et al. (1985b). The soil samples had been stored and we reanalyzed them for C, obtaining much higher results than originally published, but more in line with the other data. The N results reported here are those originally published. The second dataset was for samples that were collected from site C in 1990 (Cambardella and Elliott, 1992) and site D in 1991 (Cambardella, unpublished). As only small changes would be expected in soil organic matter in the native sod treatment between 1981/82 and 1990/91, it can be seen that the results for these two datasets are consistent with each other (Figures 4 and 5).

In the native sod treatment simulated soil organic matter levels increased slightly from 1970 to 1992, reflecting the change from an equilibrium established with grazing prior to 1940, to ungrazed conditions after 1940. The simulated and actual soil organic matter C and N levels in the native sod treatment were lower on site D than on site C reflecting the higher sand content on site D. While there had been some calibration of grazing intensity in order to match the simulated and field soil organic matter data, the fact that simulation results agreed fairly closely with the field data for the native sod plots on both sites for the 1981/82 and 1990/91 sampling dates indicate that CENTURY correctly predicted the effect of soil texture on native grassland soil organic matter levels. The model predicted C/N ratios of 11.9 and 12.5 for sites C and D respectively, compared to measured C/N ratios of 11.6 and 11.1. Thus, on site D, the C levels were slightly overestimated, while N levels were slightly underestimated.

For the 1981/82 and 1990/91 datasets the simulated soil organic C (Figure 4) and N (Figure 5) agreed fairly closely with the field data for all treatments. Differences between simulated and observed data were mostly less than the standard error of the difference between treatments (SED) calculated from the field data. The largest discrepancy occurred for the plow treatment in 1981/82. In general, there was more differentiation between tillage treatments than was predicted by the model. Coefficients of determination (r^2) for simulated versus observed data were 0.86 and 0.69 for soil organic C and N, respectively. In both cases the slopes of the regression lines were not significantly different ($p > 0.05$) from 1.0 and the intercepts were not significantly different ($p > 0.05$) from 0.0.

The CENTURY model predicted that the average annual effect on decomposition rate was 1.66, 2.06, and 2.34 times greater than the native sod treatment for no-till, stubble mulch, and plow, respectively. The field data suggest the decline in soil organic matter had initially been rapid and then stabilized, whereas the model predicted a more gradual and continual decline.

C. Physical Separation of Soil Organic Matter

The physical separation of soil organic matter fractions showed that POM accounted for 39% of the total soil organic C in the native sod treatment (Cambardella and Elliot, 1992; Table 3). Almost all of the differences in amounts of soil organic matter in the four treatments could be accounted for by changes in the size of the POM fraction. After twenty years of wheat-fallow management the C content in this fraction had been reduced by approximately 750, 1000, and 1100 g C/m^2 for the plow, stubble mulch, and no-till treatments respectively. N contents had dropped by about 50 g N/m^2. Examination of the POM by scanning electron microscope indicated that the POM fraction was composed mostly of root fragments in various stages of decomposition. It contained 47% lignin and 18% cellulose. Most of the POM has a density less than 1.85 g/cm^3.

The size-density fractionation of macroaggregates revealed that the fine-silt sized particles (2-20 μm) which have a density of 2.07-2.21 g/cm^3 contained 10-20% of the total soil organic C and 8-28% of the total

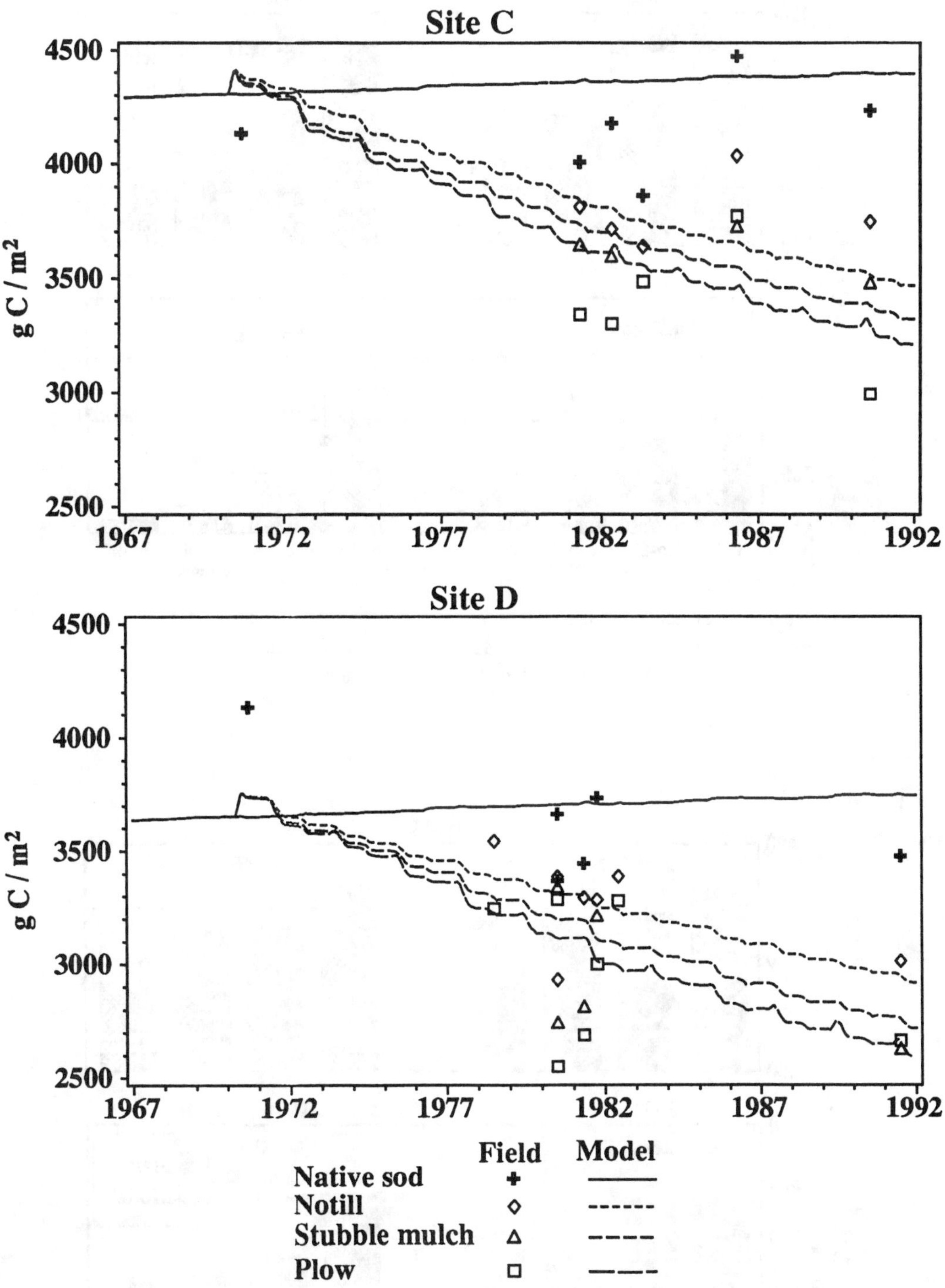

Figure 3. The effect of tillage treatment on simulated and measured soil organic carbon to 20 cm depth for Sidney sites C and D.

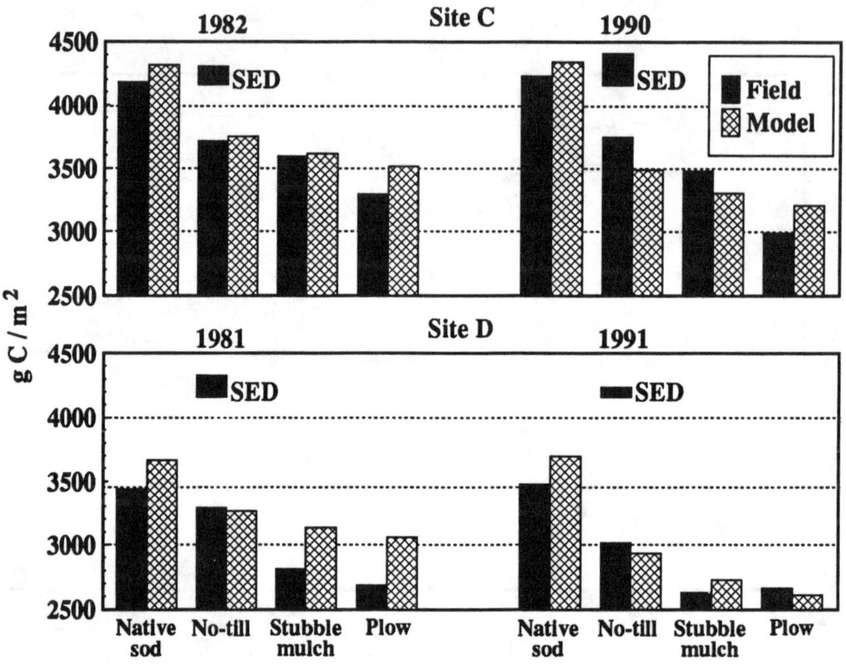

Figure 4. The effect of tillage treatment on simulated and measured soil organic carbon to 20 cm depth for the 1981/82 and 1990/91 datasets. SED = standard error of the difference for field data treatment means.

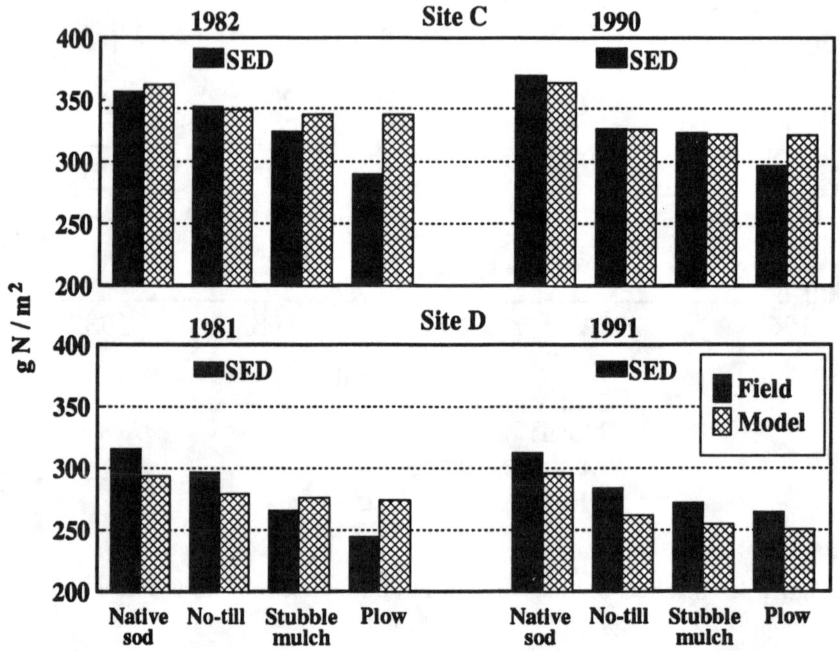

Figure 5. The effect of tillage treatment on simulated and observed soil organic nitrogen to 20 cm depth for the 1981/82 and 1990/91 datasets. SED = standard error of the difference for field data treatment means.

Table 3. Partitioning of soil organic carbon and nitrogen by physical separation for Sidney site C

Treatment	Native sod	No-till	Stubble mulch	Plow
		g C/m^2		
Total C	4237	3729	3494	3055
POM C	1671	926	681	563
ELF C	458	647	603	624
		g N/m^2		
Total N	366	326	325	295
POM N	107	40	53	48
ELF N	28	81	52	25

(Modified from Cambardella and Elliot, 1992 and 1994.)

soil N (Cambardella and Elliot, 1994; Table 3). The proportion of N mineralized in this fraction (4.7%) was much greater than for intact aggregates (2.1%), suggesting a labile, but physically protected pool which has been termed the enriched labile fraction (ELF).

IV. Discussion

A. Nitrogen Balance

The largest discrepancy between field data and model results was in the prediction of mineral N during the fallow period. The accumulation of mineral N in the soil profile was underestimated by amounts ranging from about 2 to over 20 g N/m^2. In the simulations the low mineral N levels then reduced crop N uptake, and at sites A and B, in the absence of N fertilizer, crop yields. There is field evidence from sites C and D which indicates a positive N budget in the no-till treatment (Lamb et al., 1985b) suggesting an unaccounted for N input.

The discrepancy between the CENTURY model's prediction of mineral N and the field measurements cannot be ascribed to a gross underestimation of mineralization rates in the surface soil because the model gave reasonable predictions of the decline in soil organic N in the surface 20 cm. Nor can the difference be explained by incorrect estimation of the effect of cultivation on mineralization rates because the problem occurred with no-till as well as the tilled treatments. Three processes are suggested which could contribute to the simulated N deficiency.

First, the CENTURY model is parameterized to simulate soil organic matter dynamics in the surface 20 cm of soil only. Although the concentration of organic matter is usually lower in soil layers below 20 cm, there may still be a considerable reservoir of organic N, which, if mineralized, would be measured as nitrate in deep soil samples and would be available for plant uptake by wheat. We calculated that up to 1.5 g $N/m^2/y$ could be mineralized from lower depths. This would be sufficient to make up the deficit for crop production, but would not match the measured accumulation of nitrate in the profile during the fallow period.

The second process which could have contributed to the simulated N deficiency is an overestimation of volatile N losses. Volatile N losses due to nitrification are estimated as being 5% of gross mineralization, originally calibrated to balance the N budget in native grasslands, with volatile N losses being approximately equal to external N inputs. However, volatile losses from senescing herbage were ignored. Recent research (Zachariassen, 1993) has shown that volatilization of ammonia from native grasslands is greatest at the end of the season when the herbage is senescing. If this pathway of N losses was accounted for in the grassland simulations, the volatile losses associated with nitrification could be reduced by 40%. Hence, the simulated volatile N losses with wheat-fallow management would be reduced by up to 0.4 g $N/m^2/yr$. The simulated equilibrium soil organic matter level in native grasslands is very sensitive to the rate of N loss because net primary production is limited by N. Hence, further field research and model testing are necessary to establish the amounts and mechanisms of N losses in native grasslands.

Table 4. Partitioning of soil organic carbon and nitrogen between pools by the CENTURY model for Sidney site C

Treatment	Native sod	No-till	Stubble mulch	Plow
	g C/m^2			
Metabolic	14	0	7	5
Structural	294	17	50	65
Active	146	62	87	103
Slow	2769	1975	1822	1719
Passive	1485	1473	1471	1471
	g N/m^2			
Metabolic	0.4	0.0	0.2	0.1
Structural	2.0	0.1	0.3	0.4
Active	12.3	13.3	15.9	18.0
Slow	156.5	118.9	114.5	112.1
Passive	199.6	198.7	198.7	198.8

Clearly, neither of the above mechanisms could account for all of the discrepancy between simulated and measured mineral N, and we suggest that higher than predicted rates of non-symbiotic N_2 fixation may have occurred in the wheat-fallow plots, especially in the no-till treatment. Combined with the positive feedback of having more N cycling through the plants and crop residue these three mechanisms could account for a large part of the discrepancy between model and field results.

B. Soil Organic Matter

The CENTURY model, within a reasonable degree of error, simulated the changes in soil organic C and N when native sod was converted to a wheat-fallow system on sites C and D. There was an indication that the model did not sufficiently differentiate between the tillage treatments. In particular, the soil organic matter level in the plow treatment declined at a faster rate than was predicted by the model. An increase in the factor for the effect of cultivation on decomposition rates seems justified for the tillage practices with the greatest degree of soil stirring, such as plowing and use of a field cultivator.

The laboratory results indicated that the POM fraction accounted for the majority of the soil organic matter initially lost as a result of the cultivation of grassland soils. This is analogous to the CENTURY model's prediction that the change in size of the slow pool would account for almost all of the difference between treatments (Table 4). However, the size of the slow pool is much larger than the size of the POM fraction. The present structure of the CENTURY model has flows to the slow pool from both the structural pools and the active pool. However, material derived from the structural pools is very different in physical form and chemical composition from material derived from the active and surface microbe pools. The former flow, representing the decomposition of highly lignified material is very likely to correspond to particulate soil organic matter which has a high lignin content. POM would thus provide a relatively simple laboratory technique for initialization and validation of changes in one of the model's pools. The enriched labile fraction, which is physically protected from decomposition in the aggregate structure, may possibly account for most of the latter part of the slow pool.

A modification of the model with the slow pool split into these two parts, but with no changes to parameter values indicated that the structural derived part of the slow pool (Table 5) underestimated the quantities of C and N in POM. Testing at more sites and further calibration of the parameters controlling C and N partitioning and decay rates will be necessary before the model can explicitly represent the POM fraction.

Table 5. Partitioning of soil organic carbon and nitrogen to the slow pool of the CENTURY model for Sidney site C

Treatment	Native sod	No-till	Stubble mulch	Plow
	-----	-----	g C/m^2 -----	-----
Structural derived	775	500	427	388
Active derived	1991	1473	1393	1330
	-----	-----	g N/m^2 -----	-----
Structural derived	41	26	23	21
Active derived	115	93	92	91

V. Summary and Conclusions

The CENTURY soil organic matter model (Parton et al., 1987; Parton et al., 1988) integrates the effects of climate and soil driving variables with agricultural management to simulate C, N, and water dynamics in the soil-plant system, with an emphasis on long-term soil organic matter dynamics. A new agroecosystem version of the CENTURY model has been designed for the simulation of complex agricultural management systems.

Validation of the effects of tillage on soil organic matter dynamics using results from a long-term tillage experiment near Sidney, Nebraska (Fenster and Peterson, 1979) showed that the CENTURY model closely predicted the changes in soil organic C and N when native sod was converted to wheat-fallow management. However, the model greatly underestimated the accumulation of inorganic N in the soil during the fallow period. Further research investigating the N cycle in the field and its representation in the CENTURY model is necessary.

The comparison of data for physical fractionations of soil organic matter with the CENTURY model's conceptual, kinetically defined pools revealed that the changes in the total amount of soil organic matter could be mostly accounted for by changes in particulate organic matter and the slow pool respectively. However, the predicted size of the slow pool was much larger than the measured amount of particulate organic matter. The fraction of the slow pool that is derived from the structural pool is likely to correspond to particulate organic matter. The model needs to be revised with the slow pool subdivided into pools for particulate organic matter and protected microbial products.

References

Broder, M.W., J.W. Doran, G.A. Peterson, and C.R. Fenster. 1984. Fallow tillage influence on spring populations of soil nitrifiers, denitrifiers, and available nitrogen. *Soil Sci. Soc. Am. J.* 48:1060-1067.

Brown, R.E., J.L. Havlin, D.J. Lyon, C.R. Fenster, and G.A. Peterson. 1991. Long term tillage and nitrogen effects on wheat production in a wheat fallow rotation. *Agronomy Abstracts 1991 Annual Meetings American Society of Agronomy*. 326.

Burke, I.C. C.M. Yonker, W.J. Parton, C.V. Cole, K. Flach, and D.S. Schimel. 1989. Texture, climate, and cultivation effects on soil organic matter content in U.S. grassland soils. *Soil Sci. Soc. Am. J.* 53:800-805.

Cambardella, C.A. and E.T. Elliott. 1992. Particulate soil organic-matter changes across a grassland cultivation sequence. *Soil Sci. Soc. Am. J.* 56:777-783.

Cambardella, C.A. and E.T. Elliott. 1993. Methods for physical separation and characterization of soil organic matter fractions. *Geoderma* 56:449-457.

Cambardella, C.A. and E.T. Elliott. 1994. Carbon and nitrogen dynamics of soil organic matter fractions from cultivated grassland soils. *Soil Sci. Soc. Am. J.* 58:123-130.

Cole, C.V. 1987. A conceptual framework for analysing agricultural and environmental concerns, p. 5-14. In: A.S. Phillips, and G.E. Schweitzer (eds) *Agricultural Development and Environmental Research: American and Czechoslovak Perspectives*, National Academy Press, Washington, D.C.

Elliott, E.T. and C.A. Cambardella. 1991. Physical separation of soil organic matter. *Agric. Ecosyst. Environ.* 34:407-419.

Fenster, C.R. and G.A. Peterson. 1979. Effects of no-tillage fallow as compared to conventional tillage in a wheat-fallow system. *University of Nebraska, Agric. Exp. Sta. Res. Bull. 289.* 28pp.

Follet, R.F. and G.A. Peterson. 1988. Surface soil nutrient distribution as affected by wheat-fallow tillage systems. *Soil Sci. Soc. Am. J.* 52:141-147.

Goering, H.K. and P.J. Van Soest. 1970. A system for forage fiber analysis including equipment, reagents, methods, and some analysis application. USDA-ARS Agric. Handb. 379. U.S. Gov. Print. Office, Washington, D.C.

Greb, B.W. 1979. Reducing drought effects on croplands in the west-central Great Plains. *USDA Agric. Info. Bull. No. 420.* 31 pp.

Haas, H.J., C.E. Evans, and E.F. Miles. 1957. Nitrogen and carbon changes in Great Plains soils as influenced by cropping and soil treatments. *USDA Tech. Bull. 1164.*

Jenkinson, D.S. 1990. The turnover of organic carbon and nitrogen in soil. *Phil. Trans. Royal Soc. Lond.* B 329:361-368.

Jenkinson, D.S. and J.H. Rayner. 1977. The turnover of soil organic matter in some of the Rothamsted classical experiments. *Soil Sci.* 123:298-305

Lamb, J.A., G.A. Peterson, and C.R. Fenster. 1985a. Fallow nitrate accumulation in a wheat-fallow rotation as affected by tillage system. *Soil Sci. Soc. Am. J.* 49:1441-1446.

Lamb, J.A., G.A. Peterson, and C.R. Fenster. 1985b. Wheat fallow tillage systems' effect on a newly cultivated grassland soils' nitrogen budget. *Soil Sci. Soc. Am. J.* 49:352-356.

Metherell, A.K., L.A. Harding, C.V. Cole, and W.J. Parton. 1993. CENTURY Soil Organic Matter Model Environment Technical Documentation. Agroecosystem Version 4.0. Great Plains System Research Unit Technical Report No. 4. USDA-ARS, Fort Collins, Colorado. 245 pp.

Meyers, T.P., B.B. Hicks, R.P. Hosker, J.D. Womack, and L.C. Satterfield. 1991. Dry deposition inferential measurement techniques-II. Seasonal and annual deposition rates of sulfur and nitrate. *Atmos. Environ.* 25A:2361-2370.

Mielke, L.N., J.W. Doran, and K.A. Richards. 1986. Physical environment near the soil surface of plowed and no-tilled soils. *Soil Till. Res.* 7:355-366.

National Atmospheric Deposition Program (NRSP-3)/National Trends Network. 1991. NADP/NTN Coordination Office, Natural Resource Ecology Laboratory, Colorado State University, Fort Collins, CO 80523

Nelson, D.W. and L.E. Sommers. 1980. Total nitrogen analysis of soil and plant tissues. *J. Assoc. Off. Anal. Chem.* 63:770-780

Parton, W.J., D.S. Schimel, C.V. Cole, and D.S. Ojima. 1987. Analysis of factors controlling soil organic matter levels in Great Plains grasslands. *Soil Sci. Soc. Am. J.* 51:1173-1179.

Parton, W.J., J.W.B. Stewart, and C.V. Cole. 1988. Dynamics of C, N, P and S in grassland soils: a model. *Biogeochemistry* 5:109-131.

Parton, W.J., J.M.O. Scurlock, D.S. Ojima, T.G. Gilmanov, R.J. Scholes, D.S. Schimel, T. Kirchner, J-C. Menaut, T. Seastedt, E. Gaand, J.I. Kinyamario. 1993. Observations and modelling of biomass and soil organic matter dynamics for the grassland biome worldwide. *Global Biogeochemical Cycles* 7:785-809.

Smika, D.E. and G.A. Wicks. 1968. Soil water storage during fallow in the central Great Plains as influenced by tillage and herbicide treatments. *Soil Sci. Soc. Am. Proc.* 32:591-595.

Snyder, J.D. and J.A. Trofymow. 1984. A rapid accurate wet oxidation diffussion procedure fore determining organic and inorganic carbon in plant and soil samples. *Commun. Soil Sci. Plant Anal.* 15:487-597

van Veen, J.A. and E.A. Paul. 1981. Organic carbon dynamics in grassland soils. I. Background information and computer simulation. *Can. J. Soil Sci.* 61:185-201.

Zachariassen, J. 1993. Ammonia exchange above grassland canopies. Ph.D. dissertation. Colorado State University, Fort Collins, Colorado.

CHAPTER 23

Management of Isoelectric Soils of the Humid Tropics

G. Uehara

I. Introduction

Mohr (1930), a Dutch soil scientist surveying the soils of Indonesia during colonial times, was struck by the diversity of soils he saw there. To organize his thoughts and observations he categorized the soils into young, mature, and senile soils. In doing so, Mohr was attempting to group soils into distinct behavioral and performance classes. On the Island of Java he witnessed young soils on landscape undergoing constant rejuvenation by volcanic eruptions. Under the warm and humid conditions of the tropics he could see evidence of these soils turning into mature, productive soils. But Mohr was most intrigued by the senile and impoverished soils occurring on the oldest landscapes. These soils, now known as Ultisols and Oxisols have been intensely studied in the past 25 years. The purpose of this paper is to summarize what we now know about Mohr's senile soils of the humid tropics and offer glimpses into what the future holds for them.

II. A Textbook Example

In order to understand and appreciate the objects of Mohr's curiosity, it may be useful to compare and contrast a typical senile soil from the humid tropics with a soil from the midwestern United States. Data for two such soils are presented in Tables 1 and 2. While one may argue that the soils are not typical for those regions, no one would claim that the soils and regions could be reversed. Any soil scientist reviewing the data in Tables 1 and 2, will, without hesitation, recognize them for what they are. What even the inexperienced eye will see are:

1. the high calcium levels in the midwestern soil, and
2. the high acidity of the surface horizons of the soils from the humid tropics.

But calcium levels and acidity can vary greatly over short distances even in the midwestern United States. While soil calcium and pH tend to be higher in the midwest than in the humid tropics, those are not the attributes that distinguishes the soils from the two regions. The telltale sign of a senile soil resides in the way soil pH measured in water and salt solution (1NKCl) varies with depth. In Table 1, pH in 1NKCl is depressed relative to the value in water for samples taken near the soil surface, but the reverse is true in the deeper horizons so that the difference in pH determined in 1NKCl and water is negative near the surface and positive in the deeper layers. It suffices to say at this point that the sign of this pH difference or delta pH corresponds to the sign of the net electric charge on the soil particles. Net electric charge implies that a particle has both positive and negative charge and can have a net zero charge when they occur in equal amounts. When net charge is zero, delta pH is also zero.

In the midwestern United States, or for that matter in all of North America, the net charge on soil particles is almost always negative. Exceptions to this are so rare that there is no benefit in measuring pH in 1NKCl to ascertain the existence of net positive charge. Even in the humid tropics, net positive charge

ISBN 1-56670-117-1/95/$0.00+.50

Table 1. Selected chemical data for an isoelectric soil from the humid tropics

Depth	Organic carbon	Extractable bases					Ext. iron	pH		
		Ca	Mg	Na	K	Sum		H_2O	KCl	ΔpH
cm	%	--------------cmol/kg---------------					%			
0-38	3.88	0.5	0.6	0.1	0.1	1.3	24.5	5.2	4.2	-1.0
38-48	2.43	trace	trace	0.1	0.1	0.2	32.5	4.7	4.9	-0.3
48-76	1.58	trace	0.6	trace	0.6	0.6	26.2	5.0	5.1	+0.1
76-100	1.10	trace	trace	trace	---	---	26.1	4.8	5.3	+0.5
100-155	0.66	0.2	0.1	0.1	0.5	0.5	23.5	4.8	5.4	+0.6
155-230	0.16	trace	trace	trace	0.1	0.1	20.5	4.9	5.3	+0.4

(From USDA, 1975; p 692.)

Table 2. Selected chemical data for a soil from the midwestern U.S.

Depth	Organic carbon	Extractable bases					Ext. iron	pH
		Ca	Mg	Na	K	Sum		
cm	%	------------------cmol/kg-------------------					%	H_2O
0-18	2.35	13.9	3.4	0.1	0.5	17.9	0.6	5.7
18-28	1.95	13.8	4.2	0.1	0.4	18.5	1.1	5.8
28-43	1.42	14.3	5.8	0.1	0.4	20.6	1.1	5.7
43-51	0.97	14.6	6.4	0.1	0.4	21.5	1.2	5.8
51-64	0.68	15.0	6.8	0.1	0.4	22.3	1.2	5.7
64-74	0.45	14.7	6.6	0.1	0.3	21.7	1.2	5.7
74-89	0.34	14.3	6.6	0.1	0.3	21.3	1.2	5.7
89-115	0.21	14.1	5.8	0.1	0.3	20.6	1.2	5.7

(From USDA, 1975; p. 660.)

is rare and almost always occurs only in the deep, subsoil. The data in Table 1, therefore, represent a textbook example of a soil from the humid tropics. The analyses contain the full range of characteristics of soils from that ecological zone, and is also useful in explaining the behavior of soils that have not yet acquired all of the same characteristics, but show early symptoms of senility. What we now know, and Mohr probably did not, is that the young and mature soils of the world differ from their older counterparts in the humid tropics mainly in their surface charge characteristics.

III. Surface Charge and Soil Vigor

Infertile and unproductive soils occur throughout the world, so why should we fault soils of the humid tropics for their impoverished state? The answer may lie in our belief that only a rich soil can support the immense biomass and species diversity one finds in tropical rainforests. We discovered too late that the richness of the rainforest resides not in the soil, but in the litter and living canopy. Even today its verdant appearance continues to entice its exploitation.

We now know that Mohr's senile soils are impoverished because they have lost negative charge and gained positive charge with age. The high levels of positively charged calcium ions in the midwestern soil is there because the negatively charged soil particles prevent their escape. As soils age and become senile, they lose negative charge and cations including calcium, magnesium, potassium, and sodium, on the one hand, and gain positive charge and anions such as phosphate and sulfate ions on the other.

In most young and mature soils, the charge on the particle arises from defects in the interior crystal structure. A common defect occurs when a trivalent aluminum ion substitutes for a tetravalent silicon atom during mineral synthesis. The difference in ionic charge is expressed as a negative surface charge which is in turn neutralized by cations from the weathering environment. All things being equal, the surface charge has greater affinity for a divalent cation such as calcium than for a monovalent one such as potassium or sodium. And because calcium is abundant in the earth's crust there is a strong bias for its retention by soils.

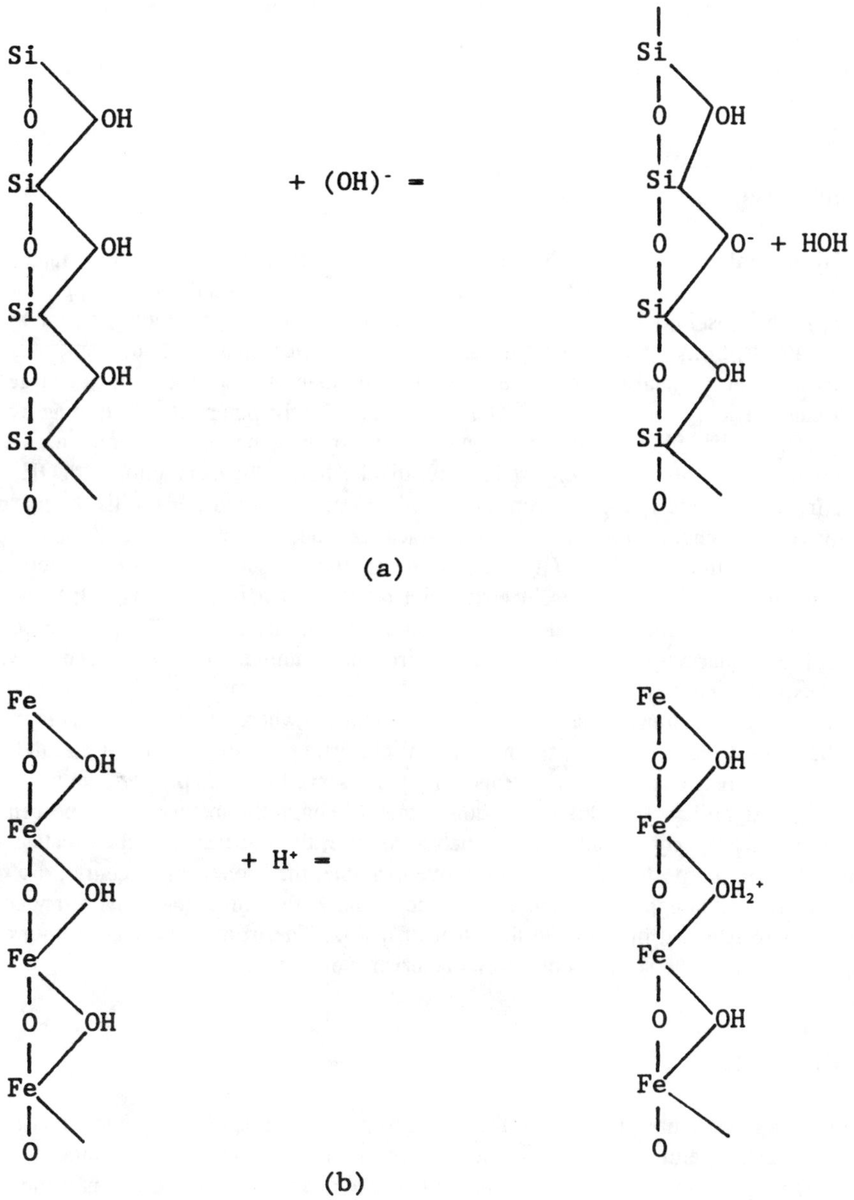

Figure 1. Surface charge originating from adsorption of potential determining hydrogen (H^+) and hydroxyl (OH^-) ions.

So long as there are weatherable minerals to replenish cations and other nutrients consumed by plants, a soil can retain its negative charge and vigor. Under warm and humid conditions, however, the supply of weatherable minerals is soon depleted and even the stable, negatively charged clay minerals begin to decompose. In the end, only the most resistent and insoluble minerals like the oxides of silicon, iron, and aluminum remain. Very few soils ever reach this advanced stage of weathering and in the cooler temperate climates they never do, but in the equatorial regions, weathering is accelerated and some soils do attain senility as illustrated by the data in Table 1. Many more tropical soils are not far behind and show early signs of old age.

IV. Surface Charge of Oxides

Unlike the clay minerals whose surface charges originate from internal defects and are permanent and almost always net negative, the charge on oxides arises from adsorption of potential determining ions (Uehara and Gillman, 1981). The most important potential determining ions in soils are the hydrogen and hydroxyl ions. The hydrogen and hydroxyl ions have little effect on the surface charge of minerals with permanent, negative charge, but are crucial to, and govern the sign and magnitude of the surface charge and surface potential of oxides. The oxides, however, differ in their affinity for potential determining hydrogen (H^+) and hydroxyl $(OH)^-$ ions. Silicon dioxide or quartz, for example, has great affinity for $(OH)^-$ at pH's normally encountered in soils, and is rendered net negatively charged by its reaction with OH^- as illustrated in Figure 1. In contrast, the oxides of iron and aluminum are rendered net positively charged by their preference for H^+. The chemical parameter that characterizes quartz and iron and aluminum oxides is the pH at which equal quantities of H^+ and OH^- are adsorbed so that the surface acquires net zero charge. This all important parameter is called the isoelectric point or point of net zero charge. It turns out that the isoelectric points of quartz and the oxides of iron and aluminum are near pH 2 and 8 respectively. This means that at pH 5.0, quartz is negatively charged and iron and aluminum oxides are net positively charged. As one would expect, quartz and the oxides of iron and aluminum interact strongly owing to their opposite charges. This interaction is more apparent in drier environments where the iron oxide is more likely to be red-colored hematite ($\alpha Fe_2 0_3$). There, a soil material consisting of 98 percent quartz and two percent hematite can be bright red. Under the microscope, one can observe fine hematite particles coating the coarse quartz grains. The significance of this interaction is that, although the material is 98 percent quartz, the mixture takes on the property of hematite. In such a mixture, the mineral with the greatest surface area controls soil behavior and performance. In the above example, the properties of quartz, are masked by a thin coat of hematite. Although the particle is 98 percent quartz, the mineralogy, chemistry and physics of its surface are more like that of hematite than that of quartz. The iron oxide is not only skin deep, the particle begins to acquire the surface charge characteristics of hematite.

V. Organic Matter

If it were not for organic matter, the soils of the humid tropics would be even more impoverished. In some ways organic matter is chemically very much like quartz. Both quartz and organic matter have isoelectric points in the pH range of two to three. This means that like quartz, organic matter is negatively charged in the pH range usually found in soils. But here, their similarities end.

Unlike quartz, organic matter possesses high surface area per unit weight, so much so that when organic matter meets iron or aluminum oxide, organic matter coats the oxide particle. Here again, as with the quartz-iron oxide mixture, the mixture's chemical properties are determined by its outer coating. This is clearly illustrated in the soil data for the humid tropics. The data reveals that with increasing soil depth the sum of bases ($\Sigma\ Ca^{++} + Mg^{++} + K^+ + Na^+$) decreases as soil pH increases. This is contrary to what every student of soil science learns, namely that soil pH should increase with increasing sum of bases. And for most soils outside the humid tropics that rule applies, but the soils of humid tropics operate under a different set of rules.

These rules were first described over 60 years ago by a Swedish chemist (Mattson, 1928) who probably knew very little about the soils of the humid tropics. Mattson worked with mixtures of silica and alumina gels and discovered that the isoelectric point of his mixtures decreased with increasing silica-alumina ratio. Moreover, when a mixture of specified silica-alumina ratio and isoelectric point was leached with water, the pH of the suspended mixture always shifted to the pH of the isoelectric point. Mattson (1932), called this phenomenon "isoelectric weathering." We now know that organic matter, quartz and the oxide of iron and aluminum are isoelectric materials, and if they are leached with rain water as they are in the rainforests, the pH in any depth of the soil profile is governed by the material's isoelectric point. The humus-rich surface horizons are acid because the isoelectric point of humus is acid. Similarly, the iron oxide-rich, humus-poor material in the deepest horizons is higher in pH because the isoelectric point of iron oxide is higher. And, as one would expect the middle horizons display intermediate pH's corresponding to intermediate isoelectric points. But this does not explain the higher base levels in the acid surface layers than in the near- neutral subsoil. The simple explanation is that the acid organic matter is net negatively charged and it is able to adsorb positively charge basic cations, whereas the subsoil material although higher in pH is net positively charged and therefore has no affinity for bases. The humus-rich, surface layer is net negatively charged because the pH of material in that layer lies above its isoelectric point, and the material in the deep subsoil is net positively charged because the ambient pH lies below the material's isoelectric point.

With this introduction, we are now ready to explore management options for isoelectric soils of the humid tropics.

VI. Soil Management

The purpose of soil management is to manipulate soil behavior so that a soil performs in a prescribed way. How, for example, should an isoelectric soil be managed so that it behaves as a reservoir rather than a conduit for nutrients? Unlike the fertile soils of the temperate regions which have a large reserve of nutrients, the isoelectric soils of the humid tropics are not only impoverished, they have little capacity to retain nutrients that are applied to them.

One measure of a soil's capacity to retain nutrients is its cation exchange capacity (CEC). This quantity is the product of specific surface (surface area per kilogram of soil material) and surface charge density (electric charge per unit surface area). CEC can therefore be expressed as:

$$CEC\left(\frac{cmol}{kg}\right) = S\left(\frac{m^2}{kg}\right)\ \sigma\left(\frac{cmol}{m^2}\right) \tag{1}$$

where S is the specific surface and σ is the surface charge density. The values for montmorillonite and kaolinite are given in Table 3.

Table 3. Cation-exchange capacity [CEC], specific surface (S), and surface charge density (σ) of two common soil minerals montmorilonite and kaolinite

Mineral	CEC $\frac{cmol}{kg}$	S $\frac{m^2}{kg}$	σ $\frac{cmol}{m^2}$
Montmorillonite	100-140	$6 - 8 \times 10^5$	$1.6 - 1.8 \times 10^{-3}$
Kaolinite	6-20	$3 - 10 \times 10^3$	2.0×10^{-3}

Table 3 illustrates the large difference in CEC between montmorillonite and kaolinite. These differences are mainly due to their specific surfaces, and one can attribute the lower CEC of kaolinite to its lower specific surface. Quartz, for example, has a negligible CEC because it occurs as coarse particles and presents a small surface area per unit weight. For that reason, a small amount of finely divided iron oxide can coat a large amount of quartz.

The CECs of organic matter and the oxides of iron and aluminum have not been included in Table 3 because as isoelectric materials, their CECs vary with pH.

It may be helpful at this point to compare and contrast the nature of surface charge on a mineral like montmorillonite, the principal clay mineral in the soil represented in Table 2, and the surface charge for isoelectric materials.

In isoelectric materials, the surface charge density (σ) varies with the surface potential (Φ). The surface charge density and surface potential may be expressed as

$$\sigma = \left(\frac{2n\epsilon RT}{\pi}\right)^{1/2} \sinh \frac{ZF}{2RT} \Phi \qquad (2)$$

and

$$\Phi = \frac{RT}{F} \ln \frac{H^+}{H^+_o} \qquad (3)$$

where:

σ = surface charge density in coulombs per meter^2 (C/m^2)
n = electrolyte concentration in kilomoles per meter^3 ($kmol/m^3$)
ϵ = dielectric constant of water in coulombs^2 per joule meter [C^2/Jm]
R = gas constant in joules per kilomole °K (J/kmol°K)
T = absolute temperature (°K)
Z = counter ion valence
F = Faraday constant in coulombs per kilomole (C/kmol)
Φ = surface potential (volts)
H^+ = hydrogen ion activity, and
H^+_o = hydrogen ion activity at the isoelectric point.

Equation 3 can be written in terms of pH

$$\Phi = \frac{RT}{F} 2.303 \, (pHo - pH) \qquad (4)$$

and combined with equation 2 to give

$$\sigma = \left(\frac{2n\epsilon RT}{\pi}\right)^{1/2} \sinh Z \, (1.15) \, (pHo\text{-}pH) \qquad (5)$$

where pHo is the isoelectric point.

Equations 2 and 5 represent the surface charge characteristics of the soils in Tables 2 and 1, respectively. In Table 2 and Equation 2, the surface charge density is a constant and therefore not subject to change by management. But if we examine equation 5, we note that the sign of surface charge can be negative, zero or positive depending on whether soil pH is above, equal to, or below the isoelectric point. Since soil pH can be readily adjusted, the sign and magnitude of surface charge of isoelectric soil materials can also be easily altered. The error of the past was that isoelectric soils were treated as though they were constant surface charge soils.

VII. Soil Rejuvenation

The high performance of most intensively cultivated soils are sustained by replacing nitrogen, potassium, and phosphorus removed from them. It has always been thought that the nutrient-deficient soils of the humid tropics might also benefit from fertilizers. Unfortunately, they have not responded to fertilizers as most agronomists would have liked. While this has been used as an argument to discourage further clearing of tropical rainforests, it has also blocked efforts to salvage abandoned or underutilized deforested lands. Examples of such lands can be seen on the Islands of Kalimantan, Sulawesi, and Sumatra in Indonesia. By transforming such lands into productive farms, the pressure to clear more rainforests can be lessened.

In order to understand the causes for the failure of isoelectric soils to respond to conventional soil management practices, we need to recall what Sante Mattson learned about these materials 60 years ago. Mattson (1932) discovered that when isoelectric materials were washed with distilled water (just as rainforest soils are washed with rain water), the pH always shifted towards the isoelectric point. In Equation 4, this corresponds to the convergence of pH and pHo. When this condition is reached, the surface potential attains a value of zero, and the surface charge density also acquires a net zero value. Isoelectric soil display all the symptoms of old age when they reach this state. The most common symptoms of a senile soil are low CEC and severe nutrient deficiency.

What is less well known is that in isoelectric soils, chemical impoverishment goes hand-in-hand with physical stability. By physical stability we mean the capacity of a soil material to maintain good tilth even with repeated cultivation. This state is reached when repulsive forces between like charged particles disappear at the isoelectric point, and soil particles begin to adhere to each other to form stable aggregates. In rare instances, the aggregates harden into gravel-sized particles, but the usual case is a situation in which isoelectric weathering produces soils with agronomically desirable, physical characteristics. Soil physical characteristics are as important to sustainable agriculture as the nutrient and chemical status of soils, and in the long term, it is the former attribute that will be sought by future farmers seeking new agricultural lands.

VIII. Managing Soil Acidity and Fertility

In the past, the shifting cultivator could always clear more land when soil fertility or pests and weeds lowered productivity to unacceptable levels. If shifting cultivators were denied that option, what can they do as sedentary farmers to sustain productivity indefinitely? This was one of the questions researchers of the Soil Management Collaborative Research Support Program set out to answer in West Sumatra, Indonesia (Arya, 1990; Colfer, 1987). The research site was located in a transmigration area with mean annual rainfall and temperature of 2500-3000 mm and 28 °C respectively. The transmigrants were mostly Javanese and Sundanese families from the Island of Java. Having farmed the fertile soils of Java, the transmigrants were unprepared for the low fertility and unresponsiveness of the soils to fertilizer inputs.

When confronted with this problem, the researchers hypothesized that water stress caused by restricted root development prevented crops from performing at their expected levels. It was further hypothesized that a chemical barrier of toxic aluminum prevented roots of acid sensitive crops from proliferating into and exploiting the large store of subsoil moisture. Lime which was so ineffective when applied on the surface, dramatically improved crop performance when incorporated 30 cm into the soil. Deep incorporation of lime not only eliminated water stress, but also enabled roots to intercept nutrients that would otherwise have been lost through leaching.

In Hawaii, where highly weathered soils have been cultivated and studied for nearly a hundred years, crop yields have increased more dramatically in the isoelectric soils of the high rainfall areas than in the younger, more fertile irrigated soils of the semi-arid areas. Like their counterparts in the humid tropics, the isoelectric soils of Hawaii are notoriously deficient in phosphorus, and often require heavy doses of phosphorus fertilizers before benefits to crops are observable. This situation stems from the strong affinity of positively charged iron and aluminum oxides for negative phosphate ions. With repeated, heavy application of phosphatic fertilizers, the surface of the iron and aluminum oxides acquires a net negative charge. This negative charge enables the soil to attract and retain positively charged nutrient ions which would otherwise be leached below the root zone. This increase in cation retention by phosphated isoelectric soils has been observed in the field by Gillman and Fox (1980). Earlier, Mekaru and Uehara (1972) had shown in laboratory studies that cation retention capacity of isoelectric soil materials could be increased several fold by addition of phosphate ions. In terms of Equation 5, phosphate absorptions increases net negative charge by lowering the isoelectric point. This suggest that when added to isoelectric soils, phosphorus is not only a plant nutrient, but acts as an amendment to increase surface potential and net negative charge. Phosphorus, in short, is a rejuvenating agent for senile, isoelectric soils, but is too valuable and expensive to be used for this purpose.

There is a cheaper and more abundant material to replace phosphorus as an amendment. This material is calcium silicate and is available as slag from old steel mills, or today, as fly ash from coal burning power plants. Calcium silicate rejuvenates and improve productivity of isoelectric soils by (1) neutralizing soil acidity and detoxifying aluminum, (2) lowering phosphorus fixation by iron and aluminum oxides, (3) increasing the cation retention capacity of iron and aluminum oxides and (4) increasing the silica content of plant tissue, and thereby rendering corps less susceptible to pest damage and lodging. Calcium silicate is regularly applied to sugarcane fields in Hawaii (Plucknett, 1971) and to rice fields in Korea and Japan.

IX. Summary and Conclusion

Unlike the majority of the world's soils which consist of minerals whose surface charge is constant and negative, the surface charge on the minerals in most humid tropics soils is variable and pH-dependent, and can be net positive, net zero or net negative. The pH corresponding to net zero charge is known as the isoelectric point. The charge on the mineral particle is net positive below the isoelectric point and net negative above it. Most metal oxides commonly found in soils of the humid tropics possess isoelectric points between pH 7 and 9. At soil pH's normally encountered in the humid tropics, the oxides are highly positively charged. In contrast soil organic matter with isoelectric points near pH 2 is net negatively charged and interacts strongly with the metal oxides. The isoelectric point of the metal oxide-organic matter complex is intermediate between the interacting components, and becomes more acid with increasing organic matter content. Since organic matter generally decreases with depth, soil pH often increases with depth even though the sum of bases decreases in that direction. The sum of the bases decreases with depth in many soils of the humid tropics because the charge on the soil particles becomes increasingly positive with depth and diminishing organic matter content. An important feature of these soils is that soil pH is controlled principally by the isoelectric point.

The isoelectric point can be readily changed by addition of soil amendments. Like organic matter, phosphate fertilizers can also lower the isoelectric point to render the inorganic minerals more negatively charged. But phosphorus is too expensive to be used in this way. A cheaper and more plentiful amendment is calcium silicate which is a waste product of coal-burning power plants.

The primary reason for investing in the rejuvenation of the soils of the humid tropics is to prevent further encroachment of humans into the rainforests. The alternative to slash and burn agriculture is sedentary farming on rejuvenated soils now either under-utilized or abandoned. The cost of rejuvenation may be high, but should be markedly lower than the ecological price of losing an equal area of rainforests.

The potential for high productivity of isoelectric soils resides in their outstanding physical characteristics. When these characteristics are combined with high soil fertility (which they now lack), adequate rainfall and year-long growing season, the soils of the humic tropics can be as productive as any in the world.

References

Arya, L.M. 1990. Properties and processes in upland acid soils in Sumatra and their management for crop production. Sukarami Research Institute for Food Crops; Padang, Indonesia.

Colfer, C.J. 1987. Social science and soil management: An anthropologists role in the Tropsoils project. Farming Systems Support Project Newsletter 5:15-18. University of Florida, Gainesville.

Gillman, G.D. and R.L. Fox. 1980. Increase in the cation exchange capacity of variable charge soils following superphosphate applications. *Soil Sci. Soc. Am. J.* 44:934-938.

Mattson, S. 1928. The electrokinetic and chemical behavior of the aluminosilicates. *Soil Sci.* 25:289-311.

Mattson, S. 1932. The laws of soil colloidal behavior: 1X. Amphoteric reactions and isoelectric weathering. *Soil Sci.* 34:209-240.

Mekaru, T. and G. Uehara. 1972. Anion adsorption in Ferruginous tropical soils. *Soil Sci. Soc. Am. Proc.* 36:296-300.

Mohr, F.C.J. 1930. Tropical soil farming processes and the development of tropical soils. National Geographical Survey of China, Peiping.

Plucknett, D.L. 1971. The use of soluble silicates in Hawaiian agriculture. University of Queensland Papers 1:203-223.

Uehara, G. and G.P. Gillman. 1981. *The mineralogy, chemistry, and physics of tropical soils with variable charge clays.* Westview Press, Boulder, Colorado.

U.S. Department of Agriculture, Soil Conservation Service, Soil Survey Staff. 1975. Soil taxonomy. USDA Agric. Handbook No. 436. Govt. Printing Office, Washington, D.C.

CHAPTER 24

The Significance of Greenhouse Gas Emissions from Soils of Tropical Agroecosystems

John M. Duxbury

I. Introduction

Assessments of climate forcing created by greenhouse gases evolved from soils have for the most part focussed on carbon dioxide (CO_2), methane (CH_4), and nitrous oxide (N_2O) because they are considered to be the most important soil derived greenhouse gases and because their effects can be reasonably well quantifed with present knowledge. Soil sources of nitric oxide (NO) and ammonia (NH_3) have potentially important, but less well understood, effects on climate forcing through their impacts on atmospheric chemistry and dispersal of nitrogen over landscapes. In general, we have a good understanding of the processes and controls on the generation of all of these gases in soils and reasonable knowledge of their emissions from soils at small scales. Emission estimates are less certain at regional and larger scales. Other greenhouse gases evolved from soils include carbonyl sulfide (COS) and carbon disulfide (CS_2), but little is known about their source strengths.

In this paper, I discuss greenhouse gas emissions from soils of tropical agroecosystems within the contexts of total agricultural sources and anthropogenically created radiative forcing of climate. I do this to provide a perspective of the relative importance of both agriculture and soils of the tropics to the climate change issue. I also comment on some of the uncertainties and gaps in our knowledge of greenhouse gas emissions from soils of the tropics.

II. Contributions of Agriculture and Managed Soils to Radiative Forcing of Climate

A. Relative Radiative Forcing Created by Agricultural and Non-Agricultural Sectors

The relative impact of different anthropogenically generated greenhouse gases on radiative forcing of climate can be assessed by multiplying the global warming potential (GWP) index of a gas by its source strength (Shine et al., 1990; Isaken et al., 1992). The GWP index combines the radiative capacity and atmospheric residence time of a gas with a time frame of analysis and expresses the result relative to CO_2 , which always has a value of 1. Such analyses are, of course, limited by uncertainties in both the radiative effects and source strengths of gases and usually only include CO_2, CH_4 , N_2O, chlorofluorocarbons (CFCs), and sometimes ozone (O_3). They may or may not include indirect effects of these gases. The issue of indirect effects of CH_4 is especially important to assessment of agricultural contributions to climate forcing because agriculture is a major contributor to anthropogenic emissions of this gas (Duxbury et al.,1993). In 1992, the Intergovernmental Panel on Climate Change (IPCC) withdrew an earlier value for the GWP associated

ISBN 1-56670-117-1/95/$0.00+.50

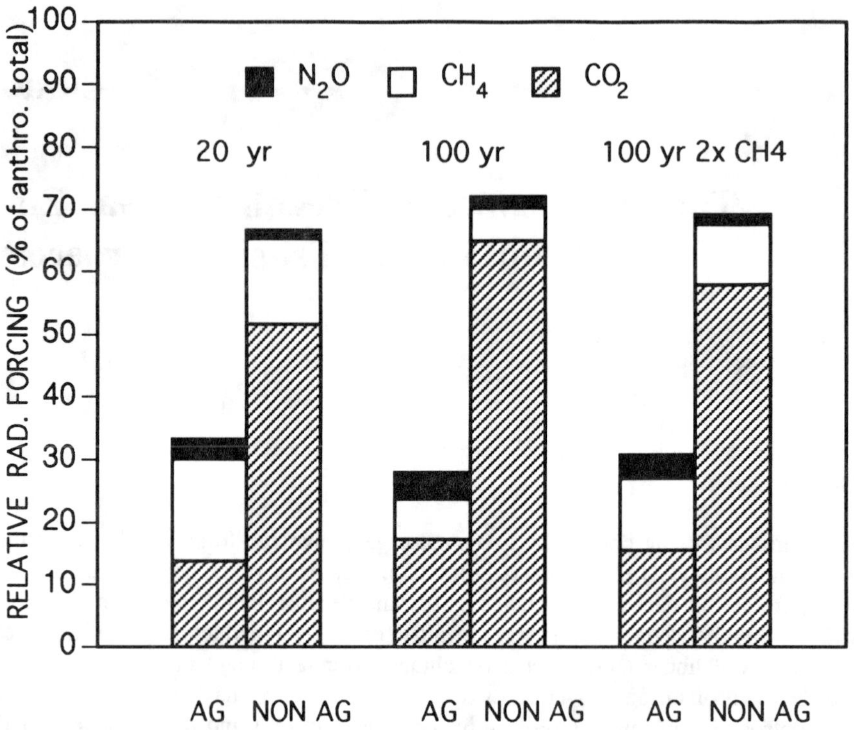

Figure 1. Relative global radiative forcing created by agricultural and non-agricultural sectors for 1990 estimated emissions of CO_2, CH_4, and N_2O. (Calculated from data in Watson et al., 1992.)

with the indirect effects of CH_4, citing insufficient knowledge (Isaksen et al., 1992). It is known, however, that the indirect effects of CH_4 are positive, i.e. they increase the GWP of this gas, and it is possible that cumulative indirect effects may generate a GWP similar to that of the direct effect.

The relative contributions of agricultural and non-agricultural sectors to radiative forcing of climate created by 1990 estimated anthropogenic emissions of greenhouse gases are shown in Figure 1. The analysis only includes CO_2, CH_4, and N_2O for the reasons stated above and is based on IPCC estimates of source strengths and GWP values (Watson et al., 1992; Isaksen et al., 1992). Two analysis time frames, 20 and 100 y, were used and the doubled CH_4 scenario represents the case where the indirect GWP of CH_4 is assumed to be equal to it's direct GWP. These analyses show that the combined effects of agriculture and clearing of land for agricultural use contribute between 28-33% of the radiative forcing generated by present emissions of the three gases considered. Carbon dioxide emissions from fossil fuel combustion are the dominant source of radiative forcing created by the non-agricultural sector. Emissions of CO_2 and CH_4 are both important contributors to radiative forcing generated by agriculture, while N_2O is less important.

B. Agricultural Activities Contributing to Radiative Forcing of Climate

Table 1 shows 1990 estimated emissions of greenhouse gases from agricultural sources broken down by tropical and other agroecosystem types and by soil and non-soil sources. This data was coupled with GWP indices to compare the relative radiative forcing created by tropical agroecosystems with that from other agroecosystems (Figure 2a) and that created by soil sources with non-soil sources (Figure 2b) using the methodology described in the preceding section. It can be seen from Figure 2a that 80% of the radiative forcing created by agriculture arises in tropical agroecosystems. Of the three gases considered, CO_2 emissions associated with agricultural development are a major component of radiative forcing attributable to tropical agroecosystems. Methane emissions from ruminant animals and rice paddies are also important, but N_2O emissions are of minor importance. It is likely, however, that emissions of N_2O from agriculture

Table 1. Estimated emissions of greenhouse gases from agricultural sources in 1990

Gas	Total	Temperate	Tropical	Soil	Non-soil
	----------	----------	Tg yr^{-1}	----------	----------
CO_2	6527	660[a]	5867	1173	5354
CH_4	198	57	141	60	138
N_2O	5.3[b]	2.6[c]	2.6[c]	4.7	0.6

[a] Fossil fuel use on farm and for manufacture of farm chemicals; [b] using the high end of the range given by IPCC for emissions from fertilizer (Watson et al., 1992); [c] based on equal use of fertilizer N by countries in temperate and tropical regions.

have been considerably underestimated (see later discussion) and this picture may change as better information becomes available. About one-third of the anthropogenic radiative forcing attributable to agriculture is associated with soil sources, while the bulk is derived from loss of forest biomass (CO_2) and CH_4 emissions from ruminant animals and animal wastes. The soil sources of CO_2 and CH_4 are essentially all in the tropics and half of the N_2O emissions are estimated to come from the tropical soils.

III. Emissions of Greenhouse Gases from Soils of the Tropics

A. Carbon Dioxide

It is well known that conversion of land from it's natural state to agriculture generally leads to losses of soil organic matter, that the soil organic matter content reaches an equilibrium level under a given set of management practices, and that factors such as climate, residue return, manure additions, and tillage influence the equilibrium level. It may take up to 50 years for the organic matter content of soils in temperate climates to reach a new equilibrium level following a change in management, but this time period is much shorter in the tropical environment. Most of the loss of organic matter from soils occurs quickly following conversion of natural ecosystems to agriculture so that on a global scale only newly cleared lands are considered to be significant sources of CO_2 (Houghton et al.,1983).

Large scale clearing of lands for agriculture is presently occuring in the tropics. Carbon emissions to the atmosphere from tropical deforestation are currently estimated to be within the range of 1.1-3.6 x 10^{15} g C yr^{-1} (Houghton, 1991); the IPCC uses a value of 1.6 x 10^{15} g C yr^{-1} (Watson et al., 1992). Loss of soil organic matter is thought to contribute up to 30% of the carbon loss with the bulk coming from the trees (Houghton et al., 1983). For comparison, the present global rate of emission of CO_2 from fossil fuel combustion is 6 x 10^{15} g C yr^{-1} (Watson et al., 1992). Loss of soil organic matter from soils of the tropics could be equivalent, at most, to 8% of the global fossil fuel source (using the IPCC estimate for total C loss created by deforestation).

Most recent discussions of soil carbon changes following conversion of lands in the tropics to agriculture utilize data gathered by agricultural scientists over the last 50 years or so. Unfortunately, these data, which were gathered mainly to assess the fertility of the surface soil layer, are not well suited for assessment of CO_2 release from soils. Their use for this purpose has led to an overestimation of CO_2 release from soils following conversion of land to agriculture because of:

- extrapolation of measured losses in the surface soil, usually <20 cm, to a depth of 1m,
- failure to recognize that comparisons of soil organic matter made on a % C basis did notconsider increases in soil bulk density that usually occured with agricultural use; consequently different quantities, and effectively different depths, of soil were compared, and
- failure to consider soil erosion, which removes carbon rich surface soils and hence organic matter from study sites. Eroded soil is either transported to aquatic environments or redistributed over land-scapes. Although the fate of carbon in eroded materials is uncertain, it is inappropriate to simply equate soil organic matter loss with CO_2 production.

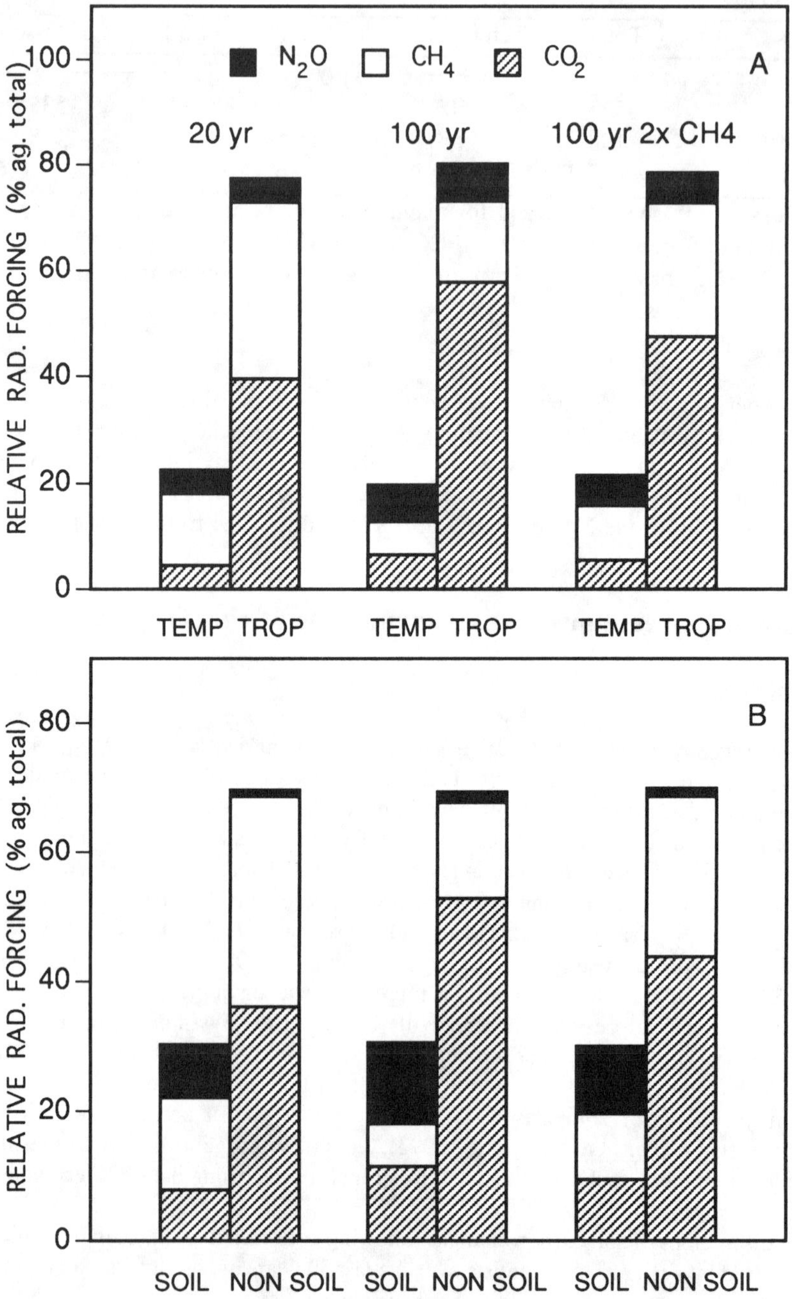

Figure 2. Agricultural sources of radiative forcing; a) tropical agroecosystems versus other agroecosystems and b) soil versus non-soil sources. (Calculated using data in Table 1 and GWP values from Isaken et al., 1992.)

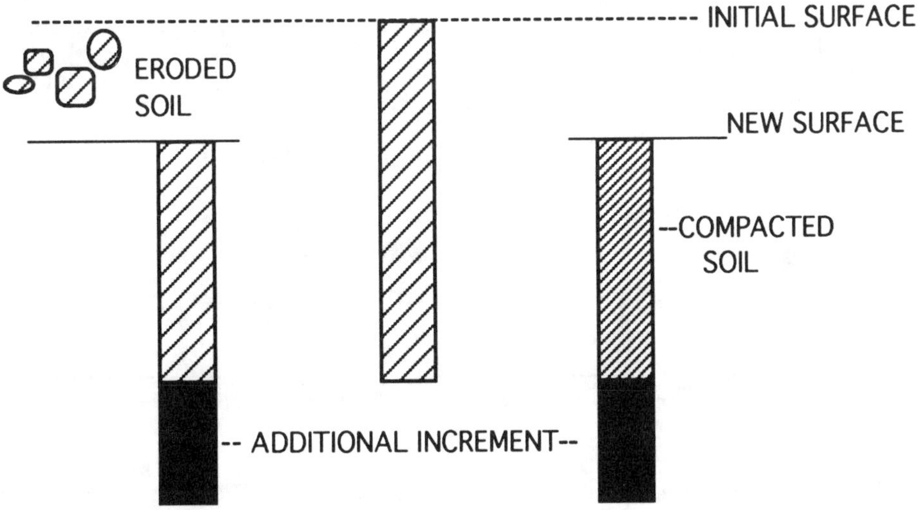

Figure 3. Effects of erosion and compaction of soil on the zone and mass of soil sampled.

Declines in soil organic matter levels following conversion of land to agriculture are mostly restricted to surface soils and there often is little or no loss below 20 cm (e.g. Juo and Lal, 1979; Vitorello et al., 1989; Krebs, 1975). There is therefore little justification for applying declines in soil organic matter measured in surface soils to soils at lower depths in the profile. Moreover, there is evidence that the carbon content of long-term (50+ yr) pastures is equivalent to that in forests (Brown et al.,1984; Lugo et al., 1986). Thus, Houghton et al.'s (1983) calculation of the soil contribution to CO_2 release following deforestation, which used minimum values for soil organic carbon content in the top 1m of agricultural soils of 50% and 25% of natural ecosystem levels for pasture and cropland, respectively, undoubtedly overestimates soil organic carbon loss.

The effects of compaction of soil and soil erosion on effective sampling depth are illustrated in Figure 3. In both cases, sampling to a fixed depth, using the surface as the reference point, leads to inclusion of soil from deeper in the profile. Inclusion of this additional depth increment, with a lower organic matter content, automatically reduces the percent organic C in the soil sample from the agroecosystem relative to that from the natural ecosystem. This problem was noted by Nye and Greenland (1964) but has largely been overlooked in subsequent research. Comparisons of soil organic matter levels have almost always been made on a % basis, which is inadequate for any purpose. Changes in soil organic C content are properly determined by measuring the total quantity of organic C in a soil profile. If this is done for the surface layer only, compaction would of course increase the measured quantity of organic C in the sample.

Quantitative measurements of soil compaction when highly weathered soils of forested tropical ecosystems in Tanzania were converted to agriculture are summarized in Table 2, which was compiled using data reported by Allen (1985). Of the 24 comparisons (0-30 cm depth) between forest and agricultural sites, 14 of 15 soils with an initial bulk density < 0.80 g cm^{-3}, showed increases of at least 30% in agricultural use. In contrast, 9 soils with initial bulk densities > 0.80 g cm^{-3} were within $\pm 10\%$ of their original value, and 4 had lower bulk densities in agricultural use. As would be expected, soils that had low initial bulk densities (<0.8 g cm^{-3}) and high initial organic carbon levels ($>5\%$) showed the greatest increase in bulk density and the largest apparent decline in organic matter when converted to agriculture.

Further evidence for the extent of soil compaction is found in the study of Juo and Lal (1977) who reported that bulk densities of the 0-15 cm depth of soil under managed fallows and agricultural crops were 1.4-1.6 times their initial value of 1.00 g cm^{-3} three years after clearing a secondary forest in Nigeria. In the absence of soil erosion, the effects of the various systems on soil carbon content vary widely with the method used to assess change in soil carbon (Table 3). When comparisons are made on a % C basis, soil carbon decreases (-9 to -37 %) for all systems except the guinea grass fallow where it increases by 16%. In contrast, soil carbon contents increase (+1 to +59 %) for all systems when the total mass of C in the 0-15 cm soil depth is compared. However, neither of these two methods correctly assesses changes in soil

Table 2. Effect of conversion of tropical forests to agriculture on bulk density and carbon content of highly weathered soils[a]

Initial value[b]	Number of comparisons	Frequency distribution of ratio Ag soil/Forest soil			
		Bulk density			
(g cm^{-3})		0.9-1.0	1.0-1.25	1.25-1.5	>1.5
<0.8	15	0	1	7	7
>0.8	9	4	5	0	0
		Organic carbon content			
(%)		0.25-0.5	0.5-0.75	0.75-1.0	>1.0
>5	18	13	5	0	0
<5	19	4	4	4	7

[a] Prepared from data in Allen, 1985; [b] for 0-30 cm soil depth.

Table 3. Effect of analysis method on assessment of soil C change (0-15 cm depth) three years after clearing a secondary forest in Nigeria[a]

System	Soil bulk density (gm cm^{-3})	Soil C content (%)	% change in soil carbon: %C	% change in soil carbon: Change in mass of C in 0-15 cm depth	% change in soil carbon: Corrected for subsoil increment
Cleared forest	1.00	1.74	-	-	-
Bush regrowth	1.38	1.44	-17	+14	0
Guinea grass	1.37	2.01	+16	+59	+45
Pigeon pea	1.49	1.57	-10	+35	+16
Maize + residue	1.46	1.58	-9	+33	+15
Maize - residue	1.59	1.10	-37	+1	-21

[a] Calculated from data in Juo and Lal, 1977.

carbon, and they represent the two extremes. The data is not available to properly assess carbon change in these systems, but the correct approach can be illustrated by making several assumptions. First, the soil carbon content of the subsoil increment included in the sample because of soil compaction can be calculated if it is assumed that there has been no change in soil C content for the regrowing bush. In this event, subsoil C is 0.65%. If it is further assumed that subsoil C does not change over time, the contribution of this to the total mass of C in the 15 cm deep soil sample can be calculated for each of the systems. This quantity of C is then subtracted from the total mass of C in the sample to obtain the mass of C in soil corresponding to the initial sample, i.e. the soil collected immediately after forest clearing. The corrected soil C content is then compared with that in the initial soil sample. When this approach is followed, soil C content decreases by 21% for maize without residue return but increases (+15 to +45%) in the other systems and results are intermediate between those obtained by the other two methods.

Soil erosion rates in tropical agroecosystems vary from <10 to several hundred t soil ha^{-1} yr^{-1} depending on climate, slope, soil type, amount and continuity of vegetative cover, tillage, and clearing methods (Cassel and Lal, 1992; Lal, 1990; El-Swaify et al.,1982; Sheng, 1982). Erosion of soil preferentially moves low density materials, i.e. those with high organic content. Enrichment factors of 2 to 5 have been reported for the organic matter content of eroded materials (Lal, 1976). The significance of erosion to soil carbon measurements depends on erosion intensity, the time since conversion of land to agriculture, the depth of mixing of soil, and the C content of soil below the initial sampling depth. As a simple illustration, 40% of the carbon in a soil having a bulk density of 1.25 g cm^{-3} and sampled to a depth of 20 cm is lost in a 50 yr period at an erosion rate of 10 t ha^{-1} yr^{-1} when the carbon content of eroded material is enriched by a factor of 2. If the original carbon levels in the soil were 2% for the top 20 cm and 0.5% below 20 cm, the measured level of soil carbon after 50 years would be 1.06% (assuming no mixing of the two soil layers).

A simulation model for soil organic matter decomposition and rainfall erosion and its application to soils of the tropics has been described by Bouwman (1989). Application of the model to two tropical soils suggested that soil organic C losses in 50 yr from a 80 cm deep profile would be between 31-50% with no erosion and that erosion would increase C loss to 40-70% depending on slope (5 or 10%). The model may, however, overestimate mineralization of soil organic matter because all of it was considered to be biologically active, whereas there is strong evidence for a biologically inactive or passive pool of organic matter in soils, on this kind of time scale (Balesdent et al., 1988; Duxbury et al., 1989). Omission of a refractory C pool would cause the importance of decomposition to be elevated relative to erosion, especially when 50-54% of the organic matter in the profile was in the top 15 cm of soil.

Although erosion lowers the organic matter content of soils, this loss of carbon cannot be equated with CO_2 release. There are no studies to my knowledge of the fate of organic matter in eroded soils, but since erosion generally moves soils to aquatic environments or to lower and wetter positions in the landscape it should lead to preservation of organic matter. It is not inconceivable that agricultural development, through acceleration of soil erosion and subsequent accuumulation of organic matter in former subsoils, actually leads to carbon storage rather than to loss, although such a hypothesis is usually greeted with enthusiastic skepticism. Support for the conclusion that soil erosion is a carbon storage process, at least on geological time scales, comes from the fact that 99% of the organic carbon at the earth's surface is in sedimentary rocks (Zehnder, 1982).

B. Methane

The predominant agricultural sources of CH_4 are ruminant animals (80 Tg CH_4 yr^{-1}) and flooded rice (60 Tg CH_4 yr^{-1}), with smaller amounts associated with animal waste (28 Tg CH_4 yr^{-1}) (Watson et al., 1992; Safley et al., 1992). The estimated global annual flux of CH_4 from flooded rice is probably less certain than that from ruminant animals because few measurements have been made under actual production conditions in Asia, where water management is highly variable (Neue et al., 1990) and where 90% of the world's rice is grown (USDA, 1990). In contrast, animal scientists have long studied CH_4 generation by ruminant animals, especially cattle, because the 8-15% of digestible energy that is converted to CH_4 represents an inefficiency in animal nutrition (Leng, 1991).

Recent estimates of annual emissions of CH_4 from flooded rice range from 25-170 Tg CH_4 yr^{-1} (Holzapfel-Pschorn and Seiler,1986; Schütz et al., 1989; Neue et al., 1990; Watson et al.,1992). The IPCC (Watson et al., 1992) recently reduced its estimate from 110 to 60 Tg CH_4 yr^{-1} based on recent data from India (Parashar et al., 1991 and 1992), but this change should be considered speculative due to the limited quantity of flux data gathered in the Indian studies. The data base for annual flux estimates for CH_4 emissions from rice fields is summarized in Table 4 and it can be seen that estimated emissions vary widely in all locations. The highest reported values are from China, while the limited data from India is towards the low end of the range. It is clearly important to identify the reasons for the wide range in emissions and to derive global estimates in a systematic way based on controlling variables. Water status is likely to be an important variable (Neue et al., 1990). Emissions have also been found to be lower from acid soils (Tsuruta, personal communication) and higher when crop residues are returned (Schutz et al., 1989; Yagi and Minami, 1990) or when green manures are used (Lauren and Duxbury, 1993; Lauren et al., 1994). The latter finding illustrates that conflicts may arise between the desire to utilize practices that promote agricultural sustainability and the desire to mitigate greenhouse gas emissions.

Using the most recent IPCC estimates of CH_4 source strength (Watson et al., 1992), USDA statistics for ruminant animal numbers in different countries (USDA, 1990), and data from Safley et al. (1992) for CH_4 emissions from animal waste, it can be calculated that 71% of CH_4 emissions associated with agriculture occur in the tropics (Figure 4). It should be noted that cattle/buffalo account for about 70% of the emissions from ruminant animals, and that emissions per unit of digestible energy vary with feed quality, being about 8% with high protein feeds and 15% with high fiber-low protein feeds, which are commonly used in the tropics (Leng, 1991). Methane emissions per animal unit will therefore vary with both quantity and quality of feed; such adjustments were not included in the present calculation. Furthermore, CH_4 emissions from biomass burning, which are estimated to be 40 Tg CH_4 yr^{-1} (Watson et al., 1992) were also not considered. With these caveats, emissions of CH_4 from flooded rice (the soil contribution), are 36% of the agricultural total and 50% of the tropical total.

Table 4. Estimated emissions of methane from flooded rice

Country/Region	Estimated annual flux[a]	Comments
	g CH_4 m^{-2} yr^{-1}	
India	8-23	Variable water
China	14-18	One crop; one study
	55-170	Two studies
Thailand	4-75	Low from acid soils
Developed countries	5-77	Eight studies

[a] Prepared from data in Watson et al., 1992; data from Thailand is from Tsuruta, personal communication; data from developed countries include Cicerone, 1992 and Lauren et al., 1992.

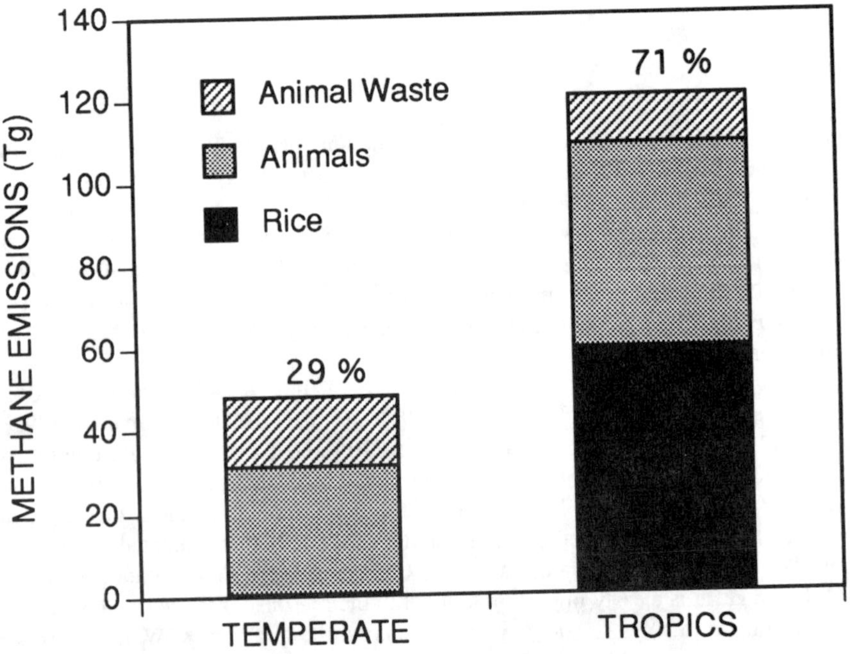

Figure 4. The distribution of agricultural sources of CH_4 between temperate and tropical environments.

While flooded, predominantly anaerobic soils are sources of CH_4, aerobic soils are a sink for atmospheric CH_4 (Steudler et al., 1989; Whalen and Reeburgh, 1990; Mosier et al, 1991). Estimates of the global soil sink strength are quite uncertain because of the limited number of measurements (Schimel et al., 1993) and incompletely understood interactions with the soil nitrogen cycle that appear to reduce CH_4 oxidation (Steudler et al., 1989; Mosier et al., 1991). A recent estimate suggests that the global soil sink strength could be about 10% of the tropospheric chemical sink or 50 Tg CH_4 yr^{-1}(Summerfeld et al., 1993). Measurements in tropical ecosystems suggest that agricultural development may reduce CH_4 uptake by as much as an order of magnitude (Goreau and de Mello, 1988; Keller et al., 1990), but tropical pastures can sometimes be a small source of CH_4 (Goreau and de Mello, 1988). Clearly, more research is needed to establish the global strength of the soil sink and to define the impact of agricultural management on this. Without the estimated soil sink, the atmospheric concentration of CH_4 would be increasing at about 1.5 times it's current rate.

Table 5. Estimated global source strengths of ammonia and N oxides

	-Global source strength-		---Soil source strength---		Other Ag.
	Total	Anthropogenic	Total	Anthropogenic	sources
	Tg N yr^{-1}				
N_2O[a]	10.2	2.4	7.8	2.1	<0.2
NO[b]	68	40	20	?	?
NH_3[c]	75	50	19	9	32

[a] From IPCC (1992), using the upper value given for the effect of fertilizer addition; [b] from Logan (1983) and Davidson (1991); [c] from Schlesinger and Hartley (1992).

Table 6. Estimated annual inputs into the terrestrial N-cycle

Process	Amount
	(10^9 kg)
Background N fixation	110
New fixation	
Fertilizer manufacture	90
Biological N fixation by legume crops	40
Mobilized by human activities	
Fossil fuel combustion	25
Biomass burning	40
Flux from cleared lands	20
TOTAL	325

C. Nitrogen Oxides and Ammonia

Oxides of N (N_2O and NO), and NH_3 are all evolved from soils, although under somewhat different conditions. Soils are the major source of N_2O and an important source of NO (Table 5). Anthropogenic emissions of N oxides are associated with fertilizer additions and enhanced mineralization of soil organic N in agricultural ecosystems, while anthropogenic emissions of NH_3 (Table 5) arise from animal manures (Bouldin et al., 1984) and soils fertilized with NH_4^+-N sources, especially rice paddies (Freney et al., 1990). The impact of N_2O on radiative forcing is direct, but the impacts of NO and NH_3 are indirect and more difficult to assess. Nitric oxide participates in a variety of reactions that affect troposopheric levels of O_3, which is a greenhouse gas, and OH radicals, which initiate oxidative destruction of CH_4 (Williams et al., 1992). Volatilization of NO and NH_3 leads to unintended fertilization of landscapes surrounding sources which, coupled with leaching of NO_3^- from cropland, undoubtedly causes further anthropogenic emissions of N_2O and also NO. Such effects have not been evaluated, nor are they considered in the IPCC budget for N_2O (Watson et al., 1992). This budget is quite fragmentary and ignores major anthropogenic perturbations to the terrestrial N cycle that are largely related to agriculture, and which in aggregate are thought to have increased annual inputs of fixed N by a factor of three (Table 6). It is therefore probable that the imbalance in global N_2O budgets (Robertson, 1993) is largely caused by an underestimation of anthropogenically induced emissions of N_2O from soils.

Fewer measurements of N_2O emissions from soils have been made in tropical than in temperate regions and they have been less intensive, both in terms of the frequency of measurement and duration of studies. Moreover pastures are the only agroecosystem that have received much attention (Matson and Vitousek, 1990; Goreau and de Mello, 1988). Although emissions of N_2O in tropical environments are less well characterized than those in temperate environments, the importance of tropical regions has been amply demonstrated (Matson and Vitousek, 1990).

Nitrous oxide is formed in small amounts during oxidation of NH_4^+ and under anaerobic conditions during denitrification. Emissions of N_2O arising from nitrification are quantitatively dependent upon the extent of nitrification and tend to be higher, per unit of N nitrified, when soils are wet. In natural ecosystems, nitrification is dependant on the extent of N cycling within the soil plant system, which is greater in tropical than in temperate ecosytems (Matson and Vitousek, 1990). Consequently, N_2O emissions are higher in tropical forests and grasslands than in their temperate ecosystem counterparts. Additions of ammonium fertilizer to agricultural ecosystems create an anthropogenic source of N_2O derived from nitrification. Emissions of N_2O arising from denitrification are highly variable because N_2O is not the terminal product of denitrification. Hence, N_2O emissions to the atmosphere resulting from denitrification represent the balance between the rates of further reduction and diffusion to the atmosphere. Emissions of N_2O are, in general, higher in wetter environments, but highest fluxes are often found under conditions that are marginal for denitrification.

The importance of the different sources of N_2O needs to be considered when making measurements of N_2O emissions. Where nitrification is the source of N_2O, fluxes to the atmosphere follow predictable patterns. The pattern of nitrification within a year regulates N_2O fluxes on longer time scales, while temperature and moisture provide short-term controls through effects on the rate of nitrification and on the solubility of N_2O over diel cycles (Duxbury, 1986). Under these circumstances, flux measurements made over a 24h period once a week could adequately characterize N_2O emissions. In contrast, where denitrification is the source, N_2O flux patterns are not as predictable because they occur as a series of short-term events following rainfall. Under these conditions, a sampling scheme oriented to denitrification events is required to adequately characterize N_2O emissions. The importance of measurement frequency is illustrated by the fact that 27% of the annual emission of N_2O from an alfalfa field in NY occured over two consecutive days (Duxbury et al., 1982). Unfortunately, many measurements of N_2O emissions in the tropics have been inadequate to determine annual emission values (although this is often done), because of inappropriate sampling schedules and short duration of studies.

IV. Rice Based Cropping Systems

Rice based cropping systems are perhaps the most complex with respect to emissions of trace gases from soils and the rice paddy is very difficult to manage in terms of nitrogen efficiency. A conceptual diagram of the effect of water management on emissions of CH_4, NH_3, and N_2O from a paddy rice based cropping system is shown in Figure 5. Under flooded conditions, large emissions of CH_4 occur but no N_2O is evolved because of the intense reducing conditions. However, N_2O and perhaps also NO will be evolved during flooding and after draining soils. Attempts to reduce CH_4 fluxes by water management will undoubtedly lead to higher emissions of N_2O and tradeoffs between the two gases will need to be evaluated. Additionally, high floodwater pH (caused by depletion of dissolved CO_2 by algal photosynthesis) leads to volatilization of fertilizer N as NH_3. Volatilized NH_3 is returned to the earth's surface in wet and dry deposition and can also be absorbed by plant canopies. The extent of NH_3 redistribution over landscapes and the impact of this on greenhouse gas fluxes is unknown. Furthermore, the rice paddy would seem to be an ideal environment for generation of COS and CS_2 but only one study of emissions of these gases has been reported (Kanda et al., 1992).

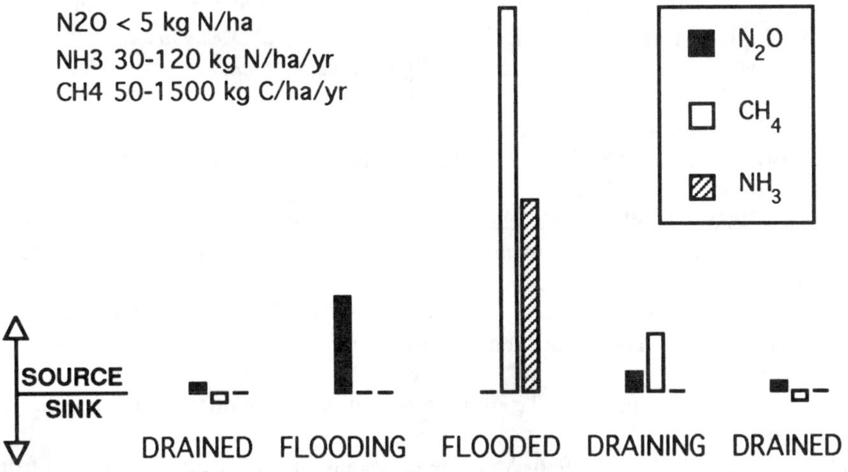

Figure 5. A conceptual diagram (not to scale) of the effect of water management on emissions of CH_4, NH_3, and N_2O from a paddy rice based cropping system. Values represent measured emissions per rice crop for CH_4 and NH_3 and an estimate of potential emissions for N_2O.

Acknowledgements

This paper is a contribution from Cornell University's Agricultural Ecosystems Program, which is funded by a special grant from USDA-CSRS.

References

Allen, J.C. 1985. Soil response to forest clearing in the United States and the tropics: Geological and biological factors. *Biotropica* 17:15-27.

Balesdent, J., G.H. Wagner, and A. Mariotti. 1988. Soil organic matter turnover in long-tem field experiments as revealed by carbon-13 natural abundance. *Soil Sci. Am. J.* 52:118-24.

Bouldin, D.R., S.D. Klausner, and W.S. Reid. 1984. Use of nitrogen from manure. p. 221-245. In: R.D. Hauck (ed.), *Nitrogen in crop production.* Am. Soc. Agron., Madison, WI.

Bouwman, A.F. 1989. Modelling soil organic matter decomposition and rainfall erosion in two tropical soils after forest clearing for permanent agriculture. *Land Degredation and Rehabilitation* 1:125-140.

Brown, S., A. Glubczynski, and A.E. Lugo. 1984. Effects of land use and climate on the organic carbon content of tropical forest soils in Puerto Rico. p. 204-209. In: *New forests for a changing world.* Soc. Amer. Foresters, Bethesda, MD.

Cassel, D.K. and R. Lal. 1992. Soil physical properties of the tropics: Common beliefs and management restraints. p. 61-89. In: R.Lal and P.A. Sanchez (eds.), *Myths and science of soils of the tropics.* Soil Sci. Soc. Am. Special Publication no. 29, Soil Sci. Soc. Am., Madison WI.

Cicerone, R.J., C.C. Delwiche, S.C. Tyler, and P.R. Zimmmerman. 1992. Methane emissions from California rice paddies with varied treatments. *Global Biogeochem Cycles* 6:233-248.

Davidson, E. A. 1991. Fluxes of nitrous oxide and nitric oxide from terrestrial ecosystems. p. 219-235. In: J.E. Rogers and W.B. Whitman (eds.), *Microbial production and consumption of greenhouse gases: methane, nitrogen oxides and halomethanes.* Am. Soc. Microbiol., Washington D.C.

Duxbury, J.M. 1986. Advantages of the acetylene method of measuring denitrification. p. 73-91. In: R.D. Hauck et al. (ed.), *Field measurement of dinitrogen fixation and denitrification.* Soil Sci. Soc. Am., Special Publication no. 18, Soil Sci. Soc. Am., Madison WI.

Duxbury, J.M., D.R. Bouldin, R.E. Terry, and R.L. Tate III. 1982. Emissions of nitrous oxide from soils. *Nature* 298:462-464.

Duxbury, J.M., L.A. Harper, and A.R. Mosier. 1993. Contributions of agroecosystems to global climate change. p.1-18. In: L.A. Harper, A.R. Mosier, J.M. Duxbury, and D.E. Rolston (eds.), *Agroecosystem effects on radiatively important trace gases and global climate change*. Am. Soc. Agron. Special Publication no. 55, Am. Soc. Agron., Madison WI.

Duxbury, J.M., M.S. Smith, and J.W. Doran. 1989. Soil organic matter as a source and a sink of plant nutrients. p. 33-67. In: D. C. Coleman, J.M. Oades, and G. Uehara (eds.), *Dynamics of soil organic matter in tropical ecosystems.* Univ. of Hawaii, Honolulu, HI.

El-Swaify, S.A. and E.W. Dangler. 1982. Rainfall erosion in the tropics: a state-of-the-art. p. 1-25. In: W. Kussow et al. (eds.), *Soil erosion and conservation in the tropics.* Am. Soc. Agron. Special Publication no. 43, Am. Soc. Agron. Madison, WI.

Freney, J.R., A.C.F. Trevitt, S.K. De Datta, W.N. Obcemea, and J. G. Real. 1990. The interdependence of ammonia volatilization and denitrification as nitrogen loss processes in flooded rice in the Phillipines. *Biol. Fert. Soils* 9:31-36.

Goreau, T.J. and W.Z. de Mello. 1988. Tropical deforestation: Some effects on atmospheric chemistry. *Ambio* 17:275-281.

Holzapfel-Schorn, A. and W. Seiler. 1986. Methane emissions during a cultivation period from an Italian rice paddy. *J. Geophys Res.* 91:11803-11814.

Houghton, R.A. 1991. Tropical deforestation and atmospheric carbon dioxide. *Climatic Change* 19: 99-118.

Houghton, R.A., J.E. Hobbie, and J.M. Melillo. 1983. Changes in the carbon content of terrestrial biota and soils between 1860 and 1980: a net release of CO_2 to the atmosphere. *Ecol. Monographs* 53:235-262.

Isaksen, I.S.A., V. Ramaswamy, H. Rodhe, and T.M.L. Wigley. 1992. p. 51-67. In: J.T. Houghton, B.A. Callander, and S.K Varney (eds.), *Climate Change 1992.* Cambridge Univ. Press, Cambridge.

Juo, A.S.R. and R. Lal. 1977. The effect of fallow and continuous cultivation on the chemical and physical properties of an alfisol in Western Nigeria. *Plant and Soil* 47:567-584.

Juo, A.S.R. and R. Lal. 1979. Nutrient profile in a tropical alfisol under conventional and no-till systems. *Soil Sci.* 127:168-173.

Kanda, K.I., H. Tswuta, and K. Minami. 1992. Emission of dimethyl sulfide, carbonyl sulfice and carbon disulfide from paddy fields. *Soil Sci. Plant Nutr.* 38:709-716.

Keller, M., M.E. Mitre, and R.F. Stallard. 1990. Consumption of atmospheric methane in soils of central Panama: effects of agricultural development. *Global Biogeochem. Cycles* 4:21-27.

Krebs, J.E. 1975. A comparison of soils under agriculture and forests in San Carlos, Costa Rica. p. 381-390. In: F.B. Golley and E. Medina (eds.), *Tropical ecosystems-trends in terrestrial and aquatic research.* Springer Verlag, N.Y.

Lal, R. 1976. Soil erosion problems on an alfisol in Western Nigeria and their control. *IITA Monograph 1*, Ibadan, Nigeria.

Lal, R. 1990. *Soil erosion in the tropics: principles and management.* McGraw Hill, N.Y.

Lauren, J.G. and J.M. Duxbury. 1993. Methane emissions from flooded rice amended with a green manure. p. 183-192. In: L.A. Harper, A.R. Mosier, J.M. Duxbury, and D.E. Rolston (eds.), *Agroecosystem effects on radiatively important trace gases and global climate change.* Am. Soc. Agron. Special Publication no. 55, Am. Soc. Agron., Madison WI.

Lauren, J.G., G.S. Pettygrove, and J.M. Duxbury. 1994. Methane emissions associated with a green manure amendment to flooded rice in California. (In press.)

Leng, R.A. 1991. Improving ruminant production and reducing methane emissions from ruminants by stragic supplementation. EPA/400/1-91/004, USEPA, Office of Air and Radiation, Washington, D.C.

Logan, J.A. 1983. Nitrogen oxides in the troposphere: a global perspective. *J. Geophys. Res.* 88:10785-10807.

Lugo, A.E., M.J. Sanchez, and S. Brown. 1986. Land use and organic carbon content of some subtropical soils. *Plant and Soil* 96:185-196.

Matson, P.A. and P.M. Vitousek. 1987. Cross system comparisons of soil nitrogen transformations and nitrous oxide flux in tropical forest ecosystems. *Global Biogeochem. Cycles* 1:163-170.

Mosier, A.R., D. Schimel, D. Valentine, K. Bronson, and W.J. Parton. 1991. Methane and nitrous oxide fluxes in native, fertilized, and cultivated grasslands. *Nature* 350:330-332.

Matson, P.A. and P.M. Vitousek. 1990. Ecosystem approach to a global nitrous oxide budget. *BioScience* 40:667-672.

Neue, H.U., P. Becker-Heidman, and H.W. Scharpenseel.1990. Organic matter dynamics, soil properties, and cultural practices in rice lands and their relationship to methane production. p. 457-466. In: A.F. Bouwman (ed.), *Soils and the geenhouse effect*. John Wiley, N. Y.

Nye, P.H. and D.J. Greenland. 1964. Changes in the soil after clearing tropical forest. *Plant and Soil* 21:101-112.

Parashar, D.C., J. Rai, P.K. Gupta, and N. Singh. 1991. Parameters affecting methane emissions from paddy fields. *Indian J. of Radio and Space Physics* 20:12-17.

Parashar, D.C., A.P. Mitra, and S.K. Sinha. 1992. Methane budget from Indian paddy fields. p. 57-69. In: K. Minami (ed.), *Proc. of CH_4 and N_2O Workshop: CH_4 and N_2O Emissions from Natural and Anthropogenic Sources and their Reduction Research Plan*. Nat. Inst. of Agro-Environ. Sci., Tsukuba, Japan.

Robertson, G.P. 1993. Fluxes of nitrous oxide and other nitrogen trace gases from intensively managed landscapes. p. 95-108. In: L.A. Harper, A.R. Mosier, J.M. Duxbury, & D.E. Rolston (eds.), *Agroecosystem effects on radiatively important trace gases and global climate change*. Am. Soc. Agron. Special Publication no. 55, Am. Soc. Agron., Madison WI.

Safley, L.M., M.E. Casada, J.W. Woodbury, and K.F. Roos. 1992. Global methane emissions from livestock and poultry manure. EPA/400/1-91/048, Office of Air and Radiation, U.S. EPA, Washington, D.C.

Schimel, J.P., E.A. Holland, and D. Valentine. 1993. Controls on methane flux from terrestrial ecosystems. p.167-182. In: L.A. Harper, A.R. Mosier, J.M. Duxbury, and D.E. Rolston (eds.), *Agroecosystem effects on radiatively important trace gases and global climate change*. Am. Soc. Agron. Special Publication no. 55, Am. Soc. Agron., Madison WI.

Schlesinger, W.H. and A.E. Hartley. 1992. A global budget for atmospheric NH_3. *Biogeochemistry* 190-211.

Schütz, H., W. Seiler, and R. Conrad. 1989. Processes involved in the formation and emission of methane in rice paddies. *Biogeochemistry* 7:33-53.

Sheng, T.C. 1982. Erosion problems associated with cultivation in humid tropical hilly regions. p. 27-39. In: W.Kussow et al. (eds.) *Soil erosion and conservation in the tropics*. Am. Soc. Agron. Special Publication no. 43, Am. Soc. Agron. Madison, WI.

Shine, K.P., R.G. Derwent, D.J. Wuebbles, and J.-J. Morcrette. 1990. Radiative forcing of climate. p. 47-68. In: J.T. Houghton et al. (eds.) *Climate change: the IPCC scientific assessment*. Cambridge Univ. Press, Cambridge.

Steudler, P.A., R.D. Bowden, J.M. Melillo, and J.D. Aber. 1989. Influence of nitrogen fertilization on methane uptake in temperate soils. *Nature* 341:314-316.

Summerfeld, R.A., A.R. Mosier, and R.C. Musselman. 1993. CO_2, CH_4, and N_2O flux through a Wyoming snowpack and implications for global budgets. *Nature* 361:140-142.

USDA. 1990. *Agricultural statistics 1990*. U.S. Govt. Printing Office, Washington, D.C.

Vitorello, V.A., C.C. Cerri, F. Andreux, C. Feller, and R.L. Victoria. 1989. Organic matter and natural carbon-13 distribution in forested and cultivated oxisols. *Soil Sci. Soc. Am. J.* 53:773-778.

Watson, R.T., H. Rhode, H. Oeschger, and U. Siegenthaler. 1990. Greenhouse gases and aerosols. p. 7-40. In: J.T. Houghton et al. (eds.), *Climate change: the IPCC scientific assessment*. Cambridge Univ. Press, Cambridge.

Watson, R.T., L.G. Meira Filho, E. Sanhueza, and T. Janetos. 1992. Sources and sinks. p. 25-46. In: J.T. Houghton, B.A. Callander, and S.K Varney (eds.), *Climate change 1992*. Cambridge Univ. Press, Cambridge.

Whalen, M. and W. Reeburgh. 1990. Consumption of atmospheric methane by tundra soils. *Nature* 346:160-162.

Williams, E.J., G.L. Hutchinson, and F.C. Fehsenfeld. 1992. NO_x and N_2O emissions from soil. *Global Biogeochem. Cycles* 6:351-388.

Yagi, K. and K. Minami. 1990. Effect of organic matter application on methane emission from some Japanese paddy fields. *Soil Sci. Plant Nutr.* 36:599-610.

Zehnder, A.J.B. 1982. The carbon cycle. p. 83-109. In: O. Hutzinger (ed.), *The handbook of environmental chemistry , vol. 1 part B: The natural environment and the biogeochemical cycles*. Springer-Verlag, N.Y.

CHAPTER 25

Agricultural Activities and Greenhouse Gas Emissions from Soils of the Tropics

R. Lal and T.J. Logan

I. Introduction

Land area of the world is 13.5 x 10^9 hectares, or 29% of the earth's surface. About 11% of the land area in the world is cultivated, and 22% is grassland. Tropics, defined as the part of the world between the Tropic of Cancer and the Tropic of Capricorn, comprise some 4.9 x 10^9 hectares or 37.2% of the land. The sub-tropics, regions north and south of the tropics to about 35° latitude, comprise some 25% of the land surface.

World soils vary greatly in soil organic matter (SOM) content ranging from 4 to 6% (weight basis) organic carbon (C) in the top 15 cm of some fertile soils to less than 0.5% in others. The accumulative amount of organic C stored in world soils is far more than in atmosphere, fresh waters, or world biota (Bohn, 1978; Stevenson, 1982).

Estimates of total reserves of organic C in the world soils are highly variable. The total C reserve according to these estimates is 3.62 x 10^{17} g. However, the data by Post et al. does not contain all land uses, e.g. pastures, permanent crops, etc.

Estimates of total C reserves in world soils vary by a factor of 4 with a range from 700 X 10^{15} g to 3000 x 10^{15} g. The most commonly referred estimate is about 1500 x 10^{15} g (Buringh, 1984; Schlesinger, 1984), implying thereby that C stored in the soil surface is at least twice as much as that contained in the atmosphere, and three times that in the biomass. The wide variation observed in these estimates is due to the lack of reliable data on soil C content for major soils, and ambiguity in the area of these soils.

II. Soils of the Tropics and their Organic Matter Content

Soils of the tropics are as diverse and varied as those of the world (Table 1). Total area of soils of the tropics is 4.9 x 10^9 ha out of the world total of 13.5 x 10^9 ha or 37.2%. Organic matter content in soils of the tropics is important for several reasons. One, in soils with low activity clays, organic matter becomes an important sink or source of plant nutrients through its effect on cation exchange capacity (CEC). Two, soils of the tropics are generally cultivated by resource-poor farmers who rely on plant nutrients supplied by soil organic matter. Three, by-products of mineralization of organic matter (e.g., CO_2, CH_4, N_2O) have important global impact on atmospheric concentration of these gases. This third reason is the theme of this report.

Some researchers agree that there are no major differences in quantity or quality of SOM in soils of the tropics vs. those of temperate regions (Sanchez et al., 1989). However, the rate of accumulation and distribution of SOM, and the rate of SOM mineralization vary widely depending on temperature and moisture regimes, activity and species diversity of soil flora and fauna, the rate of supply of fresh organic material/biomass, etc.

ISBN 1-56670-117-1/95/$0.00+.50

Table 1. Soil distribution in the tropics

Soil order	Total		Humid/sub-humid		Semi-arid tropics	
	Area (10^6 ha)	%	Area (10^6 ha)	%	Area (10^6 ha)	%
Alfisol	800	16.1	694	34.0	106	3.6
Ultisols	550	11.1	52	2.5	498	17.1
Oxisols	1100	22.2	188	9.2	912	31.2
Aridisols	900	18.1	520	25.5	380	13.0
Entisols	400	8.1	272	13.3	128	4.4
Inceptisols	400	8.1	66	3.2	334	11.5
Vertisols	131	1.6	100	4.9	31	1.1
Mollisols	78	1.6	58	2.8	20	0.7
Other	600	12.1	93	4.6	507	17.4
Total	4959	100.0	2043	100.0	2916	100.0

(Modified from Buringh, 1979.)

Table 2. Mean organic matter contents in 61 soils from the tropics and 45 soils from the temperate region

Depth (cm)	Tropical soils	Temperate soils	Significance	CVC %	
				Tropics	Temperate region
0-15	1.68	1.64	NS	53	64
0-50	1.10	1.03	NS	57	69
0-100	0.69	0.62	NS	59	75

(From Sanchez et al., 1982.)

Table 3. Mean total carbon and nitrogen reserves of soil orders in tropical and temperate regions

Region	Soil order	No. of profile samples	Total C (kg C m^{-2})	
			0-15 cm depth	0-100 cm depth
Tropics	Oxisols	19	3.8a	11.3a
	Alfisols	13	2.9a	6.4b
	Ultisols	18	2.1b	6.4b
Temperate region	Mollisols	21	3.3a	10.1a
	Alfisols	16	2.8ab	5.8b
	Ultisols	8	2.4b	4.2b

(From Sanchez et al., 1982.)

Processes governing dynamics of SOM in soils of the temperate climate have been studied extensively over the last 50 years (Stevenson, 1982). In contrast, however, dynamics of SOM in soils of the tropics have been studied for a few soils only. On the basis of available literature, it is now evident that soils of the tropics do not necessarily contain lower organic matter contents than soils of the temperate region (Post et al., 1985; Kimble et al., 1989). Sanchez et al. (1982) compared SOM content from 61 randomly chosen soils of the tropics with that from 45 soils of the temperate region. Their data indicate no significant differences in total C between soils of the two regions (Table 2). These authors did not observe any significant differences in SOM contents between Alfisols from the tropics vs. Alfisols from the temperate region. Similar observations were reported for Ultisols and Mollisols (Table 3). Similar conclusions were arrived at by Post et al. (1982) based on the analyses of over 3000 soil profiles. All other factors remaining the same, SOM is greatly influenced by rainfall, temperature, clay content, supply of biomass (Spain et.

Table 4. Organic carbon pool in soils of the tropics (10^{12} g)

Soil order	Area (10^6 ha)	o-30 cm depth	0-50 cm depth	0-100 cm depth
Alfisols	800	47.7	67.3	95.6
Andisols	9	1.2	1.8	3.0
Aridisols	900	13.7	20.9	34.6
Entisols	400	19.4	23.1	29.4
Inceptisols	400	26.1	36.2	47.6
Mollisols	78	8.7	11.6	16.9
Oxisols	1100	72.4	93.6	133.0
Ultisols	550	29.5	43.8	65.7
Vertisols	131	3.5	4.8	8.2
Others	600	26.7	36.4	52.0
Total	4968	248.9	339.5	486.0

Estimates are based on mean C content and bulk density from the published data of several soil profiles in each soil order. Eswaran et al. (1993) estimated that soils of the tropics contain about 506 Pg of carbon.

al., 1983; Anderson and Swift, 1983), and internal drainage of the soil. The ratio of temperature to precipitation (T/P) is strongly and inversely correlated with SOM (Theng et al., 1989).

III. Organic Carbon Pool in Soils of the Tropics

Calculations by Post et al. (1982) show that C reserves in soils of different tropical biomes are 2 x 10^{15} g in desert bush, 129 x 10^{15} g in woodland and savanna, 22 x 10^{15} g in very dry forest, 24 x 10^{15} g in dry forest, 60 x 10^{15} g in moist forest, and 7.8 x 10^{15} g in wet forest. An extrapolation of the data by Post et al. (1982) would estimate C reserve in arable lands of the tropics at 47 x 10^{15} g.

Knowing the area of major soil groups, soil C profile and bulk density, it is possible to estimate total C stored in soils of the tropics. The data shown in Table 4 is an estimate of the carbon pool in major soil orders of the tropics. It is apparent that about 4.9 x 10^9 ha of soils in the tropics contain about 486 Pg of C in the top 1 m depth. Eswaran et al. (1993) estimated that soils of the tropics contain about 506 Pg of C. The C pool most susceptible to mineralization and other processes is generally in the top 30 cm or 50 cm of soil profile. Carbon pool estimated in the upper layers is about 2.5 x 10^{17} g for 0-30 cm depth, and ~3.4 x 10^{17} g for 0-50 cm depth. The relative distribution of organic C in different soil orders is as follows: Oxisols (27.4%), Alfisols (19.7%), Ultisols (13.5%), Inceptisols (9.8%), Aridisols (7.1%), Entisols (6.0%), Mollisols (3.5%), and Vertisols (1.7%), and remainder (11.3%) in miscellaneous soils.

IV. Current Land Use in the Tropics

There are four predominant agricultural land uses in the tropics, e.g. arable, permanent/perennial crops, pastures and forest/woodlands (Table 5). Out of the total land area of 4.9 x 10^9 ha, 8.6% is for arable, 1.0% permanent crops, 25.1% pastures, and about 38.2% forest/woodland. The remainder (27.0%) is under miscellaneous uses, e.g. urban, lakes and rivers, swamps, etc.

Changes in land use from beginning to the end of the 20th century indicate that the area of tropical rainforest has decreased by 41.6% over the 76 year period ending in 1976, and is expected to decrease by an additional 35.8% of the existing area in 1976 by the year 2000 (Blasco and Achard, 1990). Conversion of tropical rainforest is considered to be a major factor in emission of greenhouse gases (Houghton et al., 1987).

Another relevant statistic in the arable land use is the net cultivated area under rice paddy. Of the total world area of about 0.146 x 10^9 ha, 89.9% of the area under rice paddy lies in Asia (Table 6). Of 0.131 x 10^9 ha of rice paddy in Asia, 31.7% is in India, 24.7% in China, 7.9% each in Indonesia and Thailand,

Table 5. Landuse in the tropics (10^6 ha)

Region	Arable land	Permanent crops	Permanent pastures	Forest and woodland	Other lands
Africa	130.8	14.2	637.6	659.3	786.4
Asia	191.4	20.6	138.3	283.8	288.9
Central America and the Caribbean	21.7	3.8	56.7	43.8	42.9
Pacific	12.0	0.7	106.5	68.8	53.3
South America	62.5	12.1	286.9	149.0	94.3
Total	418.4	51.4	1226.0	1867.5	1320.5

The data is compiled from FAO Year Book (1989). However, the data is compiled for only tropical countries for each region.

Table 6. Rice cultivation in the World

Region	Area (10^6 ha)	Production (10^6 Mg)
A. World Regions		
Africa	5.5	10.5
NC America	1.8	9.3
South America	6.9	17.2
Asia	131.1	463.9
Europe	0.4	2.1
Oceania	0.1	0.8
World total	145.8	503.8
B. Principal Countries		
China	32.4	179.4
India	41.5	107.5
Indonesia	10.3	43.6
Japan	2.1	12.9
Pakistan	2.1	4.8
Philippines	3.5	9.5
Thailand	10.3	21.3
Vietnam	5.9	18.1
U.S.A.	1.1	7.1

(From FAO, 1989.)

4.5% in Vietnam, 2.7% in Philippines, and 1.6% each in Japan and Pakistan. Several countries of south Asia grow 2 or 3 crops of rice every year.

V. Carbon Dynamics in Different Land Use Systems in the Tropics

Examples of land use systems that influence C dynamics in soils of the tropics are described in the following sections.

A. Deforestation

Estimates of total forest reserves of the tropics and the rate of deforestation vary widely. UNESCO (1978) estimated tropical forest reserves at about 3.5 x 10^9 ha out of the total land area of about 4.9 x 10^9 ha. O'Keefe and Kristoferson (1984) estimated the total forest reserves of the tropics at 3.16 x 10^9 ha. FAO

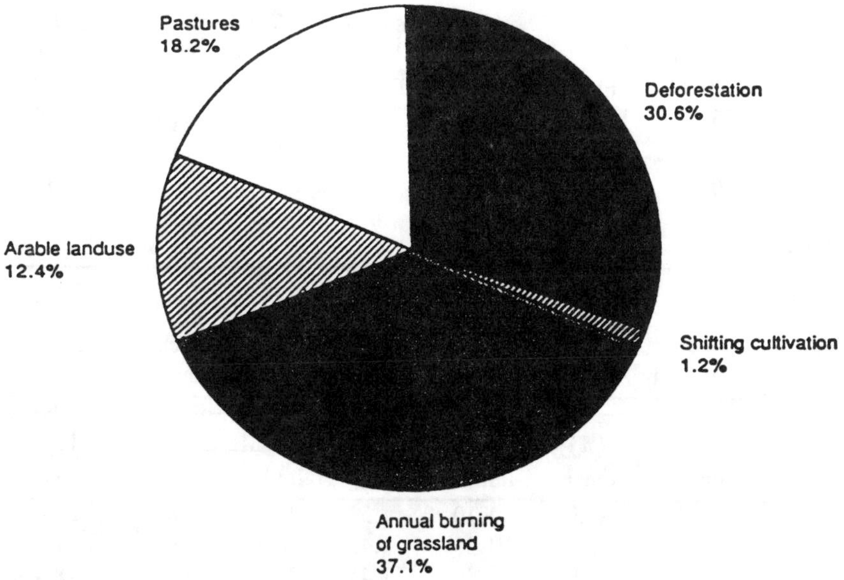

Figure 1. Relative (%) C emission rate from agricultural activities; total emission equals 0.5×10^{12} kg C yr^{-1}.

(1981) estimated that forest reserves of the tropics are 2.94 x 10^9 ha. Tropical forest resources (closed forest, open woodlands, forest fallow and shrubland) comprise about 1.9 x 10^9 ha or 38% of the total land area of the tropics (Table 5).

The most recent estimate of the rate of tropical deforestation is 20 x 10^6 ha yr^{-1} (WRI, 1990-91; Buringh, 1984; Houghton, 1990). There are other estimates ranging from 12 to 17 x 10^6 ha yr^{-1}. The rate of loss of C following deforestation depends on many factors, e.g. antecedent level of C in soil, method of deforestation, land use, cropping system, tillage methods, etc. Total C loss from the time of deforestation to the time when SOM reaches a steady state with the new land use may be as much as 60% in the top 0-5 cm layer and 30% in the 5-10 cm layer (Cunningham, 1963). Several studies on dynamics of SOM following deforestation have been conducted in western Nigeria. Field experiments conducted by Lal (1976a,b;)and Ghuman and Lal (1991) showed a rapid decline in organic C content within the first S years. The rate of decline is, however, somewhat less with conservation tillage and agro-forestry (Lal, 1989). Several long-term experiments conducted in Nigeria (Lal, 1976a,b, 1982; 1987a,b) showed 70% decline in organic C content of the soil in plowed treatment compared with 20% decline in the no-till treatment in 15 years of cropping. In another experiment in Nigeria, Lal et. al. (1980) observed that organic C contents in the top 5cm declined by about 50% in 12 months, and by about 57% in 18 months. In a study based on mechanized cultivation, Lal (1985) observed that organic C content in a no-till system decreased by 20% in 6 years for 0-10 cm layer. Aina (1979) observed that organic C content in 0-10 cm depth of the soil under 15-25 year secondary forest regrowth was about 2.6% compared with 0.6% in soil cultivated for 10 years. In Australia, several workers have demonstrated that soil disturbance increases mineralization of organic C, especially the disturbance that leads to disruption of micro-aggregates (Dalal and Mayer, 1986; Roberts and Chan, 1980). Cropping systems experiments conducted in Peru in the upper Amazon basin also showed rapid decline in C during first few years of cropping (Sanchez and Salinas, 1981).

The losses of C from tropical and temperate zone ecosystems following deforestation are very similar, except that the rate of loss due to mineralization is faster by a factor of 4 to 5 in the tropics than in the temperate climates (Figure 1). The organic C content of the surface 30-cm layer is reduced by about 50% in soils of both regions, in about 10 years in tropical climate compared with 50 years in temperate regions.

Percent decrease in soil C varies from 4 to 54% per year. Literature examples of C loss by continuous cultivation of newly deforested land by conservation tillage and conventional tillage systems are summarized

Table 7. Loss of organic carbon with continuous intensive cultivation in 10 years following deforestation with no-till and agroforestry

Depth (cm)	--Organic C (%)-- Initial	Final	Bulk density -----Mg m^{-3}----- Initial	Final	Total soil C -----Mg ha^{-1}---- Initial	Final	Carbon emission in 10 yr Mg ha^{-1}
0-10	2.5	1.5	1.10	1.40	27.5	21.0	6.5
10-25	1.4	1.0	1.25	1.45	26.3	21.8	4.5
25-50	0.9	0.8	1.30	1.45	29.3	29.0	0.3
Total					83.1	71.8	11.3

Table 8. Loss of organic carbon with continuous intensive cultivation using plow-based mechanized systems in 10 years following deforestation

Depth (cm)	Organic C (%) Initial	Final	Bulk density -----Mg m^{-3}----- Initial	Final	Total soil C -----Mg ha^{-1}----- Initial	Final	Carbon loss Mg ha^{-3}
0-10	2.5	0.5	1.10	1.5	27.5	7.5	20.0
10-25	1.4	0.4	1.25	1.45	26.3	8.7	17.6
25-50	0.9	0.3	1.30	1.45	29.3	10.9	18.4
Total					83.1	27.1	56.0

Assuming no erosion preventive/control measures are applied.

in Tables 7 and 8, respectively. It seems that the rate of C loss from soil may be as low as 1.13 Mg ha^{-1} yr^{-1} from conservation tillage system to 5.6 Mg ha-l yr^{-1} with conventional tillage. The subsequent loss in C is practically negligible under both systems, because soil C reaches a steady state level. It is further assumed that conservation tillage and agro-forestry systems are adopted on 25% of the cleared land, and conventional tillage and plow-based systems are adopted on 75% of the land area.

On the basis of the assumptions outlined above, we estimate that soils from newly cleared land release as little as 89.7 x 10^{12} g C yr^{-1} and as much as 218.8 x 10^{12} g C yr^{-1} (Table 9). Median estimate of C emission from tropical deforestation is 154.3 x 10^{12} g C yr^{-1}.

It is estimated that 1.0 x 10^9 ha out of a total land area of 4.9 x 10^9 ha in the tropics is still forested. Assuming that a total of 2 x 10^9 ha have been deforested since the dawn of civilization, the total C released from soils supporting tropical forests from beginning of settled agriculture till now is 3.5 Mg ha^{-1} x 2 x 10^9 ha, or 7 x 10^{15} g. The low estimate is 2.3 x 10^{15} g and the high estimate 13.2 x 10^{15} g C.

B. Shifting Cultivation

Shifting cultivation is an important farming system of the tropics (NRC, 1993; Sanchez et al., 1990). It is estimated that about 300 million people, mostly in the humid and sub-humid tropics, practice shifting cultivation. Shifting cultivators annually cut about 25 x 10^6 ha of new forest and use an additional 125 x 10^6 ha already cut for a 5-year rotation cycle. The traditional cycle, when land shortage is not a problem, is about 15-20 years. So the annual loss of forest due to shifting cultivation is assumed to be about 25 x 10^6 ha.

The loss of C from soil profile under shifting cultivation is drastically less compared with loss of C under intensive cultivation. Large trees are kept standing and the biomass is returned to the soil surface. However, shifting cultivators rarely, if at all, use any fertilizers. Therefore, most crop nutrients are made available from mineralization of residue and soil C pool. The data on C dynamics in soils under shifting cultivation obtained from West Africa by Nye and Greenland (1960) showed that the loss of C in 100 years ranges

Table 9. Carbon emission from tropical soils after deforestation

Particulars	Low estimate	High estimate
1. Area deforested (10^6 ha y^{-1})	20	40
2. Mean C in 0-50 cm depth (Mg ha^{-1})	83.1	100.6
3. Deforestation and management by conservation effective measures (19^6 h yr^{-1})	5	10
4. Deforestation and management by soil depletive measures 10^6h yr^{-1})	15	30
5. Rate of C emission with conservation effective measures (Mg ha^{-1} yr^{-1})	1.13	2.08
6. Rate of C emission with soil depletive practices (Mg ha^{-1} yr^{-1})	5.6	6.60
7. Total C emission with conservation effective measures (10^6 Mg ha^{-1} yr^{-1})	5.65	20.8
8. Total C emission with soil depletive measures (10^6 Mg ha^{-1} yr^{-1})	84.0	198.0
9. Total C emission from tropical deforestation (10^9 kg C y^{-1})	89.7	218.8

from 20% for a soil with 12-year fallow cycle to 45% for a soil with 4 year fallow cycle. An average loss of soil C under this system is assumed to be about 16.6 Mg ha^{-1} for the 12 year cycle and 37 Mg ha^{-1} for the 4 year cycle over 100 years (Nye and Greenland, 1960). That being the case, the annual loss of C due to shifting cultivation is about 0.27 Mg ha^{-1} x 25 x 10^6 ha yr^{-1} or 6.25 x 10^{12} g C yr^{-1}. The range of C emission due to shifting cultivation in the tropics may be 3.75 x 10^{12} g C yr^{-1} to 9.18 x 10^{12} g C yr^{-1}. It is difficult to estimate the total C emission due to shifting cultivation because some of the land has reverted back to fallow for restoration.

C. Fire in Tropical Savannas and Grassland

Tropical grasslands and savannas occupy about 1.5 x 10^9 ha and contain about 4 x 10^{13} g C m^{-2} (Schlesinger, 1984). The maximum surface soil temperature during fire can be extremely high --> 500 °C (Lal, 1987a,b). However, soil temperature below 1 cm depth is only slightly increased. Effects of burning on soil C have been extensively studied in the tropics.

An experiment conducted in Thailand indicated that burning increased total C pool in the 13 to 30-cm depth, probably due to contamination by ash (Kyuma and Pairintra, 1982). Assuming the loss of organic C due to burning is limited to the top 0-2 cm, a realistic loss can be about 0.5 Mg ha^{-1} yr^{-1} with a low value of 0.3 and a high of 0.8 Mg ha^{-1} yr^{-1}. Assuming that only 25% of grassland burns every year, the annual emission of C due to burning of tropical grassland, therefore, is 0.5 Mg ha^{-1} yr^{-1} x 375 x 10^6 ha or 187.5 x 10^{12} g yr^{-1}. The low and high estimates range from 112.5 x 10^{12} g to 275.6 x 10^{12} g. The annual burning of tropical grassland and savannas over the history of earth has been, therefore, a major source of C emission into the atmosphere, not only due to the release of C in the biomass but also due to combustion and mineralization of SOM.

D. Arable Land Use and Permanent Pastures

Total cultivated land in the tropics is about 418 x 10^6 ha (Table 5). With some exceptions, most of this land is used by subsistence or semi-commercial farmers with none or low level of off-farm input. Fertility depletion is a norm due to intensive cropping and resultant soil degradation. Most arable land has already lost at least 50% of its C reserve. Over the centuries, therefore, the total emission of C into the atmosphere by arable land use in the tropics may have been as much as 40 Mg ha^{-1} or a total emission (since the cultivation of tropical land began) of 40 Mg ha^{-1} x 418.4 x 10^6 ha, or 16.7 x 10^{15} g C. The range of C loss may be 10.0 x 10^{15} g C to 24.5 x 10^{15} g C.

Table 10. Total and per hectare population of livestock (cattle) and small ruminants (sheep and goats)

Region	Permanent pasture (10^6 ha)	Total population (10^6) Livestock	Small ruminants	Head per hectare Cattle	Small ruminants	Total animal herds
Africa	637.6	154.9	347.1	0.25	0.55	0.80
Asia	138.3	220.1	253.8	1.60	1.84	3.44
C America/ Caribbean	56.7	55.6	20.9	1.00	0.37	1.37
Pacfic	106.5	56.7	41.6	0.53	0.39	0.92
S America	286.9	147.1	60.1	0.51	0.21	0.72
Total	1226.0	634.4	723.5	0.52	0.59	1.11

Small ruminants is total of sheep and goats.
(Compiled from FAO, 1989.)

The present rate of emission from these lands is, however, low because the SOM has attained the equilibrium level. However, because of low-input agriculture, an equivalent of about 10 kg N ha^{-1} yr^{-1} net is absorbed by crops from mineralization of SOM. There is some additional contributions to crops from biological N fixation and atmospheric inputs, resulting in a loss of about 0.1 to 0.2 Mg C ha^{-1} yr^{-1} with an average value of 0.15 Mg ha^{-1} yr^{-1}. The total loss of C from arable land due to low input or subsistence farming is, therefore, estimated at 0.15 Mg C ha^{-1} yr^{-1} x 418.4 x 10^6 ha, or 62.8 x 10^{12} g C yr^{-1} with a range of 37.7 x 10^{12} g to 92.3 x 10^{12} g. This loss of C, can, however, be mitigated by addition of organic and inorganic fertilizers and amendments.

Livestock numbers in Africa increased from 295 million animal units compared with 219 million people in 1950 to 520 million animal units compared with 515 million people in 1983 (Brown & Wolfe, 1985). Total number of livestock in the tropics is 634.4 million, with an additional 723.5 million small ruminants (Table 10). The largest concentrations of animals is in Asia, with 3.44 animals ha^{-1}. The average per capita livestock and small ruminants in the tropics is 0.52 and 0.59, respectively. Globally, livestock contribute from 65 to 100 x 10^{12} g of CH_4 per year (Lerner et al., 1988). Of this, approximately 57% is from cattle, 19% is from dairy cows and 10% from sheep (Burke and Lashof, 1989). Aside from CH_4 and CO_2 emissions by livestock, the rate of C emission from excessively grazed pastures is assumed to be about 50% of that from arable land use, e.g. 0.05 to 0.1 Mg C ha^{-1} yr^{-1} with a mean value of 0.075 Mg C ha^{-1} yr^{-1}. For a total land area of 1226 x 10^6 ha, total C emissions from tropical pastures is estimated at 0.075 Mg C ha^{-1} yr^{-1} x 1226 x 10^6 ha, or 92 x 10^{12} g C yr^{-1}. The range of C emission from tropical pastures ranges from 55.2 x 10^{12} g C yr^{-1} to 133.4 x 10^{12} g C yr^{-1}. It seems that indiscriminate use of tropical grassland and pastures may contribute significantly more C to the atmosphere than tropical deforestation. In comparison with arable land, pasture land has lost about 25% of its C over the period of human occupation. This would mean a loss of 20 Mg C ha^{-1} for the total pastureland of 1226 x 10^6 ha. The total loss of C from pastureland over the period of human occupation is estimated at 24.5 x 10^{15} g C, with a range of 14.7 x 10^{15} g to 36.0 x 10^{15} g.

E. Cultivation of Upland Crops in Rotation with Rice Paddy

Out of about 146 x 10^6 ha of paddy rice cultivated in the world, 131 x 10^6 ha is in Asia. Literature surveys indicate paddy soils contain 25 to 40% more C than soils growing upland crops. However, there is little quantitative data regarding the amount of C loss during the period of oxidation when upland crops are grown following paddy cultivation. The loss during the 8-month period may be as much as 10% of organic C in the plowed layer, or about 3 Mg ha^{-1} yr^{-1}. Some C may, however, be assimilated back into the profile during paddy cultivation presumably at about 1 Mg ha^{-1} yr^{-1}. The net loss of C due to cultivation of paddy land for upland crops may be 2 Mg ha^{-1} yr^{-1}, with a range of 1.2 Mg ha^{-1} yr^{-1} to 3.0 Mg ha^{-1} yr^{-1}. The total loss of C from growing upland crops in rotation with paddy, therefore, is 1 Mg ha^{-1} yr^{-1} x 90 x 10^6 hectares in the tropics or 90 x 10^{12} g C yr^{-1} with a range of 50 x 10^{12} g C ha^{-1} yr^{-1} for low value to a high value of 130 x

10^{12} g C ha^{-1} yr^{-1}. Emission of this C is not necessarily related to CH_4 emission for rice paddies due to anaerobic decomposition of straw and other biomass. There have been several attempts to estimate CH_4 produced in rice paddies. Holzapfel-Pschorn and Seiler (1985) estimated worldwide CH_4 emission by rice paddies of 33-83 x 10^{12} g yr^{-1} or 10-20% of the global atmospheric CH_4 budget. Burke and Loshof (1989) estimated global CH_4 emission of 60-170 x 10^{12} g from rice paddies. Estimates of flux rates of CH_4 from rice paddies range from 4-27 mg m^{-2} hr^{-1} (EPA, 1990). Assuming an average rate of 15 mg m^{-2} hr^{-1} for a 150-day variety, total CH_4 production from 131 x 10^6 ha of rice paddies in the tropics may be 70.7 x 10^{12} g CH_4 per season with a range of 18.9 x 10^{12} g to 127.3 x 10^{12} g CH_4 per season.

F. Carbon Flux From Tropical Peat Lands

Peat soils, soils with 35% organic C or more and a minimum depth of 50 cm, occupy some 240 x 10^6 ha worldwide. An estimated 32 x 10^6 ha occur in the tropics, of which 20 x 10^6 ha are in the coastal lowlands of southeast Asia (Beek et al., 1980). A large proportion of peat soils is in Indonesia (Driessen et al., 1975).

Reclamation of peat soil for growing crop production involves drainage that leads to subsidence and rapid oxidation of organic material. The process of drying and shrinkage is accelerated in bare drained soils. Although rate of subsidence may be 2-5 cm yr^{-1}, it is difficult to assess loss of C from subsidence because the latter involves compaction from a low initial density (0.1-0.3 Mg m^{-3}) to a high density (0.4-0.6 Mg m^{-3}). A possible method to assess C emission from peat soils is through measurement of fluxes of CH_4 and CO_2. There is little or no data on flux measurements from these soils in the tropics. Experiments conducted in sub-arctic regions (Moore and Knowles, 1987) show that emission rates of 10-20 g C m^{-2} yr^{-1} for CO_2 and 0.1-0.6 g C m^{-2} yr^{-1} for CH_4 are common. Rate of CO_2 evolution from sub-arctic peat lands in Canada were measured by Moore (1986). Daily rates ranged from 1 to 10 mg CO_2 m^{-2} and decreased to less than 2 mg m^{-2} in winter. Annual rates of CO_2 evolution were 300-600 g m^{-2}. In West Virginia, rates of CO_2 production from surface peat ranged from 3.2 to 20 μ mol (mol C)$^{-1}$ hr^{-1} (Yavitt et al., 1987). Assuming that flux rates in the tropics are 4 times that of the sub-arctic region (Jenkinson and Ayanaba, 1977), it is possible that C emissions from drained peatland in tropics are 60 g C m^{-2} yr^{-1} for CO_2 and 2 g C m^{-2} yr^{-1} for CH_4. Most peat soils in Asia have already been developed, and are intensively used for production of vegetable and horticultural crops. Consequently, annual C emission from peat soils of the tropics is estimated at 62 g C m^{-2} yr^{-1} x 32 x 10^{10} m^2, or 2.0 x 10^{12} g C yr^{-1}.

VI. Total Carbon Emission Due to Agricultural Activities

Total emission of C due to agricultural activities in the tropics is shown in Table 11. The C flux ranges from about 0.3 x 10^{15} g C yr^{-1} to about 0.73 x 10^{15} g C yr^{-1} with a realistic value of 0.5 x 10^{15} g C yr^{-1}. Since soils of the tropics contain about 4.86 x 10^{15} g C, the annual emission of 0.5 x 10^{15} g is about 0.1% yr^{-1} of the total C reserve, with a range of about 0.06% yr^{-1} to 0.15% yr^{-1}. The relative loss of C from different agricultural activities is in the order annual burning of grassland > tropical deforestation > pastures > arable land use >shifting cultivation > peat soils (Figure 1).

Total release of C from soils of the tropics to date is estimated at about 45.1 x 10^{15} g C with a range of 27.0 x 10^{15} g to 73.7 x 10^{15} g. Relative emission due to different activities is shown in Figure 2. The rate of C emission from world soils has been estimated at 0.8 x 10^{15} g C yr^{-1} by Schlesinger (1984), implying thereby that 62.5% of C released from world soils comes from tropical regions covering only 38% of the total land area. Schlesinger estimated that cumulative transfer of C from world soils since prehistoric times may have been 40 x 10^{15} g. The present estimate of 64.7 x 10^{15} g is from soils of the tropics alone.

While present data of annual C emission compare favorably with those of Schlesinger (1984), these estimates are drastically lower than those reported by Buringh (1984). Buringh's estimates of C emission from world soils of 4.6 x 10^{15} g C yr^{-1} seem to be too high. His estimates of cumulative transfer of C from world soils since prehistoric times of (537 x 10^{15} g) also seem to be an overestimation. In contrast to Buringh, Revelle (1966) estimated that the CO_2 released by cultivation of virgin soils in the last 100 years is 30 x 10^{15} g.

Table 11. C emission from soils of the tropics due to agricultural activities

Agricultural activity	C emission (10^{12} kg yr^{-1})			Total emission to date (10^{15} kg)		
	Low	Realistic	High	Low	Realistic	High
1. Deforestation	89.7	154.3	218.8	2.3	3.5	13.2
2. Shifiting cultivation[a]	3.8	6.3	9.2	-	-	-
3. Annual burning of grassland[b]	112.5	187.5	275.6	-	-	-
4. Arable landuse	37.7	62.8	92.3	10.0	16.7	24.5
5. Pastures	55.2	92.0	133.4	14.7	24.5	36.0
6. Rotation with paddy[c]	0.05	0.09	0.13	-	-	-
7. Peat soils	1.50	2.0	3.00	0.3	0.4	0.6
Total	300.5	505.0	732.4	27.3	45.1	74.3

[a]Total emission of C due to shifting cultivation over the period of human history is included within the estimate of deforestation.
[b]Total emission due to natural or anthropogenic fire in grasslands over the period of human history is included within the estimate of arable landuse and pastures.
[c]Total estimates of C emission from rice paddies are included within arable landuse.

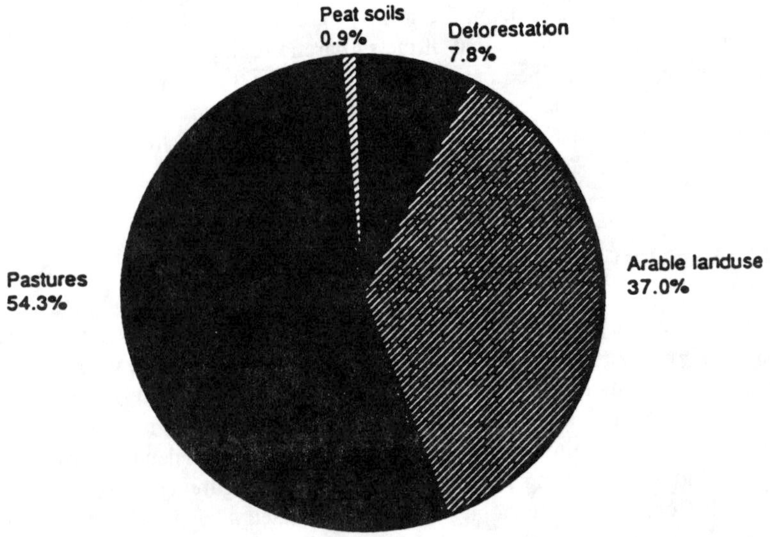

Figure 2. Relative (%) C emission rate from different landuses in the tropics; total emission equals 45.1×10^{12} kg C yr^{-1}.

Estimates of C emission from soil presented in this report do not take into account potential increases in global temperature. Increase in global temperature is likely to accelerate the rate of mineralization and enhance C emission from soils of the tropics.

Table 12. Nitrogen fertilizer use (urea and $NaNO_3$) in the tropics during 1989-90

	Fertilizer use (1000 Mg)	Total land area of arable land and permanent crops (10^6 ha)	Fertilizer use (kg ha^{-1})
Africa	613.0	145.0	4.2
Asia	13563.5	212.0	64.0
Central America and the Caribbean	1416.0	25.5	55.5
Pacific and Oceania	122.6	12.7	9.7
South America	1182.0	74.6	15.8
Total	16897.1	469.8	36.0

• Asia Includes 40% of land in India, 33% in China, 35% in North Korea, and 50% in South Korea.
• Central America includes 50% of land in Mexico.
• Pacific and Oceania includes 25% of land in Australia.
(Adapted from FAO-Supply/demand statistics, 1990.)

Table 13. Change in fertilizer consumption in developing countries

	Consumption (10^3 Mg)		
	1988	1989	% increase yr^{-1}
Africa	0.836	0.897	7.3
Latin America	3.801	3.881	2.1
Southeast Asia	32.030	32.76	2.3
Pacific Region	0.017	0.020	17.6

(Adapted from FAO, 1989.)

VII. N_2O Emission Due to Agricultural Activities

Emission of N_2O from soils is an important factor responsible for increase in atmospheric concentration of this gas. Emission of N_2O can be due to changes in N in SOM, inorganic fertilizers or organic manures applied to agricultural land. Global estimates of N_2O emission are hard to make, because there is little, if any, experimental data available on gaseous emissions under different land use systems. Global emission of N_2O from all sources is estimated at $15 \pm 7 \times 10^{12}$ g N_2O-N yr^{-1}.

In view of the scanty data available, an alternate approach would be to estimate N_2O emission as a percentage of the total fertilizer applied. Total nitrogenous fertilizer consumed in the tropics during 1989-90 was 16.9×10^6 Mg on 469.8×10^6 ha of arable land and permanent crops (Table 12). Therefore, the rate of fertilizer use is estimated at 36.0 kg ha^{-1}, which varies widely from 4.2 kg ha^{-1} in Africa to 64.0 kg ha^{-1} in Asia (Table 13).

Estimates of annual loss of N by N_2O emission range from 1 to 40 kg N ha^{-1} (Groffman and Tiedje, 1989). Robertson and Tiedje (1986) estimated high rates of 800 mg N cm^{-2} hr^{-1} from soil supporting tropical rainforest. N_2O emission rates are much higher from organic soils and may range from 50 to 165 kg N_2O-N ha^{-1} yr^{-1} (Terry et al., 1981). Sahrawat and Keeney (1986) reported N_2O emission rates ranging from 0.3 ng N_2O-N m^{-2} s^{-1} (0.10 kg N_2O-N ha^{-1} yr^{-1}) to 135 ng N_2O-N m^{-2} s^{-1} (42.5 kg N_2O-N ha^{-1} yr^{-1}). A few studies conducted in the tropics (Engelstad and Russel, 1975; Gould et al., 1986) have shown that denitrification rate can be 20% of fertilizer applied can be denitrified. De Datta et al. (1991) estimated loss of urea from rice to be 43 to 64% of the applied N. Loss of N_2O was 0.1% of the applied N. Lindau et al. (1990) measured N_2O flux of 4 g ha^{-1} d^{-f} from $(NH_4)_2$ SO_4 plots and 122 g ha^{-1} day^{-1} from plots treated with KNO_3. Mosier et al. (1991) measured N_2O flux from temperate zone grassland at 1.8-3.0 g N ha^{-1} d^{-1} in unfertilized plots vs. 3.6-6.3 g N ha^{-1} d^{-1} in fertilized treatments. Duxbury et al. (1982) estimated N_2O emission of 1 kg N_2O-N ha^{-1} yr^{-1} in Florida. With this assumption, the loss of N_2O-N from tropical land is estimated at 3.6 x 10^{12} g N_2O-N yr^{-1}. In addition, experimental data indicate emission of 1 x 10^{-5} to 1 x 10^{-2} g N2O-N m^{-2} hr^{-1} (Sahrawat and Keeney, 1986) from natural land use systems. Assuming an average flux of 2 x 10^{-5} g N_2O-N m^{-2} hr^{-1}, flux from natural ecosystems over land area of 3.6 x 10^9 ha amounts to 6.3 x 10^{12} g N_2O-N yr^{-1}. Therefore, total N_2O flux from tropical ecosystem amounts to 9.9 x 10^{12}g N_2O-N yr^{-1}. These estimates of N_2O flux may increase because the fertilizer demand in the tropics is rapidly increasing (Table 13). The most rapid expansion is expected in Africa. This means higher losses of applied fertilizers as N_2O with potentially higher risks.

VII. General Conclusions and Researchable Priorities

World soils contain about 1500 x 10^{15} g C in the top 1-m depth. The C pool in world soils is about twice as much as in the atmosphere, and three times as much as in world's biota. Soils of the tropics contain about 486 x 10^{15} g C in the top l-m depth, or about 32.4% of C stored in world soils. Estimates of the C stored in soils of the tropics amount to 145 x 10^{15} g in the top 30-cm depth and 340 x 10^{15} g in the top 50-cm depth. Most soils in the tropics contain 40 to 100 Mg C ha^{-1} in the top 30 m depth.

Out of the total land area of 4.8 x 10^9 ha in the tropics, 8.6% is arable, 1.0% is under permanent/plantation crops, 25.1% is under pasture, 38.2% is forest/woodland, and the remaining 27.1% is under miscellaneous uses. Forest reserves of the tropics are estimated at about 3 x 10^9 ha, with annual rate of deforestation at about 20 x 10^6 ha. Shifting cultivation is practiced on an additional 25 x 10^6 ha.

Principal agricultural activities relevant to C emission from soils include deforestation, shifting cultivation, natural or anthropogenic annual burning of the grassland, arable land use, pasture, and cultivation of upland crops in rotation with rice paddy. Annual rate of C emission from tropical deforestation is estimated at 154.3 x 10^{12} g C yr^{-1}. Rate of C emission due to shifting cultivation in the tropics is 6.25 x 10^{12} g C yr^{-1}. Fire in tropical savannas and grasslands is responsible for emission of 187.5 x 10^{12} g C yr^{-1}. Emission of C from arable lands in the tropics due to subsistence or semi-commercial farming is 62.8 x 10^{12} g C yr^{-1}, and the rate of emission from pastures is 92 x 10^{12} g C yr^{-1}. Cultivation of upland crops in rotation with rice paddies leads to emission of 90 x 10^{12} g C yr^{-1}.

Total C emission due to agricultural activities is 0.60 x 10^{15} g C yr^{-1} with a range of 0.3 x 10^{15} to 0.8 x 10^{15} g C yr^{-1}. Since soils of the tropics contain about 486 x 10^{15} g C in the top 1 m, the annual emission of 0.6 x 10^{15} g is equivalent to about 0.1% of the total C reserve. It is estimated that 62.5% of the C emission form world soils is contributed by soils of the tropics. Cumulative C emission from soils of the tropics since pre-historic times is estimated at 44.7 x 10^{15} g C with a range of 27.0 x 10^{15} g to 44.7 x 10^{15} g.

Oxidation of peat soils in the tropics is another important source of C. Peat soils occupy about 32 x 10^6 ha in the tropics. When drained for cultivation, peat soils can contribute as much as 2 x 10^{12} g C yr^{-1}. Over the last 200 years, total C emission from peat is estimated to be 0.4 x 10^{15} g C.

Annual rate of nitrogenous fertilizer use in the tropics is 36 kg ha^{-1}, with a range of 4.2 kg ha^{-1} in Africa to 64.0 kg ha^{-1} in Asia. It is estimated that 0.7-2.1 kg ha^{-1} yr^{-1} of N_2O-N (with an average of 1.0) is lost from fertilizers and manure applied to arable lands and pastures. This would amount to a total of 3.6 x 10^{12} g N_2O-N yr^{-1}. In addition, an emission rate of 2 x 10^{-5} g N_2O-N m^{-2} hr^{-1} is assumed for four principal land uses in the tropics over the total area of 3.6 x 10^9 ha. Therefore, the N_2O-N flux from all land uses amounts to 6.3 x 10^{12} g N_2O-N yr^{-1}. The total N_2O-N flux from tropical ecosystems amounts to 9.9 x 10^{12} g N_2O-N yr^{-1}.

Estimates of C and N_2O emission presented in this report are first approximations. There is an urgent need to obtain reliable data of C balance under a range of land use systems in major soils and principal ecoregions of the tropics. These data should be obtained from existing or newly established long-term field experiments (>10 years) in Africa, Asia, Central America, South America, and the Pacific regions. However, there are few, if any, on-going long-term experiments currently in progress. There is a need to establish such projects in principal ecoregions of the tropics.

Recent increases in the concentrations of radiatively active gases in the atmosphere and the concern about their effects on global climate has stimulated considerable interest regarding sources and sinks of these gases. The role of world soils as a regulator of C in the atmosphere has not been given the attention it deserves.

Among world soils, specific contribution of soils of the tropics to emission of greenhouse gases is the least understood. Attempts to model these emissions are greatly hindered by the paucity of reliable data on:

(i) the extent and ecological/geographical distribution of soil resources, (ii) C and N profiles of major soils under different land uses, (iii) dynamics of C and N due to change in land use, e.g. from forest or grassland to arable land, plantations, or pastures, (iv) fluxes of CO_2, CH_4, and N_2O from soils as influenced by seasonal changes, land use changes, and soil management practices.

Whatever meager data exist, they are incomplete, obsolete, often inaccessible, and not obtained for the purposes of calculating C or N budgets. The information on C and N profiles for principal soils in major tropical biomes to at least 1-m depth is essential to estimate C and N reserves. Most available data are for the top few centimeters and only for a small fraction of the vast arrays of soils in the tropics. There have been some studies on deforestation effects on soil properties in the tropics. Rarely, however, have the effects been measured in terms of C and N balance to 1-m depth under different land use, farming systems, or methods of soil surface management following deforestation.

There is an urgent need to initiate long-term experiments on benchmark soils to study dynamics of C and N in soils of the tropics. These experiments to be conducted for about 5 years should be located on at least 3 major soils (Oxisols, Alfisols, Ultisols) in 2 principal biomes (rainforest, savanna). Detailed measurements should be made on annual changes in C and N status of soil to 1-m depth for arable land use, pastures, plantations and traditional farming. Measurements of C dynamics and gaseous emissions from intensive rice-wheat systems in south Asia should also be a priority.

Sequestering atmosphere C into soil reservoir, and reversing the trend is an important researchable priority. C can be sequestered into the soil as humus. There are several promising land use and farming systems that enhance humus content of the plow layer. Those farming systems that would increase humus content of the sub-soil horizon (30-100 cm deep) would be more effective in increasing soil's C pool. Humus stored in sub-soil horizons is not easily mineralized. Another important strategy is to increase micro-aggregation in the soil. C blocked within stable micro-aggregates is not easily accessible even to micro-organisms. Those farming systems and agricultural practices of soil and crop management that enhance micro-aggregation would also increase C pool of the soil.

Acknowledgement

Financial support received from USEPA-Climate Change Division, Washington, D.C. is gratefully acknowledged. Support and help received from Mr. Ken Andrasko is greatly appreciated. Mr. J. McLaugholin was involved during initial phases of the project.

References

Aina, P.O. 1979. Soil changes resulting from long-term management practices. *Soil Sci. Soc. Am. J.* 43:173-177.

Beek, K.J., W.A. Blokhuis, P.M. Driessen, N. van Breeman, R. Brinkman, and L.J. Pons. 1980. Problem soils: their reclamation and management. ILRI Publication 27, Wageningen, Netherlands, 72 pp.

Blasco, F. and F. Achard. 1990. Analysis of vegetation changes using satellite data. In: A F. Bouwman (ed.), *Soils and the Greenhouse Effect*, J. Wiley and Sons :303-325.

Bohn, H. L. 1976. Estimate of organic C in world soils. *Soil Sci. Soc. Am. J.* 40:468470.

Bohn, H. L. 1978. Organic soil C and CO_2. *Tellus* 30:472475.

Brown, L.R. and E.C. Wolf. 1985. Restoring African soils. World Watch Paper 65. World Watch Institute, Washington, D.C.

Buringh, P. 1979. Introduction to the study of soils in tropical and sub-tropical regions. Centre for Agric. Publishing and Documentation, Wageningen, The Netherlands, 124 pp.

Buringh, P. 1984. Organic C in soils of the world. In: G.M. Woodwell (ed.), *The Role of Terrestrial Vegetation in the Global C Cycle: Measurement by Remote Sensing*. J. Wiley & Sons, NY: 91-109.

Burke, L. and D.A. Leshof. 1989. Greenhouse gas emissions related to agriculture and land use practices. US-EPA Report, Washington, D.C., 30 pp.

Cook, L. 1939. A contribution to our information on grass burning. *S. Afr. J. Sci.*, 36:270-282.

Cunningham, R.K. 1963. The effect of clearing a tropical forest soil. *J. Soil Sci.* 14:334-345.

Dalal, R.C. and R.J. Mayer. 1986. Long-term trends in fertility of soils undercontinuous cultivation and cereal cropping in southern Oueensland. II. Total organic C and its rate of loss from the soil profile. *Aust. J. Soil Res.* 24:281-292.

De Datta, S.K., R.J. Buresh, M.I. Samson, W.N. Obcemea and J.G. Real. 1991. Direct measurement of NH_4 and denitrification fluxes from Urea applied to rice. *Soil Sci. Soc. Am. J.* 55:543-548.

Driessen, P.M., M. Soepraptohardjo and L.J. Pons. 1975. Formation, properties, reclamation and agricultural potential of Indonesian ombrogenous lowland peats. In: K.M. Schallinger (ed.), *Peat in Agriculture and Horticulture*, Proc. Intl. Symposium, Intl. Peat Society Comm. III, Special Publ. 205:67-84.

Duxbury, J.M., D.R. Bouldin, R.E. Terry, and R.L. Tata. 1982. Emissions of N_2O from soils. *Nature* 298:462-464.

Edwards, D. C. 1942. Grass Burning. *Emp. J. Expt. Agric.* 10:219-231.

Engelstad, O.P. and D.A. Russel. 1975. Fertilizers for use under tropical conditions. *Adv. Agron.* 27: 175-208.

Eswaran, H., E. Van Den Berg, and P. Reich, 1993. Organic C in soils of the world. *Soil Sci. Soc. Am. J.* 57: 192-194.

EPA. 1990. Greenhouse gas emissions from agricultural systems. Summary Report, Vol. 1: IPPC, Washigton, D.C.

FAO. 1981. Tropical forest resources assessment project. Forest resources of tropical Africa. Part I, Regional Analysis. FAO, Rome, Italy.

FAO. 1989. Production Year Book, Rome, Italy.

FAO. 1990. Fertilizer Statistics, Rome, Italy.

Ghuman, B.S. and R. Lal. 1991. Land clearing and use in humid Nigerian tropics. II: Soil chemical properties. *Soil Sci. Soc. Am. J.* 55:184-188.

Gould, W.D., C. Hagedorn and R.G.L. McCready. 1986. Urea transformations and fertilizer efficiency in soil. *Adv. Agron.* 40:209-238.

Groffman, P.M. and J.M. Tiedje. 1989. Denitrification in north temperate forest soils: spatial and temporal patterns at the landscape and seasonal cycles. *Soil Biol. Biochem.* 21:613-620.

Holzapfel-Pschorn, A., and W. Seiler. 1985. Contribution of CH_4 produced in rice paddies to the global CH_4 budget. *Biometeorology* 9:53-61.

Houghton, R.A. 1990. The global effects of tropical deforestation. *Env. Sci. Technol.* 24:414-22.

Houghton, R.A., R.D. Boone, J.R. Eruci, J.E. Hobbie, J.M. Melillo, C.A. Palm, B.J. Peterson, G.R. Shaver, G.M. Woodwell, B. Moore, D.L. Skole, and N. Myers. 1987. The Flux of C from Terrestrial Ecosystem to the Atmosphere in 1980 due to Changes in Land use: Geographic Distribution of the Global Flux. *Tellus* 398:122-139.

Kimble, J.M., T. Cook, and H. Eswaran, 1990. Organic matter in soils of the tropics. p. 250-258. In: Proc. Symp. on Characterization and Role of Organic Matter in Different Soils. 14th Int. Cong. Soil. Sci., Kyoto, Japan.

Kyuma, K. and C. Pairintra. 1982. Shifting cultivation. Ministry of Sci. Technology and Energy, Thailand, 219 pp.

Lal, R. 1976a. No-tillage effects on soil properties under different crops in western Nigeria. *Proc. Soil Sci. Soc. Amer.* 40:762-768.

Lal, R. 1976b. Soil erosion on Alfisols in western Nigeria. I. Effects of slope, crop rotation and residue management. *Geoderma* 16:363-375.

Lal, R. 1982. Effects of 10 years of no-tillage and conventional plowing on maize yield and properties of a tropical soil. p. 111-117. In: Proc. 9th ISTRO Conf., Osijek, Yugoslavia.

Lal, R. 1985. Mechanized tillage systems effects on properties of a tropical Alfisol in watershed cropped to maize. *Soil and Tillage Res.* 6:149-162.

Lal, R. 1987a. Conversion of tropical rainforest: agronomic potential and ecological consequences. *Adv. Agron.* 39: 173-264.

Lal, R. 1987b. *Tropical Ecology and Physical Edaphology, United Kingdom.* John Wiley & Sons, 732 pp.

Lal, R.D. De Vleeschauwer, and R.M. Nganje. 1980. Changes in properties of a newly cleared Alfisol as affected by mulching. *Soil Sci. Soc. Am.* J. 44:827-833.

Lerner, J.E., E. Matthews and I. Fung. 1989. Methane emission from animals: A global high resilution database. *Global Biogeochemical Cycles* 2:139-156.

Lindau, C.W., R.D. De Launa, W.H. Patrick, Jr., and P.K. Bollich. 1990. Fertilizer effects on dinitrogen, nitrous oxide and methane emission from lowland rice. *Soil Sci. Soc. Am. J.* 54: 1789-1794.

Moore, T.R. 1986. CO_2 evolution from sub-arctic peatlands in eastern Canada. *Arctic and Alpine Res.* 18: 189-193.

Moore, T.R. and R. Knowles. 1987. Methane and CO_2 evolution from subarctic fens. *Can. J. Soil Sci.* 67:77-81.

Mosier, A., D. Schimel, D. Valentin, K. Bronson, and W. Parton. 1991. Methane and nitrous oxide fluxes in native, fertilized and cultivated grassland. *Nature* 350:330-332.

National Research Council. 1993. *Sustainable Agriculture and Environment in the Humid Tropics.* National Academy Press, Washington, D.C, 702 pp.

Nye, P.H., and D.J. Greenland. 1960. *The Soil Under Shifting Cultivation. Commonwealth Bureau of Soils,* Hapenden, England Tech. Comm. No. 51.

O'Keefe, P. and L. Kristofferson. 1984. The uncertain energy path and the Third World development. *Ambio* 13:168-170.

Post, W.M., W.R. Emanuel, P.J. Zinke, and A.G. Stangenberger. 1982. Soil C pools and world life zones. *Nature* 298:156-159.

Revelle, R. 1966. Atmospheric CO_2. p 111-136. In: Restoring the Quality of Our Environments. Rep. Env. Publication Panel. President Sci. Adv. Council, The White House.

Roberts, W.P. and K.Y. Chan. 1990. Tillage induced increases in CO_2 loss from soil. *Soil and Tillage Res.* 17:143-151.

Robertson, G.P. and J.M. Tiedje. 1988. Deforestation alters denitrification in a lowland tropical rainforest. *Nature* 336:756-759.

Sahrawat, K.L. and D.R. Keeney. 1986. N_2O emission from soils. *Adv. Soil Sci.* 4:103-148.

Sanchez, P.A., C.A. Palm, and T.J. Smyth. 1990. Approaches to mitigate tropical deforestation by sustainable soil management practices. *Developments in Soil Sci.* 20:211-220.

Sanchez, P.A., C.A. Palm, L.T. Scott, E. Cuevas, and R. Lal. 1989. Organic input management in tropical agroecosystems. p. 125-152. In: D.C. Coleman, J.M. Oades and G. Uehara (eds.), *Dynamics of Soil Organic Mater in Tropical Ecosystems.* Univ. of Hawaii Press, Honolulu.

Sanchez, P.A. and I.G. Salinas. 1981. Low-input technology for managing Oxisols and Ultisols in tropical America. *Adv. Agron.* 34:279 406.

Sanchez, P.A., D.E. Bandy, J.H. Villachica, and I.I. Nicholaides. 1982. Amazon Basin soils: management of continuous crop production. *Science* 216:821-827.

Schlesinger, W. H. 1984. Soil organic matter: a source of atmospheric CO_2. In: G.M. Woodwell (ed.), *The Role of Terrestrial Vegetation in the Global C Cycle: Measurement by Remote Sensing.*" J. Wiley & Sons, NY.

Stevenson, F.I. 1982. *Humus Chemistry: Genesis, Composition, Reactions.* J. Wiley & Sons, NY, 443 pp.

Terry, E.R., R.L. Tate and J.M. Duxbury. 1981. N_2O emissions from drained, cultivated organic soils of South Florida. *J. Am Pollution Control Assoc.* 31:1173-1176.

Theng, B.K.G., K.R. Tate, and P. Sollins. 1989. Constituents of organic matter in temperate and tropical soils. p. 5-31. In: D.C. Coleman, J.M. Oades, and G. Uehara (eds.), *Dynamics of Soil Organic Matter in Tropical Ecosystems.* Univ. Hawaii Press L 5-31.

UNESCO. 1978. Tropical forest ecosystems: A state of knowledge report, UNESCO, Paris.

WRI. 1990-91. World Resources Report, World Resources Institute, Washington, D.C.

Yavitt, J.B., G.E. Land, and R.K. Wieder. 1987. Control of C mineralization to CH_4 and CO_2 in anaerobic, sphagnum-derived peat from Big Run Bog, West Virginia. *Biogeochemistry* 4:141-157.

CHAPTER 26

Economic and Resource Impacts of Policies to Increase Organic Carbon in Agricultural Soils

Aziz Bouzaher, Derald J. Holtkamp, Randall Reese, and Jason Shogren

I. Introduction

Recently there has been widespread documentation of increasing levels of CO_2 and other greenhouse gases in the atmosphere (Houghton et al., 1990). The balance of carbon stored in the atmospheric, terrestrial, and oceanic pools is beginning to shift disproportionally toward the atmosphere leading to predictions of global climate change. It is estimated that agricultural soils store 1.5×10^{15} kilograms of carbon, twice the amount held in the atmosphere (Post et al., 1990). The agricultural activities carried out on any particular tract of land may have a significant capacity to affect the amount of carbon stored in the soil. Policies designed to encourage or compel the adoption of practices or land use patterns that promote the buildup of soil organic carbon may affect reduced emissions of carbon gases as well as potential economic costs to producers and consumers. Depending upon their design and implementation, policies may have considerable regional and national impacts on agricultural profitability, land use patterns, soil erosion, and the use of pesticides and fertilizer. The purpose of this chapter is to evaluate these impacts and to provide meaningful measures with which to evaluate both environmental and economic outcomes. Economic impacts are evaluated with two modeling systems: the Resource Adjustment Modeling System (RAMS) and the Basic Linked System (BLS). The work summarized here is part of the U.S. EPA Climate Change Program and involves the integration of RAMS and BLS with a soil organic formation model called CENTURY. Soil organic formation models simulate long-term changes in soil organic nitrogen and carbon. For this analysis CENTURY is used to examine the impacts of alternative management practices on soil organic carbon. The CENTURY runs are performed by Aqua Terra Consultants. Some key CENTURY results are reported in this paper. These results are reported to demonstrate economic and environmental tradeoffs produced by the integrated approach. The carbon results are controversial and the CENTURY model is being carefully reviewed.

RAMS is a regional, comparative static, linear programming model of the crop production sector. The model objective is maximization of net returns to crop production. RAMS is designed for disaggregated analysis of agricultural and environmental policy and the results are used to estimate short-run adjustments in producer decisions about crops grown, input use, and cropping systems employed. The RAMS study region includes the major corn and sorghum producing areas of the United States. The strength of RAMS is in the extensive detail about agricultural practices with environmental ramifications.

The Basic Linked System of applied general equilibrium models (Fisher et al., 1988; Abkin, 1985) is a large-scale computer simulation of the world agricultural economy. Unlike RAMS, BLS simulates the long-run behavior of agricultural production and commodity markets at the national and world levels and specifies interactions with nonagricultural markets. The rules governing these markets are consistent with microeconomic principles of general equilibrium and accommodate various policy regimes.

Two types of policies are evaluated in this paper. The first type, involving region-specific targeting of crop production practices, is evaluated by RAMS. The second type of policy, involving national shifts in land use patterns prompted by the Conservation Reserve (CRP) and Wetlands Reserve (WRP) Programs,

ISBN 1-56670-117-1/95/$0.00+.50

is evaluated by BLS. BLS is better suited to analyze national policies, such as the CRP and WRP, for which international trade effects may be significant. While the scope of BLS is broad, it does not have RAMS's detail for specific production practices. RAMS is thus better suited for evaluation of targets on specific production practices.

Regionally targeted production practices include winter cover crops and conservation tillage. Winter cover crops are expected to increase soil organic carbon by expanding annual biomass production while conservation tillage is expected to increase soil organic carbon by minimizing soil disturbances. Shifts in land use patterns through CRP may have substantial impact on soil carbon levels even though increasing soil carbon levels was not a policy objective when the CRP was first authorized. The WRP examined here is intended as a carbon sequestering measure. The program targets bottomland suitable for hardwood tree growth, which has the potential for significant carbon benefits.

This paper discusses the RAMS and BLS models and their linkages to CENTURY. The policies and their implementation in the models are described and are followed by a review of the economic results of the policy analyses and key soil organic carbon results from CENTURY. Unfortunately, only partial soil organic carbon results were available at publication for the CPR scenarios, and none were available for the WRP scenario. Conclusions are presented at the end to stimulate discussion and further research.

II. Models

A. RAMS

The Resource Adjustment Modeling System (RAMS) was developed in 1990 and 1991 by the Center for Agricultural and Rural Development (CARD). The system is geographically delineated by producing areas (PAs), hydrological areas representing aggregated subareas defined by the U.S. Water Resources Council (USWRC, 1970). The areas are small enough that the assumption of homogenous production across the area can reasonably be made. RAMS is interfaced with the CENTURY model through areas of crop production by crop, tillage, and rotation, and they are passed from RAMS to CENTURY.

Crop production is characterized in the model by activities that specify crop rotation, tillage, contour management, winter cover crops, and irrigation. Each activity is defined by a combination of these five dimensions.

Cover crops are included only for a subset of all crop sequences within each rotation. The subset depends on the geographic location of the PA. In general, the eligible sequences in any PA depend upon the time period in which cover crops can grow and on the scarcity of moisture. In some cases there is simply not enough time or moisture to establish and get a reasonable growth of the cover crop. The RAMS study area is segregated into three regions for the purpose of defining cover crops. The northern one-third of the producing areas is considered unsuitable for cover crops and therefore the cover crop alternatives are excluded in those regions. The middle and southern thirds have cover crops defined but the subset is smaller in the middle third. Southern PAs typically have longer, and warmer, growing seasons allowing substantial growth of a winter cover crop. Some western PAs simply do not have enough water to support a commodity and a winter cover crop. The growing period for winter cover crops is greater following crops with early harvests and/or succeeding crops with late planting dates. (Cruse, 1992).

Winter cover crops not only improve the soil carbon content of the soil, they reduce soil erosion, fix nitrogen available to the subsequent crop in the case of hairy vetch (*Vicia villosa*), cost money to establish and, because of competition for moisture, reduce yields in the subsequent crop. Winter cover crop activities are modeled by adding new production activities with adjustments to each of these variables defined for the original RAMS activities. Because each of the variables is defined for rotations in RAMS, adjustments are weighted according to the share occupied by each crop sequence in the rotations.

Data needed to estimate cost adjustments are obtained from two sources. Machinery usage and costs are estimated from CARD budgets. Seeding rates are estimated from field data (Cruse, 1992) and prices for seed are obtained from *Agricultural Prices* (NASS, 1990).

Estimates of erosion reductions vary according to crop sequence and tillage. Reductions are estimated from two sources. When available, percentage differences in the universal soil loss equation (USLE) C factors for land treated with a winter cover crop, versus untreated, are obtained from the Soil Conservation Service Field Office Technical Guides (SCS, 1991) and used to adjust the C factor used in the original calculation of erosion for the activity without winter cover. The alternative is to use rainfall erosion index curves, which measure the percentage of annual erosion occurring at each point in time to estimate the percentage of annual erosion occurring between harvest and planting of commodity crops (Wischmeier and Smith, 1965). This value is then adjusted by an estimate of the percent reduction in erosion due to the cover crop during that same time period. For example, if the rainfall erosion index curves indicate that 30 percent of annual erosion occurs between harvest and planting and that cover crops will reduce erosion by 50 percent between the harvest and planting dates, then the erosion reduction is 15 percent (50% of 30%). Normal planting and harvest date are taken from Burkhead et al. (1968).

Nitrogen carryover from hairy vetch, which is a nitrogen fixing legume, is estimated from reported field data. Carryover amounts of 25 pounds per acre in the middle region and 35 pound in the south are limited to use by the crop immediately succeeding the cover crop. Nitrogen in excess of the succeeding crop's demand is assumed lost.

Yield adjustment factors are estimated from field data when available. In cases where relevant field data are unavailable, yield impact estimates are based on information about water competition taken from the literature (Cruse, 1992). The outcomes of the model include the indicators used to evaluate the economic costs and benefits as well as the production patterns that are fed to the CENTURY model. Primary economic indicators include net returns to production, acreage planted, tillage practices, rotations, erosion, and fertilizer and herbicide use.

B. BLS

The BLS is characterized as a system of applied general equilibrium models. The term *applied general equilibrium* means that all behavior in the model derives from standard microeconomic assumptions about how markets work together. All commodities, markets, and regions are linked through prices and specific accounting rules: prices and quantities must be such that all markets clear and global commodity markets balance. BLS is also flexible enough to accommodate several specific agricultural, trade, and economic policies and to account for changes over time. Thus, when policies such as the CRP and WRP scenarios are included in the model, the impacts of those policies projected by BLS reflect (1) the domestic linkages among several agricultural markets and the nonagricultural market; (2) the dynamic effects in production, consumption, and trade over time; (3) the feedback from international markets, and (4) the influences of several national policies.

The U.S. submodel specifically accounts for 23 commodities, but this discussion is limited to the seven program crops -- corn, wheat (*Tritium aestivum*), sorghum, barley (*Hordeum vulgare*), oats (*Avena sativa*), rice (*Oryza sativa*), and cotton (*Gossypium hirsutum*) -- and soybeans (*Glycine max*) plus livestock. Results generated include annual projections of supply and utilization tables that identify production (including acreage seed, waste, and stocks), retail prices, and consumer raw material prices.

III. Policies Evaluated with RAMS: Region-specific Targeting of Production Practices

The baseline for the RAMS model is the 1990 growing season. Commodity program parameters for the 1991 programs are used to reflect the new commodity programs of the 1990 farm bill. Input prices for 1989 are used. Crop inputs such as seed and fertilizer are often purchased in the previous year for tax purposes or because discounts are offered. Crop areas are calibrated to 1990 data. Commodity prices are estimated as averages for the 1990 calendar year, using projected prices for the later months. Yields for 1990 are also estimated.

A. Alternatives

Targets for adoption of no-till and reduced tillage practices, together labeled *conservation tillage* are derived by altering the criteria for defining highly erodible land for conservation compliance purposes. Targeting of the most erosive land offers a dual benefit of reducing erosion and increasing storage of carbon in agricultural soils. Targets for planting of winter cover crops are based upon baseline areas of land planted to crops with early harvest dates, namely small grains and silage, followed by crops not seeded in the fall.[1] The objective is to select and promote situations where winter cover crops have the most opportunity to fix carbon. Targeting both conservation tillage and winter cover crops is accomplished by including a constraint in RAMS to force selection of activities meeting the appropriate criterion.

1. Conservation Tillage Targets

Estimating the distribution of highly erodible land for modeling conservation compliance in RAMS for the baseline involved calculating of erodibility index[2] (EI) values for land in the 1982 National Resource Inventory (NRI). Land with an erodibility index (EI) greater than 8 is considered highly erodible. Six *highly erodible land groups* (HELG) based on ranges of the estimated EI values are defined. Highly erodible land groups 1 through 5 enclose the range of EI from 9 to 43 in increments of 6. Land with an EI greater than 43 is included in the sixth HELG.

The USLE C and P factors are indices of the crop grown and management practices used. Values for both coefficients are estimated for all of the production activities in RAMS and the product of the two is labeled the CP factor. Values of the CP factor for each activity in RAMS are evaluated for each group of highly erodible land. If $EI < 1/CP$, where EI is the average of the range for each HELG, the practice meets the conservation compliance requirements for that HELG. For the present analysis, the sum of land falling into one of the six highly erodible land groups is used as the target for conservation tillage. This is modeled by constraining the sum of no-till plus reduced till to be equal to the amount of highly erodible land. The constraint may be satisfied by devoting cropland to either no-till or reduced tillage systems. Because the minimum EI for land to be considered highly erodible is reduced from 8 to 2, the amount of land considered highly erodible increases.

Four runs are performed to evaluate the sensitivity of the results to the size of the target and to different assumptions about relative yields. A single run is performed, assuming relative yields between tillage practices are essentially equal[3] for each of two targets based on EI values of 8 and 2. In addition, the same runs are performed assuming a 10 percent reduction in no-till yields and a 5 percent reduction for reduced till. The lower yields reflect yield losses that might be expected during the initial adoption period when producers are learning to use the conservation tillage systems (Cruse, 1992). Experience has shown that once producers gain sufficient management skill, yields equivalent to those with conventional tillage systems are attainable.

2. Winter Cover Crops

Cover crop activities are created by inserting winter cover crops into previously defined production activities in RAMS. The cropping sequences are defined, in part, by a crop rotation that is merely a set of sequences in which crops are to be grown. Therefore, adding winter cover crops to a cropping system does not increase the amount of land required, only the intensity of its use.

To establish a target for winter cover crops the following question is posed. How many hectares could be devoted to cover crops grown between small grains or silage and crops not seeded in the fall without altering the mix of cropping systems used under the baseline? These crop sequences are targeted because they provide the best opportunities for the establishment and growth of winter cover crops. Small grains include barley, oats, and spring and winter wheat. Silage includes both corn and sorghum silage. The area in the appropriate sequences is calculated from baseline results for each producing area in RAMS. These

[1]Crops seeded in the fall circumvent the need for establishing a winter cover crop. Fall seeded crops modeled include winter wheat, legume hay, and nonlegume hay.

[2]EI is equal to RKLS/T for water erosion. R, K, L and S are universal soil loss equation (USLE) coefficients and T is the theoretical amount of soil loss that if exceeded will lead to losses in productivity.

[3]While yields in RAMS were estimated to reflect differences for different tillage practices, the estimation procedure used produced very similar yields for each of the four tillage practices.

values are used as the winter cover crop target. About 5 percent of the cropped area in the RAMS study region is planted to these sequences under the baseline.

IV. Policies Evaluated with the BLS: National Shifts in Land Use Patterns

Three alternative CRP scenarios and a targeted WRP scenario are analyzed using BLS to project their likely long-term economic impacts on the agriculture sector of the United States. The four scenarios are labeled CRP1, CRP2, CRP3, and WRP1. The three CRP scenarios each reflect different assumptions about the size of future CRP programs and about alternative uses of CRP land. Specifically, two alternative CRP proposals—a 16.2 million hectare CRP (CRP2) and a 20.2 million hectare CRP (CRP3)—are compared with a baseline scenario consisting of a 7.0 million hectare CRP (CRP1), considered a likely outcome after current CRP contracts expire. A 2.02 million hectare WRP targeted to bottomland capable of supporting hardwood tree growth is also analyzed and the results are compared with those obtained under the 7.08 million acre CRP baseline.

The CRP and WRP scenarios are implemented in BLS as additions (reductions) in hectares planted to the program crops and soybeans as land is retired from (enters) the reserve program being analyzed. In a particular simulation year, the model determines an initial estimate of hectares planted to a crop, then the addition or reduction from the reserve program is made. Other variables in the model then adjust to this final area planted and the hectares harvested of each crop are finally determined.

A. Baseline - CRP1

All CRP and WRP scenarios assume that the present CRP program reaches its goal of 16.2 million hectares by 1995. In the baseline scenario (CRP1) contracts begin to expire in 1996 according to historical sign-ups. Coverage would ultimately be maintained on 7.08 million hectares to include (a) 1.04 million tree hectares, (b) 3.36 million hectares of environmentally sensitive grassland, and (c) 2.63 million hectares of additional grassland. This means that by 2005, 7.08 million hectares will have entered or remained in the CRP, and will remain there through the end of the study period. The remaining 9.11 million hectares return to production according to patterns indicated by historical sign-ups.

B. Alternatives

1. CRP2 and CRP3

In CRP2, all contracts in the current 16.2-million hectare program are renewed indefinitely. It thus reflects current land use patterns. CRP3 is an expanded CRP of 20.2 million hectares. It is modeled as CRP2 with an additional 10 million acres removed from crop production over the 10- year period from 1996 to 2005 at the rate of 404,858 hectares per year. Land is removed from production of the program crops and soybeans according to the average proportion of area reductions in a particular crop attributed to CRP in the last five years of the current CRP (1991-1995) as projected in the *FAPRI 1992 U.S. Agricultural Outlook*. Roughly 25 percent of the new 4.04 million hectares is assumed to come from other crops.

Because no specific constraint exists on the total number of acres available to U.S. crop production in the model, there is no internal mechanism to determine the way in which CRP land will be used as CRP contracts expire -- in other words, what proportion of old CRP land will be planted to wheat, what proportion to corn, and so on. Similarly, the BLS cannot determine which crops will surrender land to a new CRP or WRP. It is thus necessary to supply estimated area changes to the model for each crop exogenously. For the CRP scenarios, this distribution is based on the USDA estimates of area reductions from historical sign-ups.[4] While there are slight differences in the distribution of cropland in each CRP, wheat land makes up the largest portion in each program, about 30 percent. Corn total and other feed grains (sorghum, oats, and barley) makes up about 18 percent. Cotton acreage is only 4 percent.

[4]These reductions are outlined in the FAPRI 1992 Outlook and based on estimates provided by USDA.

2. WRP1

WRP1 is a 2.02 million hectare reserve of wetlands consisting predominantly of drained bottomland currently planted to agricultural crops and is run in conjunction with CRP1. The potential crop-specific area of the reserve is estimated using a national database of hydric soils for the United States obtained from the 1982 National Resource Inventory (NRI) and SOILS5 (SCS, 1985; SCS/ISSL, 1989). Land in the database is ranked according to USDA estimated easement and restoration costs (Heimlich et al., 1989) and the least costly 2.02 million hectares are selected for the program.[5] About 40 percent of these acres are planted to soybeans and 28 percent to corn. The remainder comes mostly from wheat and other acreages. Not surprisingly, most of the land projected to enter the reserve is located in the Mississippi Delta and Southern states. Most of the remainder comes from the Midwest.

V. Empirical Results

Estimates produced by RAMS for the region-specific targets on conservation tillage and winter cover crops are aggregated and summarized at the RAMS study region which covers the major corn and sorghum producing areas in the United States. Crop areas and production practice outcomes from RAMS and BLS are passed to the CENTURY model used to predict soil organic carbon levels in yearly increments to the year 2030 by Aqua Terra Consultants (Donigian, et al., 1993). Presently, only results for alternative levels of CRP, and the targeted levels of conservation tillage (with no yield adjustments for conservation tillage) and winter cover crops are available. However, further study is needed before these conclusions can be confirmed, especially with respect to the conservation tillage and CRP scenarios. Soil organic carbon levels in 2030 are estimated assuming the crop production activities, estimated in RAMS for the baseline and each policy alternative, are practiced starting in the base year, 1990, and continuing every year through 2030.

A. Results from RAMS

1. Region-Specific Targeting of Production Practices under the Baseline

Under the baseline, net returns to crop production defined as gross revenue less variable costs on all land in the RAMS study region is $19.165 billion or $218.67 per hectare. A total of 87,644,000 hectares are in crops. The major crop is corn for grain grown on 27.81 million hectares or 32 percent of the total area in crops. Other principal crops include soybeans (17.89 million hectares), legume hay (12.59 million hectares), nonlegume hay (8.66 million hectares), winter wheat (5.22 million hectares) and spring wheat (4.17 million hectares).

Nearly 70 percent of the crops are grown with conventional tillage. Forty-six percent (40.81 million hectares) is plowed in the spring and 24 percent (20.48 million hectares) is plowed in the fall. Reduced tillage is practiced on 27 percent (24.01 million hectares) of the cropped area and no-till on nearly 3 percent (2.3 million hectares)[6]. The dominant rotation in the RAMS study region is a corn-soybean rotation (CRN SOY). Twenty-four percent of the area is devoted to this rotation. Six other rotations are practiced on 5 to 9 percent of the total area. They include continuous corn (CRN), continuous legume hay (HLH), continuous nonlegume hay (NLH), corn-corn-soybean (CRN CRN SOY), corn-soybean-winter wheat (CRN SOY

[5]Land only in states bordering or east of the Mississippi River was included because it was believed that very little bottomland in the western states had the suitable climate to support hardwood growth. All land planted in rice and all land in Florida were excluded from the database because of difficulty in determining the hydric nature of these soils. These exclusions had relatively little impact on the final make-up of the WRP.

[6]Because of conservation compliance provisions introduced in the 1985 farm bill, the area in reduced tillage and no-till systems has increased considerably subsequent to 1990 which is the base year for this analysis.

Table 1. Average net returns, baseline versus targeted levels of winter cover crops

Baseline	Cover crop	Change from baseline
(dollars per hectare)		(percent)
218.67	216.08	1.18

WWT) and summer fallow-spring wheat (SMF SWT). None of the cropland is treated with winter cover crops under the baseline.

Average fertilizer application rates for the macro nutrients are 61.52 kg per hectare of nitrogen, 38.68 kg per hectare of potassium, and 26.6 kg per hectare of phosphorous. In total, 5.40 billion kg of nitrogen, 3.40 billion kg of potassium, and 2.31 billion kg of phosphorous are applied. In all cases, the units are pure nutrient equivalents.

2. Region-Specific Targeting with Winter Cover Crops

Consistent with the target, a total of 4.601,291 hectares or approximately 5 percent of all crop acres are treated with winter cover crops. Of those, 2,467,174 hectares are planted to rye (*Secale cereale*) and 2,134,117 hectares to hairy vetch. When winter cover crops are forced to be grown, average net returns are reduced by 1.18 percent or $2.59 per hectare (Table 1). While establishing the cover crop in the fall costs money and yields of the crop planted the following spring are typically reduced, the costs are partially offset by nitrogen savings provided by hairy vetch. No other economic benefits are considered. Establishment costs for cover crops range from $34.58 to $54.34 per acre and costs associated with lost yields are estimated at $12.35 to $37.05 per hectare with 1991 commodity prices. Therefore, total costs per acre for establishing cover crops ranges from $46.93 to $91.39 per acre. If no other adjustments in crop mix, tillage, and rotations, for example, are made in response to the constraint, we would expect the average net returns per acre to decrease by about 5 percent of the establishment costs plus lost revenue from lower yields or by $2.35 to $4.57 per hectare. The value estimated falls at the lower end of this range. While some savings are realized in nitrogen costs on the acres planted to hairy vetch, the low-end estimate probably also reflects adaptive behavior on the part of producers; that is, choosing the least-cost cover crops, changing the crop mix, choosing alternative tillage and rotation practices, and so forth.

Changes in the crop mix ranged from a 8.81% increase in hectares of wheat grown to a 2.65% reduction in the area of soybeans grown (Table 2). In general, the areas of the crops targeted for cover crops either remained constant or experienced small increases. Increases are for corn silage (1.03%), oats (0.44%), sorghum silage (0.17%), and winter wheat. Reductions occurred in the area planted to the two major crops, corn (0.59%) and soybeans, and to nonlegume hay (0.17%).

Employment of conventional tillage with fall plowing decreases by 4 percent while conventional tillage with spring plowing increases by 2.2 percent (Table 3). This outcome is largely due to the inconsistency between fall plowing and establishing winter cover crops. The combination is simply prohibited in RAMS. Other adjustments included a 0.33 percent decrease in the use of reduced tillage. The major shift in crop rotations is a movement from planting a corn-soybean-winter wheat (CRN SOY WWT) rotation without cover crops in the baseline to one with winter cover crops (Table 4). Under the baseline 6 percent of total cropland is planted to the rotation. With the constraint on cover crops, 5.47 percent of the acres are planted to the same rotation but only 0.52 percent without cover crops, 1.27 percent with a rye winter cover and 3.68 percent with an hairy vetch winter cover. The other major rotations with cover crops are corn silage-soybean-3 years of legume hay (CSL SOY HLH HLH HLH), (1.17%) and sorghum-winter wheat (SRG WWT), (1.24%). The list of major rotations in the baseline, with the exception of corn-soybean-winter wheat, remains unchanged after the cover crop constraint is introduced.

Fertilizer application rates for all three macronutrients decline when winter cover crops are forced into the solution (Table 5). Nitrogen rates decline by 1.68 percent or 1.03 kg per hectare. In total, nearly 90.72 million fewer kg of nitrogen are applied. The principal cause is likely the nitrogen supplied by the hairy

Table 2. Hectares of crops grown, baseline versus targeted levels of winter cover crops

Crop	Baseline	Cover crops	Change from baseline
	(hectares)		(percent)
Barley	2113407	2113407	0.00
Corn Grain	27813294	27649836	-0.59
Corn Silage	1582254	1598567	1.03
Cotton	497162	497162	0.00
Legume Hay	12599064	12600846	0.01
Nonlegume Hay	8666514	8651934	-0.17
Oats	1949852	1958507	0.44
Sorghum Grain	1981928	2148602	8.41
Sorghum Silage	23442	23483	0.17
Soybeans	17895492	17420954	-2.65
Summer Fallow	2738553	2738553	0.00
Sunflower	424813	424813	0.00
Spring Wheat	4178668	4178669	0.00
Winter Wheat	5210447	5669514	8.81

Table 3. Tillage practices, baseline versus targeted levels of winter cover crops

	Baseline	Cover crops	Percentage change from baseline
	(hectares)		(percent)
Conventional tillage /fall plow	20518191	19698390	-4.00
Conventional tillage /spring plow	40838985	41737680	2.20
Reduced tillage	24020753	23941940	-0.33
No-till	2296739	2296743	0.00

vetch. Given about 2.4 percent of cropland is treated with hairy vetch with maximum[7] nitrogen savings of 61.2 kg per hectare in the middle region and 87.0 kg per hectare in the south, we might expect a savings of between 1.45 and 2.13 kg per hectare. The lower value estimated indicates that adjustments, such as the decrease in soybean acreage, may have offset some of the gains made by the hairy vetch. Lesser reduction in the amount of potassium (0.41%) and phosphorous (0.03%) are experienced. A reduction in soil erosion of 2.72 percent or 0.13 kg per hectare is also estimated (Table 6). Most of the savings is probably attributable to the winter cover crops but the shift from fall plowing to spring plowing under the conventional tillage systems might have contributed to the savings as well. Some of those savings are offset by the smaller area treated with reduced tillage systems.

With targeted levels of winter cover crops grown every year, soil organic carbon levels observed in 2030 are 34 percent higher than in 1990. The level observed is 7 percent higher in 2030 than if the baseline cropping systems are continued to 2030.

[7]Because all of the nitrogen fixed by the hairy vetch is not always demanded by the succeeding crop and because any excess is not carried over to the next year, the values represent maximum savings of nitrogen.

Table 4. Major rotations, percentage of total hectares, baseline versus targeted levels of winter cover crops

Crop rotation	Baseline	Cover crops
	(percent of total hectares)	
Continuous CRN	3.39	3.43
CRN CRN SOY	2.88	2.80
CORN SOY	9.67	9.43
CRN SOY WWT	2.43	0.21
SMF SWT	2.20	2.20
Continuous HLH	3.41	3.15
Continuous NLH	3.41	3.38
CRN SOY WWT (rye cover)	0.00	0.51
CSL SOY HLH HLH HLH (rye cover)	0.00	0.47
SRG WWT (rye cover)	0.00	0.50
CRN SOY WWT (vetch cover)	0.00	1.49

Key:
CRN=corn for grain
CSL=corn for silage
HLH=legume hay
NLH=nonlegume hay
SMF=summer fallow
SWT=spring wheat
SOY=soybeans
SRG=sorghum for grain
WWT=winter wheat

Table 5. Fertilizer applications rates, baseline versus targeted levels of winter cover crops

Fertilizer	Baseline	Cover crops	Change from baseline
	(kg per hectare)		(percent)
Nitrogen	61.46	60.43	-1.68
Potassium	38.72	38.56	-0.41
Phosphorous	26.63	26.62	-0.03

Table 6. Soil erosion rates, baseline versus target levels of winter cover crops

Baseline	Cover crops	Change from baseline
	(mt per hectare)	(percent)
4.09	3.98	-2.72

3. Region-Specific Targeting with Conservation Tillage

When no yield adjustment is assumed, average net returns per acre actually rise by \$1.65 per hectare (0.76%) for the low target (EI = 8) and \$10.03 per hectare (5.35%) for the high target (EI = 2) (Table 7). This outcome is a manifestation of the relative costs and yields in RAMS. While, as mentioned in the previous section, yields are nearly equal across tillages, the costs of no-till and reduced tillage systems are lower. Lower machinery costs more than offset the higher chemical costs in these systems. Consequently, they are very attractive alternatives in RAMS. In the baseline the dilemma these circumstances create is

Table 7. Net returns per acre, baseline versus targeted levels of conservation tillage (dollars/hectare)

		Tillage targets			
	Baseline	Low-w/ yield Adj	High-w/ yield Adj	Low-wo/ yield Adj	High-wo/ yield Adj
Dollars per hectare	218.67	220.32	230.35	211.21	211.43
Percentage change from Baseline		0.76%	5.35%	-3.41%	-3.31%

handled with flexibility constraints,[8] which force the model to calibrate to observed historical patterns of tillage. Modeling higher conservation tillage targets demands that these constraints be dropped in favor of even higher levels of no-till and reduced tillage. Consequently, returns increase relative to the baseline. If yields with the conservation tillage systems are adjusted downward then net returns per acre fall by $7.46 (3.41%) per hectare for the low target and $7.24 (3.31%) per hectare for the high target. Clearly the lower costs associated with no-till and reduced tillage systems are not sufficient to offset the lower yields.

The pattern of shifts in crop acreages is rather cloudy and only a few generalizations are apparent. Soybeans, sorghum grain, and barley are the only crops to show increases from the baseline for both low and high targets under both yield assumptions (Table 8). Furthermore, the areas for these three crops increase more with the higher target and when lower yields for no-till and reduced tillage are assumed. For soybeans, with no yield adjustments, areas increase by 112,550 (0.63%) and 211,740 (1.18%) hectares for the low and high target. With the yield adjustment, increases are 382,186 (2.14%) and 261,538 (1.46%) hectares. For barley the respective changes are 52,651 (2.5%), 52,631 (2.5%), 123,886 (5.87%) and 124,291 (5.88%) hectares. For sorghum grain the changes are 106,882 (5.39%), 240,080 (12.11%), 554,000 (11.32%) and 282,986 (14.25%) hectares. Nonlegume hay and legume hay are the only crops to experience consistent area losses. Just as with soybeans, sorghum grain, and barley, the changes are larger for the higher target and when lower yields for no-till and reduced tillage are assumed. The changes from the baseline in area planted to legume hay with no yield adjustment are 8,502 (0.07%) and 221,457 (1.76%) hectares for the low and high targets. With the yield adjustment, the areas decrease by 129,555 (1.03%) and 333,198 (2.65%) hectares for the low and high targets. The corresponding changes for nonlegume hay are 0.221 (2.56%), 1.25 (14.43%), 0.63 (7.32%), and 1.47 (16.98%) million hectares. The outcomes suggest that no-till and reduced tillage systems favor soybeans, sorghum for grain, and barley relative to other crops. It is not surprising that increases in these crops come at the expense of nonlegume and legume hay because the small seed of these crops benefit more from a well-prepared seed-bed. One other interesting shift is the increase in summer fallow with the higher target. When no yield adjustments are made, 0.48 (17.17%) million more hectares are summer fallowed and with yield adjustments, 0..36 (13.33%) million. Because no vegetation is grown during the fallow period, such an increase may have important implications for soil organic carbon levels.

The most striking outcomes are the percentage increases from the baseline in the area treated with no-till. The increase is as high as 905.37 percent or 20.76 million hectares for the high target with no yield adjustment (Table 9). Other increases for no-till in descending order are 9.23 (401.42%) million hectares for the high target with yield adjustments, 1.58 (69.13%) million hectares for the low target without yield adjustments, and 1.58 (47.91%) million hectares for the low target with yield adjustments. Increases for reduced till with no yield adjustments are 1.62 (6.77%) and 1.78 (7.46%) million hectares for the low and high targets. With yield adjustments the increases are 2.10 (8.8%) and 13.36 (55.64%) million hectares. Clearly no-till is favored over reduced till when a higher level of either is required. Only when a 10 percent yield reduction for no-till versus a 5 percent reduction for reduced tillage systems is assumed do the absolute aree increases for reduced tillage exceed those for no-till. In all but one case, the percentage decreases in area treated with conventional tillage with fall plowing exceed those for conventional tillage with spring plowing. However, the absolute area decreases for spring plowing are always larger. With no yield

[8]Flexibility constraints are simply upper bounds on the areas of no-till and reduced tillage. A separate constraint is included for each tillage system.

Table 8. Hectares of crops grown, baseline versus targeted levels of conservation tillage

		Tillage Targets			
	Baseline	Low-w/ yield Adj	High-w/ yield Adj	Low-wo/ yield Adj	High-wo/ yield Adj
		(hectares)			
Barley	2113407	2166155	2166248	2237471	5525140
		2.50%	2.50%	5.87%	5.88%
Corn grain	27813294	27847719	27876677	27719901	69360500
		0.12%	-0.23%	-0.34%	1.00%
Corn silage	1582254	1589082	1511930	1632393	3735510
		-4.43%	-4.44%	3.17%	-4.38%
Cotton	497162	497162	497162	497162	1227560
		0.00%	0.00%	0.00%	0.00%
Legume hay	12599064	12590600	12377246	12469221	30285000
		-1.07%	-1.76%	-1.03%	-2.65%
Nonlegume hay	8666514	8444655	7415834	8032446	17766200
		-2.56%	-14.43%	-7.32%	-16.98%
Oats	1949852	1941242	2029682	1954473	504880
		-0.44%	4.09%	0.24%	4.81%
Sorghum grain	1981928	2088848	2221899	2206363	5590920
		5.39%	12.11%	11.32%	14.25%
Sorghum silage	23442	23301	23301	23360	57207
		-0.60%	-0.60%	-0.35%	-1.16%
Soybeans	17895492	18008163	18107510	18277853	44832300
		0.63%	1.18%	2.14%	1.46%
Summer fallow	2738553	2738553	3937641	2739339	7663330
		0.00%	17.17%	0.03%	13.33%
Sunflower	424813	424813	424813	428259	1057430
		0.00%	0.00%	0.81%	0.81%
Spring wheat	4178669	4178669	417393	4121361	10189000
		0.00%	-0.11%	-1.37%	-1.25%
Winter wheat	5210447	5135886	5639909	5335187	14145100
		-1.43%	8.24%	2.39%	9.95

Note: Percentage changes from baseline shown below estimated values.

adjustments, the decreases for the low and high targets are 0.44 (2.18%) and 9.07 (44.24%) million hectares with fall plowing and 2.75 (6.78%) and 13.52 (33.07%) million with spring plowing. When yield adjustments are made the decreases for the low and high targets are 1.50 (7.23%) and 9.39 (45.73%) million hectares with fall plowing and 1.50 (4.24%) and 13.2 (32.33%) million hectares with spring plowing. The mix of crop rotations employed is relatively steady. The major rotations used in the baseline are also the major rotations for each of the policy runs (Table 10). With the exception of the corn-soybean (CRN SOY) rotation, no patterns are apparent and generalizations do not seem warranted. The corn-soybean (CRN SOY) rotation increases with the higher target and is higher when yield adjustments are made.

As with winter cover crop targets, fertilizer use and erosion are reduced when conservation tillage targets are included. Without yield adjustments, nitrogen rates decreases by 0.15 (0.22%) and 0.66 (1.07%) kg per hectare for the low and high targets (Table 11). When yield adjustments are considered, nitrogen rates go down by .77 (1.25%) and 0.21 (0.34%) for the low and high targets. Total savings of nitrogen are 12.74, 57.92, 67.77, and 18.64 million kg respectively. Similar decreases are observed for the other macro nutrients, never exceeding 1.4 percent for potassium or 0.6 percent for phosphorous. Most of the savings in nitrogen can be attributed to the larger areas planted to soybeans and to more corn being grown in a corn

Table 9. Tillage practices, baseline versus targeted levels of conservation tillage

	Tillage targets (hectares)				
	Baseline	Low-w/yield adj.	High-w/yield adj	Low-wo/yield adj	High-wo/yield adj
Conventional tillage/fall plow	20518191	20071436 -2.18%	11440157 -44.24%	19034636 -7.23%	11134787 -45.73%
Conventional tillage/spring plow	40838985	38071539 -6.78%	27332033 -33.07%	39108339 -4.24%	27637322 -32.33%
Reduced tillage	24020753	25647395 6.77%	25811906 7.46%	26134610 8.80%	37386441 55.64%
No-till	2296739	3884468 69.13%	23090792 905.37%	3397221 47.91%	11516256 401.42%

Note: Percentage changes from baseline shown below estimated values.

Table 10. Major rotations, percentage of total hectares, baseline versus targeted levels of conservation tillage

		Tillage targets			
	Baseline	Low-w/ yield Adj	High-w/ yield Adj	Low-wo/ yield Adj	High-wo/ yield Adj
	(percentage of total hectares)				
Continuous CRN	3.39	3.39	3.22	3.41	3.53
CRN CRN SOY	2.88	2.92	2.89	2.44	2.59
CRN SOY	9.67	9.69	9.81	10.14	10.05
CRN SOY WWT	2.43	2.43	2.43	2.51	2.55
Continuous HLH	3.41	3.41	3.09	3.38	3.42
Continuous NLH	3.41	3.40	2.82	3.22	2.69

Key: CRN=corn for grain; HLH=legume hay; NLH=nonlegume hay; SMF=summer fallow; SWT=spring wheat; SOY=soybeans; WWT=winter wheat

Table 11. Fertilizer application rates, baseline versus targeted levels of conservation tillage

		Tillage Targets			
	Baseline	Low-w/ yield adj	High-w/ yield adj	Low-wo/ yield adj	High-wo/ yield adj
		(pounds per hectare)			
Nitrogen	61.46	61.32	60.80	60.69	61.25
		-0.22%	-1.07	-1.25	-0.34
Potassium	38.72	38.71	38.30	38.56	38.21
		-0.03%	-1.08	-0.40	-1.32
Phosphorous	26.62	26.61	26.46	26.59	26.55
		-0.07%	-0.60	-0.16	-0.26

Note: Percentage changes from baseline shown below estimated values.

Table 12. Soil erosion rates, baseline versus targeted levels of conservation tillage

	Tillage targets			
Baseline	Low-w/ yield adj	High-w/ yield adj	Low-wo/ yield adj	High-wo/ yield adj
		(mt per hectare)		
12.28	11.90	8.06	11.98	9.48
	-3.08%	-34.34%	-2.58%	-22.85%

Note: Percentage changes from baseline shown below estimated values.

soybean rotation. Substantial decreases in erosion rates are estimated to occur. With no yield adjustments, erosion rates decrease by 313 (3.08%) and 3473 (34.34%) kg per hectare for the low and high targets (Table 12). With yield adjustments the corresponding reductions are 0.11 (2.58%) tons per acre and 2308 (22.85%) kg per hectare. Total soil savings in the region are 27.49, 304.36, 21.59, 23.8, and 202.30 million metric tons. For targets on conservation tillage, projected soil organic carbon levels, for the RAMS study region as a whole, are relatively unchanged or lower compared with the levels estimated for the baseline in the year 2030. Under the baseline, soil organic carbon levels are projected to increase nearly 32 percent from 1990 to 2030. With the low and high conservation tillage targets, soil carbon increases are 31 to 32 percent.

B. Results from BLS

1. National Shifts in Land Use Patterns under the Baseline

Baseline values for acres harvested of the program crops and soybeans are given at the top of Table 13 for the years 2010 and 2030. Note that over time, hectares harvested of all these crops except cotton are anticipated to increase. Total area harvested of these crops rises 8 percent from 107.69 million hectares in 2010 to 115.79 million hectares in 2030 due to underlying assumptions in the model about growth in the general economy and in crop production technology. The land use pattern in the baseline, on the other hand, remains about the same over time: corn accounts for about 32 percent of all hectares in both years; wheat for 29 percent; soybeans for 23 percent; and barley, sorghum, and oats together account for about 10 percent of the total area.

Table 13. Baseline and percentage changes in production indicators to CRP1, selected years

Year	Wheat	Rice	Corn	Other grains	Soybeans	Cotton
CRP1			(million hectares)			
2010	30.4	1.5	34.5	11.1	5.4	13.3
	33.0	1.8	36.9	12.1	26.6	5.3
CRP2						
Hectares harvested			(percent change)			
2010	-8.0	-2.0	-4.0	-15.0	-2.0	-6.8
2030	-7.0	-1.6	-5.0	-18.0	1.0	-7.3
Yields						
2010	0.2	1.4	0.3	-1.3	0.1	1.0
2030	0.3	1.2	0.3	-2.6	0.2	0.9
Producer prices						
2010	4.1	1.0	5.6	5.6	3.1	1.0
2030	5.5	2.6	4.7	4.7	4.4	0.0
CRP3						
Hectares harvested			(percent change)			
2010	-10.0	-2.3	-6.0	-20.0	-5.0	-8.9
2030	-9.0	-1.8	-7.0	-28.0	-1.0	-9.5
Yields						
2010	0.3	1.6	0.4	-1.2	0.2	1.0
2030	0.3	1.3	0.4	-4.6	0.3	0.9
Producer prices						
2030	5.3	1.2	7.6	7.6	6.2	1.4
2030	6.8	3.3	6.2	6.2	7.8	-0.3
WRP1						
Hectares harvested			(percent change)			
2010	0.0	0.0	-2.0	-1.0	-4.0	-0.7
2030	0.0	0.1	-2.0	-1.0	-3.0	-0.8
Yields						
2010	0.0	0.1	0.1	0.1	0.0	0.0
2030	0.0	0.0	0.0	-0.1	0.1	0.0
Producer prices						
2010	0.8	0.2	1.4	1.4	3.4	0.3
2030	1.3	0.8	1.0	1.0	3.6	0.1

Table 14. Estimated changes from CRP1 in net producer returns, selected years

	Crop production net returns	Program payments	CRP payments	Total crop net returns	Livestock net returns	Agriculture net returns
			(million 1989 dollars)			
CRP2						
2010	1300	-2103	1035	232	-576	-345
2030	1950	-2498	1035	488	-838	-350
CRP3						
2010	1793	-2739	1519	573	-939	-366
2030	2660	-3254	1519	925	-1384	-459
WRP1						
2010	417	-373	0	44	-717	-673
2030	542	-393	0	149	-856	-708

2. National Shifts under the CRP Scenarios

Table 14 gives the percentage changes from the baseline in crop areas, yields, and crop producer prices for the years 2010 and 2030 for all scenarios. As expected, crop areas in CRP2 and CRP3 are lower throughout the study period than under CRP1. In percentage terms, hectares harvested of barley, sorghum, and oats are reduced the most, especially barley hectares. This is due mostly to the high percentage of feed-grain hectares idled by the present CRP program and the assumption that the new programs will generally reflect historic sign-ups. Of the major crops, soybean area is affected least by the increases in CRP coverage, primarily because a relatively small proportion of present soybean hectares are idled in CRP. Total area does not fall by as much as the size of the CRPs. Under CRP2, total area is only 6.76 million hectares lower than under CRP1 in 2010, even though the move from CRP1 to CRP2 ultimately removes 6.92 million hectares from production. The difference is due primarily to farmers bringing previously unplanted land into production in response to the higher crop prices induced by the CRP area reductions over time. For CRP3, the gross reduction in area of these crops due to CRP is 9.96 million hectares, but projected acreage planted actually falls by only 9.2 million hectares.

The reductions in area are slightly offset by increases in yields on harvested hectares of the major crops, except barley. The direction and magnitudes of these changes depend upon the responsiveness of farmers to changes in price expectations, the impact of changes in feed demands, and the rate of technological change (represented by time trends in BLS) for each crop. The net result is not at all surprising: production of all major crops is less in the larger CRP scenarios throughout the study period than in the 7.08 million hectare CRP1. The relative differences between scenarios (in percentage terms) are seen most dramatically in feed grains. The smaller supply of these crops leads to higher producer prices for all major crops.

Beef production increases in both scenarios relative to CRP1 while production of pork, milk, poultry, and eggs falls. The increases in beef production are slightly higher under CRP3, but in both scenarios, the increases are on the order of only 1 to 2 percent. The increases reflect improved demand for beef and veal relative to CRP1. Pork production falls by 2.5 to 4 percent under CRP2 and by 4.2 to 5.8 percent under CRP2. The drops in poultry and egg production are relatively small in both scenarios, and milk production is barely changed compared with milk production under CRP1. All livestock producer prices increase in both scenarios but more so under CRP3 than under CRP2.

Consumption of wheat and coarse grain products falls under both scenarios by 1 to 2 percent, reflecting retail price increases of 2 to 5.5 percent in CRP2 and 3.6 to 7.4 percent in CRP3. Consumption of beef changes very little in either scenario even though retail prices for beef and veal increase slightly. In both scenarios, per capita consumption of other animal products (primarily pork, poultry, and fish) decreases by 1 to 2 percent. Consumption of dairy products, cotton, and tobacco are relatively unchanged between scenarios.

Projected changes in net production returns and government costs are given in Table 14 in constant dollar terms. Changes in net returns to production are estimated outside of the BLS, but are based upon recent (1989) data and the trends suggested by the BLS runs. In general crop producers gain, livestock producers lose, and net government costs fall relative to CRP1. The results indicate that in the long run, however, the industry as a whole does not gain from the larger programs relative to the 7.08 million hectare program.

Higher grain prices in CRP2 and CRP3 boost net returns from crop production, excluding government payments, and lower net returns to livestock production relative to CRP1. Higher grain prices also lower government transfer payments for the crop programs (price supports, deficiency payments, etc.), more than offsetting the estimated increases in CRP payments. Thus, total government payments actually fall relative to CRP1. Crop producers benefit substantially from the larger CRP scenarios. The net gains to crop producers through transfer payments and production returns do not, however, fully compensate for declines in net returns to livestock production. The net result for the agriculture sector is a decline in industry net returns of $350 to $460 million per year. By comparison, the average U.S. net farm income from 1989 to 1991 was about $47 billion (ERS, 1992). So these annual losses would probably amount to less than 1 percent of net farm income.

Estimated annual consumer expenditures are greater under the larger scenarios (Table 14). For CRP2, the estimated increase in 2030 is up to $6.3 billion (in 1989 terms) and up to $8.8 billion in 2030 for CRP3. To put these numbers in perspective, 1989 personal consumption expenditures are about $367 billion; so the increases in estimated consumer expenditures are likely to be less than 2 percent of total expenditures.

For the soil organic carbon analysis of the CRP scenarios, 1986 is considered the base year. Two alternatives are examined. The first alternative corresponds with the BLS baseline (CRP1) in that 44 percent of the CRP land is returned to production as the original contracts expire and 56 percent remains idled. The second alternative, corresponding with CRP2, assumes that all of the land enrolled in the CRP is re-enrolled upon expiration of the original contract. A third case that does not correspond with any of the BLS runs but provides a benchmark of comparison assumes that CRP never existed and that all of the land remained in production. Compared with this benchmark, soil organic carbon levels are 12 percent higher under the assumptions corresponding with the BLS baseline and 4 percent higher for CRP2. The higher value for the BLS baseline is a result of incorporating surface organic matter into the soil when the CRP land is converted back to cropping.[9]

3. National Shifts under the WRP Scenario

The largest impacts on crop production due to the WRP are for soybeans and corn. Indeed, the results show that area of these crops falls about 4 percent for soybeans and 2 percent for corn. Other feed-grain area also falls 1 percent. Cotton area falls by less than 1 percent while the other crops are relatively unaffected. Yields for all crops change very little, if at all. Soybean producer prices increase by about 3 percent and corn prices by about 1 percent. Other feed-grain prices can be expected to follow the changes in corn price, but wheat and rice prices rise by only small margins. Production of pork and poultry fall by 1 to 2 percent, while production of other livestock products is relatively unaffected. Consequently, producer prices for pork and poultry increase by about 1 to 2 percent while other livestock prices increase only slightly.

With respect to net farm income, much the same story holds for the WRP scenario as for the CRP scenarios: crop producers gain, government payments fall, and livestock producers absorb most of the costs to agriculture due to higher feed costs. The estimated cost of obtaining easements on WRP land is $1.4 billion, but this represents one-time payments based on the discounted present value of expected net returns to that land if it remained in production. There is thus no change in annual reserve program payments relative to CRP1, which is assumed to run in conjunction with the WRP. Because there are no additional reserve program payments, the annual projected costs to agriculture of WRP are much higher than under CRP2 and CRP3, about $700 million versus $350-$400, even though the reserve is smaller. The increase in annual consumer expenditures for WRP1 is given in Table 13. The largest increase relative to CRP1 is $2.3 billion in 2030.

Soil carbon results for the WRP scenario are not available at the time of publication.

VI. Conclusions

This paper presents an economic evaluation of several policies designed to promote the build-up of organic carbon in agricultural soils. It is part of a broader, integrated effort to evaluate both economic

[9]CENTURY evaluates soil organic values in the top 20 cm of the soil profile and does not account for residue stored on the soil surface.

and environmental outcomes. Two economic models are employed to evaluate two types of policies. RAMS is used to evaluate region-specific targeting of conservation tillage and winter cover crops while BLS is used to evaluate changes in land use patterns caused by the CRP and WRP. Both RAMS and BLS are linked to CENTURY, a soil organic formation model, which estimates soil organic carbon levels over time under alternative assumptions about agricultural practices and land use.

For the targets on conservation tillage and winter cover crops, the results indicate that the economic cost to producers is not overwhelming. Measured in terms of changes in net returns to crop production, decreases are never more than 3.5 percent. The changes are sufficiently small that relatively minor changes in the assumptions about yields can produce increases in net returns. But, if increases in soil organic carbon are the only environmental indicators considered, only the targeted winter cover crop alternative is successful. In general, forcing higher levels of conservation tillage does not improve soil organic carbon levels in the top 20 centimeters of the soil profile. More benefits may be observed if the organic matter stored at the surface of the soil is added to that in the top 20 centimeters of soil. Other environmental benefits, however, are evident. For the targeted level of winter cover crops, nitrogen application rates are reduced by about 1.5 percent. Smaller but positive reductions are also observed for potassium and phosphorous. In addition, decreases in erosion per acre of nearly 3 percent occur. The results for the targeted levels of conservation tillage are more striking. Reductions of up to 34 percent in erosion rates are estimated. Results for fertilizer applications are similar to those for the targets on winter cover.

Under the alternative CRP and WRP scenarios, producers and consumers are worse off than under the baseline scenario. Therefore, 12 and 4 percent increases in soil organic carbon stored, under the CRP1 and CRP2 alternatives relative to the no CRP case, must be weighed against the economic costs. In assessing the impacts of the CRP and WRP policies, it is important to remember that the assumptions about the make up of each alternative reserve program—the amount of land idled from each crop—play an important role in determining the relative impacts within agriculture, for instance whether corn producers benefit more than wheat producers, or whether pork production is affected more heavily than beef production. Our assumptions about the distribution of cropland within the scenarios should reasonably approximate the most likely outcomes. The overall effects on agriculture, however, should not depend so much on the particular assumptions made within each scenario, and the trends reflected in the results should be fairly robust.

Economic analyses from two divergent models are presented in this paper. Each provides useful information about policy impacts. But individually neither can provide results with a broad range of temporal and geographic characteristics. Linking RAMS and BLS is one strategy that would provide opportunities to obtain a more comprehensive list of economic indicators. While both RAMS and BLS are linked to CENTURY, linking the two economic models together is more difficult. Several issues and obstacles must be resolved. Both models have several common parameters (e.g., commodity prices and crop acreages) that are either estimated or used as input. Decisions must be made about which parameters to pass and at what level of aggregation. Furthermore, inconsistencies regarding the fundamental assumptions underlying each model must be resolved.

References

Abkin, M. 1985. The Intermediate United States Food and Agriculture Model of the IIASA/FAP Basic Linked System: Summary Documentation and User's Guide. WP-80-30. Laxenberg, Austria: International Institute for Applied Systems Analysis.

Burkhead, C.E., J.W. Kirkbride, and L.A. Losleben. 1968. *Usual Planting and Harvesting Dates.* USDA Handbook no. 283. Washington, D.C.: U.S. Government Printing Office.

Cruse, Richard. 1992. Personal Communication. Department of Agronomy, Iowa State University.

Donigian, A.S., A.S. Patwardhan, R.B. Jackson, T.O. Barnwell, K.B. Weinrich, and A.L. Rowell. 1993. Modeling the Impacts of Agricultural Management Practices on Soil Carbon in the Central United States. Unpublished Manuscript. Mountain View, California: Aqua Terra Consultants.

Fisher, G.K., K. Frohberg, M.E. Keyzer, and K. Parikh. 1988. *Linked National Models: A Tool for International Food Policy Analysis*. Dordrecht, The Netherlands: Kluwer Academic Publishers.

Heimlich, R.E., M.B. Carey, and R.J. Brazee. 1989. Beyond Swampbuster: A Permanent Wetlands Reserve. *Journal of Soil and Water Conservation* 44:445-50.

Houghton, J.T., G.J. Jenkins, and J.J. Ephraums. 1990. Climate Change, The IPCC Scientific Assessment. Cambridge: Cambridge University Press.

Johnson, Stanley R. and FAPRI staff. 1992. *FAPRI 1992 U.S. Agricultural Outlook*. Staff Report #1-92. Ames: Food and Agricultural Policy Research Institute, Iowa State University.

National Agricultural Statistics Service (NASS). 1990. Agricultural Prices: Annual Price Summary. NASS, Washington, D.C.: U.S. Department of Agriculture,

Post, W.M., T.H. Peng, W.R. Emaniel, A.W. King, V.H. Dale, and D.L. DeAngelis. 1990. The Global Carbon Cycle. *American Science* 78:256-258.

U.S. Department of Agriculture, Soil Conservation Service (SCS). 1991. Soil Conservation Service Field Office Technical Guides. Washington, D.C.: U.S. Department of Agriculture.

U.S. Department of Agriculture. 1985. *Hydric Soils of the United States, 1985*. Washington, D.C.: U.S. Department of Agriculture.

U.S. Department of Agriculture, Soil Conservation Service (SCS) and Iowa State Statistical Laboratory (ISSL). 1989. Basic Statistics: 1982 National Resource Inventory. SB-756. Washington, D.C.: U.S. Department of Agriculture.

U.S. Department of Agriculture, Economic Research Service (ERS). 1992. *Agricultural Outlook*. Washington, D.C: U.S. Government Printing Office.

U.S. Water Resources Council (USWRC). 1970. Water Resources: Regions and Sub-regions for the National Assessment of Water and Related Land Resources. Washington, D.C.: U.S. Government Printing Office.

Wischmeier, W.H., and D.D. Smith. 1965. *Predicting Rainfall Erosion Losses From Cropland East of the Rocky Mountains—Guide for Selection of Practices for Soil and Water Conservation*. USDA Handbook no. 282. Washington, D.C.: U.S. Government Printing Office.

CHAPTER 27

Carbon Sequestration Through Tree Planting on Agricultural Lands

Bruce A. McCarl and J. Mac Callaway

I. Introduction

Policies to plant trees are a frequently discussed way of sequestering carbon and mitigating the environmental effects of greenhouse gas emissions (National Academy of Sciences, 1991; Center for Strategic and International Studies, 1991). A number of appraisers have argued that planting trees to sequester carbon is an inexpensive alternative with broad based benefits (Moulton and Richards 1990; Sedjo and Solomon 1983; Dudek and LeBlanc 1990). However, Adams et al. (1993) argued that large scale tree planting programs were more costly than the previous studies suggested. They also found major implications for the forestry sector showing the potential expansion in agricultural based timber harvest could swamp forest product markets.

This paper presents results and discussion based on an ongoing EPA sponsored project examining tree planting programs and potential policy in terms of: a) the sensitivity of cost and timber supply impacts to differences in carbon and timber yield assumptions; b) the effects of carbon sequestration losses at the time of harvesting and during the life cycle of wood products; c) the interaction between agricultural programs and tree planting policies; and d) the amount of net carbon sequestered due to changes in planting and harvesting decisions on commercial timberlands.

This project involves two distinct phases, the first involving appraisals largely within the context of the agricultural sector with a broad based examination of forestry sector implications and the second with detailed examinations of implications for both sectors. Empirical work will be reported here from the first study phase. Later we will discuss the conceptual formulation of the model for the second phase of the study.

II. Long Run Agricultural and Forestry Implications of Tree Planting

The first study phase concentrated on the agricultural sector using a long run agricultural sector model - ASM (Chang and McCarl, 1990; Chang et al., 1992) - integrated with forestry data from the Timber Assessment and Market study (Adams and Haynes, 1980). The ASM base model was set up for 1990 conditions.

The ASM represents production and consumption of 24 primary agricultural commodities, including both crop and livestock products in 63 U.S. regions. Processing of agricultural products into 36 secondary commodities is also included. A forestry component was added to ASM during the Adams et al. (1993) study by introducing tree planting possibilities on agricultural land and then entering forest product market characteristics adapted from Adams and Haynes (1980). The net effect is that the resultant integration can simulate the long run agricultural and forestry sector consequences of displacement of agricultural land use

ISBN 1-56670-117-1/95/$0.00+.50

by trees. Water resources are disaggregated in the model into surface and ground water available in each of the 63 regions. Surface water is available for a constant price up to a prespecified quantity, but pumped ground water is provided according to a supply schedule in which the unit price increases with increasing rates of withdrawal. The model assumes a large number of individuals make up both production and consumption sectors, each operating under competitive market conditions, and thus maximizes the area under the demand curves less the area under the supply curves. This area is a measure of economic welfare or net social benefit. Both domestic and foreign consumption (exports) are considered. This model structure allows projection of the effects of carbon sequestration on: 1) the regional agricultural and timber economies across the United States; 2) irrigated versus dryland cropping tradeoffs in response to regional water demand and availability; 3) producer welfare at the regional and national level, as well as consumption effects for both domestic and foreign consumers; and 4) supply of logs from the traditional forest sector and agricultural sector sequestration activity.

The resultant model is used herein to generate information on the subsidy costs required to meet various carbon suggestion targets. The carbon targets used range from 0 to 508 million megagrams (Mg). The cost estimates include: a) the cost of establishing trees, which is computed exogenously; and b) the endogenous change in land opportunity costs as agricultural commodities are displaced by trees. The specific analysis involved varying:

1. Carbon and timber yields based on differing authors estimates of these yields;
2. The assumptions that trees would be harvested or left standing; and
3. The incidence of the Farm Program.

All analyses at this stage are done under 1990 conditions with and without the 1990 farm program. The discussion herein is a summary of that in Callaway and McCarl (1992).

A. Analysis Setup

The literature contains differing estimates regarding both timber and carbon yields as a result of agriculturally based tree planting programs. Three sets of timber and carbon yields were used:

1. The yields used in Adams et al. (1993), based on Moulton and Richards (1990), but modified to reflect emerging judgements about timber yields under non-plantation conditions.
2. A revised set of yields, developed by Richards (1992), correcting previous errors in Moulton and Richards (1990), carbon yields.
3. The yields reported by Birdsey (1991), based on timber yields used in the 1989 RPA assessment (USDA Forest Service, 1990).

Most studies of carbon sequestration assume that trees, once planted, will not be harvested. This is plausible only if a small amount of land is afforested so that harvest is irrelevant or the program assures the land will be a permanent timber reserve. If a permanent reserve is not assured, then the cost of carrying the trees will lead to harvest. When trees are harvested, carbon sequestration will be affected for three reasons:

- carbon in the soil, limbs, roots, understory, etc. will dissipate soon after harvest;
- carbon in wood products will be released as those products undergo processing, use and disposal;[1] and
- carbon sequestration on commercial timberlands may be reduced due to reduced timber holdings in response to market forces.

In the empirical work, we consider losses of 17-22% depending on region, where the carbon in the roots and the above ground woody debris is lost based on Birdsey's (1991) estimates.

B. Analysis Results

The results discussion is organized by harvesting assumption where both no harvesting and harvesting optional cases were run. The first results shown are those from the No Farm Program Case.

[1] Phase 1 of the study does not deal with wood products post-harvest losses.

Table 1. Changes due to carbon sequestration - no harvesting, no harvest losses, no farm programs

	Total carbon sequestered (million Mg carbon/yr)					
	39	77	154	308	463	617
Adams et al. yields						
Cost/Mg[a]	13.52	14.23	14.87	19.48	26.92	38.97
Hectares/trees planted[b]	6.03	11.5	22.9	44.7	66.1	89.1
Net social benefit[c]	-419	-860	-1786	-3894	-6849	-10905
Carbon prod. cost[d]	429	903	1889	4947	10257	19799
Revised Moulton and Richards yields						
Cost/Mg carbon	17.40	17.59	18.61	35.50	53.66	100.71
Hectares/trees planted	5.9	11.9	24.3	74.9	100.0	125.2
Net social benefit	-515	-1067	-2219	-4919	-8686	-14262
Carbon prod. cost	552	1117	2363	6620	13522	27254
Birdsey Yields						
Cost/Mg carbon	18.03	22.43	27.36	56.56	121.72	727.12
Hectares/trees planted	10.3	19.1	37.6	75.8	115.3	157.3
Net social benefit	-442	-1086	-2601	-7595	-17902	-312053
Carbon prod. cost	572	1424	3475	14365	46372	369319

[a] Dollars; [b] million hectares;[c] million dollars; [d] in dollars.

1. No Harvest No Farm Program

The no harvest cases adopt the assumption that trees planted on agricultural land grow indefinitely following the assessment assumption in Moulton and Richards, 1990; and Parks and Hardie, 1992. Table 1 contains results for the various carbon scenarios[2]. The data given in the table are: 1) cost per megagram (Mg) - the marginal cost of producing the last megagram of carbon associated with a specific carbon target; 2) hectares of trees planted - the amount of land, in million hectares, that it takes to achieve this sequestration; 3) net social benefits - the change in net social benefit measured as the difference between the base case agricultural producers' and consumers' surplus without any tree planting and the same quantity associated with a specific carbon target (a negative indicates value cost to society, but note that the benefits of reduced atmospheric carbon are not factored in); and 4) carbon production cost - a measure of the marginal cost of having the trees planted, which equals the marginal cost of carbon times the carbon times the carbon target level. An example of the nature of the results can be found by examining the marginal cost of sequestering the 39th Mg/yr of carbon. In particular using the yields from Adams et al. (1993) the sequestration of this volume would cost society $13.52/Mg/yr, would require 6.03 million hectares, costing 429 million in subsidization and reducing welfare by 419 million without counting the benefits of less atmospheric carbon.

The marginal cost estimates include both the annualized establishment cost and the opportunity costs of land due to the displacement of crops by trees. At low average levels of carbon sequestration, the marginal cost is dominated by establishment costs; however, as the carbon target increases more and more land is drawn out of agricultural production, and the opportunity cost component dominates.

[2] Note a range of sequestration targets is used here in identifying the scenarios as the target level of sequestration has not been determined and is an ongoing subject of debate.

The different yield estimates do not have a very pronounced effect on the marginal cost estimates at the 31.8 or 63.5 million Mg/yr carbon targets. Furthermore, at these low levels the opportunity costs found herein are close to those obtained by Moulton and Richards (1990) and others although at higher targets the results here are consistent with Adams et al. (1993) is higher costs.[3] However, starting at the 127 million Mg/yr carbon target, the yield assumptions begin to have a significant effect on the results. In particular, the marginal costs and land area requirements estimated using Birdsey's yields begin to increase rapidly becoming about twice the next highest estimate for the 254 million Mg/yr carbon three times at 380.9 Mg and by 507.9 Mg/yr the Birdsey based estimate ($726.57/Mg) is 13 times higher.[4] The marginal cost curve for carbon associated with the revised Moulton and Richards (Richards 1992) yields also lies above the curve based on Adams et al. (1993). However, the differences are not nearly so great.

Table 1 also presents estimates for the changes in net social benefits and carbon production costs. For the 31.8 million Mg/yr level, the welfare cost estimates range from $515 million/yr to $442 million/yr, while the corresponding range of carbon production cost estimates is from $429 million/yr to $572 million/yr. To achieve the 63.5 million Mg target would cost society about twice as much in terms of lost welfare and carbon production costs. After that point, the costs estimated using Birdsey's yields are again substantially higher than those estimated with the remaining sets of yields. Three final points stand out in the results. First, between zero and the 254 Mg/yr target, a doubling in the target results, roughly, in a doubling of both the opportunity and carbon production costs. Beyond that point, costs begin to increase more rapidly largely due to the opportunity costs. Second, as the carbon target increases, the carbon production costs rise more rapidly than do the welfare costs. Third, the welfare cost estimates give the minimum level of benefits which the rest of society must gain from sequestering this quantity of carbon.

2. Harvesting Allowed Cases - No Farm Program

Here, we assume that farmers can harvest trees. Thus the planting decision is based on both stumpage price and carbon subsidy considerations. We also introduce carbon harvest losses. There are three confounding effects introduced by such a setup. The first is that by introducing carbon losses, more land is needed to sequester an equivalent volume driving up the cost. Second, the tree establishment and agricultural land opportunity cost are partially offset by revenues from harvesting trees. Third, the more trees harvested the lower the harvesting revenues since the timber prices fall, thus the harvesting revenue offset falls as with increases in tree planting, and sequestering targets.

Estimates of the changes caused by the harvest scenario are presented in Table 2. At the lowest first few target levels, all timber is harvested and the sequestration costs are reduced relative to the no harvest case. For example, using the revised Moulton and Richards yields, the carbon production cost estimates are $339 million/yr and $966 million/yr for the lowest and highest scenarios under no harvest case, the corresponding cost estimates are $552 million/yr and $1,117 million/yr. However, at the higher carbon target levels, the carbon production costs in Table 2 become much closer to the no harvest case.

The effects on the forestry sector are significant (Table 3). Under existing conditions, combined annual harvests of sawtimber and pulpwood equal 177 million cubic meter/yr. The results show estimated agricultural harvest levels can increase to as much as 78.5% of this total. The dramatic increase in harvest is accompanied by sharp decreases in stumpage prices. These stumpage price decreases range from about 20 to 25% for wood products for the 31.8 million Mg/yr carbon target but fall as much as 48% across the scenarios. Increasing carbon targets lead to a U-shaped response in timber harvested and timber prices. This occurs since dropping timber prices, and increasing agricultural opportunity costs increase the carbon price which makes the harvest time carbon losses unattractive and this reduces the proportion harvested.

[3] The range of estimates for 31.8 million Mg just about spans the estimates developed by Moulton and Richards (1990) - $13.52/Mg, by Richards (1992) - $16.77/Mg, and Parks and Hardie (1992) - $18.22/Mg, for the same case.

[4] The fact that it costs so much more and takes more land to sequester carbon using Birdsey's yields is due primarily to the lower yields associated with Birdsey's work. On average, Birdsey's carbon yields are about one-half the magnitude of those in Moulton and Richards (1990).

Table 2. Effects of carbon sequestration - with harvest possible with harvest losses, no farm program

	Total carbon sequestered (million Mg carbon/yr)					
	39	77	154	308	463	617
Adams et al yields						
Cost/Mg[a]	6.99	10.75	14.83	19.43	27.07	40.04
Hectares/trees planted[b]	6.8	13.8	26.0	47.0	69.6	90.6
Net social benefit[c]	-129	-413	-1296	-3386	-6426	-10620
Carbon prod. cost[d]	222	683	1883	4933	10313	20338
Revised Moulton and Richards yields						
Cost/Mg	10.67	15.20	18.58	25.71	36.88	55.42
Hectares of trees	7.01	14.12	27.60	51.51	77.19	101.72
Net social benefit	-194	-629	-1749	-4459	-8293	-14070
Carbon prod. cost	339	966	2359	6529	14047	28151
Birdsey Yields						
Cost/Mg	12.77	22.44	27.99	56.28	121.72	--[e]
Hectares of Trees	12.3	22.7	40.5	77.2	115.2	--
Net Social Benefit	-230	-798	-2317	-7473	-17872	--
Carbon Prod. Cost	405	1425	3541	14293	46369	--

[a] Dollars; [b] million hectares; [c] million dollars; [d] in dollars;[e] this solution is infeasible.

One important factor not incorporated into the carbon cost calculation is the potential reduction in carbon grown on existing timberland. Given the reductions in forest product prices and commercial harvest, existing timberland owners would have substantial incentives to undertake actions like: harvest existing stands earlier than usual; reduce the level of management intensity in existing and newly regenerated stands; and move lands out of forestry. The second phase of this project embodies an effort to develop a dynamic forest and agricultural sector model that will account for such factors.

C. Effects of the Farm Program and Farm Program/Sequestration Policy Tradeoffs

The model was also run under existing farm programs and under farm program reductions. Table 4 contains the results of the current farm program run for the revised Moulton and Richards and Birdsey yield. The entries largely have the same conceptual basis as in Table 1. The new government farm payments row, is the amount of money that the Federal Government pays for farm programs. The sixth row entry, total payments, is the sum of the carbon production cost and the farm program payments. It represents the financial cost the Government would have to pay to farm bill and carbon sequestering payments. The results show that the sequestration consequences are virtually identical to the non-farm program supply curves shown above and do not merit much more discussion.

However the results do reveal that there are policy tradeoffs between tree planting and farm programs which merit examination.

1. Trading off Farm Program Elimination with Tree Planting

Comparisons of the with and without farm program runs in Tables 1 and 4 indicates that current farm programs cost tax payers (and benefit farmers by) about $8.2 billion/yr. Current farm programs cost society

Table 3. Timber effects (million m^3) - optional harvests, high harvest losses, no farm program

	Total carbon sequestration (million Mg carbon/yr)						
	0	39	77	154	308	463	617
Adams et al. Yields							
Commercial harvest[a]	177	156	139	128	132	129	144
Agricultural harvest[a]	0	55	97	131	119	128	82
Total harvest[a]	177	211	236	259	251	257	226
Sawtimber price[b]	18,204	14,977	10,395	7,737	8,918	8,355	11,806
Pulp price[b]	6,024	4,628	4,588	4,588	4,588	4,588	4,588
Moulton and Richards yields							
Commercial harvest	177	151	135	126	131	137	144
Agricultural harvest	0	67	109	139	120	101	83
Total harvest	177	218	244	265	251	238	227
Sawtimber price	18,204	13,726	9,612	7,455	9,076	10,675	12,137
Pulp price	6,024	4,588	4,588	4,588	4,588	4,588	4,588
Birdsey yields							
Commercial harvest	177	159	140	144	152	177	177
Agricultural harvest	0	48	93	83	63	0	0
Total harvest	177	207	233	227	215	177	177
Sawtimber price	18,204	15,348	10,662	11,831	14,077	18,202	18,204
Pulp price	6,024	4,712	4,588	4,588	4,628	6,024	6,024

[a] Million cubic meters; [b] $ per 1000 cubic meter.

about $1.9 billion/yr in terms of lost welfare. Thus, if current farm programs were abolished and no programs were put in their place, government budget exposure would be reduced by $8.2 billion/yr, and societal welfare would increase by $1.9 billion/yr.

Now, consider a policy that would eliminate farm programs and plant trees until the welfare costs of planting trees equaled the welfare costs of farm programs. Such a policy could afford to sustain a $1.9 billion welfare loss. To find what size sequestration program this would be, one can refer to Table 1, and look for the size program (by yield source) which satisfies this criterion. In all yield cases, the tree planting policy sequesters between 63.5 and 127 million Mg. For example, in the case of the revised Moulton and Richards yields, a 63.5 million Mg/yr program costs society approximately $1.1 billion/yr, while a 127 million Mg/yr program costs society about $2.3 billion/yr. The corresponding land area requirements and marginal cost per megagram estimates can also be interpolated from the table. The interpolated estimates for the program levels and carbon production costs are:

Yield case	Carbon target (million Mg/yr)	Carbon prod. cost ($million/yr)
Revised Moulton and Richards	110	2018
Birdsey	98	2526

Table 4. Changes due to carbon sequestration - no harvesting, no harvest losses under farm program

	Total carbon sequestered (million Mg carbon/yr.)			
	39	77	154	308
Revised Moulton and Richards yields				
Cost of Carbon/Mg	17.38	17.38	19.75	25.96
Hectares	6.2	13.0	26.1	49.5
Change in net social benefits[a]	-497	-1036	-2086	-3835
Carbon prod. cost	552	1103	2508	6595
Gov. farm pymts.	8234	8222	8108	7067
Total pymts.	8786	9325	10616	13662
Birdsey Yields				
Cost of carbon/Mg	19.39	21.73	29.81	63.53
Hectares	10.4	20.0	38.1	75.7
Change in net social benefits[a]	-451	-1102	-2400	-4106
Carbon prod. cost	616	1380	3785	14453
Gov. farm pymts	8171	8141	7891	4502
Total pymts.	8787	9521	11676	18955

[a] Welfare change is measured in relation to estimated welfare under current farm programs. Note the units are as defined in Table 2.

These results mean that farm programs could be replaced with tree planting programs that could sequester from 98 to 109.8 million Mg of carbon/yr at the same welfare cost to society as current farm programs[5].

On the other hand, if one targets for equal government cost, one finds that a program in the range of 127-254 million Mg is consistent with a carbon production cost equal to the farm program cost ($8.2 billion), depending on the yield assumption. However, societal welfare is lower with such a policy than under the current farm bill by an amount equal to the net social benefit for the carbon sequestration program less the $1.9 billion welfare cost of current farm programs. Notice that these welfare costs can be high. For example, under the revised Moulton and Richards yields, the application of this rule leads to a program slightly bigger than 254 million Mg, but the additional social cost would exceed $3.0 billion/yr (i.e., $4.919 billion/yr versus $1.9 billion/yr. The interpolated estimates for the program levels and carbon production costs are.

Yield case	Carbon target (million Mg/yr)	Welfare cost ($million/yr)
Revised Moulton and Richards	283	3881
Birdsey	182	2867

2. Trading off Farm Program Reduction with Tree Planting

A second policy approach involves fixing the size of the carbon sequestration target one wants to achieve and then finding the percentage reduction in the size of the current farm program that will make society as a whole or government cost no worse off. The calculations for welfare compensation indicate the following for the two sets of yields used in this part of the analysis:

[5] Note this would have quite different distributional consequences. Since the incidence of tree payments and farm program payments would be quite different.

1. **Revised Moulton and Richards Yields**. For the 31.8 million Mg/yr program, a reduction of about 5% in current farm program target prices would hold welfare constant at current farm program levels. The combined program cost for this option lies between $4.1 and $2.4 billion/yr. For a 63.5 million Mg program it would take more than a 10% reduction in current farm program while a 127 million Mg program would require about a 15% reduction in target prices. Programs larger than 127 cannot be matched even by farm program elimination.
2. **Birdsey Yields.** For a 31.8 million Mg program, it would take a reduction in farm program target prices of greater than 5% to hold welfare constant at farm program levels. For a 70 MTC program a reduction of 20% or more is required. Program levels greater than 70 cannot be matched by farm program reductions.

If, on the other hand, policy could be revised to hold the sum of tree subsidy and farm program payments constant, there is a tradeoff between farm program payments and carbon sequestrations cost. By uniformly reducing program target prices, the amount of carbon that can be sequestered at a given cost level increases. At the same time, the farm program cost falls, so does the welfare cost to society in terms of the change in net benefits. For example, using the revised Moulton and Richards yields, going from a 2% to a 5% reduction in program target prices, almost doubles the amount of carbon that can be sequestered (208.3 million Mg vs. 120.4 million Mg) at the same level of cost. The size of government payments to farmers under existing programs falls from about $6 billion/yr to $3 billion yr, while payments for carbon increase by an offsetting amount to hold the total fiscal cost of the two sets of programs constant at $8.2 billion/yr. In the process, welfare losses just about double, increasing from about $1.5 to 3 billion/yr.

D. Summary and Conclusion

The empirical analysis shows six major things. The costs of sequestration continue to be higher particularly at the levels above 63.5 million Mg than the estimates developed before Adams et al. (1993). Second, the uncertainty in yields between the Birdsey (1991) and Richards (1992) estimates is a significant factor in the program cost estimation. Third, the costs appear to be yet higher than the Adams et al. (1993) costs under Richards (1992) revised estimates and there is potential for them to rise yet more if one subscribes to Birdsey's (1991) estimates. Fourth, the inclusion of harvest losses makes these costs even higher. Fifth, the harvest scenario has major implications for the forest sector. Sixth, the farm program and tree subsidy programs are two different ways of transferring money to the agricultural sector and can be used in a substitute fashion to achieve a targeted levels of funds transfer or welfare.

III. Toward a More Detailed Forest Sector Appraisal

The purpose of the second, ongoing project phase is to incorporate tree growth dynamics and to project the response of private timberland owners. Timberland owners, in the face of a large agricultural planting program, could reduce the size of their inventory holdings harvesting more timber sooner and replanting fewer harvested stands or change managements regimes. Furthermore, Haynes, Alig and Moore (1992) found dynamic considerations dampened the changes in stumpage prices.

Consideration of the above items requires adoption of a dynamic framework which depicts both the forest and agricultural sectors. This led to the conceptualization and ongoing development of the Forest and Agricultural Sector Optimization Model (FASOM -- Callaway et al., 1993). FASOM is a dynamic, open, nonlinear programming model of the United States forest and agricultural sectors. It is being developed for EPA to evaluate the welfare and market impacts of alternative policies for sequestering carbon in trees. The model is also being designed to help aid in the appraisal of a wider range of forest and agricultural sector policies. The principal characteristics of FASOM are it:

1. includes a detailed forest sector representation. The forest sector is depicted as open to trade, producing sawlogs, pulpwood, and fuelwood from both hardwood and softwood sources. Forest lands are dissaggregated regionally by species, age cohort of trees, cover type, site condition, management regime, land suitability, and private non-industrial or industrial ownership. Public forest harvest policy is included as an exogenous influence. In addition an inventory projection component is used to describe the current sector as well as project future tree growth, wood yields, carbon

sequestration quantities, and management actions. This is based on the Forest Service's Timber Assessment Database model and Birdsey's (1991) carbon sequestration data.

2. includes a detailed agricultural sector. The model simulates the production of 36 primary crop and livestock commodities and 39 secondary, or processed, commodities with competition for land, labor, and irrigation water in an open economy. The model also includes farm program policies. The agricultural model follows the existing ASM structure (Chang and McCarl, 1990) in all aspects except that the regions are aggregated.
3. is a regionally-based model. Eleven regions are included: Pacific Northwest-West, Pacific Northwest-East, Pacific Southwest, Rocky Mountains, Northern Plains, Southern Plains, Lake States, Corn Belt, South Central, Northeast, and Southeast.
4. covers a 120 period. The model incorporates the 120 year period with explicit accounting on a decade by decade basis plus terminal conditions for in process timber stands.
5. is based on a dynamic, price-endogenous, spatial equilibrium market structure. Prices for agricultural and forest sector commodities are endogenously determined in domestic and foreign markets given product demand and factor supply functions. FASOM is dynamic in that it solves jointly for the multi-market equilibrium in each product and factor market included in the model, over time. The agricultural sector and the forest product sectors are unified as drawing from a common land base. The structure of the two sectors will be based on those modeled in the existing TAMM (Adams and Haynes, 1980) and ASM (Chang and McCarl, 1990) models.
6. includes welfare accounting. Changes in total welfare, and the distribution of welfare is computed. Welfare is accounted for agricultural producers, timberland owners, consumers of agricultural products and purchasers of stumpage.
7. incorporates expectations of future prices. Farmers and timberland owners are able to foresee the consequences of current behavior (when trees are planted) on future stumpage and agricultural product prices. That information is used in determining sectoral land allocation.
8. accounts for potential changes in land use. Land can shift between sectors over time, based on future prices. For example, tree planting policy can cause upward pressure on agricultural prices and downward pressure on stumpage prices, thus there is the potential to convert forest land into agricultural land.
9. includes detailed carbon accounting. Accounting is present for carbon accumulation in existing forest stands, as well as in regenerated and afforested stands. Post harvest losses are also included covering losses in non-merchantable carbon pools associated with harvested stands and carbon decay over time in wood products.
10. is easily modifiable. The model system will be developed in the GAMS modeling system (Brooke, Kendrick, and Meeraus, 1988) which allows easy model modification for policy evaluation.

A. FASOM Solution Information

The FASOM solution information depicts a multi-period simulation of economic activity. Timber planting and harvest decision variable solutions will tell where and when existing stands are harvested. In turn the decision to regenerate, idle or move land between sectors will occur. As such the FASOM solution will provide national and regional information by decade on:

- Consumer and producer welfare
- Agricultural production and prices
- Forestry harvest and replanting by species
- Forest product and prices
- Land and forest asset values
- Land utilization by sector and intersectoral movement
- Carbon sequestration amounts and costs.

B. Anticipated Policy Applications

The initial motivation behind FASOM was to develop a model to evaluate alternative carbon sequestration policies in an economic framework including projections of the reaction of consumers and producers.

Subsequently, it became clear that FASOM could also be used to evaluate the consequences of a wide range of forest and agricultural policies, not just those intended to promote carbon sequestration. FASOM potentially could be used to evaluate a wide ranging set of policies such as:

- Agricultural lands based tree planting programs
- Substitution of tree planting subsidies for farm program payments
- Significant changes in forest and agricultural product trade policies
- Reforestation and forest management programs
- Changing harvest levels on national forest land
- Increases in paper recycling and wood processing technology
- Climate change induced alterations in forest and agricultural product yields
- Changes in erosion limits on agricultural lands.

References

Adams, D., R. Adams, J. Callaway, Ching-Cheng Chang, and B. McCarl. 1993. Sequestering carbon on agricultural land: a preliminary analysis of social costs and impacts on timber markets. *Contemporary Policy Issues* 11(1):76-87.

Adams, D. and R. Haynes. 1980. The 1980 timber assessment market model. *Forest Science*. 26(3). Monograph 22.

Birdsey, R. 1991. Prospective changes in forest carbon storage from increasing forest area and timber growth. USDA Forest Service Technical Publication. Washington, D.C. In press.

Brooke, A., D. Kendrick, and A. Meeraus. 1988. *GAMS: a user's guide*. The Scientific Press, Redwood City, California.

Chang, Ching-Cheng and B. McCarl. 1990. *Scope of ASM: the U.S. agricultural sector model*. Texas A&M University, College Station, Texas. Draft Report.

Chang, Ching-Cheng, B. McCarl, J. Mjelde, and J. Richardson. 1992. Sectoral implications of farm program modifications. *American Journal of Agricultural Economics* 74(1):38-49.

Callaway, J.M. and B.A. McCarl. 1992. Social costs and economic impacts of tree planting on agricultural land. Report prepared for Catherine Benham, Energy Policy Branch, United States Environmental Protection Agency.

Callaway, Mac, B. McCarl, R. Alig, and D. Adams. 1993. Forest and agricultural sector optimization model: conceptual model description. Final Report to United States Environmental Protection Agency, Climate Change Division. February 1993. RCG/Hagler, Bailly, Inc. PO Drawer O, Boulder CO.

Center for Strategic and International Studies. 1991. *Economic effects of alternative climate change policies*.

Dudek, D. and A. LeBlanc. 1990. Offsetting new CO_2 emissions: a rational first greenhouse policy step. *Contemporary Policy Issues* 8(3):29-42.

Haynes, R., R. Alig, and E. Moore. 1992. Alternative simulations of forestry scenarios involving carbon sequestration options: investigation of impacts on regional and national timber markets. USDA Forest Service, PNW Station. Draft report.

Moulton, R. and K. Richards. 1990. *Costs of sequestering carbon through tree planting and forest management in the United States*. USDA Forest Service, GTR WO-58. Washington, D.C.

National Academy of Sciences. 1991. *Implications of greenhouse warming*, Washington, D.C.

Parks, P.J. and I.W. Hardie. 1992. Least cost forest carbon reserves: cost-effective subsidies to convert marginal agricultural lands to forest. Draft paper. Center for Resource and Environmental Policy Research, Duke University, Chapel Hill, NC.

Richards, K.. 1992. Derivation of carbon yield figures for forestry sequestration analysis. Draft paper Prepared for Office of Economic Analysis, U.S. Department of Energy. January 8, 1992.

Sedjo, R.A., and A.M. Solomon. 1983. Climate and forests. in N.J. Rosenberg, W.E. Easterling III, P.R. Corsson, J. Darmstadter, (eds.), *Greenhouse warming: abatement and adaptation*, Resources for the Future, Washington, D.C.

USDA Forest Service. 1990. *An analysis of the timber situation in the United States, 1989-2040*. General Technical Report, RM-199. USDA Forest Service, Rocky Mt. Forest and Range Experiment Station, Ft. Collins, CO.

CHAPTER 28

The Role of Forest Management in Affecting Soil Carbon: Policy Considerations

R. Neil Sampson

I. Introduction

As interest grows in policy options to reduce greenhouse gas emissions and increase terrestrial C sequestration through management of forest ecosystems, there has been increasing interest in the role of forests and forest policy. Much of this interest has centered on the amount of C that could be stored in the relatively stable woody parts of trees, and the growth and yield estimates developed by foresters over the years have been extended to provide estimates of the total C sequestered by tree and forest growth (Birdsey, 1992).

The data for forest soils, however, and how those soils react under various forest management regimes throughout the forest growth, death, and renewal cycles, are more limited. In spite of the fact that forest soils may store as much or more C than the total stored by trees on many forest sites, Birdsey (1992) utilized fairly broad assumptions to simulate soil C dynamics in connection with his forest C tables. Major studies have evaluated the biomass contained in forest ecosystems, but restricted their consideration to the above-ground C in the system (Cost et al., 1990)

Given the importance of soil C in determining the effectiveness of various forest ecosystems and their management in addressing greenhouse gas concerns, the lack of data on forest soil C dynamics is a serious problem. Additional policy efforts to encourage actions known to enhance soil C are warranted, as are improved methods of gathering and reporting soil organic matter (SOM) data for lands where forest projects are being contemplated.

In evaluating the impact of forest management activities and policies on SOM, it is assumed that, in soils where other factors stay generally comparable, SOM:

- increases with cooler soil conditions;
- increases with moister soil conditions;
- increases with improved soil fertility; and,
- decreases with additional disturbance and aeration (Buckman, et al., 1969).

II. Expansion of Forest Areas

Since forest cover is an excellent way to provide shade and protection to topsoils, as well as a source of root dieback and litterfall to build SOM, expanding forests onto lands that are biologically suited, but not now forested, is one way to increase SOM. This is being proposed in a variety of policies aimed largely at reducing greenhouse gas emissions.

ISBN 1-56670-117-1/95/$0.00+.50

A. Carbon Offset Forests

The Framework Convention on Climate Change signed during the United Nations Conference on Environment and Development (UNCED) commits developed countries to adopt policies that can help limit emissions of greenhouse gases or enhance sinks and reservoirs. The "joint implementation" provision allows industrial countries to work towards their own objectives by reducing emissions in other countries. This could result in major tree planting programs financed by developed countries, but executed elsewhere.

In the U.S., the Energy Policy Act of 1992 (P.L. 102-486) establishes a program for the voluntary reporting of information on greenhouse gas emissions and reductions, with the latter including forest management practices and tree planting. Thus, while there is no requirement as yet, U.S. policy is also encouraging industry to look to forests as one way of mitigating greenhouse gas emissions.

The Applied Energy Services Corporation (AES) of Arlington, VA, launched the first carbon offset project in 1988 with a tree planting and care project in Guatemala designed to offset the carbon emissions of a 180 MW coal-fired plant in Connecticut. The project seeks to increase biomass yields through soil conservation techniques, conservation of forest biomass through fire prevention, and educational programs. It will involve nearly 40,000 farm families and the planting of approximately 52 million trees over a ten-year period (Trexler et al, 1989). If successful, such a program would have a significant impact on SOM in the project area.

In January 1993, the U.S. Environmental Protection Agency (EPA), U.S. Agency for International Development (AID) and U.S.D.A. Forest Service launched a cooperative effort entitled "Forests for the Future." The program, which awaits funding from Congress, will result in the negotiation of carbon offset projects in a number of countries including Mexico, Russia, Guatemala, Indonesia, and Papua New Guinea (U.S. EPA, 1993). A similar program is being discussed in Germany at this time (Hasenkamp, 1993) and in The Netherlands, the FACE (Forests Absorbing Carbon Emissions) Foundation, set up by the Dutch Electricity Generating Board, is planning to reforest 1,000 hectares per year in the tropics to offset CO_2 emissions (Kinsman et al., 1993). Carbon offset forestry seems to hold great promise as one method to manage atmospheric CO_2, due mainly to its cost-effectiveness, associated environmental benefits, and direct impact on local peoples where tree planting projects can build employment and economic activity.

With the policy push it is already getting from the U.S. and other countries, there is little doubt that many more projects will emerge. In many cases, however, predicted results are highly uncertain. Forest growth models reflect past climate and CO_2 conditions, not what might occur in the future, so projects with a 50-100 year proposed life span face significant uncertainty in terms of environmental conditions. On most of the proposed sites, so little is known about SOM conditions prior to tree planting that it is difficult to accurately analyze what these projects will actually achieve in terms of increasing C sinks.

B. Afforestation of Understocked and Non-Commercial Forests

While it is common practice throughout most of the United States to reforest lands that are harvested for timber, there are still major gaps to be filled. In the South, particularly on the small private ownerships, adequate reforestation (either through planting or natural regeneration) has been estimated at only 50% (Mixon, 1988). Policies to encourage adequate reforestation following harvest, such as Virginia's Seed Tree Law, would help close this gap.

There are about 86 x 10^6 ha in the United States classified as "other" forest land, and these lands lack the data-gathering that has been common on the commercial timberlands. Aside from state-by-state estimates of their extent, there are few data upon which to make an assessment of their potential. These are defined as forest lands that are not expected to produce economic wood crops (>1.4 m^3 of wood per ha per year) or which have been set aside for some purpose which precludes their use as timberland (Waddell, et al., 1989). About half of them are in Alaska, where there are limited opportunities for added C sequestration. Many of the remaining lands, however, are well adapted to growing trees. A privately-funded program entitled "Heritage Forests," administered by AMERICAN FORESTS, a national conservation group, provides financial support to public agencies to cost-share the restoration of forest ecosystems on these lands. The program, entering its third year, is growing rapidly, with projects ranging from restoration of longleaf pine ecosystems in Florida to riparian area reforestation in Arizona with willows, chokecherries, and cottonwoods (Tikkala et. al., 1992).

The limit to restoring forest ecosystems on many of these lands is mainly a matter of budget constraints. When viewed in the context of C sequestration, instead of in the context of producing merchantable timber, these lands offer significant opportunities. As with most afforestation efforts, other environmental benefits such as reduced soil erosion, improved watershed and groundwater conditions, and wildlife habitat enhancement all spring from improving forest cover and restoring a more natural species balance. Whether SOM will be greatly enhanced may depend on the status of the soil today. Soils under brush fields may be at or near normal SOM levels, even though the standing biomass on the site is far below what it might be under trees.

C. Conversion of Marginal Crop and Pasture Lands

There is a national pool of 47 x 10^6 ha of privately-owned crop and pastureland which rates as marginal, using the criteria of the Department of Agriculture, and which is biologically capable of supporting tree growth (Parks et al., 1992) This land pool, three times as large as the original target of the Conservation Reserve Program contained in the 1985 Farm Bill, presents a major opportunity for expanding the area of forest land in the United States.

The current use of these marginal lands is roughly divided between cropland and pasture. They are also roughly divided between land where softwoods or hardwoods will be most well adapted.

On some of these lands, conversion to timber production under current economic conditions would be profitable, as well as possible. Of the 25.3 x 10^6 ha adapted primarily to softwoods, Parks (1992) estimates that 9.5 x 10^6 ha would be economically more profitable in trees than in their current marginal crop or pasture use. On the 21.6 x 10^6 ha best suited to hardwoods, the lack of forest economic data prevents such analysis, but it is logical to assume that part of those lands would be economically competitive in forest, as well.

Experience with the CRP demonstrated, however, that farmers are reluctant to plant trees for a variety of reasons, including the fact that trees limit their flexibility in using the land in future years (Esseks et al., 1992). USDA field agency personnel advising farmers did not always explain the opportunities for tree planting and commercial harvest after the expiration of the 10-year CRP contract (Esseks et al., 1992). Thus, while it may be feasible, and even profitable, to convert selected marginal lands to trees, it will not come easily, or without cost and program effort on the part of federal and state governments.

D. Estimating Carbon Gain and Storage

In all of the types of afforestation listed in the previous sections, policymakers need to have reasonably accurate estimates of the amount of C storage likely to ensue as a result of program investments. Several analysts have given the subject of C sequestration through tree planting projects a fairly intensive review (Trexler et al., 1992; Trexler et al., 1993; Winjum et al., 1993; Sampson, 1992; Sampson, 1993). It seems readily apparent, however, that the weakest data and analysis in these efforts is associated with the potential to change SOM. Improved data on current SOM status in proposed projects, as well as data on rates of stable SOM buildup under a growing forest, are badly needed.

Carbon is stored in newly-planted forests at different rates over time, depending on the growth characteristics of the species involved, the quality of the site, the type of soil and climate regime present, and the condition of the site prior to the planting. These wide variations in factors make generalizations both difficult and dangerous. Table 1 shows several examples of the application of Birdsey's (1992) calculations to potential tree planting projects.

As is readily apparent in Table 1, a sound analysis of an afforestation project for carbon sequestration needs to establish the initial carbon condition of the sites involved, as well as the rate of SOM change and the maximum SOM accumulation likely to result during the life of the project. Considering only the change in standing biomass from the trees will, in many cases, result in a significant omission. If the goal is maximum net C sequestration per unit of cost, it is not the total carbon on the site at the end of the growth period that is important in terms of project feasibility, it is the *net* increase in stored carbon. In the case of the spruce/fir reforestation project on cutover forest, SOM is estimated to decline for 10-15 years following timber harvest, before renewed growth begins to rebuild it. In this example, SOM is estimated to return to original levels at age 55. If this is the case, assuming that costs for planting and other management were roughly equal, this site would be a poor choice for a 50-year carbon offset tree planting project because its net gain is only 1/3 to 1/4 of the other options.

Table 1. Net Gains in Soil and Above-Ground C Predicted for Afforestation Projects.

Location/Species/Site		Initial C content	C content at age 55	Net C increase
		(t C ha^{-1})		
Southern pine plantation on cropland	Trees	0	157	157
	Soil	25	74	49
	Total	25	231	206
Lake States pine plantation on cropland	Trees	0	208	208
	Soil	54	119	65
	Total	54	327	273
Northeast spruce/fir plantation on cropland	Trees	0	70	70
	Soil	61	146	85
	Total	61	216	155
Northeast spruce/fir planting on cutover forest	Trees	9	72	63
	Soil	161	161	0
	Total	170	233	63

(Calculated by the author based on tables developed by Birdsey, 1993.)

Northern sites contain most of the forest system's stored carbon in the soil, while southern sites store a larger percentage in the tree biomass. While fast tree growth has led to many proposals for tree planting in the warmer southern climates, consideration of SOM changes under a tree planting project may lead to different conclusions. For the examples in Table 1, an investment in the Lake States plantation would result in a C payback almost 30% larger than an equal investment in the southern pine example, if the costs per tree are the same.

Location may also make an enormous difference in the long-term stability of the carbon store, with the soil carbon being less dynamic and liable to losses than the above-ground biomass. A full consideration of both tree growth and SOM may lead to the conclusion that some of the best investments in new carbon storage will be found on northern temperate soils that have been cropped, with the organic matter somewhat degraded. These will provide larger "net" carbon gains than cutover lands where the soil carbon is still fairly well intact. Obviously, on crop or pasture sites where soil organic matter is so degraded that tree growth is stunted and/or commercial fertilizers are required, many of those gains would be lost.

III. Management of Existing Forests

The management of existing forests is clearly tied to the ability of the forest to build up and retain SOM. With over 195 x 10^6 ha of timberland in the United States, half of which comes under some form of public or private management, even a modest change in management methods could add up to a significant impact (WRI, 1992). There are substantial biological and economic opportunities to increase timber growth on existing timberlands. Many of those opportunities are associated with tree planting to improve the stocking rates on lands that are currently understocked (Haynes, 1993). Such investments, in addition to adding to C storage in the timber itself, would also result in increasing SOM levels, it would appear.

A. Extending Rotation Length

Forests often live for lengthy periods when there is little or no major disturbance to affect growth or SOM, then go through a period of intense disturbances, due either to natural events such as windstorms or fires, or human intervention, usually in the form of timber harvests. Most of the opportunities to use forest policy as a means of increasing and maintaining SOM are found in efforts to alter the frequency or nature of those major disturbance events.

While foresters have worked out growth models that estimate a forest's "economic maturity," those models reflect the nature of the economic climate and the market for wood products, as well as the growing rate and health of the forest itself (Hyde, 1980). Thus, a calculus that includes the value of keeping major C stores intact may lead to a different policy or management decision about when and how to harvest a particular stand (Harmon et al., 1990).

Lengthening rotation period reduces the number of major disturbances the forest will experience, lessens soil alterations, reduces the opportunity for soil erosion, and helps maintain SOM stores. Policies about forest rotation length are most likely to affect the public multiple-use timberlands, as public pressure on issues such as protection of biodiversity, retention of remaining old growth forest stands, and retention of amenity, recreational, and watershed values causes foresters to think beyond the basic wood-production calculus to the more complex art of ecosystem management (Sample, 1991).

In addition to its impact on SOM, lengthened rotations generally result in larger trees being harvested, which changes the product mix that is likely to result from the timber crop (Row and Phelps, 1992). With older, larger trees, more of the wood will go into longer-lived products, adding to the value of the forest management system in lengthening the time when C is terrestrially-bound.

B. Managing Fire Intensity

Fire is a normal part of most forest and range ecosystems. In most of these systems, fires have historically burned on a fairly regular cycle, recycling the carbon and nutrients stored in the ecosystem and strongly affecting the species that inhabited each place (Pyne, 1982). On most grasslands, Southern pinelands, and Western ponderosa pine forests in the United States, regular ground fires (less than 25-year frequency) kept flammable fuels reduced, and favored the development of large, fire-resistant species such as pine and larch, while limiting the less fire-resistant species such as juniper and fir (Kilgore, 1987). A similar fire regime in the Eastern deciduous forest favored oak and hickory over maple and other shade-tolerant species (Cronin, 1983).

The arrival of European settlers, settled agriculture, and professional forest management significantly changed that regime. The native American practice of regularly igniting fires as part of hunting or grazing management strategies was eliminated. In many regions, heavy livestock grazing removed the carpet of fine fuels that helped carry ground fires. Farming and irrigation broke up the grass fuels that had previously covered mountain valleys and carried wildfires from one forested mountain range to the next (Pyne, 1982). A landscape formed under wide-ranging and fairly frequent wildfires was dramatically altered.

In the early years, settlers and loggers undertook massive deforestation as land was cleared for agriculture and logged to provide fuel and building materials for a rapidly growing population. In the sixty years between 1850 and 1910, almost 190 million acres of forest were cleared for agricultural use (MacCleery, 1993). This clearing, along with the wasteful lumbering practices of the 19th Century, was accompanied by annual wildfire levels that were estimated at 40-50 million acres per year in the early 1900's (MacCleery, 1993). Included in these annual conflagrations were millions of acres of "cutover lands" where the residual brush, secondary forest, and soil humus were repeatedly burned, either through wildfires, through land clearing, or as part of the purposeful burning of the "piney woods" that southern settlers adopted from the native Americans (Fedkiw, 1989).

The enormity of the damage to forests, plus the loss of life and property due to catastrophic burns, led to a major upsurge in public support for improving forest management and fire protection. The American Forestry Association, started in 1875 as the first national citizens conservation organization, crusaded at the federal level for improved forest protection. By 1910, when massive fires charred northern Idaho and western Montana, public sentiment was so aroused that the federal government began to move aggressively to create the programs needed to protect forests from wildfires. The Weeks Law of 1911 and the Clarke-McNary Act of 1924 created the basis for a federal-state cooperative program which, when added to the efforts of the private timber protective associations created by the forest products industry, began to provide an effective level of protection (USDA Forest Service, 1952).

And those efforts bore fruit which was hailed widely as significant progress in resource management. On millions of acres of cutover lands, forest regeneration was able to survive as re-burning was stopped. New forests begin to emerge throughout the Lake States and the South. Losses dropped dramatically. In western Washington, for example, modern wildfires are estimated to consume only about 1/10 of the biomass each year as did their prehistoric predecessors (Fahnestock and Agee, 1983).

But there were unintended and unforseen impacts from the fire control success. Obviously, the most effective fire suppression was achieved in reducing the incidence and spread of low and moderate intensity fires, so those kinds of fires became increasingly uncommon (Arno and Brown, 1991). In fire-dependant forest ecosystems that depended on those burns for affecting both forest composition and SOM, biomass that would normally have been burned every few years has continued to build up, either in the form of older plants or an increasing amount of dead woody material decomposing slowly on the ground. Nutrient and C recycling was altered, as were soil nutrient, pH, and SOM levels.

The result is that millions of acres of forest and range now support the result of several decades of fire-less plant growth and ecosystem development. In south-central Oregon alone, there are some 100,000 hectares of lodgepole pine stands that are decadent and seriously susceptible to attack from pine beetle and catastrophic wildfire (Little and Shainsky, 1992). Under these conditions, a combination of dry weather conditions and ignition will almost certainly trigger a high-intensity fire which will, in spite of control efforts, become impossible to contain and run its course until changing fuel or weather conditions can slow its progress.

Dead wood on or near the ground has built up to unnaturally high levels, leading to hotter ground fires and more combustion of fine roots and soil organisms in the topsoil (Borchers et al., 1990). In a survey of the literature on the relationship between fire and SOM, Johnson (1992) finds a significant soil carbon loss from high-intensity wildfires, and concludes that carbon loss increases with fire severity. Fire affects the soil biological community directly by immediately killing or injuring organisms, and indirectly by longer-term influences over such processes as plant succession, soil organic matter transformations, and microclimate (Borchers et al., 1990). Clearly, soil organisms undamaged by a fire and dependent on above ground organisms are affected by the severity of the fire above ground.

In addition to losses of productivity, many of these sites suffer from significant losses of biodiversity, important soil, carbon, and nutrient cycle shifts, and major shifts in hydrologic regimes (McNabb et al., 1990). Instead of periodic light-intensity burns, which have been shown to increase SOM levels through improved nutrient recycling, these lands are now subject to high-intensity burns that will deplete SOM even further (Johnson, 1992).

An effective fire management policy that successfully converts fire-dependent forest ecosystems back to a more frequent, less-intense fire regime would lead to increased SOM levels in most of those forest soils, it appears. Such a policy must now begin with major investments in fuel removal, thinning of live trees, and careful prescribed burning to get the forest landscape back into a more complex, fire-tolerant condition, so that normal ignitions will not develop into disastrous fires so easily. The most difficult areas are those that are inter-laced with homes, towns, farms and roads, where edge forests need to be cleaned of excess fuels on a regular basis.

A major policy question arises over the amount of smoke and air pollution such a fire program would generate. The need to meet smoke management guidelines is a constraint to the use of prescribed burning as a management tool today, particularly in areas where smoke tends to hang in populated valleys, and any attempt to significantly increase the use of fire will be impossible unless the public is willing to tolerate the accompanying smoke.

In terms of SOM, additional research is needed to improve the understanding of SOM dynamics in the post-fire years. Does SOM decline rapidly after a stand-replacement fire, as the increased soil temperature caused by canopy removal and dark surfaces increases decomposition rates? Or is the rapid decomposition of roots from fire-killed trees, aided by the fertilization from mineral salts in the ash, responsible for some increase in the stable SOM fractions, as well as a significant flux of CO_2? What is the role of charcoal formation within soil profiles, and how does it affect long-term soil C levels? Answers to questions such as these would greatly assist in the development of predictive models to illustrate the temporal aspects of C dynamics in relationship to the management of wildfires.

C. Changing Harvest Methods

Timber harvest is, with the exception of intense wildfire and catastrophic weather events such as hurricanes or massive blowdowns, the most disruptive event in the forest life cycle. Attention to the effects of forest harvest methods on SOM (as well as total C balances) will lead to methods that:

- Leave some canopy cover to shade the soil and keep soil temperatures reduced;
- Leave foliage and small branches on-site to minimize nutrient export;
- Burn slash carefully, and leave large woody debris intact as a C reservoir for the ecosystem; and,
- Minimize soil disturbance and movement, either through mechanical activities or erosion, to prevent export of SOM or increased decomposition due to aeration.

The practice of clear-cut harvesting has come under fire for decades in the United States, with negative public reaction to its appearance and effect on the forest common to all forest regions. Foresters trying to defend clearcutting as a necessary and useful silvicultural practice have, for the most part, been unsuccessful in convincing the public on those points. After years of trying to gain public acceptance, the Forest Service has, in 1992, declared a new policy that will seek to minimize the use of clearcutting as a harvest method wherever other methods are available (Robertson, 1992).

This should be a positive policy change in terms of SOM and the effects of forest harvest upon it. Particularly in its most extreme forms, where forests were not only clearcut, but the slash, stumps, and debris were piled with a brush rake-equipped tractor and burned, the destruction of C stores, both in the debris and the soil, was maximized. Although little evidence of site deterioration has been found, in part due to the fact that most such sites have only been through one or two rotations, it seems inconceivable that interrupting the normal C cycle in such an aggressive fashion would go without impact.

New forms of timber harvest, whether based on the "New Forestry" being promoted in the Pacific Northwest, or the all-aged stand management being practiced in some of the pine and mixed-hardwood systems, seem designed to do far less disruption to C stores during the harvest itself. Most of these methods leave more shade to keep soils cool, and more woody debris to keep ecosystem structure intact (Franklin, 1989). Critics contend that these methods raise costs, and that they may result in "high grading" the forest by removing good trees and leaving the subsequent forest full of mediocre specimens if the selection of harvest trees is not done very carefully (Gibney, et al., 1990). In spite of the critics, public opposition to clearcutting seems certain to prevail and, with the added pressure to keep forests as a positive sink for atmospheric CO_2, the need to find other, less-C-destructive methods seems overwhelming.

IV. Forest-Derived Substitutes for Fossil Fuels

Fossil fuel burning is recognized as the major forcing factor in the buildup of greenhouse gasses (Chandler, 1990). Thus, any means of substituting woody biomass for fossil fuels results in a net reduction of fossil fuel burning and, consequently, a net reduction in the amount of net CO_2 added to the atmosphere. There are many policy issues involved in producing woody biomass as an energy feedstock, and converting it into useful energy (Wright et al., 1993). Many of those, such as conversion technologies, are beyond the scope of this paper. Many of the production technologies, however, have significant impact on the question of SOM, and the retention of soil fertility and productivity on the sites involved.

A. Biomass Feedstocks from Forest Management

As described above, there are many forests where the buildup of dead and dying trees is a serious problem that needs immediate attention (Schmidt et al., 1993). In many of these forests, however, the lack of a market for small trees, less-valuable species, or partially rotted material means that foresters have few, if any, economic ways of supporting the needed forest improvements. Where there is an economic market for wood as an energy feedstock, far greater management options are available.

Policies that encourage co-generation at industrial facilities, that maintain electric rates at a level that provides incentives, and that raise the price of fossil fuel burning through pollution restrictions, all tend to help create a market for wood as an energy source.

New harvesting technologies, largely around the removal, transport, storage, and burning of whole trees, may make wood a more competitive material in some areas (Wright et al., 1993). Caution must be exercised, however, or whole-tree technology could result in major and destructive nutrient export from many forest sites. The effect of such nutrient transport could be to lower SOM levels and adversely affect site productivity and C sequestration capacity. Whole-tree handlers that strip small limbs and foliage at the harvest site, or that remove deciduous trees after leaf-fall, may help minimize that risk.

B. Short Rotation Woody Crops

Planting cropland to woody crop plants that grow rapidly, can be harvested on a 4-12 year cycle, and re-sprout after cutting, has been intensively studied as a means of meeting energy needs, reducing fossil fuel usage, and building new income-producing options for farmers (Wright et al., 1992). Because of the intensive management needs such as maintenance of high fertility rates and weed control, the technology is much more related to agriculture than to forestry.

Where properly done, however, it appears that SOM levels under woody crops could be maintained at levels as high or higher than for ordinary cultivated crops on the same soils. This would arise due to the additional shading of the soils, the lack of annual plowing or deep cultivation, the maintenance of high fertility, and the decomposition of leaf litter from the deciduous species.

Additional research to quantify the SOM dynamics under these systems will help managers and policymakers develop these systems in ways that are most positive in sequestering C in the system, as well as producing biomass to replace fossil fuels.

C. Conservation Trees

Windbreaks and shelterbelts represent a major opportunity to improve SOM levels over vast areas of U.S. cropland. It has been estimated that there are 27.3 x 10^6 ha of cropland with wind erosion rates in excess of tolerable levels (Brandle et al., 1992). On those lands alone, 2.25 x 10^6 km of field windbreaks are needed to provide erosion protection. Those windbreaks would occupy 1.4 x 10^6 ha, but research indicates that the resulting yield increases on the remaining area would more than offset the land taken up by the trees (Brandle et al., 1992).

The implications for SOM of such a program would be enormous. Soils suffering wind erosion lose the finer soil fractions first, meaning a selective removal of SOM (Sampson, 1981). Protection from wind reduces soil drying as well as erosion, leading to higher moisture and nutrient levels, both important in building and maintaining higher SOM levels under crop management. In the land covered by the trees themselves, higher SOM levels, coupled with the standing biomass in the shelterbelt, will add to C storage.

In addition to soil erosion and SOM protection, windbreaks and shelterbelts produce indirect savings due to reduced fuel and fertilizer usage, as well as potential wood for energy production. Where windbreaks take up to 6% of the field out of cultivation, while raising yields on the remainder enough to compensate for the production losses, energy savings from reduced fuel and fertilizer usage will amount to approximately 0.6 t C ha^{-1} yr^{-1} (Brandle et al, 1992). In Germany, it has been shown that, where windbreaks are managed for sustainable wood production, the average annual energy production for the first 60 years is 0.6 t C ha^{-1} (Burschel et al., 1993).

But windbreaks and shelterbelts are not increasing in popularity with American farmers. Large machinery, center-pivot sprinkler systems, and herbicide usage all make tree borders problematic, and as a result, millions of windbreaks and shelterbelts have been removed in recent years (Sampson, 1981). Policies that encourage their re-establishment will need to address, in addition to financial incentives for planting and maintaining the trees, some larger changes in the technology utilized in production agriculture. Whether such changes will emerge as agriculture seeks more sustainable technologies and methods is an important aspect for the future of SOM and soil quality in those regions.

D. Scientific Landscaping

In urban situations, properly placed urban trees can have a significant impact on atmospheric carbon buildup through energy conservation. Studies in the U.S. indicate that the daily electrical usage for air-conditioning could be reduced by 10-50% by properly located trees and shrubs (US EPA, 1992). Savings of 1,351-1,665 kWh per year for a 137 m^2 house have been recorded (McPherson et al., 1991). On the other side of the calendar for energy conservation, properly placed trees can also reduce winter heating costs by 4 to 22 percent (De Walle, 1978).

With energy conservation as a major goal, tree planting programs should focus first upon those trees (including species selection and location) that provide direct energy benefits to buildings. The next priority for planting would be trees that would provide maximum shade for parking lots, streets, and other dark-surfaced areas. The lowest priority would be to plant those open areas where trees could "fill in" the open spaces that, while not directly shading or protecting buildings, would help reduce the urban heat island effect by modifying albedo and wind patterns as part of the total urban forest. Sampson et al. (1992) have proposed a goal for a 10-year program aimed at increasing the canopy cover by 10% on residential lands, and 5-20% on other urban lands in the U.S. They estimate that the effect of such an improvement program on U.S. urban forests could result in sequestration of 2-5 x 10^6 t C/yr in trees and soils, and a 7-29 x 10^6 t C/yr reduction in C emissions due to energy conservation from improved shading, increased evapotranspiration, and reduction of the urban heat island, along with wintertime heat savings.

The implications for SOM are highly variable in urban situations, where soils are seldom normal in terms of structure due to mixing and compaction during construction, and where most leaf litter is removed before it has the opportunity to decompose (Moll et al., 1992) In these intensively managed situations, additional policy efforts to encourage the recycling of yard waste and tree litter using methods such as home composting provide one answer to a growing waste disposal problem, while providing an opportunity to retain much-improved SOM levels in urban soils.

IV. Summary and Conclusions

The maintenance of high levels of SOM under a variety of forest situations is desirable and possible, provided that policies to encourage sound forest management practices are in place. Trees and forests in healthy, productive ecosystems can be an important element in maintaining high levels of SOM which can, in turn, be a significant element in increasing the amount of terrestrial C as one way to reduce the impact of greenhouse gas emissions.

In those situations where trees and forests are managed specifically as a terrestrial sink for C, the status of SOM is a critical component that must not be overlooked, either in analysis or in management. In the tree planting-C sequestration projects that seem certain to grow from increased federal attention to enhancing C sinks, federal cost-sharing rules should require soil C data to be gathered at the start of the project, then monitored periodically through the life of the project in order to gain additional insight into the actual functioning of the C cycle during these reforestation efforts.

Timber harvest methods that minimize soil disturbance, aeration, heating, and drying (such as the replacement of clearcuts with a form of harvest that leaves a minimal canopy) will result in less C loss during the harvest cycle. Pressure on foresters to develop such methods, particularly on the public lands where pressures are high to retain visual and other environmental qualities, will help reinforce the development of these methods. Research is needed, however, to demonstrate the extent to which SOM levels can be affected.

Some of the most serious threats to proper maintenance of SOM under forest conditions today exist in those areas where fire regimes have been seriously altered or eliminated. Restoring fire, or replacing its effects upon ecosystems, is a major challenge for land managers and public policymakers. Silvicultural efforts to replace fire's effects with vegetative manipulation may be adequate in some cases, but major questions remain in regard to such impacts as the role of fire in affecting nutrient cycling, soil pH, charcoal formation, and heat-pulse effects on germination and growth characteristics in some species. Fire, long seen as the nemesis of a healthy forest, may in fact be essential. If that proves to be true, finding ways to re-introduce it in a landscape that is dramatically fragmented and changed by civilization, with a population that abhors smoky, polluted air, may be one of the major undertakings in natural resource policy in the coming years.

Scientific understanding of SOM dynamics under various forest management regimes, and in response to various forest disturbances, is less well-understood than the dynamics of forest tree growth, so analysis to date has tended to deal with SOM dynamics largely in a theoretical way. More research and interchange of information between soil scientists, foresters, and policy analysts is needed to improve this area of knowledge.

If another national tree planting program such as CRP is initiated, or the CRP itself is extended, several policy options to increase its effectiveness should be considered. Those include:

- Efforts to link tree plantings to forward-contracting of wood sales, preferably to energy or co-generation plants where non-commercial thinnings, deformed stems, and other material could be converted to energy as an offset against fossil fuel usage;

- Special training of USDA field staff, to assure that they recognize tree planting opportunities, understand the risks and benefits involved, and are familiar with the potential to turn appropriate tree planting efforts into sustainable local development opportunities;

- An attempt to focus the program's incentives on soils where SOM levels have been depleted, but where trees are adapted and could grow and restore SOM with a minimum of purchased inputs needed to subsidize tree growth; and,

- Efforts to lengthen the program's effect, through some form of longer-term contracting or easement, so that landowners could reasonably plan for the duration of the tree crop.

References

Arno, S.F. and J.K. Brown. 1991. Overcoming the paradox in managing wildland fire. *Western Wildlands*, Spring, 1991, Vol. 17:2. p. 40-46.

Birdsey, R.A. 1992. *Carbon Storage and Accumulation in United States Forest Ecosystems*. Gen Tech Rep WO-59. Washington, D.C: USDA Forest Service. 51 pp.

Birdsey, R.A. (In press). Carbon Storage in United States Forests. In: R. Neil Sampson and Dwight Hair (eds.), *Forests and Global Change, Volume II: Opportunities for Improving Forest Management*, Washington, D.C. American Forests.

Borchers, J.G. and D.A. Perry. 1990. Effects of Prescribed Fire on Soil Organisms. p. 143-157. In: J.D. Walstad, S.R. Radosevich, and D.V. Sandberg (eds.), *Natural and Prescribed Fire in Pacific Northwest Forests*. Corvallis, OR: Oregon State University Press.

Brandle, J.R., T.D. Wardle, and G.F. Bratton. 1992. Opportunities to Increase Tree Planting in Shelterbelts and the Potential Impacts on Carbon Storage and Conservation. p. 157-176. In: R.N. Sampson and D. Hair (eds.), *Forests and Global Change, Volume 1: Opportunities for Increasing Forest Cover*. Washington, D.C: American Forests.

Buckman, Harry O. and Nyle C. Brady. 1969. *The Nature and Properties of Soils*. London: MacMillan Co. 653 pp.

Burschel, P., E. Kürsten, and B.C. Larson. 1993. Die Rolle von Wald und Forstwirtschaft im Kohlenstoffhaushalt - Eine Betrachtung für die Bundesrepublik Deutschland, Forstl. Forschungsberichte München, Nr. 126, (short version in English)

Chandler, W.U. 1990. *Carbon Emissions Control Strategies: Executive Summary*. Washington, D.C: World Wildlife Fund. 42 pp.

Cost, N.D., J.O. Howard, B. Mead, W.H. McWilliams, W.B. Smith, D.D. Van Hooser and E.H. Wharton. 1990. *The Forest Biomass Resource of the United States*. Gen Tech Rep WO-57. Washington, D.C: USDA Forest Service. 21 pp.

Cronin W. 1983. *Changes in the Land: Indians, Colonists, and the Ecology of New England*. New York: Hill and Wang. 241 pp.

DeWalle, D. R. 1978. Manipulating urban vegetation for residential energy conservation. p. 267-283. In: *Proceedings of the 1st national urban forestry conference*; November 13–16, 1978; Washington, D.C: USDA Forest Service.

Esseks, D., S.E. Kraft, and R.J. Moulton. 1992. Land Owner Responses to Tree Planting Options in the Conservation Reserve Program. p. 23-40. In: R.N. Sampson and D. Hair (eds.), *Forests and Global Change, Volume 1: Opportunities for Increasing Forest Cover*, Washington DC: American Forests.

Fahrenstock, G.R. and J. Agee. 1983. Biomass Consumption and Smoke Production by Prehistoric and Modern Forest Fires in Western Washington. *Journal Forestry* 81:653-657.

Fedkiw, J. 1989. *The Evolving Use and Management of the Nation's Forests, Grasslands, Croplands, and Related Resources*. Gen. Tech. Rep. RM-175. Washington, D.C: USDA Forest Service. 67 pp.

Franklin, J.F. 1989. Toward a New Forestry, *American Forests*, 95:11-12, pp. 37-44.

Gibney, D.R. and K.F. Wenger. 1990. Letters to the Editor, *American Forests* 96:3-4, pp. 2-5.

Harmon, M.E., W.K. Ferrell, and J.F. Franklin. 1990. Effects on Carbon Storage of Conversion of Old-Growth Forests to Young Forests, *Science* 247:4943, pp. 699-702.

Hasenkamp, K.P. 1993. Large-Scale Afforestation as a Contribution to Warding off a Climate Catastrophe, Testimony before a non-public hearing held by the Enquete Commission on "Protecting the Earth's Atmosphere," March 15, 1993, Bonn: German Bundestag.

Haynes, R.W. 1993 (In press). Forest Sector Changes from Increasing Forest Area and Timber Growth, in: R.N. Sampson and D. Hair (eds.), *Forests and Global Change, Volume II: Opportunities for Improving Forest Management*. Washington DC: American Forests.

Hyde, W.F. 1980. *Timber Supply, Land Allocation, and Economic Efficiency*, Washington, D.C: Resources for the Future. 224 pp.

Johnson, D.W. 1992. Effects of Forest Management on Soil Carbon Storage, p. 83-120. In: J. Wisniewski and A.E. Lugo (eds.), *Natural Sinks of CO_2*. Dordrecht, Holland: Kluwer Academic Publishers.

Kilgore, B.M. 1987. *The Role of Fire in Wilderness: A State-of-Knowledge Review*. Gen Tech Rep INT-220. Ogden, UT: USDA Forest Service, Intermountain Research Station.

Kinsman, J.D. and M.C. Trexler. 1993. Terrestrial Carbon Management and Electric Utilities. p. 546-560. In: J. Wisniewski and R.N. Sampson (eds.), *Terrestrial Biospheric Carbon Fluxes: Quantification of Sinks and Sources of CO_2*. Dordrecht: Kluwer Academic Publishers.

Little, S.N. and L.J. Shainsky. 1992. *Distribution of Biomass and Nutrients in Lodgepole Pine/Bitterbrush Ecosystems in Central Oregon*. Res. Pap. PNW-RP-454. Portland, OR: USDA Forest Service, Pacific Northwest Research Station. 22 pp.

MacCleery, D.W. 1993. *American Forests: A History of Resiliency and Recovery*, FS-540, U.S. Department of Agriculture Forest Service. 59 pp.

McNabb, D.H. and K. Cromack, Jr. 1990. Effects of Prescribed Fire on Nutrients and Soil Productivity. p. 125-142. In: J.D. Walstad, S.R. Radosevich, and D.V. Sandberg (eds.), *Natural and Prescribed Fire in Pacific Northwest Forests*. Corvallis, OR: Oregon State University Press.

McPherson, E.G. and G.C. Woodward. 1990. Cooling the urban heat island with water- and energy-efficient landscapes. *Arizona Review* Spring 1990:1–8.

Mixon, J.W. 1988. What we are doing in Georgia and the South Today. p. 23-25. In: *Proceedings of the Policy and Program Conference on the South's Fourth Forest: Alternatives for the Future*. USDA Forest Service Misc. Pub. No. 1463. Washington, D.C.

Moll, G. and S. Young. 1992. *Growing Greener Cities*, Venice, CA: Living Planet Press. 126 p.

Parks, P.J., S.R. Brame, and J.E. Mitchell. 1992. Opportunities to Increase Forest Area and Timber Growth on Marginal Crop and Pasture Land. p. 97-122 In: R.N. Sampson and D. Hair (eds.), *Forests and Global Change, Volume 1: Opportunities for Increasing Forest Cover*, Washington D.C: American Forests..

Pyne, S.J. 1982. *Fire in America: A Cultural History of Wildland and Rural Fire*. Princeton, NJ: Princeton University Press. 654 pp.

Robertson, F.D. 1992. Ecosystem Management of the National Forests and Grasslands: Letter to Regional Foresters and Station Directors, June 4, 1992. USDA Forest Service. 6 pp.

Row, C. and R.B. Phelps. 1992. Carbon Cycle Impacts of Improving Forest Products Utilization and Recycling, p. 208-219. In: A. Qureshi (ed.), *Forests in a Changing Climate*, Washington, D.C: Climate Institute.

Sample, V.A. 1991. *Land Stewardship in the Next Era of Conservation*, Milford, PA: Grey Towers Press. 43 pp.

Sampson, R.N. 1993. Increasing Forest Areas as a Carbon-Fixing Strategy, Testimony before a non-public hearing held by the Enquete Commission on "Protecting the Earth's Atmosphere," March 15, 1993, Bonn: German Bundestag.

Sampson, R.N. 1992. Forestry Opportunities in the United States to Mitigate the Effects of Global Warming. p. 157-180. In: J. Wisniewski, and A.E. Lugo. (eds.), *Natural Sinks of CO_2*, Kluwer Academic Publishers,

Sampson, R.N. 1981. *Farmland or Wasteland: A Time to Choose*, Emmaus, PA: Rodale Press. 465 pp.

Sampson, R.N., G.A. Moll, and J.J. Kielbaso. 1992. Opportunities to Increase Urban Forests and the Potential Impacts on Carbon Storage and Conservation. p. 51-72. In: R.N. Sampson, and D. Hair (eds.), *Forests and Global Change, Volume 1: Opportunities for Increasing Forest Cover*, Washington D.C. American Forests.

Schmidt, T., M. Boche, J. Blackwood, and B. Richmond. 1993. *Blue Mountain Ecosystem Restoration Strategy: A Report to the Regional Forester.* Portland, OR: USDA Forest Service Pacific Northwest Region. 12 pp.

Tikkala, B. and M. Robbins. 1992. Heritage Forests: Healing the Earth With Trees, *American Forests*, 98:11-12, pp. 17-19.

Trexler, M.C. and C.A. Haugen. 1993 (forthcoming). Keeping it Green: Global Warming Mitigation Through Tropical Forestry, World Resources Institute, Washington, D.C.

Trexler, M.C., C.A. Haugen, and L.A. Loewen. 1992. Global Warming Mitigation through Forestry Options in the Tropics, p. 73-96. In: R.N. Sampson and D. Hair (eds.), *Forests and Global Change, Volume 1: Opportunities for Increasing Forest Cover*, Washington D.C.: American Forests.

Trexler, M.C., P.E. Faeth, and J.M. Kramer. 1989. Forestry as a Response to Global Warming: An Analysis of the Guatemala Agroforestry and Carbon Sequestration Project, World Resources Institute, Washington, D.C. 66 pp.

USDA Forest Service. 1952. *Timber Resources for America's Future*. For. Res. Rep. No. 14. Washington, D.C: USDA Forest Service.

US Environmental Protection Agency. 1992. Cooling our Communities: A Guidebook on Tree Planting and Light-Colored Surfacing, 22P-2001, Washington, D.C.

US Environmental Protection Agency. 1993. *Forests for the Future: Launching Initial Partnerships*, Report of an Interagency Task Force, January 15, 1993.

Waddell, K.L., D.D. Oswald and D.S. Powell. 1989. *Forest Statistics of the United States, 1987.* Res Bull PNW-RB-168, USDA Forest Service, Pacific Northwest Research Station. 106 pp.

Winjum, J.K., R.K. Dixon and P.E. Schroeder. 1993. Forest Management and Carbon Storage: An Analysis of 12 Key Forest Nations. p. 239-258. In: J. Wisniewski and R.N. Sampson (eds.), *Terrestrial Biospheric Carbon Fluxes: Quantification of Sinks and Sources of* CO_2. Dordrecht: Kluwer Academic Publishers.

World Resources Institute (WRI). 1992. *World Resources: Toward Sustainable Development.* Oxford University Press. 385 pp.

Wright, L.L., R.L. Graham, A.F. Turhollow, and B.C. English. 1992. Growing Short-Rotation Woody Crops for Energy Production. p. 123-156. In: R.N. Sampson and D. Hair (eds.), *Forests and Global Change, Volume 1: Opportunities for Increasing Forest Cover*, Washington DC: American Forests, pp. 123-156.

Wright, L.L. and E. E. Hughes. 1993. U.S. Carbon Offset Potential Using Biomass Energy Systems. p. 483-498. In: J. Wisniewski and R.N. Sampson (eds.), *Terrestrial Biospheric Carbon Fluxes: Quantification of Sinks and Sources of* CO_2. Dordrecht: Kluwer Academic Publishers.

CHAPTER 29

The Role of Soil Management in Sequestering Soil Carbon

Mark G. Johnson

I. Introduction

Soils are an important component of the global carbon cycle and serve as a large reservoir of terrestrial carbon. The amount of carbon in any soil is a function of the soil forming factors including: climate, relief, organisms, parent material, and time. Over the centuries, humans, usually included as part of the "organisms" factor, have profoundly influenced the dynamics and sequestration of carbon in soils by their land use and management practices. These practices include cultivation, deforestation, and draining wet soils. In general, human activities have decreased the amount of carbon held in the affected soils. With the concern over increasing concentrations of greenhouse gases, humans need to consider how soil management affects greenhouse gas emissions from soil and the sequestration of carbon in soils, and to look for ways to protect and manage soil carbon. This paper examines soil management practices and their effects on greenhouse gas emissions and carbon sequestration. Included is an analysis of how management practices affect the physical and chemical environment of soil and how these in turn affect greenhouse gas emissions and the soil carbon sequestration potential.

II. Soils and the Greenhouse Gas Problem

A. Greenhouse Gases and Climate Change

Greenhouse gases (e.g., H_2O, CO_2) in the Earth's atmosphere trap a portion of the Sun's radiant energy thereby warming the Earth (Mitchell, 1989). This warming is essential to support life as we know it. The composition of the Earth's atmosphere is, however, rapidly changing in that the concentrations of a number of greenhouse gases are increasing (U.S. Congress, 1991). In particular the concentrations of CO_2, CH_4, N_2O, and chlorofluorocarbons (CFCs) (IPCC, 1990) have increased dramatically in the last 50 to 100 yr. Per molecule, methane is 25 times more effective at trapping longwave radiation than CO_2 (Mitchell, 1989) and N_2O is approximately 200 times more effective (U.S. Congress, 1991). Chlorofluorocarbons are estimated to be 10,000 times more effective than CO_2 (U.S. Congress, 1991). The primary concern is that the additional energy trapped in the Earth's atmosphere may lead to global warming and concomitantly, to a change in global climate. Human activities are largely responsible for the recent increase in greenhouse gases. While CO_2, CH_4, and N_2O have natural as well as anthropogenic sources, CFCs are strictly anthropogenic and have no natural sources or sinks.

Concerns regarding the relationship between increasing greenhouse gases and climatic change have caused scientists and environmental policy makers to focus efforts on identifying the natural sources and sinks of greenhouse gases for the purpose of characterizing the processes that control their emission and consumption. Photosynthesis provides a natural sink for CO_2 and provides a mechanism for withdrawing CO_2 from the atmosphere and storing it in above- and belowground plant parts. Similarly, soils can oxidize some

ISBN 1-56670-117-1/95/$0.00+.50

fraction of atmospheric CH_4 (Mosier et al., 1991). Plants and soils are also sources of atmospheric CO_2 due to respiration. Under some conditions soils can be a source of CH_4 and N_2O.

B. Soils and the Global Carbon Cycle

On a global-scale, carbon circulates between three very large reservoirs (oceans, atmosphere, and terrestrial systems (e.g., vegetation, soil)) and can be found in a variety of compounds in each reservoir (Houghton and Woodwell, 1989; Post et al., 1990). These reservoirs, or pools, exchange large amounts of carbon annually. A fourth reservoir, the geological reservoir, contains fossil (i.e., oil, coal, and natural gas) and mineral carbon, including carbonates, and consists primarily of inactive or non-circulating carbon. Perturbations, disturbances, or additions of carbon (e.g., fossil fuel combustion) to any of the reservoirs will have a concomitant effect on the others because of the dynamic linkage of the reservoirs. The loss of carbon from soils and wetlands is an example of loss from the terrestrial pool to the atmosphere. Similarly, erosion moves terrestrial carbon from the terrestrial pool to the oceans. Global warming is also expected to have an effect on the global carbon cycle. One projected effect is the shifting of global vegetation zones and the amount of carbon stored therein (Emanuel et al., 1985a; Emanuel et al., 1985b; Leemans, 1990; Prentice and Fung, 1990).

Globally, there are approximately 41,000 Pg (Pg = petagrams = 10^{15} grams) of active, or circulating, carbon (Bolin, 1983; Houghton and Woodwell, 1989; Post et al., 1990). Of this, the oceanic reservoir contains 38,000 Pg; the atmosphere 750 Pg; and terrestrial ecosystems about 2100 Pg of carbon (Post et al., 1990). Of the terrestrial carbon, living plants account for about 550 Pg of carbon and soils approximately 1500 Pg (Houghton and Skole, 1990). Soils are therefore the largest, non-fossil fuel, terrestrial reservoir of carbon.

At steady state, the net transfer of carbon between the global carbon pools is in equilibrium (Bolin, 1983; Post et al., 1990; IPCC, 1990). The amount of carbon fixed annually by terrestrial plants through photosynthesis ranges from 100 to 120 Pg (Post et al., 1990). Plant respiration releases approximately 40 to 60 Pg of carbon annually, and decomposition of organic residues, including soil carbon, releases approximately 50 to 60 Pg. At steady state, the amount of carbon oxidized by plant respiration and decomposition balance that fixed by photosynthesis. Through agricultural practices and various land uses, including deforestation, the oxidation of plant and soil carbon is exceeding the amount of carbon being fixed by photosynthesis and thereby contributing to the net 3 Pg annual increase in atmospheric CO_2 (Houghton and Woodwell, 1989).

C. Soils as a Source of Greenhouse Gases

Soils are a natural source of CO_2, CH_4, and to a lesser extent N_2O. Aerobic decomposition of soil organic matter produces CO_2 while anaerobic decomposition produces CH_4 and denitrification can lead to the formation of N_2O (Amundson and Davidson, 1990; Denmead, 1991). The use and management of soils affects the amount and type of greenhouse gases emitted from soil. Intensive tillage promotes decomposition of soil organic matter and production of CO_2 (Tate, 1987). Rice production is the principal source of CH_4 from soils (Crutzen, 1991) due to the anaerobic conditions created by flooding paddy soils. Other soils may be a sink for CH_4 (Steudler et al., 1989; Mosier et al., 1991; Crutzen, 1991), but recent research suggests, that nitrogen fertilization can inhibit or reduce soil CH_4 uptake (Steudler et al., 1989; Mosier et al., 1991).

D. Historical Effects of Land Use and Soil Management on the Reservoir of Carbon in Soils

Humans have used soils in a variety of ways that can greatly affect their soil organic matter content. Through management practices such as cultivation, fertilization, deforestation, moving land, construction, paving, draining or saturating, accelerating or preventing erosion, and other management practices, the effect of humans is far-reaching. During the past two centuries, soils have been a net source of greenhouse gases. Cumulative loss of carbon from vegetation and soils for the period from 1850 to 1980 is estimated to be 90 to 120 Pg (Houghton and Skole, 1990). Prior to the 1950's, temperate soils and vegetation were the primary source of terrestrial carbon emissions. Since then, however, the tropics have been the principal

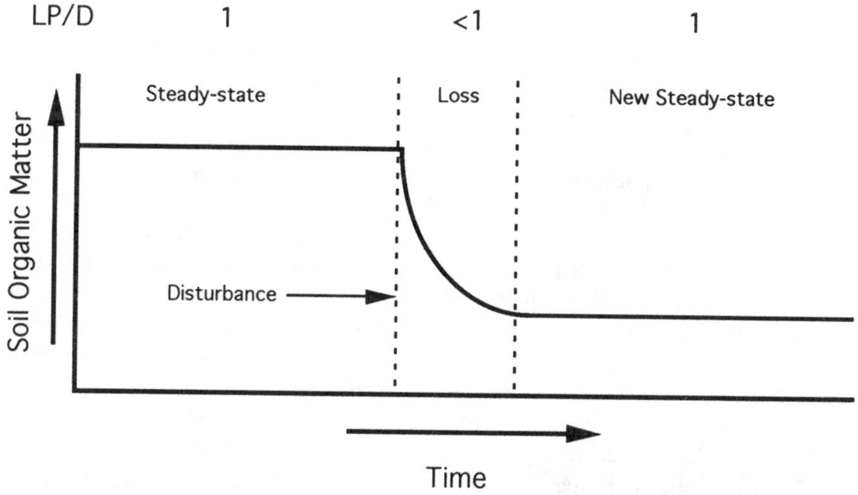

Figure 1. Conceptual model of soil organic matter (SOM) accumulation. SOM accumulates in soils when the production of plant litter (e.g., leaf, root, exudates) exceeds decomposition. When decomposition exceeds litter production, SOM is lost. At steady-state, litter production is equal to decomposition. Therefore, when the ratio of litter production (LP) to decomposition (D) is greater than 1 (LP/D > 1), SOM accumulates; when it is less than 1 (LP/D < 1) SOM is lost; and when it is equal to 1 (LP/D = 1) the system is at steady state with respect to SOM.

source of terrestrial carbon emissions to the atmosphere (Houghton and Skole, 1990) due to massive deforestation. The current net annual loss of carbon from plants and soils is estimated to be about 0.2 Pg from temperate regions and about 2 Pg from tropical regions (Houghton and Skole, 1990).

In North America, vast tracts of prairie soils have been converted to agriculture, and a large amount of native soil organic matter has been oxidized (Haas et al., 1957). The amount of organic matter lost appears to be a function of initial soil organic matter levels and the form and intensity of cultivation (Mann, 1986). Loss of soil C is most rapid in the first 20 to 30 yr of cultivation, with losses of soil carbon ranging from 30 to 50% in the uppermost soil horizons and somewhat less in the lower horizons (Lal and Kang, 1982; Schlesinger, 1986; Balesdent et al., 1988). Mann (1986) reported that soils initially high in soil organic matter lost at least 20% of their organic carbon during cultivation, while soils very low in soil organic matter actually gained carbon with cultivation due to additions of plant materials, manure, or fertilizers. Soil organic matter may also lost through erosional processes.

Figure 1 depicts the loss of soil organic matter due to disturbance, such as continuous cultivation. With the onset of disturbance, the loss of soil organic matter is characteristically faster than the accumulation rate. Over time, with other soil forming factors held constant, including management regime, held constant, the soil will reach a new steady-state condition with respect to soil organic matter. The new steady-state soil organic matter level will be a function of the intensity, duration, and extent of disturbance and how crop, forest, or other residues are managed. In general, the post-disturbance steady-state soil organic matter levels will be lower than the premanagement levels. A recent review of the literature found, however, that forest harvesting did not generally result in loss of soil organic matter if the harvested lands were replanted or allowed to naturally regenerate (Johnson, 1992). If the cleared lands were subjected to high-intensity fire, whether prescribed or wildfire, soil organic matter was lost. Similarly, if cleared lands were converted to conventional agriculture, large losses of soil organic matter would be expected.

E. Potential Effects and Feedbacks of Climate Change on Soil Carbon

One of the concerns and unknowns associated with global warming and the carbon cycle is the effect of warming on the distribution of carbon in the various reservoirs. Will increased temperatures cause a massive shift in carbon now sequestered in soils to the atmosphere thereby increasing global warming? Buol et al. (1990) projected that a 3 °C increase in mean annual surface temperatures over the next 50 years would result in an 11% decrease in soil organic carbon pools in the top 30 cm of temperate zone soils. This would result in the emission of approximately 58 Pg of carbon to the atmosphere. Additional soil carbon would also be lost from the tropical and boreal zones because of global warming. Jenkinson et al. (1991) projected that if global warming caused mean air temperatures to increase uniformly by 0.5 °C per decade over the next 60 years (3 °C total increase), that approximately 100 Pg of carbon would be evolved from soils due to the temperature increase alone. If this projection included the effects of deforestation and land use changes over the same period (the next 60 years), the projected losses of soil carbon could easily reach 10% of the global soil carbon pool, or about 150 Pg. This is roughly equivalent to carbon emissions from fossil fuels since 1850 (Houghton and Skole, 1990). The carbon loss from soils could occur, however, in a much shorter period of time. Occurrence of either of these soil carbon loss scenarios could exacerbate the amplitude and extent of global warming due to increased greenhouse gases. Increased warming could in turn, through a positive feedback mechanism, cause greater carbon emissions from soil.

III. Basic Chemistry of Soil Organic Matter

In natural systems, organic carbon (i.e., organic matter) accumulates in soils over long periods of time, but it can be lost rapidly due to disturbance, whether natural or anthropogenic. Since natural systems accumulate carbon slowly, the question is whether or not the accumulation rates, or reaccumulation rates, can be accelerated through judicious management. It is also important to know whether or not these management practices will lead to increased or decreased emissions of greenhouse gases from soils. To understand the effects of management on soil organic matter, information is required on the composition of soil organic matter and how it is processed. Additional information is also needed on how management affects the production, quality, processing, or stabilization of organic matter.

A. What is Soil Organic Matter?

Soil organic matter is the residue of plants and/or animals that remains in soil. These residues are in various stages of decomposition ranging from very slightly decomposed (e.g., recently fallen leaves) to highly decomposed material of undistinguishable origin. Since soil organic matter is derived from plants and animals, newly deposited materials will be chemically similar to the original source material. These materials consist of biomolecules containing C, H, O, N, S, P, and a variety of other elements. These elements are arrayed in numerous organic compounds including (1) carbohydrates, (2) amino acids and proteins, (3) lipids, (4) nucleic acids, and (5) lignins (Stevenson, 1986; Tate, 1987; and Tan, 1993). Each class of compounds has chemical characteristics that affect how they decompose, the extent to which they may decompose, and the potential reactivity with the other compounds, ions, and soil mineral surfaces, and their solubility in soil waters. Polysaccharides that are highly branched, for example, are less susceptible to enzymatic degradation than polysaccharides that have a low degree of branching and can therefore, more readily accumulate in soils (Tan, 1993). Polysaccharides have also been shown to adsorb on soil surfaces and to form complexes with metal cations. Both reactions tend to increase the stability of polysaccharides in soil. Polysaccharides have been shown to help form and stabilize soil aggregates (Burns, 1977 cited in Lynch and Bragg, 1985).

The stabilized or recalcitrant forms of soil organic matter are known collectively as humus. Humus is comprised of complex organic compounds (decomposition residues) that are either chemically recalcitrant (e.g., cellulose encrusted with lignin (Oades, 1988)), contain biological inhibitors (e.g., polyphenols), adsorbed on soil mineral surfaces, or chelated with metal cations (e.g., Ca^{2+}, Al^{3+}, Fe^{3+}). Recalcitrance should not be construed to mean inertness. If recalcitrant organic matter were inert, humus would continue to accumulate and the humus pool would be immense. On the contrary, a portion of the humus pool is

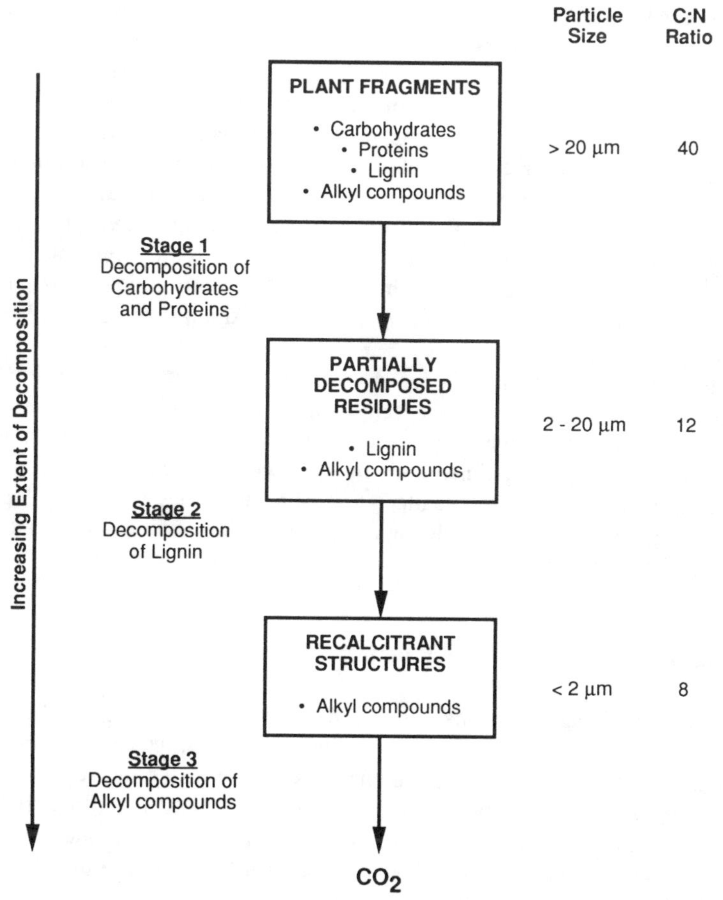

Figure 2. Three stage conceptual model of the plant litter/soil organic matter decomposition continuum. (After Baldock et al., 1992.)

mineralized annually. It seems, however, that soils have only a finite capacity to accumulate humus which is likely constrained by the extent of mineral surfaces, the chemistry of the soil, and local site conditions.

In a recent article (Baldock et al., 1992) researchers using solid state ^{13}C nuclear magnetic resonance spectroscopy (NMR) characterized the chemical constituency of organic materials along the plant litter/soil organic matter decomposition continuum. They characterized the organic matter by physically separating three particle size fractions of soil organic matter: (1) large particles (>20 μm diameter), (2) intermediate sized particles (2-20 μm diameter), and (3) the fine fraction (<2 μm diameter). The size fractions correspond to fresh plant material, partially degraded residues, and degraded residues, respectively. Baldock et al. (1992) propose three stages of decomposition (See Figure 2). In the first stage carbohydrate and protein structures are degraded, as demonstrated by a decrease in the oxygenated alkyl (O-alkyl) carbon ^{13}C NMR signal. A portion of the carbon was respired as CO_2 and the remainder was assimilated by the decomposers. The second stage of decomposition is characterized by a loss in aromatic carbon (e.g., lignin). The final stage is the loss of the most recalcitrant forms of organic carbon, alkyl carbon structures. Baldock et al. (1992) also note that the three stages of decomposition correspond to a decrease in the C:N ratio of the organic material, from a C:N ratio of plant material of about 40, to a C:N ratio of 12 for the partially decomposed residues, and finally to a C:N ratio of 8 for the most recalcitrant residues.

B. Soil Factors Affecting the Stability of Soil Organic Matter

External and internal factors affect the accumulation and stability of soil organic matter. External factors include temperature, precipitation, vegetation, and disturbance (Greenland et al., 1992). Internal factors, or intrinsic soil properties, affecting the stability of soil organic matter include: the kind and amount of clay-sized minerals, soil pH, and the abundance of base cations, in particular Ca^{2+}. Layer silicate clay minerals (e.g., kaolinite, vermiculite, montmorillonite) have a net negative surface charge due to isomorphous substitution which may provide absorption sites for soil organic matter with cationic functional groups (e.g., protonated amino groups). Because of their small size (<2 μm diameter) these clays have extremely high surface areas ranging from 10-20 $m^2 g^{-1}$ for kaolinite to more than 600 $m^2 g^{-1}$ for montmorillonite (Bohn et al., 1985). With their great surface areas and the abundance of small pores due to the stacking of individual clay units, clay minerals provide an ideal site for soil organic matter adsorption and protection from microbial decomposition (Oades, 1988). Other clay-sized hydrous oxide minerals of Al, Fe, or Si with short-range order (i.e., amorphous), such as allophane or ferrihydrite, chemically absorb soil organic matter, rendering it less susceptible to microbial or enzymatic decomposition (Stevenson, 1982; McKeague et al., 1986).

Soil pH is related to the distribution of cations in soils. At low pH (<5) Al^{3+} and Fe^{3+} can become more abundant and form stable complexes with soil organic matter. The negative effect of low pH is that potential plant toxicities to Al^{3+} that may reduce primary production and consequently the input of plant litter into soils. At higher pHs, particularly in limed soils or soils with free calcium carbonates, Ca^{2+} plays an important role in stabilizing soil organic matter. According to Oades (1988) Ca^{2+} can bridge between organic matter and clays and can form stabilizing complexes with organic matter. Calcium amendments to soil reduce decomposition and promote aggregate and organic matter stabilization.

C. How Does Soil Organic Matter Accumulate?

Soil organic matter is produced when dead plant and animal matter is deposited in or on soil but only accumulates when the rate of plant and animal detritus production exceeds the rate of decomposition. Through natural decomposition processes, these once identifiable materials are physically and chemically broken down into smaller unrecognizable pieces. A portion of the original material is respired by the decomposers and leaves the soil system as CO_2. A portion is incorporated into the decomposers to build cell walls, proteins, and other compounds, and eventually becomes substrate for other decomposers. A portion of the original material becomes immobilized, or biologically unavailable, on soil mineral surfaces or flocculated by inorganic cations. During the decomposition process, inorganic elements (e.g., N, P, S) become available for plant uptake, thereby completing the nutrient cycle.

The accumulation of soil organic matter is shown conceptually in Figure 3. In new soil parent material the accumulation phase is preceded by an initiation or organization phase during which the soil parent material is colonized by pioneering organisms, including N fixers, that create an appropriate rooting medium for higher plants. During the accumulation phase the rate (slope of the line) of soil organic matter accumulation is slow at first then accelerates (Jenny, 1984). As the soil nears its soil organic matter carrying capacity (i.e., the maximum soil organic matter level that can be maintained under the current set of soil and environmental conditions), the accumulation rate slows. Stevenson (1965) showed a rapid accumulation of organic matter in the first few years (i.e., 5-10 years) that declines slowly until soil organic matter equilibrium or steady-state levels are reached. The time to reach soil organic matter steady-state ranges from 110 yr for fine textured soils and 1500 yr for coarse textured soils (Stevenson, 1965).

The carbon carrying capacity of any soil is a function of the factors of soil formation (Jenny, 1984) and local site conditions. Cool moist conditions favor the accumulation of soil C, whereas, hot moist conditions promote decomposition. When the system is at or near steady-state the net rate of soil organic matter accumulation is approximately equal to zero.

D. Rates of Soil Organic Matter Accumulation

The rate of soil organic matter accumulation is governed by a variety of factors including climate, vegetation, intrinsic soil properties, period of time since disturbance, and whether or not the soil is under

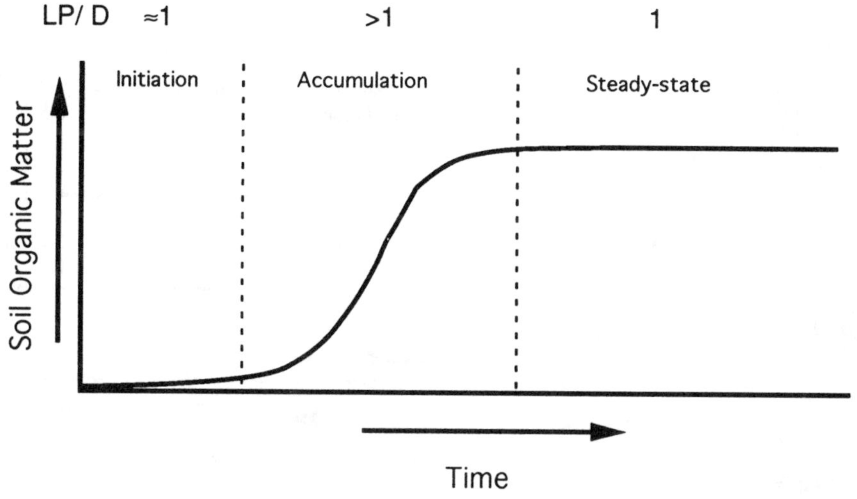

Figure 3. Conceptual model of soil organic matter (SOM) loss due to disturbance, such as continuous cultivation. SOM is lost when decomposition (D) is greater than plant litter production (LP), or LP/D < 1. Over time, with the other soil forming factors held constant, including management regime, the soil will reach a new steady-state condition with respect to SOM (LP/D = 1).

some form of management. The rate of accumulation in new soils can be very low. Schlesinger (1990) reported a long-term soil carbon accumulation rate for newly formed land surfaces (e.g., mudflows, retreating glaciers) of 2.4 g C m^{-2} yr^{-1}. In contrast, in soils that have lost some, but not all, of their native organic matter, the accumulation or reaccumulation rate may be greater. For example, Jenkinson (1991) reported carbon reaccumulation rates on the order of 25 to 50 g C m^{-2} yr^{-1} for soils that were carbon-depleted by long-term continuous agriculture, abandoned, then naturally revegetated in mixed deciduous forest. Alexander et al. (1989) reported soil carbon accumulation rates for forested soils in southeastern Alaska ranging from 29 to 113 g C m^{-2} yr^{-1}. In managed tropical systems, accumulation rates as high as 120 g C m^{-2} yr^{-1} have been reported (Lugo, 1991).

IV. Managing Soil Organic Matter

Historically, soils have not been managed to limit greenhouse gas emissions nor to sequester or store carbon. Managing soil processes may provide an opportunity to reduce greenhouse gas emissions from soils or to use soils as a net carbon sink.

A. Management Strategies

Three soil management strategies have been suggested by Johnson and Kern (1991) for managing soil organic matter to reduce greenhouse gases:

1. Manage Soils to Maintain Existing Levels of Soil Organic Matter

Globally, the reservoir of carbon in soils is vast, but can be potentially lost to the atmosphere through human activities such as conventional tillage practices and deforestation. The purpose behind this strategy is to prevent further loss of soil organic matter by implementing soil management practices, such as reduced tillage or no-tillage agriculture, that preserve and protect soil organic matter.

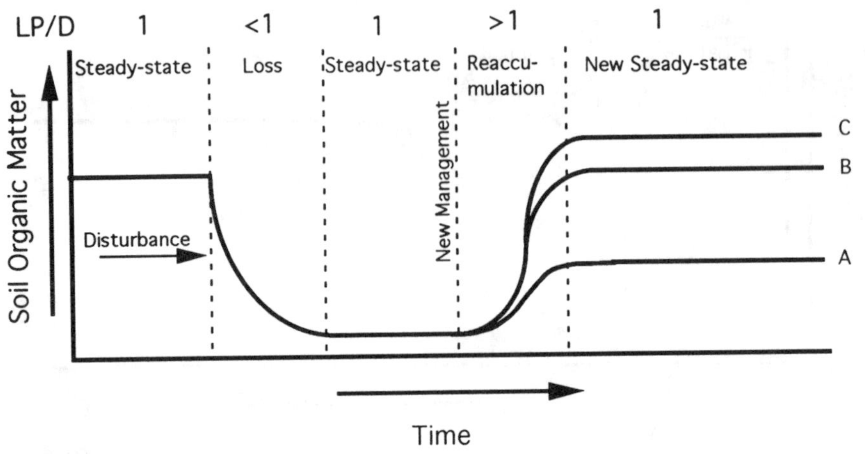

Figure 4. Conceptual model of soil organic matter (SOM) reaccumulation follow disturbance. With the onset of a new soil management regime in which soil processes are managed so that litter production (LP) exceeds decomposition (D), LP/D > 1, and SOM accumulates. If this new management regime is held in place for a sufficient period of time, a new SOM steady-state will be reached (LP/D = 1). The difference between the SOM level at the lowest point on the curve and the new steady-state level is equal to the new sequestration. Line A indicates only partial SOM reaccumulation; Line B, complete SOM reaccumulation; and Line C, SOM reaccumulation in excess of original levels.

2. Manage Soils to Restore Depleted Soil Organic Matter Levels

An opportunity exists to restore soil organic matter in soils from which it has been lost due to previous land use and management practices. For some soils, intensive management may be required (e.g., repeated fertilizer additions). For other soils, only minor management changes may be necessary (e.g., returning marginal farmland back to native vegetation) to convert a system from a net source of carbon to the atmosphere to a net sink. The long-term objective of this strategy is to restore soil organic matter to the premanagement soil organic matter carrying capacities.

3. Manage Soils to Enlarge the Soil Organic Matter Pools above Historic Carrying Capacity

The objective of this strategy is to increase the soil organic matter carrying capacity of soils through management. While conceptually simple, this strategy may be difficult to implement because the soil organic matter carrying capacity of a soil may be largely determined by factors (e.g., climate, parent material) that cannot be managed. Improving soil fertility may be one way to raise the soil organic matter carrying capacity above premanagement levels.

Figure 4 depicts the effect of soil management soil organic matter. With the onset of a new soil management regime, one in which soil processes are managed so that litter production (LP) exceeds decomposition (D), there is net accumulation of soil organic matter (LP/D > 1). If this new management regime is held in place for a sufficient period of time, a new soil organic matter steady-state will be reached (LP/D = 1). The difference between the soil organic matter level at the lowest point on the curve and the new steady-state level is equal to the new net sequestration. Two questions emerge from this management model, however. First, what is the rate of soil organic matter reaccumulation? And second, what will be the new steady-state soil organic matter level under the new management regime: less than, equal to, or more than the native (undisturbed) soil organic matter levels?

Table 1. Soil management practices for increasing C storage in soils

Soil change	Practice	Applications
Cooler soil	Mulch, shade	All soils
Wetter soil	Irrigation	Dry soils
Increase fertility	Apply fertilizer	Most soils
Increase subsoil pH	Deep liming	Acid subsoils
Fracture subsoil pans	Subsoiling	Hardpan soils
Deeper rooting	Al tolerant cultivars	Acid subsoils
Reduced aeration	Limited tillage	All soils

(From Buol, 1991.)

B. Management Practices

Published reports provide detailed information on the effects of specific soil management practices on soil organic matter (Arshad et al., 1990; Blevins et al., 1983; Follett et al., 1987; Gebhardt et al., 1985; Johnson, 1992; Lal and Kang, 1982; Lucas et al., 1977; Lugo et al., 1986; Mann, 1986; Schlesinger, 1986; Smith and Elliot, 1990; Tate, 1987). A few examples are given here to demonstrate the potential of management to affect soil organic matter levels and to support the view that soil management does have a role to play in either maintaining, restoring, or increasing soil organic matter levels. It is not intended to be an exhaustive treatise on the subject.

A long-term soil management experiment has been reported by Jenkinson et al. (1987) that demonstrates a number of effects of management on soil C. Beginning in 1852 continuous spring barley received annual additions of farm yard manure (FYM) or no manure (Figure 6 of Jenkinson et al., 1987). In 1871 applications of FYM ceased on one third of the plots. The plots in continuous barley without FYM showed a slow steady decline in soil C levels. The plots receiving FYM show a net increase in soil C from about 3000 g C m^{-2} to about 8500 g C m^{-2} to a depth of 23 cm, for an annual increase of about 45 g C m^{-2}. The plots that received FYM from 1852 to 1871 still have soil C levels greater than the unamended soils 100 yr after cessation of the FYM additions. This experiment demonstrates the potential for soil management, albeit intensive management, to increase soil C storage and the long-term C storage potential of soils.

Buol (1991) has listed a number of soil management techniques for increasing soil organic matter (Table 1). These techniques incorporate information about the different kinds of soils, their uses, and the factors of soil formation. Uncertainties exist, however, concerning the success of widespread implementation of soil management practices because of the potential for the occurrence of concomitant negative effects. In some cases the associated negative effects may outweigh the positive benefits. For example, under certain conditions some practices, such as increasing soil fertility via nitrogen fertilization, could lead to increased primary productivity and carbon storage, but could also lead to increased N_2O emissions. Similarly, increasing the water content of soils may reduce rates of decomposition in temperate regions (Greenland et al., 1992), but may increase the potential for the production of CH_4. The implementation of any new or altered management practices should be considered on a site-by-site or region-by-region basis prior to implementation, and should be evaluated in terms of the overall effect on greenhouse gas production or conservation. The carbon, or energy, costs of fertilizer production, transportation, and application must also be included in the carbon sequestration equation.

Managing agricultural soils to conserve and store carbon requires the use of conservation tillage practices to minimize soil disturbances, and the incorporation of crop residues into the soil to increase carbon inputs to the soil (Tate, 1987). Soil temperature is positively related to decomposition. Use of mulches or cover crops can reduce soil temperatures, and thereby reduce soil carbon losses through decomposition. Lower soil temperatures can, however, also slow primary production.

One of the effects of soil and land management is on the soil fauna that decompose plant and animal detritus. These organisms are situated in food webs that sequentially process organic substrates into humus, energy, CO_2, and other component pieces (Wood, 1989). The different faunal functional groups are dependent on other organisms to provide them with substrate or to be substrate. Because of these linkages, the use of herbicides, pesticides, or other management practices that disrupt the food webs will affect organic matter cycling and soil organic matter sequestration. For example, in a decomposition study conducted in the Chihuahuan desert, Santos et al. (1981) and Santos and Whitford (1981) found that when mites were removed from the food web, bacterial-feeding nematode populations increased. Consequently, bacterial populations were reduced and rates of decomposition decreased. In contrast, eliminating the mites and bacterial-feeding nematodes increased bacterial populations and decomposition rates. This example highlights the importance of considering all the effects of any proposed soil management approach.

C. Global Potential

The current amount of global cropland is estimated to be 1.5 billion hectares (World Resources Institute, 1990) and provides an opportunity to implement management practices for carbon sequestration in soils on a vast amount of currently managed land. Conservation tillage could be implemented on many of these soils, thereby reducing soil carbon losses and converting these soils from sources of atmospheric carbon to sinks. Additionally, degraded soils could be managed to sequester additional carbon. Previous estimates of the amount of degraded land and the severity of degradation are highly uncertain. UNEP (1986, UNEP is the United Nations Environment Program) estimated the amount of degraded land to be about 2 billion hectares, or 15% of the earth's land surface area. In the tropics there is an estimated 865 million hectares of deforested and abandoned land that is potentially available for reforestation (Houghton, 1990). If reforested, these systems would withdraw approximately 1.5 Pg C a year from the atmosphere sequestering it in soils and vegetation over the next century (Houghton, 1990). Similar opportunities exist in other regions of the world and need to be considered. Efforts are underway to obtain more reliable global estimates (Oldeman et al., 1990). When the amount and condition of degraded land are known, more accurate estimates of the potential carbon sequestration can be made.

If both the land currently in cultivation and degraded lands were to accumulate carbon at 2.4 g C m^{-2} yr^{-1} (the long-term soil carbon accumulation rate estimated by Schlesinger (1990) for unmanaged, newly formed soils), they would accumulate only about 0.08 Pg C annually, or about three percent of the annual atmospheric increment (currently estimated to be 3 Pg C per year). However, if the average accumulation rate was 30 g C m^{-2} yr^{-1} (on the low end of the range of carbon accumulation rates reported in Jenkinson, 1991), then the annual accumulation rate would be in excess of 1 Pg of carbon, or a third of the current annual atmospheric increment.

V. Summary and Conclusions

Through active management, the organic matter content of most soils that have been under some form of management can be maintained or increased. The potential, therefore, to sequester carbon in soils around the world through active soil management is significant in the context of the annual global carbon cycle. Managing soils at the global-scale is by no means simple or even tractable, since many social, economic, political, and technological barriers exist. Developed countries have the tools and technology to take the lead within their own borders to develop strategies for more sustainable land use practices that conserve and sequester carbon in soils and the biosphere. Developed countries may also have opportunities to assist other countries with their soil carbon management practics.

By understanding the processes and factors that affect the sequestration of carbon in soils, soil management can be more effectively utilized to maintain soil carbon stocks or increase carbon sequestration in soils. It is unlikely that soils alone can provide a large enough sink for carbon to offset all of the observed increase in atmospheric CO_2 and other greenhouse gases. Sequestration of carbon in global soils and vegetation, coupled with reductions in fossil fuel combustion and deforestation can reduce the rate at which CO_2 is accumulating in the atmosphere, thereby reducing or slowing the potential for global warming and climate change.

References

Alexander, E.B., E. Kissinger, R.H. Huecker, and P. Cullen. 1989. Soils of southeast Alaska as sinks for organic carbon fixed from atmospheric carbon-dioxide. p.203-210. In: E.B. Alexander (ed.), *Proceedings of watershed '89: A conference on the stewardship of soil, air, and water resources.* USDA Forest Service, Alaska Region, R10-MB-77

Amundson, R.G. and E.A. Davidson. 1990. Carbon dioxide and nitrogenous gases in the soil atmosphere. *J. Geochem. Explor.* 38:13-41

Arshad, M.A., M. Schnitzer, D.A. Amgers, and J.A. Ripmeester. 1990. Effects of till vs no-till on the quality of soil organic matter. *Soil Biol. Biochem.* 22:595-599.

Baldock, J.A., J.M. Oades, A.G. Waters, X. Peng, A.M. Vassallo, and M.A. Wilson. 1992. Aspects of the chemical structure of soil organic matter materials as revealed by solid-state ^{13}C NMR spectroscopy. *Biogeochem.* 16:1-42.

Balesdent, J., G.H. Wager, and A. Mariotti. 1988. Soil organic matter turnover in long-term field experiments as revealed by carbon-13 natural abundance. *Soil Sci. Soc. Amer. J.* 52:118-124.

Blevins, R.L., M.S. Smith, G.W. Thomas, and W.W. Frye. 1983. Influence of conservation tillage on soil properties. *J. Soil Water Conserv.* 38:301-305.

Bohn, H.L., B.L. McNeal, and G.A. O'Connor. 1985. *Soil chemistry.* John Wiley and Sons, New York, NY.

Bolin, B. 1983. The carbon cycle. p. 41-45. In: B. Bolin and R.B. Cook (eds.), *The major biogeochemical cycles and their interactions.* SCOPE 21. John Wiley & Sons, New York, NY.

Buol, S.W. 1991. Soil genesis and the carbon cycle. p.25. In: M.G. Johnson and J.S. Kern (eds.), *Sequestering carbon in soils: A workshop to explore the potential for mitigating global climate change.* EPA/600/3-91/031. USEPA, Environmental Research Laboratory, Corvallis, OR. 85 pp.

Buol, S.W., P.A. Sanchez, S.B. Weed, and J.M. Kimble. 1990. Predicted impact of climatic warming on soil properties and use. p. 71-82. In: B.A Kimble, N.J. Rosenberg, and L.H. Allen, Jr. (eds.) *Impact of carbon dioxide, trace gases, and climate change on global agriculture.* ASA Spec. Publ. 53. ASA, CSSA, and SSSA, Madison, WI.

Burns, R.G. 1977. *The soil microenvironment: aggregates, enzymes and pesticides.* CNR, Lab. Chimica Terrene Conference 5(1).

Crutzen, P.J. 1991. Methane's sinks and sources. *Nature* 350:380-381.

Denmead, O.T. 1991. Sources and sinks of greenhouse gases in the soil-plant environment. *Vegetation* 91:73-86.

Emanuel, W.R., H.H. Shugart, and M.P. Stevenson. 1985a. Climate change and the broad-scale distribution of terrestrial ecosystem complexes. *Climate Change* 7:29-43.

Emanuel, W.R., H.H. Shugart, and M.P. Stevenson. 1985b. Response to Comment: Climate change and the broad-scale distribution of terrestrial ecosystem complexes. *Climate Change* 7:457-460.

Follett, R.F., S.C. Gupta, and P.C. Hunt. 1987. Conservation practices: Relation to the management of plant nutrients for crop production. p. 19-51. In: R.F. Follett, J.W.B. Stewart, and C.V. Cole (eds.) *Soil fertility and organic matter as critical components of production systems.* American Society of Agronomy Special Publication 19. American Society of Agronomy, Madison, WI.

Gebhardt, M.R., T.C. Daniel, E.E. Schweizer, and R.R. Allmaras. 1985. Conservation tillage. *Science* 230:625-630.

Greenland, D.J., A. Wild, and D. Adams. 1992. Organic matter dynamics in the soils of the tropics--from myth to complex reality. p.17-34. In: R. Lal and P.A. Sanchez (eds.), *Myths and science of soils of the tropics.* Soil Science Society of America Special Publication Number 29. Soil Science Society of America, Madison, WI.

Haas, H.J., C.E. Evans, and E.R. Miles. 1957. *Nitrogen and carbon changes in soils as influenced by cropping and soil treatments.* USDA Tech. Bull. 1164. U.S. Gov. Print. Office, Washington, D.C.

Houghton, R.A. 1990. Projections of future deforestation and reforestation in the tropics. p. 87-92. In: *Tropical forestry response options to global climate change*, Sao Paulo, January 1990, Conference Proceedings, U.S. Environmental Protection Agency, Washington, D.C.

Houghton, R.A. and D.L. Skole. 1990. *The long-term flux of carbon between terrestrial ecosystems and the atmosphere as a result of changes in land use.* Research Project of the Month - July. Carbon Dioxide Research Program, Office of Health and Environmental Research, U.S. Department of Energy, Washington, D.C.

Houghton, R.A., and G.M. Woodwell. 1989. Global climate change. *Scien. Amer.* 260:36-44.

IPCC (Intergovernmental Panel on Climate Change). 1990. In: J.T. Houghton, G.J. Jenkins, and J.J. Ephraums (eds.), *Climate change: The IPCC scientific assessment.* Cambridge University Press, Cambridge, England.

Jenkinson, D.S. 1991. The Rothamsted long-term experiments: are they still of use? *Agron. J.* 83:2-10.

Jenkinson, D.S., D.E. Adams, and A. Wild. 1991. Model estimates of CO_2 emissions from soil in response to global warming. *Nature* 351:304-306.

Jenkinson, D.S., P.B.S. Hart, J.H. Rayner, and L.C. Parry. 1987. *Modelling the turnover of organic matter in long-term experiments at Rothamsted.* INTECOL Bulletin 15:1-8.

Jenny, H. 1984. The making and unmaking of a fertile soil. p.42-55. In: W. Jackson, W. Berry, and B. Colman (ed.), *Meeting the expectations of the land.* North Point Press, San Francisco.

Johnson, D.W. 1992. Effects of forest management on soil carbon storage. *Water, Air, and Soil Poll.* 64:83-120.

Johnson, M.G. and J.S. Kern. 1991. *Sequestering carbon in soils: A workshop to explore the potential for mitigating global climate change.* EPA/600/3-91/031. USEPA, Environmental Research Laboratory, Corvallis, OR. 85 pp.

Lal, R. and B.T. Kang. 1982. *Management of organic matter in soils of the tropics and subtropics.* p. 152-178. Transactions of the 12th International Congress of Soil Science, New Delhi, India.

Leemans, R. 1990. *Possible changes in natural vegetation patterns due to a global warming.* Publication number 108 of the Biosphere Dynamics Project, International Institute for Applied Systems Analysis (IIASA), Laxenburg, Austria.

Lucas, R.E., J.B. Holtman, and L.J. Connor. 1977. Soil carbon dynamics and cropping systems. p. 333-350. In: W. Lockeretz (ed.), *Agriculture and energy.* Academic Press, San Deigo, CA.

Lugo, A.E. 1991. Soil carbon in forested ecosystems #1. p. 29-30. In: M.G. Johnson and J.S. Kern (ed.), *Sequestering carbon in soils: A workshop to explore the potential for mitigating global climate change.* EPA/600/3-91/031. USEPA, Environmental Research Laboratory, Corvallis, OR. 85 pp.

Lugo, A.E., M.J. Sanchez, and S. Brown. 1986. Land use and organic carbon content of some subtropical soils. *Plant and Soil* 96:185-196.

Lynch, J.M. and E. Bragg. 1985. Microorganisms and soil aggregate stability. *Adv. Soil Sci.* 2:133-171.

Mann, L.K. 1986. Changes in soil carbon storage after cultivation. *Soil Sci.* 142:279-288.

McKeague, J.A., M.V. Cheshire, F. Andreux, and J. Berthelin. 1986. Organo-mineral complexes in relation to pedogenesis. In: P.M. Huang, and M. Schnitzer (eds.), *Interactions of soil minerals with natural organics and microbes.* Soil Science Society of America Special Publication Number 17. Soil Science Society of America, Inc., Madison, WI.

Mitchell, J.F.B. 1989. The "greenhouse" effect and climate change. *Reviews of Geophysics* 27:115-139.

Mosier, A., D. Schimel, D. Valentine, K. Bronson, and W. Parton. 1991. Methane and nitrous oxide fluxes in native, fertilized and cultivated grasslands. *Nature* 350:330-332.

Post, W.M., T.H. Peng, W.R. Emaniel, A.W. King, V.H. Dale, and D.L. DeAngelis. 1990. The global carbon cycle. *Amer. Sci.* 78:310-326.

Prentice, K.C. and I.Y. Fung. 1990. The sensitivity of terrestrial carbon storage to climate change. *Nature* 346:48-51.

Oades, J.M. 1988. The retention of organic matter in soils. *Biogeochem.* 5:35-70.

Oldeman, L.R., R.T.A. Hakkeling, and W.G. Sombroek. 1990. *World map of the status of human-induced soil degradation: An explanatory note.* International Soil Reference and Information Center, Wageningen, The Netherlands.

Santos, P.F., J. Phillips, and W.G. Whitford. 1981. The role of mites and nematodes in early stages of burried litter deomposition in a desert. *Ecology* 62:664-669.

Santos, P.F. and W.G. Whitford. 1981. The effects of microarthropods on litter decomposition in a Chihuahuan desert ecosystem. *Ecology* 62:654-663.

Schlesinger., W.H. 1986. Changes in soil carbon storage and associated properties with disturbance and recovery. p.194-220. In: J.R. Trabalka and D.E. Reichle (eds.), *The changing carbon cycle: A global analysis.* Springer-Verlag, New York, NY.

Schlesinger, W.H. 1990. Evidence from chronosequence studies for a low carbon-storage potential of soils. *Nature* 348:232-234.

Smith, J.L. and L.F. Elliot. 1990. Tillage and residue management effects on soil organic matter dynamics in semiarid regions. *Adv. Soil Sci.* 13:69-88.

Stevenson, F.J. 1965. Origin and distribution of nitrogen in the soil. In: W.V. Bartholemew and F.E. Clark, (eds.), *Soil Nitrogen.* American Society of Agronomy, Madison, WI.

Stevenson, F.J. 1982. *Humus chemistry.* John Wiley and Sons, New York, NY.

Stevenson, F.J. 1986. *Cycles of soil.* John Wiley and Sons, Inc. New York, N.Y.

Steudler, P.A., R.D. Bowden, J.M. Melillo, and J.D. Aber. 1989. Influence of nitrogen fertilization on methane uptake in temperate forest soils. *Nature* 341:314-316

Tan, K.H. 1993. *Principles of soil chemistry* (2nd Ed.). Marcel Dekker, Inc., New York, NY.

Tate, R.L. 1987. *Soil organic matter: Biological and ecological effects.* John Wiley and Sons, Inc., New York, NY.

UNEP. 1986. *Farming systems principals for improved food production and the control of soil degradation in arid and semi-arid tropics.* ICRISAT, Hyderabad, India.

U.S. Congress. 1991. *Changing by degrees: Steps to reduce greenhouse gases.* Office of Technology Assessment, OTA-0-482 Washington, DC.

Wood., M. 1989. *Soil biology*. Chapman and Hall, Inc., New York, NY.

World Resources Institute. 1990. Food and agriculture. p. 277-290. In: Hammond, A.L. et al. (ed.) *World resources 1990-91.* A report by the World Resources Institute, Oxford University Press, Oxford, England.

The information in this document has been funded wholly (or in part) by the U.S. Environmental Protection Agency. This document has been prepared at the EPA Environmental Research Laboratory in Corvallis, Oregon through Contract 68-C8-0006 to ManTech Environmental Technology, Inc. It has been subjected to the Agency's peer and administrative review and approved for publication. Mention of trade names or commercial products does not constitute endorsement or recommendation for use.

CHAPTER 30

Management of Forest Soils

Richard G. Cline and Gregory A. Ruark

I. Introduction

Is there really such a thing as management of forest soils? Most people who think about this seem to think that there is. The authors would like to take an opposing view. This is a view we have suggested verbally on several occasions. Most of the time when it is suggested, the authors are met with sharp disagreement. Our suggestion is that man is essentially a small, modestly furry animal that has a distinct penchant for running about pursuing its own ends with little concern for the consequences of its actions. Look around and think about this. Are the authors right or not? Further, man really does not know enough about the detailed operation of most ecosystems, soils included, to be able to tinker with them without making some rather appalling mistakes. Our history and present problems seem to bear this out.

What is management, and how should we approach the term? The authors would like to suggest that we really are not very capable of managing anything as complex as a soil ecosystem. We really do not understand it well enough to do that successfully because we are not particularly capable of controlling the system and all of its functions.

What are we capable of managing? How about us? The one thing we are able to control (manage) is us. Management is all about people because people really are the only things people are truly able to control and direct effectively. When we deal with most natural systems we are basically taking advantage of the things those systems do normally.

How, then, do we propose to manage soils. Maybe the most effective strategy is to try to understand them as well as we can, let the processes that occur normally do their thing, spend our efforts directing our actions to take advantage of what soils do and manage ourselves such that we do not irreparably damage the system. That is a considerably different approach than is usual for mankind. We often behave as if we were given "dominion" (substitute authority?) over the earth and all its ecosystems and, consequently, we attempt to control all aspects of what occurs. The truth might well be that dominion really means we will be held responsible for our actions and will be required to pay for our misjudgments, and reap the benefits of good judgment in direct proportion to the quality of our decisions. We must live with the consequences of our actions, however, we seldom consider what they might be. Rather, we are motivated by a view of an objective which we strive to achieve.

The subject at hand is the soil resource and how we should manage to sustain its integrity and long-term viability. How can we evaluate the potential effects of our actions on soil resources? One approach is through use of simulation models. This can be dangerous because the models we produce are usually substantially imperfect in mirroring the effects and system processes they are intended to reflect. There are many reasons for this, not the least of which is our lack of understanding and the accompanying inability to correctly portray a system's myriad interrelationships and feedbacks in mathematical form. In consequence, these models need to be used with caution.

Models are useful because they can be made to handle many complex relationships very rapidly. Models can be constructed to be detailed and specific to the conditions of a system or locality under a specified set

ISBN 1-56670-117-1/95/$0.00+.50

of conditions. This limits their utility to that system, locality and set of conditions, and they are seldom extrapolatable for use in a general sense. Models can be constructed to reflect general system relationships. This causes them to be unusable in predicting specific responses of particular systems, given a set of specified conditions. We cannot expect models to be both specifically accurate and generally applicable.

A simple model will be used to generalize a few soil carbon processes and examine assumptions concerning their interrelationships. Then the model will be used to compare disparately different systems to see what that indicates about approaches to management and how they should change with changing conditions. If the model's structure and assumptions provide a reasonable approximation of the driving processes in the systems being considered, we should be able to speculate about the best strategies for long-term management. The model used is very general and therefore not usable for specific prediction. It is expected to be usable for making general comparisons, however.

II. The Model

The model used is a conceptual carbon allocation model published by Ruark and Blake (1991). The details of the model have been published elsewhere, so this discussion will be limited to model properties that bear on the problem at hand. Model assumptions are important, but its structure and internal specifics are not for purposes of this discussion. The model allocates annual net primary production (NPP) to above ground and below ground tree components. Several assumptions are made:

1. Annual small root mortality and production are equal, and cycle in the upper 20 cm of the mineral soil (Santantonio, 1978 and Ruark, 1993). Soil bulk density is assumed to be 1.2 Mg m^{-3}.

2. Labile SOM (soil organic matter) annual decomposition rate is baselined at 70% and varies linearly from 50% to 90% with changing precipitation.

3. Labile SOM is initially zero and increases over time when small root mortality exceeds decomposition. additionally, 15% of annual foliage is input to the labile SOM pool.

4. Increased SOM improves site quality in terms of soil water and nutrient supply to roots causing a shift in carbon allocation toward the stem with an emphasis on foliage. The effects of water and nutrition are not separated.

5. Annual carbon allocation to shoot is constrained between 50 and 80% of total NPP.

6. Empirical look up tables are used to determine annual baseline NPP and large root production based on stand age. The NPP is assumed to reach an equilibrium after age 15 years. Its value is then adjusted each year by precipitation and labile SOM levels.

7. Maximum increases due to labile SOM levels are:
   20% on NPP
   10% on carbon allocation to shoot.

8. Maximum changes to baseline values due to precipitation are:
   20% increase or decrease in SOM decomposition
   30% increase or decrease in NPP
   15% increase in carbon allocation to shoot.

Ruark and Blake's model is parameterized to reflect conditions in a slash pine ecosystem using published data from Gholz et al. (1986) and Gholz and Fisher, 1982) for comparison to model outputs. The model was constructed using the STELLA (Hi Performance Systems: Lyme New Hampshire) modeling software for the Macintosh computer.

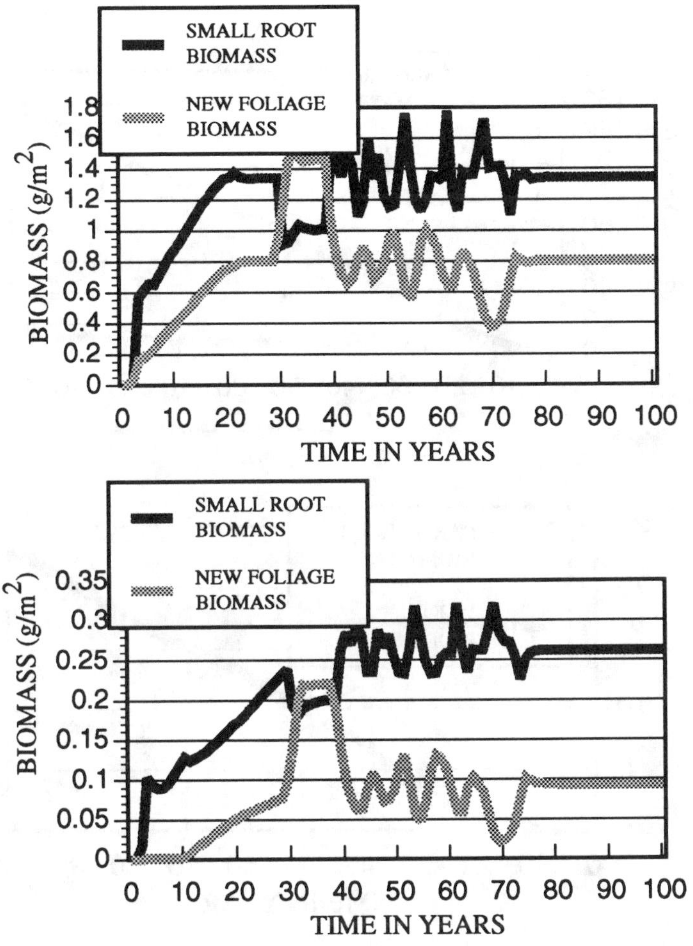

Figure 1. Slash pine (top) and pinyon-juniper (bottom) production of small roots and new foliage during a 100 year model simulation.

III. Slash Pine Application

Ruark and Blake's (1991) original formulation for a slash pine system resulted in an increase in NPP from about 1.4 g m^{-2} d^{-1}, roughly the starting growth function value, to nearly 4.1 g m^{-2} d^{-1} at 15 years when the growth function was 3.5 g m^{-2} d^{-1}. Precipitation was held constant for this period. During this time, small root production increased sharply then stabilized as did new foliage production (Figure 1). Both wood and large root production maintained fairly constant production rates throughout the 100 year simulation period (Figure 2). Large root rates of production were slightly greater early in the period then decreased slightly where as wood production was slower at first then increased as site quality improved because of accumulating labile SOM.

At year 30 precipitation was allowed to fluctuate including a 5 year period of excess rainfall from year 30 to 35 and a 5 year drought from year 65 to 70. Random fluctuations dominated the precipitation pattern from year 35 to 64. A stable average value (1.0) was used from year 70 to 100. Excess rainfall caused the model to allocate a larger proportion of resources to shoot growth and new foliage (Figure 1). Root growth decreased with most of the decrease coming from small root production. Net primary production (NPP, a function of growth, precipitation and site quality) increased and labile SOM decreased. The model indicated opposite results during the drought period: small roots and labile SOM increased, while new foliage and NPP decreased.

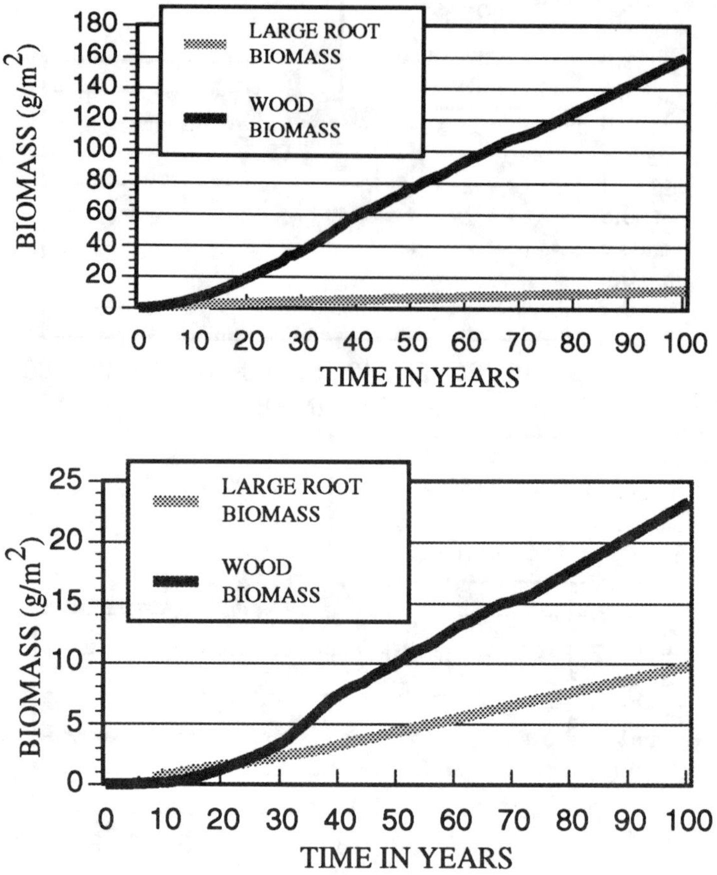

Figure 2. Slash pine (top) and pinyon-juniper (bottom) production of wood and large roots during a 100 year model simulation.

These seem like reasonable results. Periods of limited moisture put the system under stress. More resources and energy are invested in obtaining and retaining moisture hence more small roots and less foliage. There is also less moisture available to support decay processes, hence more SOM because mortality of small roots increases and organic material decomposition slows.

The model estimated foliage, fine root production and NPP within the range of values expected in a slash pine system. This does not prove that the model is fundamentally correct. Models are often formulated to reflect, empirically, the data at hand. This one is no exception to that rule even though the authors believe the underlying assumptions and feedback mechanisms designed into its structure are reasonably sound. If the model is well designed and conceived, it should be extrapolatable at least in concept. One test of the concepts is to apply it to a system that is substantially different from the one it was designed to mimic and observe the changes needed to make the outputs reflect the new system.

IV. Pinyon-Juniper Application

Pinyon-juniper (PJ) ecosystems are extensive in the western United States, and are markedly different, ecologically, from the slash pine systems of the southeastern United States. An application of the model to PJ should be a good test of its conceptual structure. PJ systems are considered semi-arid woodlands and can occur at elevations from about 90 to above 2200 m in rainfall zones averaging about 200 to over 500 mm annually (Hawkins, 1987). Some areas receive more precipitation during summer, while others receive more in winter, often as snow. The one common denominator in these systems is that evaporative demand greatly exceeds supply so surviving species must be adapted to withstand considerable moisture stress.

The primary data source for testing the model comes from Tiedmann (1987) who provided most of the data on biomass distribution for a typical system. A NPP value was derived partly from Gholz (1980) and modified according to discussions with Tiedmann (personal communication, 1993). Tiedmann indicated that his data was for a site slightly more than twice as productive as the site Gholz used in his 1980 article. The basis for soil organic matter accumulations comes from McDaniel and Graham (1992) compared with an estimate of SOM from Tiedman's (1987) soil nitrogen value back calculated to SOM using an organic carbon to SOM ratio of 1:1.7 and a SOM carbon to nitrogen ratio of 20:1 (Brady, 1974). Both estimates produced an approximate soil organic carbon percentage of 3% (SOM = 5%) for the upper 20 cm of soil which is the soil depth used in the original model.

A critical issue for use of the model was the question of how much of the SOM was in the labile fraction. Two articles from the literature were used to make an estimate using two independent calculation methods. These two approaches produced values that were close but by no means the same. One approach used data from an article by Klopatek (1987). Klopatek provided data on NO_3-N, NH_4-N, total N, organic carbon, and the C:N ratio both under the canopy and in interspaces (areas between trees with no tree canopy cover) for two mature PJ stands in Arizona. The assumption was made that the NO_3 and NH_4 nitrogen represented nitrogen in the labile fraction of the SOM. This is a weak assumption but the calculation is worth doing for the sake of attempting estimates that might be used for comparison purposes. Klopatek's data indicated that the C:N ratio for the two sites was quite variable. The C:N in the interspace was greater than under the tree canopy on one site. The reverse was true at the other. The differences were not statistically different ($P < 0.05$) so a single average value was used. An estimate of the proportion of labile material from these data suggested it might be near 0.08%. Klopatek's estimates are from a 60 day incubation experiment, however, and this might be an underestimate. Ruark (1993) in an *in situ* study of fine root decomposition, found that stable respiration rate values were not reached until 120 days.

McDaniel and Graham (1992) provided data on soil organic carbon distribution both under the canopy and in the interspace for two PJ stands in Utah. The organic carbon distribution under the canopy appeared typical of most forested soils. It was concentrated near the surface and decreased with depth. In the interspace it was low at the surface and increased to a maximum in the main rooting zone between a depth of 10 cm and a lithic contact at 20 to 35 cm. Only modest organic carbon is contributed to the interspace soils from canopy litter fall, therefore, most of the organic matter bulge in the fine earth below 10 cm should be from small root contributions throughout the rotation. The average of this estimate for the two sites was 0.2% Since the calculation from Klopatek's data was a weak one and could well be an under estimate the estimate from McDaniel and Graham seemed the best one to use. This value was just 0.1% lower than the estimate used for slash pine.

The model application to PJ required surprisingly little modification. The first and most obvious change was the site-level graphical growth function. This function was started out at .20 g m^{-2} d^{-1} and increased to .67 g m^{-2} d^{-1} at 34 years. Reasoning that the semi-arid climate should slow organic decomposition, we doubled the time to achieve maximum growth rates to 30 years instead of the 15 used in the original slash pine version. The maximum production value for the example is .73 g m^{-2} d^{-1}, or slightly more than double what Gholz (1980) observed.

The function allocating carbon to large roots in the model was observed to prevent any from going to small roots. This was not reasonable, and was indirectly a result of the much larger production rates used for the slash pine system. The large root allocation function was a graph of absolute values of carbon going to large roots against age. This graph was replaced with a product function using a similar graph against age which used a decimal fraction of below ground carbon to be allocated to large roots. This graph started out with very low values on the assumption that most resources need to go into small roots the first year or so, then large investments in the more permanent large root system are needed. The function increased quickly from 0.035 to 0.5 at age 3 years then decreased gradually to .29 at age 20 where it stabilized. This reflects the need, in a dry environment, to invest significant resources in root systems to assure survival. The model outputs for PJ versus slash pine suggested that PJ is likely to invest more of its resources in root systems (Figure 2).

These changes produced a model that nearly matched Tiedmann's measured values for a mature PJ system using a 100 year model run. Tiedmann (1987) quotes Meeuwig (1979) to the effect that 100 years is a typical length of time required to establish a mature PJ system. The decay rate of labile SOM required to achieve this result and end up with the 0.2% estimate for labile SOM was lower than that used in the slash pine model. The decay rate for slash pine was set at 0.7 of the labile SOM annually. The value used for PJ was 0.2 and is reflective of the relative rates of decay in the two systems, and the differences in SOM accumulation. Tiedmann (personal communication) indicated observation of organic materials that are identifiably 10 years old in the litter layer under PJ. Much of this material is expected to be part of the recalcitrant SOM fraction and is expected to decay more slowly. Another indication of the slow rates of decay in semi-arid environments can be seen in Progulske (1974). This publication contains a series of photographs taken 99 years apart in the Black Hills, South Dakota. One picture in particular contains a series of exposed tree roots in the foreground that are identifiable in a picture from the same location 99 years later. A rate of decay of labile SOM in the order of at least 3.5 times slower in semi-arid environments would appear quite reasonable for organic matter in contact with a soil matrix that is wet only periodically.

The model also suggested that the PJ ecosystem is likely to invest more heavily in roots, both large and small (Figures 1 and 2), but the response to moisture during high rainfall periods and drought was similar (Figure 1) for the two systems. This was a part of model design, however, and should be an expected output. This is, however, consistent with field observations of PJ where the root density can extend and occupy the interspace to the point of out competing most of the other plants that might be found in those areas.

V. Conclusions

The ease with which the model could be adjusted to give reasonable results in a markedly different environment was interesting. One can always force a model to fit reality, however, so care must be exercised when drawing conclusions. The model is quite general, and conceptual in nature, and if the assumptions are reasonable it should be useful in making comparisons, but not for projecting specific values. The suggestion is made that carbon inputs to soil labile SOM pools by small roots can be important enough to substantially dominate below ground carbon cycling. This is a premise upon which the model is based, and the results of applying it to each of the two systems do not refute the idea. The use of the model cannot demonstrate that this premise is correct either. If it is, however, and the decay rates are at least as much slower in PJ as the model implies, then management efforts to retain organic reserves on site in semi-arid environments are likely to be critically important to long-term productivity. The loss of soil organic matter and associated nutrient reserves in dry climates from erosion or other causes is likely to be more important than in more humid climates because of the much slower rates of carbon assimilation and cycling. Our management approaches should become more cautious as climatic regimes become drier and soil nutrient cycling slows. These drier systems are less capable of accumulating nutrient reserves than those in more humid areas because of productivity limitations. This suggests that, while fertilizer or other similar amendments can be helpful, they should be used in smaller increments, when they are used, because of growth and nutrient incorporation limitations. Further, if climates become much drier because of climate change, the organic reserves can be critically important to ecosystem integrity. A moister climate is likely to yield more resilient systems because of the ability of most systems to take advantage of increased nutrient cycling rates by shifting carbon allocation toward more actively photosynthesizing parts of plants with a concomitant increase in nutrient cycling via the labile SOM pool.

The differences between the slash pine and PJ systems illustrated by the carbon allocation model are basic differences that are characteristics of the systems. These characteristics are unlikely to change much without causing the systems themselves to change in some very fundamental ways. We, as managers, need to understand these differences and control our actions such that the systems are allowed to function in ways that fall within their adaptive limitations if we want them to retain their essential character and productivity. Alternatively, we can attempt, purposely or by accident, to force them to function some other way. The effects of this kind of action are unpredictable. For example, an increase in atmospheric CO_2 levels is predicted to increase production of C3 plants more than C4 plants. This is predicted even in the presence of moisture stress and low fertility (Kimball et al., 1993). If most shrubs and trees in the PJ system are C3 and most grasses are C4, the effect of this kind of production forcing might cause the grasses in many areas to be out competed and the vegetative composition of the system to change. These kinds of changes can occur surprisingly quickly in systems that are constantly under stress.

The model used in this exercise, if allowed to run for long periods, will continue accumulating large roots and wood. A reality check will suggest this is not what would happen. At some point site occupancy, nutrients, water availability or some combination of these will limit biomass accumulation. This is purposely a shortcoming of the model's design. It was not intended to look at factors such as nutrient status and limitations in detail. It should be possible to build features like this into the model's structure but it was not designed that way for this exercise. This would make model outputs for very long (multiple rotation) periods very misleading.

References

Brady, N.C. 1974. *The Nature and Properties of Soils*. MacMillan, New York.

Gholz, H.L. 1980. Structure and productivity of Juniper occidentalis in central Oregon. *The American Midland Naturalist* 103:251-261.

Gholz, H.L. and R.F. Fisher. 1982. Organic matter production and distribution in a slash pine (*Pinus elliottii*) plantation. *Ecology* 63:1827-1839.

Gholz, H.L., L.C. Hendry, and W.P. Cropper. 1986. Organic matter dynamics of fine roots in plantations of slash pine (*Pinus elliottii*) in north Florida. *Canadian Journal of Forest Research* 16:529-538.

Hawkins, R.H. 1987. Applied hydrology in the pinyon-juniper type. p. 352-359. In: Everett, R.L. (compiler), Proceedings--pinyon-juniper conference. Reno, NE, January 13-16, 1986. General Technical Report INT-215, USDA, Forest Service, Intermountain Forest and Range Experiment Station, Ogden, UT.

Kimball, J.R., J.R. Mauney, F.S. Nakayama, and S.B. Idso. 1993. Effects of elevated CO_2 and climate variables on plants. *J. Soil and Water Cons.* 48:9-14.

Klopatek, J.M. 1987. Nitrogen mineralization and nitrification in mineral soils on pinyon-juniper ecosystems. *Soil Sci. Soc. Am. J.* 51:453-457

McDaniel, P.A. and R.C. Graham. 1992. Organic carbon distributions in shallow soils of pinyon-juniper woodlands. *Soil Sci. Soc. Am. J.* 56:499-504.

Meeuwig, R.O. 1979. Growth characteristics of pinyon-juniper stands in the western Great Basin. Research Paper INT-238. USDA, Forest Service, Intermountain Forest and Range Experiment Station, Ogden, UT. 22 pp.

Progulske, D.R. 1974. Yellow ore, yellow hair, yellow pine. Bul. 616. South Dakota Agricultural Experiment Station, South Dakota State University, Brookings, SD.

Ruark, G.A. 1993. Modeling soil temperature effects on in situ decomposition rates for fine roots of loblolly pine. *Forest Science* 39:118-129.

Ruark, G.A. and J.I. Blake. 1991. Conceptual stand model of plant carbon allocation with a feedback linkage to soil organic matter maintenance. In: W.J. Dyck and CA. Mees (eds.), Long-term field trials to assess environmental impacts of harvesting. Proceedings, IEA/BE T6/A6 Workshop, Florida, USA. February 1990. IEA/BE T6A6 Report No. 5. Forest Research Institute, Rotorua, New Zealand, FRI Bulletin No. 161.

Santantonio, D. 1978. Seasonal dynamics of fine roots in mature stands of Douglas fir of different water regimes-a preliminary report. In: A. Riedacker and J. Gagnaire-Michard (ed.), Symposium on root physiology and symbiosis. Centre Nationale de Recherches Forestieres, Seichamps, France.

Tiedmann, A.R. 1987. Nutrient accumulations in pinyon juniper ecosystems--managing for future site productivity. p. 352-359. In: R.L. Everett, (compiler), Proceedings--pinyon-juniper conference. Reno, NE, January 13-16, 1986. General Technical Report INT-215, USDA, Forest Service, Intermountain Forest and Range Experiment Station, Ogden, UT.

CHAPTER 31

Towards Soil Management for Mitigating the Greenhouse Effect

R. Lal, J. Kimble, and B.A. Stewart

I. Introduction

Atmospheric concentration of CO_2 increased from about 200 ppm in the pre-glacial era to 300 ppm in 1900, and to 350 ppm in 1990 (Post et al., 1990). Total annual emission is estimated at 5.4 Gt C yr^{-1} from burning fossil fuel and 1.6 Gt C yr^{-1} from deforestation (Sarmiento and Sundquist, 1992). Out of the total release of 7.0 Gt C yr^{-1}, about 2.0 Gt C yr^{-1} is estimated to be absorbed by the oceans, and 3.2 Gt C yr^{-1} goes to increase the atmospheric concentration of CO_2. In addition to CO_2, atmospheric concentrations of other radiatively-active or greenhouse gases is also steadily increasing. Annual increase in concentration is estimated at 0.5% for CO_2, 0.8% for CH_4, 1.0% for N_2O, and 3.0% for CFCs (Bouwman, 1990; USEPA, 1990). Several agricultural activities and dynamics of terrestrial ecosystems play a major role in greenhouse gas emissions. Within terrestrial ecosystems, the role of world soils in regulating atmospheric concentrations of greenhouse gas has not been given the importance it deserves. Soils can be a major source of CO_2, CH_4 and N_2O emissions to the atmosphere. In contrast, soils can be purposefully managed to render them as an effective sink.

II. Soil Mangement for Carbon Sequestration

Hitherto, the usefulness of soil resources as the medium for plant growth as the most basic of all resources are widely recognized. In the 21st century, however, soil resources will have to be purposefully used for regulating environment problems. Among its environmental regulatory functions, using soil resources for maintenance of water and air quality are crucial. Atmospheric concentrations of greenhouse gases is an important aspect of air quality which will gain importance with time. World soil resources will be increasingly used to serve as a sink for atmospheric carbon.

Not only should total organic carbon content be increased but it should also be made inaccessible to microorganisms and buffered against human perturbations and climatic fluctuations. The objective is to increase the passive or inert fraction of the soil organic pool. This implies taking soil organic carbon out of the humification-mineralization cycle. Knowledge of soil's attributes and processes that regulate its carbon and nitrogen contents can be used to increase total and inert fraction. Consequently, there are several strategies to sequester carbon in world soils. Important among these are: (i) increase total soil organic carbon content, (ii) increase soil organic carbon content of the sub-soil horizons, (iii) increase micro-aggregation, and (iv) increase soil biodiversity (Figure 1).

ISBN 1-56670-117-1/95/$0.00+.50

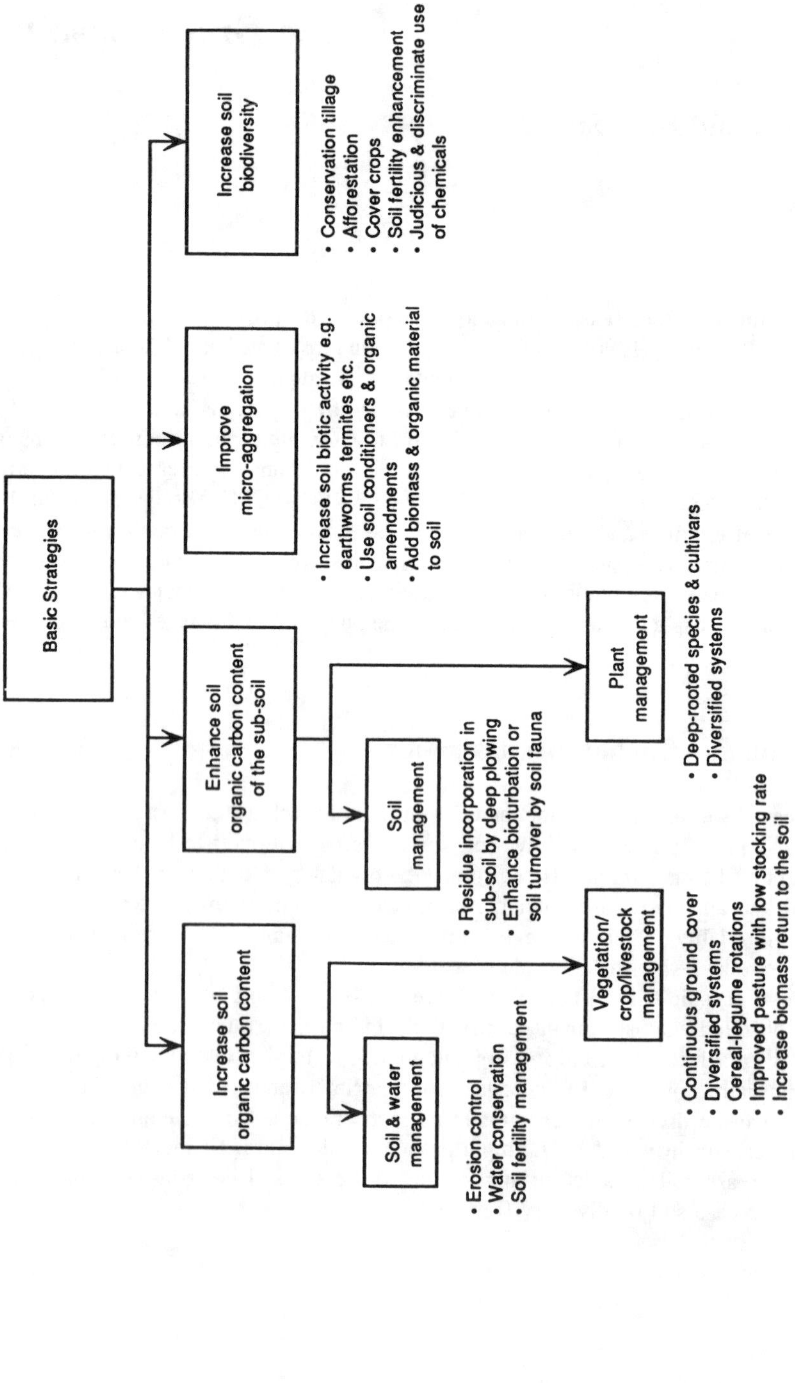

Figure 1. Strategies for carbon sequestration in soils.

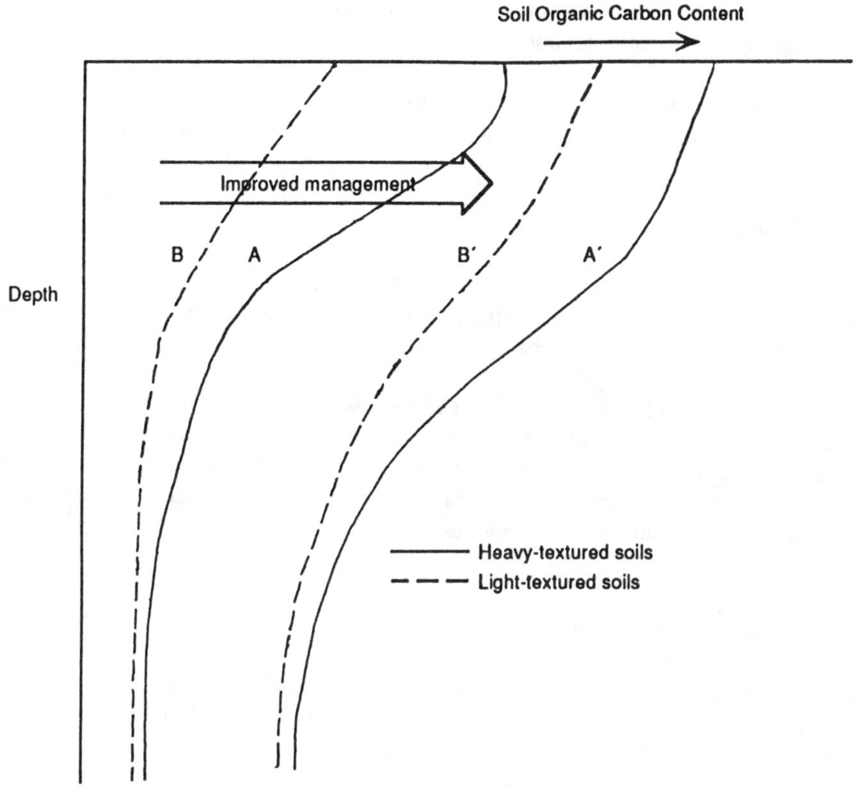

Figure 2. A schematic showing the impact of soil texture and management systems on soil carbon profile. Graph A and B refer to carbon profile with resource-based and low-input systems while A' and B' refer to science-based and high input systems..

A. Increasing Soil Organic Carbon Content

The maximum or equilibrium level of soil organic carbon content within an ecosystem depends on land use, farming/cropping system, and soil and crop management. Figure 2 shows that the objective of improved management is to enhance soil organic carbon content of the profile especially of the sub-soil layers. In general, soils reverted to natural ecosystems and under climax vegetation have more organic carbon contents than those under agricultural ecosystems. Within agricultural ecosystems, there are a wide range of soil and crop management practices that can be used to enhance soil organic carbon content. Important among soil management practices are conservation tillage (Kern and Johnson, 1991), soil fertility management to achieve a balanced nutrient supply, water management to decrease losses by runoff and leaching, and erosion control. Objectives of vegetation management are: (i) to maintain continuous ground cover, (ii) supply frequent and liberal doses of biomass and organic wastes, (iii) grow vegetation with deep root system to transfer carbon into sub-soil horizons, and (iv) apply diverse and complex cropping systems with capacity to produce large biomass.

B. Increase Micro-aggregation

Enhancement of soil structure through increasing total aggregation is an important strategy to sequester carbon in soil. Increasing total aggregation and relative proportion of micro-aggregates can immobilize large quantity of carbon, especially if increase in micro-aggregation can be achieved for the sub-soil horizons. Carbon immobilized in organo-mineral complexes to form stable micro-aggregates and clay domains is inaccessible to micro-organisms. There are several options to enhance aggregation including the use of long chain polymers and soil conditions (DeBoodt and Gabriel, 1975), enhancing activity and species diversity of soil fauna or bioturbation, and growing plant species with extensive and deep root system.

C. Enhance Soil Biodiversity

The importance of soil flora and fauna in enhancing soil organic carbon content and sequestering carbon on long-term basis cannot be over-emphasized. In addition to improving soil structure and aggregation, increase in soil biodiversity increases biomass or active carbon content that accelerate the rate of carbon turnover.

Principles of soil biology and ecology have been widely used to some extent for agronomic goals but not so for ecological and environmental goals. Soil is a living and a dynamic entity. The larger, more diverse, and highly active its faunal and floral components are the more productive and highly efficient soil is to regulate environment. Important among soil fauna is the role of macrofauna including earthworms and termites. Agronomic management of soil resources should be reconciled to enhance soil biodiversity.

III. Soil Enhancing Management Practices

Agricultural sustainability implies achieving desired levels of agronomic productivity while maintaining or enhancing soil and environmental qualities. Agronomic sustainability should be assessed in terms of soil quality by evaluating productivity in terms of yield per unit change in soil aggregation, soil organic carbon content, soil biomass carbon, etc. (Lal, 1994).

The basic principles of soil enhancing practices are known but require local adaptation and fine-tuning for specific soils and ecoregions. In general, natural ecosystems have more favorable soil characteristics (e.g., soil organic carbon content, microaggregation and total aggregation, and soil biodiversity) than agricultural ecosystems. In some soils, however, this is not true because of specific soil-related constraints. Several Oxisols, Ultisols, and Aridisols of the tropics have low productivity under natural state because of nutrient imbalance, water limitations, presence of root-restrictive layers, etc. Soil organic carbon content and structural attributes of these soils can be vastly improved by judicious management and discriminate use of science-based inputs.

Figure 3 depicts sets of cultural practices that have proven useful in enhancing soil quality. Enhancing soil quality implies improving soil's capacity to sequester carbon by erosion control, fertility management, soil structural improvement, and water management. Soil enhancing agricultural practices are often based on science-based and improved systems of soil and crop management. It is often the widespread use of resource-based and low-input agricultural systems that lead to depletion of soil resources and decline in soil quality.

IV. Research and Development Needs

Applications of soil science offers a tremendous opportunity to understand the impact of soil processes on dynamics of C and N under different land uses, and soil and crop management systems. Dynamics of C and N has been studied by soil scientists in the agronomic but not in the environmental context. Yet environmental applications of understanding the impact of soil processes on greenhouse gas emissions are tremendous.

The data on global carbon budget have indicated an imbalance of about 2.0 Gt C yr^{-1} presumably attributed to terrestrial ecosystems. It is necessary to determine the role of world soils in accounting for part

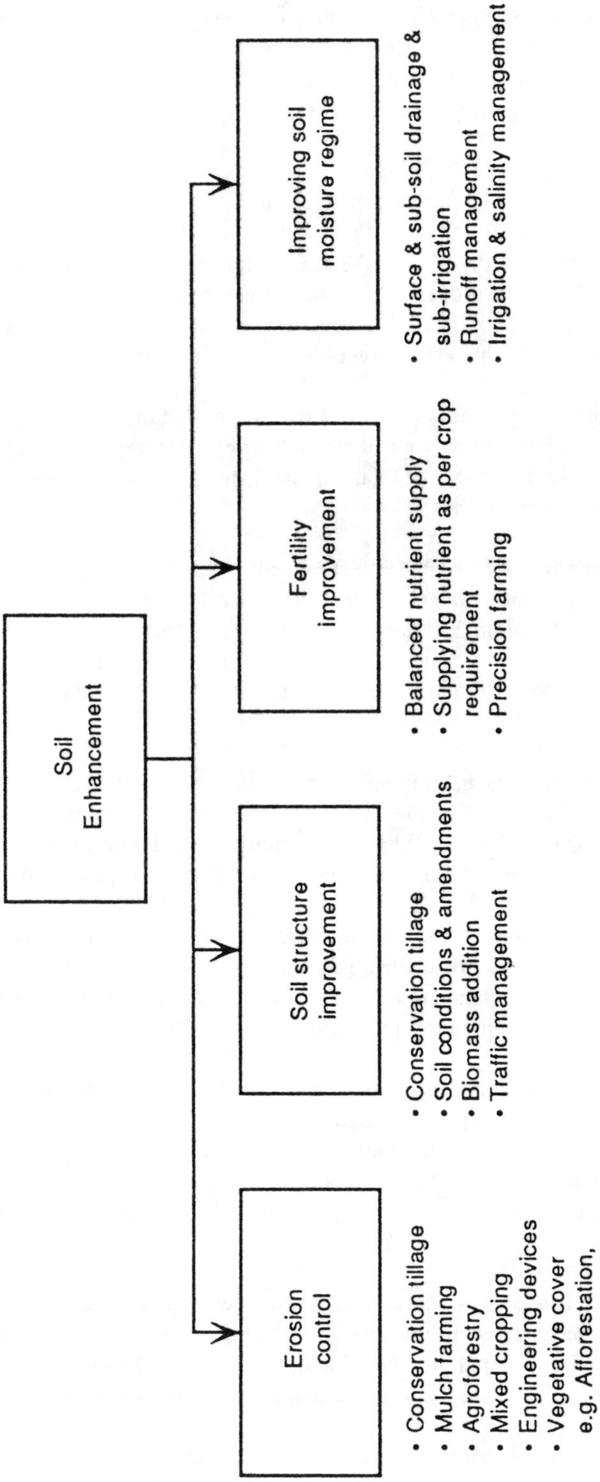

Figure 3. Soil-enhancing agricultural practices.

of this imbalance. Among several specific aspects of soil research that need to be addressed in relation to the greenhouse gas emissions are shown in Figure 4 and discussed in the following:

(a) The impact of land use and more specifically of the change in land use on type and magnitude of gaseous flux from soils needs to be studied for important ecoregions. In this connection, three ecosystems of considerable importance are tropical rainforest, tropical acid savannas, and grasslands. These ecosystems are frontiers of agricultural expansion, and little is known about the nature and magnitude of flux with change in land use.

(b) Biomass burning is an important agricultural tool, and is widely used throughout the tropics and sub-tropics. The impact of biomass burning on soil properties and crop growth has been studied (e.g., Ghuman and Lal, 1989; 1991; Ghuman et al., 1991; Lal and Cummings, 1979; Seubert et al., 1977). However, the effect of biomass burning on soil processes that impact nature and the magnitude of the greenhouse gas emissions have not been quantified. There is a need to conduct systematic study on the effects of burning on soil properties and processes that affect greenhouse gas emissions.

(c) Soil and crop management practices play a major role in greenhouse gas emissions. Yet, the impacts have not been quantified especially in tropical and sub-tropical ecoregions. There is a need to develop a systematic program to study the impact of tillage method, fertility management, and residue management systems on greenhouse gas emissions.

(i) Conservation tillage has a favorable effect on enhancement of soil organic carbon content of the surface layer. It is important to quantify the soil- and tillage-specific impact of conservation tillage systems on carbon sequesteration, and nitrogen balance. In addition to gross carbon balance as measured by changes in soil organic carbon contents, it is important to characterize pathways of carbon e.g., loss of dissolved and particulate carbon in surface runoff, leaching, etc., rate of mineralization in relation to tillage methods, etc.

(ii) Residue management goes hand-in-hand with conservation tillage and has a major impact on greenhouse gas emissions. Crop residue can be burnt, removed from the field, left on the surface as mulch, or incorporated into the soil. What is the impact of different methods of residue management techniques on greenhouse gas emissions in relation to nature and quantity of residue, soil types, and climatic conditions?

Growing in cover crop is an important aspect of residue management, and yet the impact of management of cover crop on greenhouse gas emissions is not known. What is the rate of carbon sequesteration or emission in relation to type of cover crop, mode of suppressing a cover crop for no-till system, and of the system of food crop production on greenhouse gas emissions.

(iii) Fertility management strategy has an important impact on greenhouse gas emissions. What is the C and N dynamics in relation to type, formulation, rate and method of fertilizer application? How can fertilizer use efficiency be improved to minimize losses? What is the impact of farmyard manure and of compost on greenhouse gas emissions especially in relation to ambient soil and climatic factors and method of application e.g., broadcast vs. incorporation? What is the impact of plowing under of a green manure on the greenhouse gas emissions?

(iv) There is a need to develop practical and economic methods of increasing organic carbon content of the sub-soil. Carbon sequestered in the sub-soil is neither easily disturbed by cultural practices nor susceptible to soil erosion by wind or water. Incorporation of carbon into the sub-soil may involve use of biological techniques, based on deep-rooted plants and enhancement of soil biodiversity.

(v) Crop rotations and cropping/farming systems are important tools of agronomic management. What are the impacts of cropping systems and crop combinations on greenhouse gas emissions in relation to efficiency of resource utilization? What are the impacts of multiple and mixed cropping systems vis-a-vis simple systems and monoculture on greenhouse gas emissions and carbon sequesteration? Does high Land Equivalent Ratio or high "Area Time Equivalent Ratio" also imply low gaseous emissions and high carbon sequesteration?

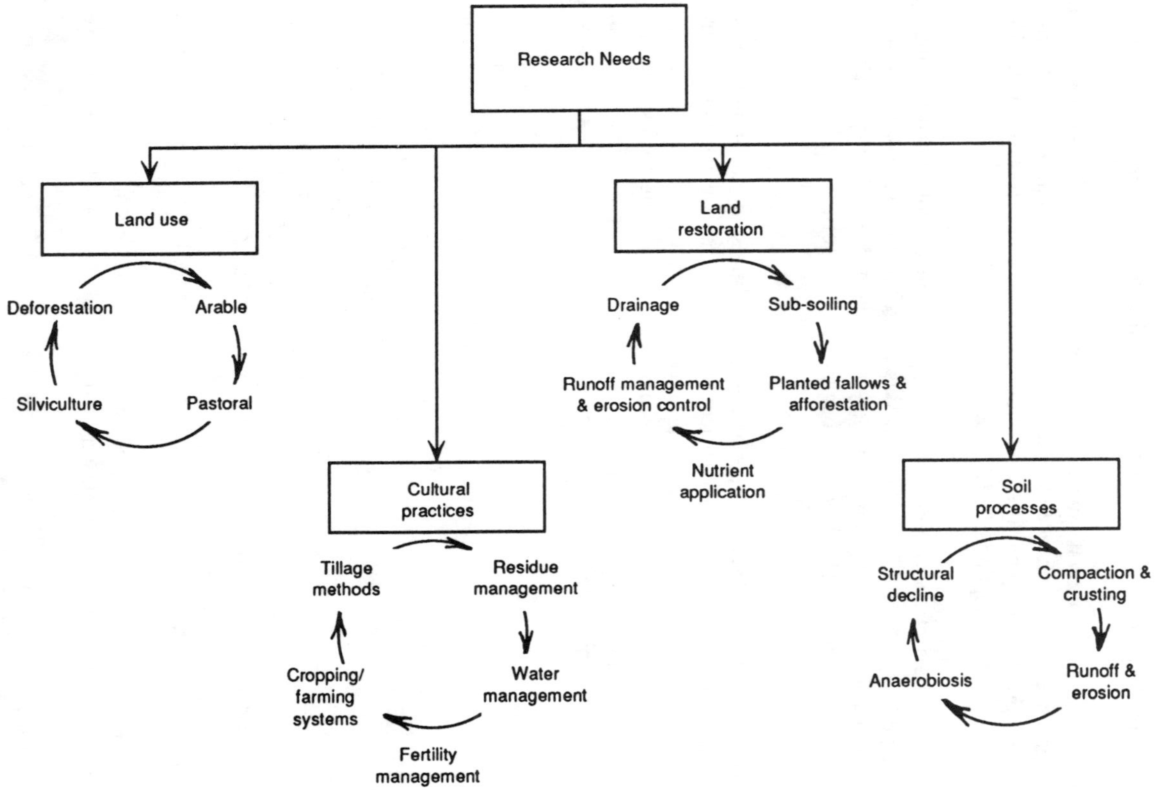

Figure 4. Research needs in evaluation of greenhouse gas emissions in relation to land use, cultural practices, soil restoration, and predominant soil processes.

(vi) Pasture or pastoral systems form an important farming system, especially in tropical and sub-tropical ecosystems. Yet, the impact of pastural systems and management on C and N dynamics in relation to greenhouse gas emissions are not known. There is a need to study the impact of traditional vs. improved pastures, fertilized vs. unfertilized systems, low vs. high stocking rate, grazing vs. stall feeding, pasture species, etc. on carbon sequesteration and greenhouse gas emissions.

(vii) Impact of soil degradative processes on greenhouse gas emissions is not known. What is the fate of particulate or dissolved carbon carried with surface runoff and eroded sediments? How does water management (e.g., drainage or irrigation) affect C and N dynamics? How does restoration of eroded and degraded land affect carbon sequesteration for different soils and agro-ecosystems.

(viii) Identification of policy consideration is crucial for widespread adoption of those land uses and soil and crop management practices which enhance carbon sequesteration and minimize gaseous emissions. Such land uses and cultural practices may be different for different soils and agro-ecosystems. However, providing necessary incentives for adoption of such practices is crucial.

(ix) It is also important to create awareness among scientists, practitioners, policy makers, and, general public regarding the importance of soils and soil processes in greenhouse gas emissions and carbon sequesteration. The importance of world soils is not widely recognized. Steps should be taken to create awareness about the role of soils in global warming through symposia, workshops, bulletins, and other popular media.

These researchable topics are not only of global importance and need immediate attention but also require a multi-disciplinary team approach representing biophysical and socio-economic and political scientists. Biophysical scientists should standardize methodology, quantify the impact of soil processes on gaseous emissions and carbon sequesteration, and develop scaling techniques to extrapolate results from small scale experiments to an ecoregion. Biophysical and social scientists should identify policy issues that encourage adoption of those land uses, farming systems, and soil and crop management techniques that sequester carbon in the soil profile.

References

Bouwman, A.F. (ed.), 1990. *Soils and Greenhouse Effect.* John Wiley and Sons, U.K., 575 pp.

DeBoodt, M. and D. Gabriels (eds), 1975. Third International Symposium on Soil Conditioning, State Univ. of Ghent, Belgium, 464 pp.

Ghuman, B.S. and R. Lal, 1989. Soil temperature effects of biomass burning in windrows after clearing a tropical rainforest. *Field Crops Res.* 22:1-10.

Ghuman, B.S. and R. Lal, 1991. Land clearing and use in the humid Nigerian tropics. II. Soil chemical properties. *Soil Sci. Soc. Am. J.* 51:184-188.

Ghuman, B.S., R. Lal, and W. Shearer, 1991. Level clearing and use in the humid Nigerian tropics. I. Soil physical properties. *Soil Sci. Soc. Am. J.* 55:178-183.

Kern, J.S. and M.G. Johnson, 1991. The impact of conservation tillage use on soil and atmospheric carbon in the contiguous United States. Man Tech Environmental Technology, Inc. USEPA-Env. Res. Lab. Corvallis, Or.

Lal, R., 1994. Methods and guidelines for assessing sustainable use of soil and water resources in the tropics. SMSS Technical Monograph 21, 78 pp. Washington, D.C.

Lal, R. and D.J. Cummings, 1979. Clearing a tropical forest. I. Effect on soil and micro-climate. *Field Crops Res.* 2: 91-107.

Post, W.M., T.H. Theng, W.R. Emannuel, A.W. King, V.H. Dale, and D.L. DeAngelis, 1990. The global carbon cycle. *Amer. Scientist* 78:310-326.

Sarmiento, J.L. and E.T. Sundquist, 1992. Revised budget for the oceanic uptake of anthropogenic carbon dioxide. Nature 356:589-593.

Seubert, C.E., P.A. Sanchez, and C. Valverde, 1977. Effects of land clearing methods on soil properties of an Ultisol and crop performance in the Amazon Jungle of Peru. *Trop. Agric.*(Trinidad) 54:307-321.

Tans, P.P., I.Y. Fung, and T. Takahashi, 1990. Observational constraints on the global atmospheric CO_2 budget. *Science*. 247:1431-1438.

USEPA, 1990. Greenhouse gas emissions from agricultural systems. Workshop Vol. I: 12-14. Dec. 1989, Washington, D.C.

Index